A comprehensive overview of food and nutrition in all stages of life

FOOD & NUTRITION THROUGHOUT LIFE

EDITED BY

Sharon Croxford

Catherine Itsiopoulos

Adrienne Forsyth

Regina Belski

Antonia Thodis

Sue Shepherd

Audrey Tierney

Routledge
Taylor & Francis Group
LONDON AND NEW YORK

First published 2015 by Allen & Unwin

Published 2020 by Routledge
2 Park Square, Milton Park, Abingdon, Oxon OX14 4RN
605 Third Avenue, New York, NY 10017

First issued in hardback 2021

Routledge is an imprint of the Taylor & Francis Group, an informa business

Cataloguing-in-Publication details are available
from the National Library of Australia
www.trove.nla.gov.au

Index by Garry Cousins
Set in 10/11.5 pt Bembo by Midland Typesetters, Australia

ISBN 13: 978-0-367-71810-7 (hbk)
ISBN 13: 978-1-74331-675-7 (pbk)

Contents

List of figures, tables and case studies

FIGURES

TABLES

CASE STUDIES

Contributors

Regina Belski

Regina is an accredited practising dietitian and senior lecturer and researcher in the Department of Dietetics and Human Nutrition at La Trobe University, Melbourne. Her work focuses on nutritional influences in the prevention and management of obesity and chronic health conditions as well as optimal nutrition for sports performance.

Kelly Bobridge

Kelly received her master's degree in nutrition and dietetics from Edith Cowan University, Perth. She has worked at Sir Charles Gairdner Hospital and at Joondalup Health Campus in Perth as a clinical dietitian. Kelly has published research investigating the relationship between fructose intake, serum uric acid and blood pressure in Australian adolescents.

Louise Brough

Louise is a senior lecturer in human nutrition at Massey University, New Zealand. Her main research focuses are nutrition during pregnancy, breastfeeding and childhood. She is especially interested in the assessment and consequences of deficiencies in micronutrients including iodine, selenium, folate, iron and vitamin D.

Jane Coad

Jane is director of the Division of Human Nutrition and Dietetics at Massey University, New Zealand. Her research focuses on maternal and infant health, and she is particularly interested in the roles of micronutrients such as iron and vitamin D and nutritional factors influencing fertility and the development of pre-eclampsia and maternal gestational diabetes.

Clare Collins

Clare is a professor of nutrition and dietetics and co-director of the Priority Research Centre in Physical Activity and Nutrition at the University of Newcastle, New South Wales. She is a fellow of the Dietitians Association of Australia. She chaired development of DAA's Best Practice Dietetic Guidelines for the Management of Overweight and Obesity for Adults. Clare has published over 160 journal articles, co-authored five books and oversaw the development of the systematic review behind the dietary guidelines for Australians. Clare is a DAA media spokesperson and is well known in Australia as a commentator on nutrition and health.

Sharon Croxford

An accredited practising dietitian and senior academic at La Trobe University, Melbourne, Sharon has worked in clinical, community and food service dietetic practice in Australia and the United Kingdom. She is a qualified chef, has written books on Turkish and Ottoman cuisine and previously ran a cooking school in Istanbul. Sharon's research interests include food and culture, and resistance to acculturation.

Adrienne Forsyth

Adrienne is a dual-qualified accredited practising dietitian and accredited exercise physiologist, and is a lecturer in dietetics and human nutrition at La Trobe University, Melbourne. Her PhD research at the

University of Wollongong, New South Wales, evaluated a community-based lifestyle intervention for primary care patients with depression and/or anxiety.

Catherine Itsiopoulos

Catherine is an associate professor and the founding head of the Department of Dietetics and Human Nutrition at La Trobe University, Melbourne. She is an accredited practising dietitian with almost 30 years' experience in academia, research and clinical practice. Catherine is recognised for her expertise in clinical trials using the Mediterranean diet in the management of cardiometabolic health.

Antigone Kouris-Blazos

Antigone is an accredited practising dietitian with over 28 years' experience. She has conducted pioneering research on the Mediterranean diet and longevity, written over 40 published papers, co-authored five university textbooks (including one on integrative medicine) and authored three books herself. Recently she turned her expertise to developing functional cookies.

Amanda Lee

Amanda is a professor in the School of Public Health and Social Work and the School of Exercise and Nutrition Sciences at the Queensland University of Technology, Brisbane. She has over 30 years' high-level experience as a 'pracademic' in nutrition and Indigenous and population health policy and practice.

Jacqueline Miller

Jacqui is a senior lecturer in nutrition and dietetics at Flinders University, Adelaide. She is a qualified lactation consultant and previously worked as a clinical paediatric dietitian specialising in infant feeding. Her research focus is on nutrition support for premature infants.

Michelle Miller

Michelle is head of discipline nutrition and dietetics at Flinders University, Adelaide, and an accredited practising dietitian. Her research expertise and publications primarily focus on clinical nutrition, with applications across aged care, rehabilitation, oncology and surgery.

Annette Murphy

Annette is a lecturer and researcher in nutrition and dietetics in the Faculty of Health and Medicine and Priority Research Centre for Physical Activity and Nutrition at the University of Newcastle, New South Wales. She is an accredited practising dietitian and has worked in community dietetics in addition to clinical dietetics in Australia and internationally.

Kerin O'Dea

Kerin is professor of population health and nutrition at the University of South Australia, Adelaide, and a nutrition scientist and public health researcher examining diet and lifestyle in the prevention and treatment of obesity, type 2 diabetes and cardiovascular diseases. She has a particular interest in the therapeutic potential of traditional diets, especially Aboriginal hunter-gatherer and Cretan Mediterranean. Kerin has held numerous senior academic and research leadership positions over the past 25 years, including Director of the Menzies School of Health Research in Darwin (2000–05), and has been active on national committees advising government on health and medical research, Indigenous health, nutrition and diabetes.

Therese O'Sullivan

Therese completed her PhD through the Queensland University of Technology, Brisbane, in the area of glycaemic carbohydrate intake and insulin resistance. She has published over twenty peer-reviewed journal articles and presented research at national and international conferences. Therese lectures at Edith Cowan University, Perth, and has worked as a dietitian in private practice and in adolescent health research through the Telethon Institute for Child Health Research (since renamed the Telethon Kids Institute), Perth.

Jane Scott

Jane is professor of public health nutrition research at Curtin University, Perth, and a fellow of the Dietitians Association of Australia. Her research focuses primarily on early infant feeding practices and the determinants of breastfeeding. She was an expert technical writer for the 2012 National Health and Medical Research Council's Infant Feeding Guidelines.

Sue Shepherd

Sue is an advanced accredited practising dietitian, recognised internationally as an expert in coeliac disease and irritable bowel syndrome. Sue is a senior lecturer at La Trobe University, Melbourne. She has authored eleven cookbooks for gluten-free and low-FODMAP diets and numerous peer-reviewed journal publications. She provides media commentary and runs a specialist gastrointestinal dietetic private practice.

Antonia Thodis

Tania is a clinical dietitian and is involved in nutrition research and teaching at La Trobe University, Melbourne. Her research interests include traditional diets, the Greek Mediterranean diet pattern, migrant health and nutrition, and ageing well in the older population. Her PhD investigated the health benefits of a plant-based Greek Mediterranean diet.

Audrey Tierney

Audrey is a senior clinical dietitian at Alfred Health and a senior lecturer and academic at La Trobe University, Melbourne. Her research interests are primarily in the areas of clinical nutrition, metabolic health and nutrigenomics.

Evelyn Volders

Evelyn is a senior lecturer at Monash University, Melbourne, and has worked as a paediatric dietitian at tertiary paediatric hospitals in the United Kingdom and Australia for over 25 years. She has extensive experience working with young children and families to improve nutrition status.

Mark L. Wahlqvist

Mark is director of the Fuli Institute and professor in food and nutrition science, Zhejiang University, Hangzhou, China; visiting professor, National Health Research Institutes, Taiwan; and emeritus professor. Monash Asia Institute and Department of Epidemiology and Preventive Medicine, Monash University, Melbourne, Australia. He has also been professor of medicine at Monash University, and professor of human nutrition at Deakin University, Australia. He has played a major role in nutrition education, science, practice and policy for almost five decades. Mark has chaired and been a member of many state, national and international committees and boards. He chaired the Food and Agriculture Organisation Centre of Excellence at Monash University and continues his role on the World Health Organization's Nutrition Advisory Panel. He is editor-in-chief of the *Asia Pacific Journal of Clinical Nutrition*.

Carol Wham

Carol is a senior lecturer in nutrition and dietetics at Massey University, Auckland. She is an experienced dietitian with expertise in nutrition screening and assessment, especially in older adults. Her present research relates to nutrition assessment and trajectories of health and social outcomes in both the New Zealand Longitudinal Study of Ageing (NZLSA) and Life and Living in Advanced Age, a Cohort Study in New Zealand (LILACS NZ). Carol is a member of the New Zealand Dietitians Board and an associate editor of *Nutrition and Dietetics*.

Alison Yaxley

Alison is a lecturer in nutrition and dietetics at Flinders University, Adelaide, and an accredited practising dietitian. Her research interests include nutrition status in older adults and healthy ageing, and she has co-authored a number of publications in peer-reviewed scholarly journals and presentations for international and national conferences.

How to use this book

PART I: FOUNDATION KNOWLEDGE

The first chapter, 'Food and nutrition basics', contains a wealth of core information that you can use as a basic introduction to food and particularly nutrition or as a reference as you read other chapters. Key concepts that reach across many of the main chapters are included in this chapter: nutrients and their roles, food sources and requirements, dietary guidelines, food guidance, anthropometry, energy balance, the glycaemic index, label reading and food safety.

'Cultures, beliefs and food habits' is the second of the background chapters and will provide you with a sound understanding of the history of and current multiculturalism in Australian and New Zealand food habits and beliefs, especially those associated with ethnicity and religion and the influence these have on our societies. As the cultural landscape continually changes, understanding the food and nutrition influences and needs of community members is key to enabling a healthy lifestyle for all.

PART II: LIFE STAGES

The eleven chapters in this part work through each of the life stages. Eating habits as described in the results of national studies are included where appropriate, current recommended nutrient intakes are discussed in the context of nutrients of interest in each stage, nutrition-related health issues are highlighted, and strategies for addressing these are included. Each of the main chapters also includes:

- case studies from real life to demonstrate application of food and nutrition interventions and to address the particular issues of population subgroups as highlighted in each of the chapters in Parts III and IV
- definitions of key terms, which are given in bold in the text
- references for additional reading
- a quiz to test your understanding
- study questions.

PART III: REGIONAL PERSPECTIVES

The three chapters in this part will add a rich depth to your knowledge of food and nutrition issues for Aboriginals and Torres Strait Islanders, New Zealand and Māori peoples and the Asia-Pacific region.

PART IV: EXTENSION KNOWLEDGE

The last four chapters, on athletes, migrants, people living with disadvantage and mental health issues will allow you to extend your knowledge of some groups within the community who may have additional food and nutrient requirements or face challenges in meeting their food and nutrition needs.

Our intention is that whatever stage you have reached in studying food and nutrition, the foundation, regional perspectives and extension knowledge chapters will provide you with a nutrition, social and cultural context so that your understanding of the complex needs of individuals and groups is enhanced.

PART I

FOUNDATION KNOWLEDGE

1

Food and nutrition basics

Sue Shepherd

Food and nutrition are related to health. In its simplest sense food provides nutrients (bioactive components) that help determine an individual's health status. For optimal health all nutrient requirements must be met, alongside taking adequate physical activity and experiencing emotional and social wellbeing.

This chapter provides an introduction to the basics when considering food and nutrition within the context of health. It can be read as a stand-alone chapter to develop a good background knowledge of food and nutrition before reading the remaining chapters; however, it can equally be used as a reference when looking into the different stages of the life cycle, the essential points of which are included in the following sections.

The chapter starts with a section on key **nutrients** in human nutrition, their **role** and **food sources**. **Dietary Guidelines** for Australia and New Zealand follow, with **recommended daily servings of foods** and **standard food and drink serve sizes** for each country. A section on **anthropometry** describes key measures used in assessing nutritional status. The concept of **energy balance** introduces energy requirements and **Nutrient Reference Values** for Australia and New Zealand, including reference body weight and heights and estimated energy, macronutrient, water, vitamin and mineral requirements by life stage and gender. **Nutrient intakes and food sources** for adults, and an introduction to the **glycaemic index** and load, **label reading** and **food safety** complete the chapter.

NUTRIENTS: THEIR ROLES AND FOOD SOURCES

Some nutrients provide energy; these include protein, fat, carbohydrate, alcohol and fibre. The energy they contribute in food is described in this chapter. Water is also a nutrient, but it does not provide any energy. Other nutrients include vitamins and minerals, many of which are described below. Vitamins are categorised according to their solubility: some are **water soluble**, and others are **fat soluble**. Minerals are categorised into two groups: **major minerals**, which are found in the body in amounts greater than or equal to 5 grams, and **trace elements**, which are found in the body in amounts less than 5 grams.

- **Water-soluble vitamins** can dissolve in water; e.g., they can leach into water when a vegetable containing them is boiled. Water-soluble vitamins include vitamin B1 (thiamin), vitamin B2 (riboflavin), vitamin B3 (niacin), vitamin B5 (pantothenic acid), vitamin B6 (pyridoxine), biotin, choline, folic acid, vitamin B12 (cyanocobalamin) and vitamin C. With the exception of vitamin B12, these are generally not stored in the body for long periods.
- **Fat-soluble vitamins** can dissolve in fat and include vitamins A, D, E and K. Generally, these can be stored in fat and/or the liver within the body for long periods of time and may pose an increased risk of toxicity. These vitamins are generally not lost from food when cooked.

- **Major minerals** are present in the body in amounts of 5 grams or greater. They include calcium (Ca), magnesium (Mg), phosphorus (P), potassium (K), sodium (Na). Chloride (Cl) and sulfur (S) are also major minerals, but there are no recommended dietary intakes (RDIs) for these.
- **Trace elements** are needed in smaller amounts by the body; however, this does not indicate their importance. They include chromium (Cr), copper (Cu), fluoride (F), iodine (I), iron (Fe), manganese (Ma), molybdenum (Mo), selenium (Se) and zinc (Zn).

Tables 1.1a and 1.1b describe the functions, signs of **deficiency**, signs of **toxicity** and rich food sources of vitamins and minerals. Note that some food sources are

- **Deficiency** in relation to nutrients is an inadequate intake of a nutrient, possibly due to inadequate dietary intake, poor absorption or uptake, altered metabolism or increased losses. It can lead to many symptoms and health conditions, varying in severity.
- **Toxicity** in relation to nutrients occurs when large doses of a nutrient are consumed usually via food or supplement or a non-food source. Toxicity levels vary according to the nutrient. Toxicity can cause many side effects.

naturally abundant in nutrients, while others are fortified; that is, they have had nutrients added to them. Fortification of foods and drinks can occur only in certain circumstances, such as those described in the Food Standards Code (FSANZ 2013: 2.1.1). The daily

Table 1.1a Vitamins: functions, signs of deficiency, signs of toxicity and rich food sources

Water-soluble vitamins	Functions	Signs of deficiency	Signs of toxicity	Rich food sources
Biotin	Co-enzyme in mitochondrial enzymes, leucine degradation	Rare, seen in unsupplemented total parenteral nutrition	Unlikely	Tomato, chard, lettuce, carrot
Thiamin (vitamin B1)	Metabolism of carbohydrate energy reactions, heart and nervous system function	Beri-beri (strong pulse, oedema, cardiac enlargement, reduced sensation in feet, absent ankle reflexes), Wernicke-Korsakoff syndrome (confusion, ataxia, reduced consciousness, memory impairment, confabulation)	Unlikely	Yeast extract, wheat germ, nuts, liver, kidney, fortified breakfast cereals (mandatory enrichment of baking flour in Australia, not in New Zealand)
Riboflavin (vitamin B2)	Energy release reactions, metabolism of carbohydrate, protein and fat, skin and eye health; conversion of vitamin B6 to active form, conversion of tryptophan to niacin	Inflammation of lining of mouth and tongue, cracks in corner of mouth, growth disturbances, normocytic anaemia	Unlikely (not seen)	Milk, cheese, yoghurt, yeast extract, liver, eggs, almonds
Niacin (vitamin B3) *Note:* Nicotinamide, the active form of niacin, is converted from the amino acid tryptophan and contributes to requirements.	Energy release reactions; metabolism of carbohydrate, protein and fat; essential for growth	Pellagra (lesions on the skin on exposure to sunlight), dermatitis, diarrhoea, delirium/dementia, inflamed tongue (glossitis)	Unlikely	Lean meat, liver, kidney, yeast, bran, peanuts, tuna, salmon, legumes, fortified breakfast cereals, eggs, mushrooms

Water-soluble vitamins	Functions	Signs of deficiency	Signs of toxicity	Rich food sources
Pantothenic acid (vitamin B5)	Component of co-enzyme A which is involved in fatty acid metabolism and oxidation of pyruvate in the citric acid cycle	Rare—only seen in unsupplemented total parenteral nutrition	Not seen	Chicken, beef, liver, egg yolk, potatoes, oats, tomatoes, broccoli, whole grains
Pyridoxine (vitamin B6)	Part of co-enzymes used in amino acid metabolism, synthesis of neurotransmitters, maintaining normal levels of homocysteine, synthesis of haem, maintenance of lymphocytes	Rare	Ataxia, photosensitivity, nausea	Meat, fish, poultry, potatoes and other starchy vegetables, fortified cereals, legumes, non-citrus fruits, banana, watermelon, prune juice, soy
Cobalamin (vitamin B12)	Part of co-enzymes used in fatty acid metabolism, DNA synthesis, synthesis of haem, normal blood and neurological function	Increased serum folate, reduced energy and exercise tolerance, fatigue, palpitations, shortness of breath, neuropathies, memory loss, visual disturbances	Not seen	Foods of animal origin—e.g., fish (including shellfish), red meat, liver, poultry, eggs, milk, milk products; fortified cereals, some edible algae and mushrooms
Total folate (as dietary folate equivalents)	Amino acid metabolism, formation of enzymes, synthesis of DNA, prevention of neural tube defect	Megaloblastic anaemia, increased homocysteine, weakness, fatigue, irritability, palpitations	Neurological abnormalities (in people with vitamin B12 deficiency)	Leafy vegetables, whole grains, peas, nuts, avocado, organ meats, yeast extract, orange juice, fortified cereals
L-ascorbic acid (vitamin C)	Healthy gums, teeth and bones, wound healing, resistance to infection, collagen formation, absorption of non-haem iron and copper, antioxidant	Scurvy (pain in extremities, oedema, skeletal and vascular lesions, death)	Gastrointestinal effects associated with acute high doses in a short timeframe	Citrus fruits, berries, guava, mango, capsicum, pawpaw, parsley, broccoli, pineapple, spinach, cabbage
Biotin	Co-enzyme in mitochondrial enzymes, degradation of leucine	Rare—seen in unsupplemented total parenteral nutrition	Unlikely	Tomatoes, chard, lettuce, carrots
Choline	Synthesis of acetylcholine (neurotransmitter) and lecithin (phospholipid), platelet activating factor, lipid and cholesterol transport	Rare—seen in unsupplemented total parenteral nutrition	Rare—hypotension, nausea, diarrhoea	Milk, liver, eggs, peanuts

continues

Table 1.1a Vitamins: functions, signs of deficiency, signs of toxicity and rich food sources *continued*

Fat-soluble vitamins	Functions	Signs of deficiency	Signs of toxicity	Rich food sources
Total vitamin A equivalents (includes retinol and beta-carotene)	Eyesight, normal growth in foetus and children, healthy skin and mucous membranes, maintenance of immune function, antioxidant	Night blindness, increased susceptibility to infection	*Retinol:* teratogenic, pregnant women advised to avoid excess intake. Those with severe protein malnutrition, chronic liver disease, hyperlipidaemia and excessive alcohol intake may be at risk of toxicity *Beta-carotene:* low toxicity; however, high intake over prolonged periods can lead to yellowing of the skin and may be harmful to smokers	Orange and green fruit and vegetables, liver, oily fish, butter, table margarine, cream, full-cream milk, egg yolk, cheese
Cholecalciferol (vitamin D)	Bone growth and remodelling, normal immune system function, normal inflammatory response, maintenance of healthy skin and muscle strength, calcium absorption in gut, maintenance of serum phosphate	Inadequate mineralisation or demineralisation of the skeleton, rickets in young children, osteoporosis and osteomalacia in adults	Unlikely from food sources or exposure to sun	Fortified milk, margarine, butter, veal, beef, egg yolks, fatty fish, fish liver oil, liver
Tocopherol (vitamin E)	Antioxidant	Peripheral neuropathy	Unlikely	Polyunsaturated plant oils, fatty meats, fish and poultry, leafy green vegetables, wheat germ, whole grains, liver, nuts, seeds
Phylloquinone (vitamin K)	Synthesis of blood clotting proteins	Rare—tendency for bleeding, increased prothrombin time	Not seen	Bacterial synthesis in digestive tract, liver, leafy green vegetables, cauliflower, milk

Table 1.1b Minerals: functions, signs of deficiency, signs of toxicity and rich food sources

Minerals	Functions	Signs of deficiency	Signs of toxicity	Rich food sources
Calcium (Ca)	Bone and teeth formation, prevent rickets and osteoporosis, muscle contraction and nerve function, blood clotting, enzyme activity	Osteoporosis (long-term deficiency), numbness and tingling in fingers, muscle cramps, lethargy, poor appetite, convulsions	Rare	Milk, cheese, yoghurt, canned salmon and sardines (when bones consumed), leafy greens, legumes, almonds, sesame seeds, ice cream, fortified soy drinks. Note that presence of oxalic acid (e.g., spinach, beans) and phytic acid (e.g., seeds, nuts, legumes) decreases ability to absorb calcium.
Magnesium (Mg)	Structure of bones, involved in more than 300 enzyme systems, control of muscle function and nerve function	Rare—may be seen in prolonged diarrhoea, neurological and neuromuscular impairment, nausea, weakness, lethargy, muscular spasms, convulsions	Diarrhoea	Wholegrain cereals and cereal products, legumes, nuts, green leafy vegetables, seafood, chocolate, cocoa
Phosphorus (P)	Bone and teeth structure, energy release factors, component of proteins and nucleic acids	Anorexia, anaemia, muscle weakness, rickets, osteomalacia, bone pain, increased susceptibility to infection, ataxia, confusion	Not seen	Meat, chicken, fish, milk, cheese, eggs, yeast extract, bran, wheat germ
Potassium (K)	Controls transmission of nerve impulse, maintenance of cell membrane differential, part of enzyme systems, role in blood pressure	Low-grade metabolic acidosis leading to demineralisation of bone, osteoporosis and kidney stones	Gastrointestinal discomfort, ulceration and perforation; arrhythmia	Nuts, fresh fruit, dried fruit, wheat bran and germ, lean meat, fish, raw vegetables, fruit juice, instant coffee, yeast extract
Sodium (Na) *Note:* Sodium is one of the few nutrients that should be limited due to its adverse effect on health.	Control of transmission of nerve impulses in conjunction with potassium, helps maintenance of water balance, necessary for transport of amino acids and glucose	Unlikely	Possible hypertension	Commercial and processed foods, salt, anchovies, monosodium glutamate, soy sauce, bottled sauces, cured meats, cheese, canned vegetables

continues

Table 1.1b Minerals: functions, signs of deficiency, signs of toxicity and rich food sources *continued*

Trace elements	Functions	Signs of deficiency	Signs of toxicity	Rich food sources
Chromium (Cr)	Enhancement of insulin action and may improve glucose levels	Rare—seen in unsupplemented total parenteral nutrition	Not seen	Meat, poultry, fish, liver, whole grains, brewer's yeast
Copper (Cu)	Assists in release of energy, helps in the production of red blood cells and absorption and transport of iron	Connective tissue abnormalities; altered immune, vascular, skeletal and central nervous system; anaemia	Liver impairment	Oysters, whole grains, legumes, nuts, organ meats, dark leafy green vegetables, dried fruits, cocoa, yeast
Fluoride (F)	Maintenance of health of bones and teeth, helps make teeth resistant to decay	Dental caries	Fluorosis (mottling of teeth enamel)	Fluoridated drinking water, tea, seafood
Iodine (I) *Note:* Iodine content of dairy foods has declined with changes in milk production practices and less use of iodophors in sanitisers.	Component of two thyroid hormones that regulate growth, development and metabolic rate	Iodine deficiency disorders, including goitre and hypothyroidism; effects on foetus including stillbirth and congenital abnormalities; mental and physical impairment	Impaired thyroid function—i.e., elevated thyroid stimulating hormone	Iodised salt, seafood, bread (with iodine fortified bread flour), dairy products, plants grown in iodine-rich soil and animals fed on those plants
Iron (Fe) *Note:* Haem (flesh) sources are absorbed more efficiently than non-haem (plant) sources.	Key component of haemoglobin and myoglobin in blood cells, component of many enzymes	Iron deficiency anaemia, fatigue, adverse pregnancy outcome, delayed development in infant, impaired cognitive function and immunity	Gastrointestinal irritation, systemic toxicity	Lean red meat, liver, kidney, heart, chicken, fish, dark green leafy vegetables, wholemeal bread, legumes, eggs, dried fruit, iron-fortified breakfast cereals
Manganese (Ma)	Cofactor for several enzymes involved in the metabolism of carbohydrate, cholesterol and amino acids; bone formation	Impaired growth, impaired reproductive function, impaired glucose tolerance, impaired skeletal development	Unlikely—muscle pain, fatigue, tremor, memory impairment, altered reflexes	Nuts, whole grains, leafy vegetables, tea
Molybdenum (Mo)	Cofactor for several enzymes	Rare—seen in unsupplemented total parenteral nutrition, genetic defects	Unclear—possible mild renal failure, weight loss	Legumes, cereals, nuts
Selenium (Se)	Antioxidant, regulates thyroid hormone	Keshan disease—cardiomyopathy (cardiac enlargement, heart failure, arrhythmia, death)	Brittle hair and nails, hair loss	Brazil nuts, seafood, meat, whole grains, fruits and vegetables (depending on soil content)
Zinc (Zn)	Wound healing; normal taste, smell and sight; component of many enzymes; hard structure of bones; male sexual maturation	Impaired growth velocity, possible pregnancy risks, impaired immune system, diarrhoea, alopecia, delayed sexual development, impotency, impaired appetite, altered taste and vision	Suppressed immune function, decreased high-density lipoprotein cholesterol	Lean meat, liver, kidney, chicken, seafood (especially oysters), milk, whole grains, wholemeal bread, legumes, nuts

requirements for each of these nutrients are described in this chapter, along with the functions of and daily requirements for macronutrients.

DIETARY GUIDELINES FOR AUSTRALIA AND NEW ZEALAND

Australian Dietary Guidelines

1. To achieve and maintain a healthy weight, be physically active and choose amounts of nutritious food and drinks to meet your energy needs.
   - Children and adolescents need to eat sufficient nutritious foods to grow and develop normally. They need to be physically active every day and their growth needs to be checked regularly.
   - Older people need to eat nutritious foods and keep physically active to help maintain muscle strength and a healthy weight.
2. Enjoy a wide variety of nutritious foods from these five food groups every day:
   - Plenty of vegetables of different types and colours, and legumes/beans
   - Fruit
   - Grain (cereal) foods, mostly wholegrain and/or high cereal fibre varieties, such as breads, cereals, rice, pasta, noodles, polenta, couscous, oats, quinoa and barley
   - Lean meats and poultry, fish, eggs, tofu, nuts and seeds, and legumes/beans
   - Milk, yoghurt, cheese and/or their alternatives, mostly reduced fat.

   And drink plenty of water.
3. Limit intake of foods containing saturated fat, added salt, added sugars and alcohol.
   a. Limit intake of foods high in saturated fat such as many biscuits, cakes, pastries, pies, processed meats, commercial burgers, pizza, fried foods, potato chips, crisps and other savoury snacks.
      - Replace high-fat foods that contain predominately saturated fats such as butter, cream, cooking margarine, coconut and palm oil with foods that contain predominately polyunsaturated and monounsaturated fats such as oils, spreads, nut butters/pastes and avocado.
      - Low-fat diets are not suitable for children under the age of two years.

   b. Limit intake of foods and drinks containing added salt.
      - Read labels to choose lower sodium options among similar foods.
      - Do not add salt to foods in cooking or at the table.

   c. Limit intake of foods and drinks containing added sugars such as confectionary, sugar-sweetened soft drinks and cordials, fruit drinks, vitamin waters, energy and sports drinks.

   d. If you choose to drink alcohol, limit intake. For women who are pregnant, planning a pregnancy or breastfeeding, not drinking alcohol is the safest option.
4. Encourage, support and promote breastfeeding.
5. Care for your food; prepare and store it safely. (NHMRC 2013: v)

Food and Nutrition Guidelines, New Zealand

Food and Nutrition Guidelines for Healthy Infants and Toddlers (Aged 0–2 years)

1. Maintain healthy growth and development of your baby and toddler by providing them with appropriate food and physical activity opportunities every day.
2. Exclusively breastfeed your baby until your baby is ready for and needs extra food—this will be at around six months of age.
3. When your baby is ready, introduce them to appropriate complementary foods and continue to breastfeed until they are at least one year of age, or beyond.
4. Increase the texture, variety, flavour and amount of food offered so that your baby receives a complementary intake of nutrients, especially iron and vitamin C, and is eating more family foods by one year of age.
5. For your baby, prepare or choose pre-prepared complementary foods with no added fat, salt, sugar, honey or other sweeteners.
6. If your baby is not fed breast milk, then use an infant formula as the milk source until your baby is one year of age.
7. Each day offer your toddler a variety of nutritious foods from each of the four major food groups, which are:
   - vegetables and fruit
   - breads and cereals, including some wholemeal
   - milk and milk products or suitable alternatives
   - lean meat, poultry, seafood, eggs, legumes, nuts and seeds.

8. For your toddler, prepare foods or choose pre-prepared foods, drinks and snacks that:
   - are low in salt, but if using salt, use iodised salt
   - have little added sugar (and limit your toddler's intake of high-sugar foods).
9. Provide your toddler with plenty of liquids each day such as water, breast milk, or cows' milk (but limit cows' milk to about 500 millilitres per day).
10. Do not give your infant or toddler alcohol, coffee, cordials, juice, soft drinks, tea (including herbal teas), and other drinks containing caffeine.
11. Purchase, prepare, cook and store food in ways to ensure food safety. (Ministry of Health 2008)

Food and Nutrition Guidelines for Healthy Children and Young People (Aged 2–18 years)

1. Eat a variety of foods from each of the four major food groups each day:
   - vegetables and fruit, including different colours and textures
   - breads and cereals, increasing wholegrain products as children increase in age
   - milk and milk products or suitable alternatives, preferably reduced or low-fat options
   - lean meat, poultry, fish, shellfish, eggs, legumes, nuts and seeds. (To reduce the risk of choking, do not give small hard foods—such as whole nuts and large seeds—until children are at least five years old.)
2. Eat enough for activity, growth and to maintain a healthy body size.
   - Eat regularly over the day: that is, have breakfast, lunch and dinner, and include in-between snacks for young children or if hungry.
3. Prepare foods or choose pre-prepared foods, snacks and drinks that are:
   - low in fat, especially saturated fat
   - low in sugar, especially added sugar
   - low in salt (if using salt, use iodised salt).
4. Drink plenty of water during the day. Include reduced- or low-fat milk every day.
   - Limit drinks such as fruit juice, cordial, fruit drink, fizzy drinks (including diet drinks), sports drinks and sports water.
   - Energy drinks or energy shots are not recommended for children or young people.
   - Do not give children less than thirteen years of age coffee or tea. If young people (thirteen years and older) choose to drink coffee or tea, limit to one to two cups per day.
5. Alcohol is not recommended for children or young people.
6. Eat meals with family, or whanau, as often as possible.
7. Encourage children and young people to be involved in shopping, growing and cooking family meals.
8. Purchase, prepare, cook and store food in ways to ensure food safety.
9. Be physically active.
   - Take part in regular physical activity, aiming for 60 minutes or more of moderate to vigorous activity each day.
   - Spend less than two hours a day (out of school time) in front of television, computers and gaming consoles.
   - Be active in as many ways as possible, for example, through play, cultural activities, dance, sport and recreation, jobs and going from place to place.
   - Be active with friends and whanau, at home, school, and in your community. (Ministry of Health 2012a)

Food and Nutrition Guidelines for Healthy Adults (Aged 19–64 years)

1. Maintain a healthy body weight by eating well and by daily physical activity.
   - Aim to do at least 30 minutes of moderate intensity physical activity on most if not all days of the week and if possible add some vigorous exercise for extra health and fitness.
2. Eat well by including a variety of healthy foods from each of the four major food groups.
   - Eat plenty of vegetables and fruits.
   - Eat plenty of breads and cereals, preferably wholegrain.
   - Have milk and milk products in your diet, preferably reduced- or low-fat options.
   - Include lean meat, poultry, seafood, eggs or alternatives.
3. Prepare foods or choose pre-prepared foods, drinks and snacks:
   - with minimal added fat, especially saturated fat
   - that are low in salt (if you use salt, choose iodised salt)
   - with little added sugar (limit your intake of high-sugar foods).
4. Drink plenty of liquids each day, especially water.
5. If you choose to drink alcohol, limit your intake.

6. Purchase, prepare, cook and store food to ensure food safety. (Ministry of Health 2003)

Food and Nutrition Guidelines for Healthy Older Adults (Aged 65+ years)

1. Maintain a healthy body weight by eating well and by daily physical activity.
2. Eat well by including a variety of nutritious foods from each of the four major food groups each day.
   - Eat plenty of vegetables and fruit.
   - Eat plenty of breads and cereals, preferably wholegrain.
   - Have milk and milk products in your diet, preferably reduced- or low-fat options. (Frail older people may require full-fat or standard milk and milk products to meet their energy requirements.)
   - Include lean meat, poultry, seafood, eggs, nuts, seeds or legumes.
3. Drink plenty of liquids each day, especially water.
4. Prepare foods or choose pre-prepared foods, drinks and snacks:
   - with minimal added fat, especially saturated fat
   - that are low in salt (if using salt, choose iodised salt)
   - with little added sugar (limit your intake of high-sugar foods).
5. Take opportunities to eat meals with other people.
6. Eat three meals every day. Nutritious snacks are recommended, especially for those who are underweight or have a small appetite.
7. Consider food safety when purchasing, preparing, cooking and storing food.
8. If choosing to drink alcohol, limit your intake.
9. Be physically active by including at least 30 minutes of moderate-intensity physical activity on most days of the week. (Ministry of Health 2013a)

Food and Nutrition Guidelines for Healthy Pregnant and Breastfeeding Women

1. Maintain a healthy body weight by eating well and by daily physical activity:
   - 30 minutes of moderate-intensity physical activity on most if not all days of the week.
2. Eat well by including a variety of nutritious foods from each of the four major food groups each day.
   - Eat plenty of vegetables and fruit.
   - Eat plenty of breads and cereals, preferably wholegrain.
   - Have milk and milk products in your diet, preferably reduced- or low-fat options.
   - Include lean meat, poultry, seafood, eggs, nuts, seeds or legumes.
3. Prepare foods or choose pre-prepared foods, drinks and snacks:
   - with minimal added fat, especially saturated fat
   - that are low in salt; if using salt, choose iodised salt
   - with little added sugar; limit your intake of high-sugar foods.
4. Drink plenty of liquids each day, especially water.
5. It is best not to drink alcohol during pregnancy.
6. Purchase, prepare, cook and store food to ensure food safety. (Ministry of Health 2006)

RECOMMENDED DAILY SERVINGS AND STANDARD SERVE SIZES OF FOODS

Australia

Recommended daily servings of foods for all Australians aged one year and beyond are shown in Table 1.2.

A standard serve of vegetables (with canned varieties, choose those with no added salt) is about 75 grams (100–350 kilojoules) or:

- ½ cup cooked green or orange vegetables (for example, broccoli, spinach, carrots or pumpkin)
- ½ cup cooked, dried or canned beans, peas or lentils
- 1 cup green leafy or raw salad vegetables
- ½ cup sweet corn
- ½ medium potato or other starchy vegetables (sweet potato, taro or cassava)
- 1 medium tomato.

A standard serve of fruit is about 150 grams (350 kilojoules) or:

- 1 medium apple, banana, orange or pear
- 2 small apricots, kiwi fruits or plums
- 1 cup diced or canned fruit (with no added sugar)

or only occasionally:

- 125 millilitres (½ cup) fruit juice (with no added sugar)
- 30 grams dried fruit (for example, 4 dried apricot halves)
- 1½ tablespoons of sultanas.

A standard serve of a grain (cereal) food (mostly wholegrain and/or high cereal fibre varieties) (500 kilojoules) is:

- 1 slice (40 grams) bread
- ½ medium (40 grams) roll or flat bread

Table 1.2 Recommended daily servings of foods for females (males) by age in Australia

Food	Energy per serve	Number of recommended daily serves by age (years)								
		1	2–3	4–8	9–11	12–13	14–18	19–50	51–70	70+
Vegetables and legumes/beans	100–350 kJ	2–3	2½	4½	5	5 (5½)	5 (5½)	5 (6)	5 (6½)	5
Fruit	350 kJ	½	1	1½	2	2	2	2	2	2
Grain (cereal) foods, mostly wholegrain and cereal fibre varieties	500 kJ	4	4	4	4 (5)	5 (6)	7	6	4 (6)	3 (4½)
Lean meat and poultry, fish, eggs, tofu, nuts and seeds, and legumes/beans	500–600 kJ	1	1	1½	2½	2½	2½	2½ (3)	2 (2½)	2 (2½)
Milk, yoghurt, cheese, and/or alternatives, mostly reduced fat	500–600 kJ	1–1½	1½	1½ (2)	3 (2½)	3½	3½	2½	4 (2½)	4 (3½)

Note: To meet additional energy needs, extra serves from the five food groups or unsaturated spreads and oils, or discretionary foods may be needed by children who are not overweight but are taller, more active or are at the upper end of their age band. An allowance for unsaturated spreads and oils for cooking, or nuts and seeds can be included in the following quantities: 4–5 grams per day for children two to three years of age, 7–10 grams per day for children three to twelve years of age, 11–15 grams per day for children twelve to thirteen years of age and 14–20 grams per day for adolescents fourteen to eighteen years of age. To meet additional energy needs for adults, extra serves from the five food groups or unsaturated spreads and oils, or discretionary foods may be needed only by those adults who are taller or more active, but not overweight. An allowance for unsaturated spreads and oils for cooking, or nuts and seeds can be included in the following quantities: 28–40 grams per day for men less than 70 years of age and 14–20 grams per day for women and older men.

Source: NHMRC (2013).

- ½ cup (75–120 grams) cooked rice, pasta, noodles, barley, buckwheat, semolina, polenta, bulgur or quinoa
- ½ cup (120 grams) cooked porridge
- ⅔ cup (30 grams) wheat cereal flakes
- ¼ cup (30 grams) muesli
- 3 (35 grams total) crispbreads
- 1 (60 grams) crumpet
- 1 small (35 grams) English muffin or scone.

A standard serve of lean meat and poultry, fish, eggs, nuts and seeds, and legumes/beans (500–600 kilojoules) is:

- 65 grams cooked lean meats such as beef, lamb, veal, pork, goat or kangaroo (about 90–100 grams raw) (weekly limit of 455 grams)
- 80 grams cooked lean poultry such as chicken or turkey (100 grams raw)
- 100 grams cooked fish fillet (about 115 grams raw weight) or 1 small can of fish
- 2 large (120 grams total) eggs
- 1 cup (150 grams) cooked or canned legumes/beans such as lentils, chickpeas or split peas (preferably with no added salt) (only to be used occasionally as a substitute for other foods in the group)
- 170 grams tofu
- 30 grams nuts, seeds, peanut or almond butter or tahini or other nut or seed paste (this amount for nuts and seeds gives approximately the same amount of energy as the other foods in this group but will provide less protein, iron or zinc).

A standard serve of milk, yoghurt, cheese (choose mostly reduced fat) and/or alternatives (500–600 kilojoules) is:

- 1 cup (250 millilitres) fresh, UHT long-life, reconstituted powdered milk or buttermilk
- ½ cup (125 millilitres) evaporated milk
- 2 slices (40 grams) or 4 × 3 × 2 cm cube (40 grams) of hard cheese, such as cheddar
- ½ cup (120 grams) ricotta cheese
- ¾ cup (200 grams) yoghurt
- 1 cup (250 millilitres) soy, rice or other cereal drink with at least 100 milligrams of added calcium per 100 millilitres. (Adapted from NHMRC 2013)

If foods from this group are not eaten the following foods contain about the same amount of calcium as a serve of milk, yoghurt, cheese or alternatives (note: the energy [kilojoules] content of some of these serves [especially nuts] is higher):

- 100 grams almonds with skin
- 60 grams sardines (including bones), canned in water

- ½ cup (100 grams) canned pink salmon with bones
- 100 grams firm tofu (check the label as calcium levels vary). (Adapted from NHMRC n.d.)

New Zealand

Recommended daily servings of foods for all New Zealanders aged one year and beyond are shown in Table 1.3.

Vegetables and fruits

- 1 medium potato, taro or kumara (135 grams)
- ½ cup of cooked vegetables (50–80 grams)
- ½ cup of salad (60 grams)
- 1 tomato (80 grams)
- 1 apple, pear, banana or orange (130 grams)
- 2 small apricots or plums (100 grams)
- ½ cup of fresh fruit salad (120 grams)
- ½ cup of stewed or canned fruit (135 grams)

Breads and cereals

- 1 roll (50 grams)
- 1 medium slice of bread (26 grams)
- 1 medium slice of *rēwena* bread
- 1 cup of cornflakes or rice bubbles (30 grams) or 2 breakfast wheat biscuits (34 grams)
- ½ cup of cooked cereal (e.g., porridge) (130 grams)
- 1 cup of cooked pasta, noodles or rice (150 grams)
- 1 cup of cassava or tapioca
- 2 plain sweet biscuits (14 grams)

Milk and milk products

- 1 cup of reduced- or low-fat milk (250 millilitres)
- 1 bottle of reduced- or low-fat yoghurt (150 grams)
- 2 slices or ½ cup of grated cheese, e.g., Edam (40 grams)

Lean meats, chicken, seafood, eggs, legumes, nuts and seeds

- 2 slices of cooked lean meat (100 grams), e.g., roast lamb, chicken, beef or pork
- ¾ cup of mince or casserole (195 grams)
- 1 medium fillet of fish or steak (100–120 grams)
- 2 chicken drumsticks or 1 chicken leg
- 1 medium paua or kina (100–120 grams)
- 1 egg
- ¾ cup of cooked dried beans (e.g., baked beans)
- ⅓ cup of nuts or seeds (50 grams)
- ¾ cup of tofu (200 grams) (Ministry of Health 2013b)

ANTHROPOMETRY

Anthropometry is the term used to describe measurements of humans. Weight and height are two measurements that many people might immediately think of regarding human measurement; however, there are many other methods for measuring. Some of the more commonly used methods are described below and are important to understand in order to be able to interpret health reports and population profiles. (For more information, see Department of Health 2010.)

Body weight

This is the total mass of all components of the body—that is, the sum of bone, protein (tendons, ligaments

Table 1.3 Recommended daily servings of foods by age in New Zealand

Food	Energy per serve	Number of recommended daily serves by age (years)					
		1[a]	2–5	5–12	13–18	19–64	65+
Vegetables	100–300 kJ	2	2	3	3+	3+	3+
Fruit[b]	180–300 kJ	2	2	2	2+	2+	2+
Breads and cereals	300–700 kJ	3	4+	5+	6+	6+	6+
Milk and milk products	350–600 kJ	2	2–3	2–3	3+	2+	3+
Lean meats, chicken, seafood, eggs, legumes, nuts and seeds	350–1200 kJ	1[c]	1[cd]	1[ce]	2[cf]	1+	1+

a Calculated from sample menu for 1–2-year-old children (Ministry of Health 2008).
b No more than one serving of fruit juice or one serving of dried fruit.
c To reduce the risk of choking, do not give small hard foods—such as whole nuts and large seeds—until children are at least five years old.
d Vegetarian preschoolers: at least one to two servings of legumes, nuts or seeds.
e Vegetarian schoolchildren: at least two servings of legumes, nuts or seeds.
f Vegetarian older children: at least three servings of legumes, nuts or seeds.

Source: Ministry of Health (2012b, 2013b).

and muscles), fat, **glycogen** and body water weights. Although weight can be useful for examining changes over time, it gives no indication of body composition, as people can be overweight without being over fat, and conversely people can be underweight but over fat.

Body mass index

Body mass index (BMI) is a ratio of height to weight; in other words, it is an equation that adjusts weight for height. It is commonly used in large epidemiological studies, including Australian and New Zealand national health surveys. BMI values are age independent and the same for both sexes. BMI is calculated using the following formula:

$$\text{BMI} = \frac{\text{weight (kilograms)}}{\text{height (metres)}^2}$$

The World Health Organization's (WHO) reference ranges have been classified to indicate healthy weight range—that is, the weight range where there are lowest risk factors for developing many chronic diseases in both sexes. These are described in Table 1.4.

There are some limitations in using BMI:

- For people aged 65 years and over, it may be appropriate to adjust BMI ranges to:
  Underweight less than 22
  Healthy weight range 22–30
  Overweight greater than 30
- BMI does not differentiate between lean tissue (muscle) mass and fat mass; therefore, it may overestimate muscular physiques, as muscle weighs more than fat.
- BMI relies on accurate recording of height.
- As there may be differences in 'normal' BMI according to ethnicity, questions may be raised as to whether different classifications should be applied for different ethnicities. The WHO (n.d.) states that the cut-offs described in Table 1.4 are appropriate for Caucasian, Asian, African-American and Polynesian populations.

Table 1.4 International BMI classifications and associated chronic disease risks for adults (18–64 years)

Classification	BMI	Chronic disease risk
Underweight	**< 18.5**	Low[a]
Normal range	**18.5–24.9**	Average
Overweight	**≥ 25.0**	
Pre-obese	25.0–29.9	Increased
Obese	**≥ 30.0**	
Class 1	30.0–34.9	Moderate
Class 2	35.0–39.9	Severe
Class 3	≥ 40.0	Very severe

a However, there is risk of increased mortality and morbidity from other causes.

Note: BMI: body mass index.

Source: Adapted from WHO (n.d.).

Glycogen is a polysaccharide that is the principal storage form of glucose in humans. It is found primarily in the liver and muscle tissue.

Waist to hip ratio

This is derived from the following equation:

$$\text{Waist to hip ratio} = \frac{\text{waist circumference}}{\text{hip circumference}}$$

Waist to hip ratio provides information about the fat distribution of individuals, but it is non-specific regarding the individual's fatness. Relationships between increased waist to hip ratio and cardiovascular disease (CVD), stroke, high blood pressure, **insulin resistance** and gallbladder disease have been reported. This is considered a good predictor of mortality in older people. Abdominal obesity cut-offs for waist to hip ratio are greater than 0.90 for males and greater than 0.85 for females.

Waist circumference

An unconverted measure of waist circumference alone may be considered superior to the waist to hip ratio or BMI as an index of general obesity. Waist circumference can be used to assess risk of developing chronic disease, as it is an indicator of internal fat deposits around the kidney, liver, pancreas and heart. Fat that is deposited predominantly around the hips and buttocks does not appear to have the same chronic disease risk. The cut-offs for waist circumference (for Caucasian men, Caucasian and Asian women) are given below:

Increased risk
- Males greater than 94 centimetres
- Females greater than 80 centimetres

Greatly increased risk
- Males greater than 102 centimetres
- Females greater than 88 centimetres

Insulin is a hormone produced by the pancreas that regulates the amount of glucose in the blood. When blood glucose is high, insulin is released, which promotes glucose uptake and storage by the cells.

Insulin resistance is a general term used to describe any defect in insulin action and is divided into abnormalities of insulin sensitivity and insulin responsiveness.

Energy density relates to the amount of energy provided per gram of food. Of the nutrients that contribute energy, fat is the most energy dense, and fibre is the least energy dense.

ENERGY BALANCE

Body weight is stable when energy consumed is equal to energy expended (energy balance) as shown in Figure 1.1. When energy consumed is greater than expended, weight increases (positive energy balance). When energy consumed is less than expended, weight decreases (negative energy balance). Estimating energy requirements is affected by many factors:

- *Age*—**Basal metabolic rate** (BMR) declines as lean body mass decreases.
- *Body composition and body size*—Taller people have more surface area, and heavier people have a higher BMR.
- *Gender*—Men generally have a higher BMR.
- *Growth*—BMR is high in people who are growing.
- *Physical activity*—Activities are clustered by intensity and vary considerably.

Energy consumed

Carbohydrate, fat and protein are the main energy-yielding nutrients. Alcohol and fibre also contribute energy. These all differ in their **energy density**, as described below. Energy is released from bonds within these nutrients as they are broken down. Water, vitamins, minerals and other bioactive compounds do not provide energy. Energy is measured in kilojoules (kJ) or kilocalories (kcal or Cal): 1 kcal ~ 4.2 kJ. The amount of energy provided by each gram of alcohol, carbohydrate, fat, fibre and protein is shown below:

Basal metabolic rate (BMR) is the amount of energy needed to support the basic processes of life—e.g., breathing, heartbeat, blinking, filtering blood. Energy expended as BMR is involuntary and supports the functions of many organs in the body, including the heart, lungs, kidneys, liver, intestine, nervous system, reproductive organs, skin and muscles.
It accounts for approximately two-thirds of energy expenditure.

- *Alcohol* 29 kilojoules
- *Carbohydrate* 17 kilojoules
- *Fat* 37 kilojoules
- *Fibre* 8 kilojoules
- *Protein* 17 kilojoules.

Energy expended

The body expends ('burns') energy via adaptive thermogenesis, basal metabolic activities, physical activity and thermic effect of food. These energy requirements differ from person to person and are affected by age, gender, weight and height.

- *Adaptive thermogenesis* is an adjustment in energy expenditure related to environmental changes—e.g., increased energy required to keep warm (shivering) or cool down (production of sweat).
- *Basal metabolism activities* support the basic processes of life.
- *Physical activity* is voluntary energy expenditure and is the most variable and changeable component of energy expenditure. The intensity, frequency and duration of physical activity will vary, and therefore the contribution to the total amount of energy expended can vary between individuals. The range often averages between 30 and 50 per cent.
- *Thermic effect of food* is estimated at 10 per cent of total energy intake and involves the breakdown

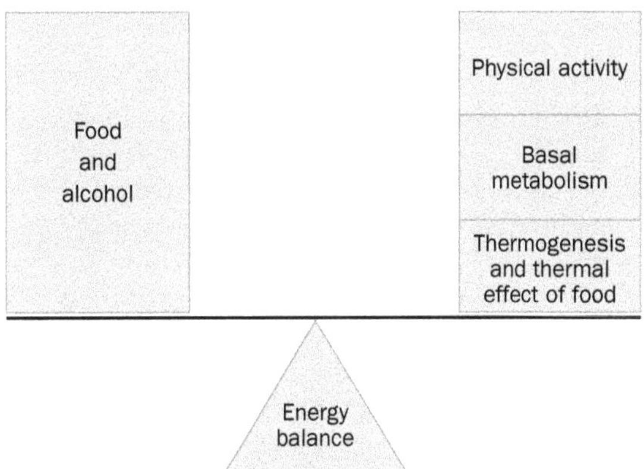

Figure 1.1 Energy balance equation

(digestion) and absorption of foods. Nutrients differ in their thermic effects:

Alcohol	15–20 per cent
Carbohydrate	5–10 per cent
Fat	0–5 per cent
Protein	20–30 per cent.

Overweight and obesity

A positive energy balance (when energy intake is greater than expenditure) can lead to weight gain. Overweight and obesity are major public health issues in Australia and New Zealand. The health implications of overweight and obesity include increased risk of many chronic diseases, such as type 2 **diabetes** mellitus (T2DM), CVD, hypertension and stroke. These are discussed in detail in Chapter 11.

In 2011–12, in Australia 69.7 per cent of males and 55.7 per cent of females were overweight or obese (ABS 2013). These results have increased by approximately 2 per cent over a four-year period. An estimated nine million Australians were overweight in 2013, and five million were obese. Obesity is the leading cause of premature death and illness in Australia. Aboriginals and Torres Strait Islanders are 1.9 times more likely to be obese than non-Indigenous Australians (Monash Obesity and Diabetes Institute 2013).

In New Zealand, **prevalence** of overweight and obesity is increasing. Thirty-one per cent of adults in 2011–12 were obese (equating to one million people), increasing by 13 per cent over fifteen years. Approximately 34 per cent of the population are overweight. And in trends similar to those seen in Australia, obesity rates are higher among Indigenous groups in New Zealand, including Pacific adults (68 per cent) and Māori adults (48 per cent) (Ministry of Health 2013c).

Diabetes has two main types. Type 1 diabetes mellitus (T1DM) develops when the pancreas is unable to produce any insulin. This typically results from the autoimmune destruction of the pancreatic beta-cells, which produce insulin. Type 2 diabetes mellitus (T2DM) is the most common form of diabetes and is characterised by a reduced production of insulin and/or an inability of the body tissues to respond fully to insulin (also known as insulin resistance). T2DM affects more than 85 per cent of the total number of people with diabetes in Australia.

Prevalence is the proportion of the population demonstrating a particular characteristic. In food and nutrition terms, it is usually associated with the food- and nutrient-related health and wellbeing of the population.

Underweight and malnourishment

A negative energy balance (when energy expenditure is greater than intake) can result in weight loss. Causes for negative energy balance can be poor appetite, intentional dieting, reduced ability to eat, reduced food availability, medication interactions, impaired digestion or absorption, increased requirements or nutrient losses and acute and/or chronic illness. Underweight, malnourishment and undernutrition are terms used interchangeably, and they exist on a spectrum. A person who is very underweight is at increased risk of reduced immunity, delayed wound healing and decreased muscle strength, mood and self-esteem. Underweight is common in the elderly, in particular those in long-care institutions such as nursing homes. Issues associated with underweight are described in Chapters 12 and 13.

NUTRIENT REFERENCE VALUES

Estimated energy requirements

The estimated energy requirement (EER) is the approximate amount of energy needed for involuntary physiological functions and voluntary movement, usually expressed as megajoules (MJ) per day. For these tables, energy is expressed as MJ, but for the following chapters kilojoules (kJ) will be used (1 MJ equals 1000 kJ). It is the amount predicted to maintain energy balance in a healthy person, based on age, gender, weight, height and level of physical activity. EERs for children, pregnant and lactating women also consider the energy needed for growth and development or milk supplies. EERs can be estimated by calculating the BMR, via the **Schofield equation**, multiplied by an activity factor (for physical activity). A factor for the thermic effect of food and adaptive thermogenesis (if applicable) may also be applied.

Reference body weights are useful as they generally describe the midpoint of a healthy weight range. The EER details are described in Table 1.5 for infants and young children (0–2 years), Table 1.6 for children and adolescents (3–18 years), and Table 1.7 for adult males and females (19 years and older). Notes regarding requirements for pregnancy and lactation are

> The **Schofield equation** is used to estimate the BMR of individuals. It is expressed in kilojoules per day. The equation is based on age, gender and body weight (kilograms). An activity factor is then applied to the BMR calculated with the Schofield equation. (Details of the equation can be found in Schofield 1985.) There are associated standard errors of estimations, so clinicians using the equation should adjust the answer upwards if a person is more muscular and downwards if more obese. The answer should also be adjusted upwards for people at the lower end of an age range and downwards for people at the higher end.

given in Table 1.8. The requirements are the same in Australia and New Zealand.

Macronutrients, water, vitamins and minerals

Australian and New Zealand governments have provided guidance for the public regarding the recommended daily intakes of essential nutrients at each life stage, according to gender. These recommendations are revised over time, and the current recommendations, introduced in 2006, are called nutrient reference values (NRVs). The NRVs comprise four different forms of recommendations, and these are described below:

- *Estimated average requirement (EAR)* is used for population nutritional assessment. The EAR daily nutrient levels are estimated to meet the requirements of half the healthy individuals in a particular life stage and gender group within a population.
- *Adequate intake (AI)* is based on the mean intake of the population who are known not to have a deficiency disease. Where the EAR is unclear or unknown, an AI is used—e.g., for fibre, sodium and potassium.
- *Recommended dietary intake (RDI)* is useful for individual assessment. The RDI provides the average daily dietary intake level that is sufficient to meet the nutrient requirements of nearly all healthy individuals (97–8 per cent) in a particular life stage and gender group. RDIs are derived from the EAR:

$$\text{RDI} = \text{EAR} + 2 \text{ coefficients of variation}$$

- *Upper limit (UL)* is used to determine upper levels likely to pose no adverse health effects for almost all of the population.

Table 1.5 Reference body weights and daily EERs for infants and young children (0–2 years) in Australia and New Zealand

Age (months)	Weight (kg), by gender		kJ (kJ/kg body weight), by gender		EER (MJ), by gender	
	Male	Female	Male	Female	Male	Female
1	4.4	4.2	2000 (455)	1800 (429)	2.0	1.8
2	5.3	4.9	2400 (453)	2100 (429)	2.4	2.1
3	6.0	5.5	2400 (400)	2200 (400)	2.4	2.2
4	6.7	6.1	2400 (358)	2200 (361)	2.4	2.2
5	7.3	6.7	2500 (342)	2300 (343)	2.5	2.3
6	7.9	7.2	2700 (342)	2500 (347)	2.7	2.5
7	8.4	7.7	2800 (333)	2500 (325)	2.8	2.5
8	8.9	8.1	3000 (337)	2700 (333)	3.0	2.7
9	9.3	8.5	3100 (333)	2800 (329)	3.1	2.8
10	9.7	8.9	3300 (340)	3000 (337)	3.3	3.0
11	10.0	9.2	3400 (340)	3100 (337)	3.4	3.1
12	10.3	9.5	3500 (340)	3200 (337)	3.5	3.2
15	11.1	10.3	3800 (342)	3500 (340)	3.8	3.5
18	11.7	11.0	4000 (342)	3800 (345)	4.0	3.8
21	12.2	11.6	4200 (344)	4000 (345)	4.2	4.0
24	12.7	12.1	4400 (346)	4200 (347)	4.4	4.2

Note: EER: estimated energy requirement.

Source: NHMRC & Ministry of Health (2006).

Table 1.6 Reference body weights and heights and daily EERs for female (male) children and adolescents (3–18 years) in Australia and New Zealand

Age (years)	Height (m)	Weight[a] (kg)	BMR[b] (MJ/day)	EER (MJ), by PAL[c]					
				1.2	1.4	1.6	1.8	2.0	2.2
3	0.94 (0.95)	13.9 (14.3)	3.2 (3.4)	3.9 (4.2)	4.5 (4.9)	5.3 (5.6)	5.8 (6.3)	6.4 (6.9)	7.1 (7.6)
4	1.01 (1.02)	15.8 (16.2)	3.4 (3.6)	4.1 (4.4)	4.8 (5.2)	5.5 (5.9)	6.1 (6.6)	6.8 (7.3)	7.5 (8.1)
5	1.08 (1.09)	17.9 (18.4)	3.6 (3.8)	4.4 (4.7)	5.1 (5.5)	5.7 (6.2)	6.5 (7.0)	7.2 (7.8)	7.9 (8.5)
6	1.15 (1.15)	20.2 (20.7)	3.8 (4.1)	4.6 (5.0)	5.4 (5.8)	6.1 (6.6)	6.9 (7.4)	7.6 (8.2)	8.4 (9.0)
7	1.21 (1.22)	22.8 (23.1)	4.0 (4.3)	4.9 (5.2)	5.7 (6.1)	6.5 (7.0)	7.3 (7.8)	8.1 (8.7)	8.9 (9.5)
8	1.28 (1.28)	25.6 (25.6)	4.2 (4.5)	5.2 (5.5)	6.0 (6.4)	6.9 (7.3)	7.7 (8.2)	8.6 (9.2)	9.4 (10.1)
9	1.33 (1.34)	29.0 (28.6)	4.5 (4.8)	5.5 (5.9)	6.4 (6.8)	7.3 (7.8)	8.2 (8.8)	9.1 (9.7)	10.0 (10.7)
10	1.38 (1.39)	32.9 (31.9)	4.7 (5.1)	5.7 (6.3)	6.7 (7.3)	7.6 (8.3)	8.5 (9.3)	9.5 (10.4)	10.4 (11.4)
11	1.44 (1.44)	37.2 (35.9)	4.9 (5.4)	6.0 (6.6)	7.0 (7.7)	8.0 (8.8)	9.0 (9.9)	10.0 (11.0)	11.0 (12.0)
12	1.51 (1.49)	41.6 (40.5)	5.2 (5.8)	6.4 (7.0)	7.4 (8.2)	8.5 (9.3)	9.5 (10.5)	10.6 (11.6)	11.6 (12.8)
13	1.57 (1.56)	45.8 (45.6)	5.5 (6.2)	6.7 (7.5)	7.8 (8.7)	8.9 (10.0)	10.0 (11.2)	11.1 (12.4)	12.2 (13.6)
14	1.60 (1.64)	49.4 (51.0)	5.7 (6.6)	6.9 (8.0)	8.1 (9.3)	9.2 (10.6)	10.3 (11.9)	11.5 (13.2)	12.6 (14.6)
15	1.62 (1.70)	52.0 (56.3)	5.8 (7.0)	7.1 (8.5)	8.2 (9.9)	9.4 (11.2)	10.6 (12.6)	11.7 (14.0)	12.9 (15.4)
16	1.63 (1.74)	53.9 (60.9)	5.9 (7.3)	7.2 (8.9)	8.4 (10.3)	9.5 (11.8)	10.7 (13.2)	11.9 (14.7)	13.1 (16.2)
17	1.63 (1.75)	55.1 (64.6)	5.9 (7.6)	7.2 (9.2)	8.4 (10.7)	9.6 (12.2)	10.8 (13.7)	12.0 (15.2)	13.2 (16.7)
18	1.63 (1.76)	56.2 (67.2)	6.0 (7.7)	7.3 (9.4)	8.5 (10.9)	9.7 (12.5)	10.9 (14.0)	12.1 (15.6)	13.3 (17.1)

a Reference weights taken from Kuczmarski et al. (2000) (see also FNB:IOM 2002).

b BMRs estimated using Schofield (1985) equations for weight, height, and age groups three to ten and ten to eighteen.

c PALs incorporate relevant growth factor for age. They correspond to the following activities: 1.2: bed rest; 1.4: very sedentary; 1.6: light activity; 1.8: moderate activity; 2.0: heavy activity; 2.2: vigorous activity.

Note: BMI: body mass index; BMR: basal metabolic rate; EER: estimated energy requirement; PAL: **physical activity levels**. EERs were calculated using BMRs predicted from weight, height and age. The height and weight to age ratios may differ markedly in some ethnic groups. In this case, if BMI is in the acceptable range, it would be more relevant to use body weight as the main guide to current energy needs.

Source: Adapted from NHMRC & Ministry of Health (2006).

Physical activity level (PAL) is a factor relating to activity level that is applied to the BMR in order to estimate total energy expenditure. Realistically, PALs range from 1.20 to 2.20 (as PALs greater than 2.50 are difficult to sustain for long periods). A PAL of 1.20 equates to bed rest, while a PAL of 2.20 equates to heavy occupational work or being very active. PALs of 1.75 and above are consistent with good health.

Table 1.7 Reference body weights and heights for BMI of 22 and daily EERs for adult females (males) (19+ years) in Australia and New Zealand

Age (years)	Height (m)	Weight (kg)	BMR (MJ/day)	EER (MJ), by PAL[a]					
				1.2	1.4	1.6	1.8	2.0	2.2
19–30	1.5 (1.6)	49.5 (56.3)	5.2 (6.4)	6.1 (7.7)	7.1 (9.0)	8.2 (10.3)	9.2 (11.0)	10.2 (12.9)	11.2 (14.2)
	1.6 (1.7)	56.3 (63.6)	5.6 (6.9)	6.6 (8.3)	7.7 (9.7)	8.8 (11.0)	9.9 (12.4)	11.1 (13.8)	12.2 (15.2)
	1.7 (1.8)	63.6 (71.3)	6.0 (7.4)	7.2 (8.9)	8.4 (10.3)	9.6 (11.8)	10.8 (13.3)	12.0 (14.8)	13.2 (16.3)
	1.8 (1.9)	71.3 (79.4)	6.5 (7.9)	7.7 (9.5)	9.0 (11.1)	10.3 (12.6)	11.6 (14.2)	12.9 (15.8)	14.2 (17.4)
	1.9 (2.0)	79.4 (88.0)	7.0 (8.4)	8.4 (10.1)	9.7 (11.8)	11.1 (13.5)	12.5 (15.2)	13.9 (16.9)	15.3 (18.6)
31–50	1.5 (1.6)	49.5 (56.3)	5.2 (6.4)	6.3 (7.6)	7.3 (8.9)	8.4 (10.2)	9.4 (11.4)	10.4 (12.7)	11.5 (14.0)
	1.6 (1.7)	56.3 (63.6)	5.5 (6.7)	6.5 (8.0)	7.6 (9.4)	8.7 (10.7)	9.8 (12.1)	10.9 (13.4)	12.0 (14.8)
	1.7 (1.8)	63.6 (71.3)	5.7 (7.1)	6.8 (8.5)	8.0 (9.9)	9.1 (11.3)	10.3 (12.7)	11.4 (14.2)	12.5 (15.6)
	1.8 (1.9)	71.3 (79.4)	6.0 (7.5)	7.2 (9.0)	8.3 (10.4)	9.5 (11.9)	10.7 (13.4)	11.9 (14.9)	13.1 (16.4)
	1.9 (2.0)	79.4 (88.0)	6.2 (7.9)	7.5 (9.5)	8.7 (11.0)	10.0 (12.6)	11.2 (14.2)	12.5 (15.8)	13.7 (17.3)
51–70	1.5 (1.6)	49.5 (56.3)	4.9 (5.8)	6.0 (7.0)	6.9 (8.2)	7.9 (9.3)	8.9 (10.4)	9.8 (11.5)	10.9 (12.7)
	1.6 (1.7)	56.3 (63.6)	5.2 (6.1)	6.2 (7.3)	7.3 (8.6)	8.3 (9.8)	9.3 (11.1)	10.4 (12.3)	11.4 (13.6)
	1.7 (1.8)	63.6 (71.3)	5.4 (6.5)	6.5 (7.8)	7.6 (9.1)	8.7 (10.4)	9.8 (11.7)	10.7 (13.1)	12.0 (14.4)
	1.8 (1.9)	71.3 (79.4)	5.7 (6.9)	6.9 (8.3)	8.0 (9.6)	9.1 (11.1)	10.3 (12.4)	11.4 (13.8)	12.6 (15.2)
	1.9 (2.0)	79.4 (88.0)	6.0 (7.3)	7.2 (8.8)	8.4 (10.2)	9.6 (11.7)	10.8 (13.2)	12.0 (14.7)	13.2 (16.1)
>70	1.5 (1.6)	49.5 (56.3)	4.6 (5.2)	5.6 (6.3)	6.5 (7.3)	7.4 (8.3)	8.3 (9.4)	9.3 (10.4)	10.2 (11.5)
	1.6 (1.7)	56.3 (63.6)	4.9 (5.6)	5.9 (6.7)	6.9 (7.8)	7.8 (8.9)	8.8 (10.0)	9.8 (11.2)	10.8 (12.3)
	1.7 (1.8)	63.6 (71.3)	5.2 (6.0)	6.2 (7.1)	7.2 (8.3)	8.3 (9.5)	9.3 (10.7)	10.3 (11.9)	11.4 (13.1)
	1.8 (1.9)	71.3 (79.4)	5.5 (6.4)	6.6 (7.6)	7.7 (8.9)	8.7 (10.2)	9.8 (11.4)	10.9 (12.7)	12.0 (14.0)
	1.9 (2.0)	79.4 (88.0)	5.8 (6.8)	6.9 (8.1)	8.1 (9.5)	9.2 (10.8)	10.4 (12.2)	11.5 (13.5)	12.7 (14.9)

a PAL ranges from 1.20 (bed rest) to 2.20 (very active or heavy occupational work). PALs of 1.75 and above are consistent with good health. PALs below 1.40 are incompatible with moving around freely or earning a living. PALs above 2.50 are difficult to maintain for long periods.

Note: BMI: body mass index; BMR: basal metabolic rate; EER: estimated energy requirement; PAL: physical activity levels. A BMI of 22 is approximately the midpoint of the WHO (1998) healthy weight range (BMI 18.5–24.9). The original Schofield (1985) equations from which this table was derived used 60 years and above as the upper age category. For people aged 51–70 years, the estimates were derived by averaging those for the adults (31–50 years) and older adults (more than 70 years).

Source: Adapted from NHMRC & Ministry of Health (2006).

Table 1.8 Daily additional energy requirements for pregnant and lactating females (all ages) in Australia and New Zealand

Stage	Additional energy (MJ)
Pregnancy	
2nd trimester	1.4
3rd trimester	1.9
Lactation	2.0–2.1

Source: Adapted from NHMRC & Ministry of Health (2006).

RDIs, EARs and AIs are recommendations for avoidance of deficiency diseases. ULs relate to the avoidance of toxicity or other symptoms related to overconsumption. The following section gives RDIs and EARs for macronutrients and RDIs for vitamins and minerals unless there is no RDI, in which case AIs are used.

Macronutrients

Macronutrients are needed by the body in large quantities and include protein, fat, carbohydrate, dietary fibre and water.

Acceptable macronutrient distribution range (AMDR)

The Australian Dietary Guidelines recommend that the spread of energy from protein, fat and carbohydrate for the maintenance of health and prevention of chronic diseases should be as follows:

- 45–65 per cent from carbohydrate (approximately 240 grams per day for women and 300 grams for men)
- 20–35 per cent of total energy intake from fat (approximately 60 grams per day for women and 80 grams for men)
- 15–25 per cent from protein (approximately 95 grams per day for women and 120 grams for men). (NHMRC 2013)

Protein

Protein is an essential macronutrient that is broken down into small building blocks called amino acids and used to manufacture body proteins for structural and functional roles. Structural roles include building connective tissue, hair, skin, nail and body organs including the lungs, heart and digestive tract, cell membranes and muscles. Functional roles include production of enzymes or transport proteins. The NRVs include recommendations as to the amount of protein necessary for health according to life stage and gender, as shown in Table 1.9.

Table 1.9a Daily AIs for protein for infants (0–12 months) in Australia and New Zealand

Age (months)	Grams (g/kg body weight) of protein
0–6	10 (1.43)
7–12	14 (1.60)

Table 1.9b Daily EARs and RDIs for protein for children and adolescents (1–18 years) in Australia and New Zealand

Gender	Age (years)	Grams (g/kg body weight) of protein, by NRV	
		EAR	RDI
All	1–3	12 (0.92)	14 (1.08)
	4–8	16 (0.73)	20 (0.91)
Male	9–13	31 (0.78)	40 (0.94)
	14–18	49 (0.76)	65 (0.99)
Female	9–13	24 (0.61)	35 (0.87)
	14–18	32 (0.62)	45 (0.77)

Table 1.9c Daily EARs and RDIs for protein for adults (19+ years) in Australia and New Zealand

Gender	Age (years)	Grams (g/kg body weight) of protein, by NRV	
		EAR	RDI
Male	19–30	52 (0.68)	64 (0.84)
	31–50	52 (0.68)	64 (0.84)
	51–70	52 (0.68)	64 (0.84)
	>70	65 (0.86)	81 (1.07)
Female	19–30	37 (0.60)	46 (0.75)
	31–50	37 (0.60)	46 (0.75)
	51–70	37 (0.60)	46 (0.75)
	>70	46 (0.75)	57 (0.94)

Table 1.9d Daily EARs and RDIs for protein for pregnant and lactating females (14–51 years) in Australia and New Zealand

Stage	Age (years)	Grams (g/kg body weight) of protein, by NRV	
		EAR	RDI
Pregnancy (2nd and 3rd trimesters)	14–18	47 (0.82)	58 (1.02)
	19–30	49 (0.80)	60 (1.00)
	31–51	49 (0.80)	60 (1.00)
Lactation	14–18	51 (0.90)	63 (1.10)
	19–30	54 (0.88)	67 (1.10)
	31–51	54 (0.88)	67 (1.10)

Note: AI: adequate intake; EAR: estimated average requirement; NRV: nutrient reference value; RDI: recommended dietary intake.

Source: NHMRC & Ministry of Health (2006).

Fat

Fats are the most concentrated source of energy available from food, with each gram of fat contributing 37 kilojoules. In addition to being an energy source, fats are a component of lipoprotein (in cholesterol). They are categorised according to the number of double bonds within the carbon chain:

- *Saturated fat* has no double bonds; i.e., it has the maximum number of hydrogens attached to the carbon backbone.
- *Monounsaturated fat* has one double bond; i.e., it lacks two hydrogen atoms.
- *Polyunsaturated fat* has two or more double bonds; i.e., it lacks four or more hydrogen atoms. The location of the first double bond gives the polyunsaturated fat its omega number; e.g., omega-3 fats, also referred to as n-3 fats, have the first double bond at the third carbon from the methyl end, omega-6 (n-6) fats have the first double bond at the sixth carbon, and so on.

Some fats are targeted in the NRVs because of their beneficial functions, including the following:

- *Omega-3 (n-3) alpha-linolenic acid* decreases inflammation, blood clotting and cell membrane fluidity. It is an essential fatty acid as the body cannot synthesise it and it must be obtained from food.
- *Omega-3 (n-3) eicosapentaenoic acid (EPA), docosahexaenoic acid (DHA) and docosapentaenoic acid (DPA)* are long-chain omega-3 fats which reduce triglyceride levels, increase breakdown of fibrin and reduce blood pressure. They are considered essential fatty acids—the body can make only small amounts from alpha-linolenic acid.
- *Omega-6 (n-6) linoleic acid* optimises brain function, normal growth and development. It is an essential fatty acid as the body cannot synthesise it and it must be obtained from food.

The NRVs include recommendations for the amount of total fat and types of polyunsaturated fats necessary for health according to life stage and gender, as shown in Table 1.10.

Carbohydrate

Carbohydrates provide energy, and each gram contributes 17 kilojoules. Carbohydrate is used for energy storage and, when in the form of glucose, is the only energy supply to the brain. The NRVs do not provide recommendations for carbohydrate for any life stage except as an AI for infants (Table 1.11), as it is not known if infants can adequately rely on gluconeogenesis—that is, generate enough glucose from other energy sources for their needs. The fact that levels have not been set for other age groups does not reflect carbohydrate's importance as a key nutrient in the diet, as discussed throughout this text.

Fibre

Fibre is found in plant materials and consists of molecules that resist digestion and absorption in the small intestine, including polysaccharides (such as cellulose, hemicellulose, resistant starch and pectin), oligosaccharides (such as inulin and fructooligosaccharides) and lignin. Fibre can be beneficial by assisting with one or all of the following functions: laxation (passing a motion), decreasing blood cholesterol and improving blood glucose levels. Rich sources of fibre include oats, barley, psyllium, chia seeds, citrus fruit, legumes, whole grains, cereal fibres (for example, bran, vegetables and fruit) particularly with the skin intact. Individual requirements for fibre vary according to life stage, as described in Table 1.12. Note that no AI has been set for fibre intake for infants (aged from birth to twelve months).

Water

Water is a nutrient that does not provide any energy; however, it is needed by the body for functioning. Water contributes 50–80 per cent of human body weight; hence, the demands are high. The body is not capable of producing enough water to sustain its needs, so water is considered an essential nutrient. Individual requirements for water vary according to many metabolic demands, and AIs are described for various life stages in Table 1.13.

Vitamins and minerals

The function, food sources, signs of deficiency and of excess for essential vitamins and minerals were described at the beginning of this chapter. From these it is possible to understand the vital roles that vitamins and minerals perform in human function. The amount of each nutrient required at each stage of the lifespan varies, as it also does with gender. The NRVs for vitamins and minerals across various life stages are described in Table 1.14.

Table 1.10a Daily AIs for polyunsaturated fats for infants (0–12 months) in Australia and New Zealand

Age (months)	Type of fat	Amount of fat (g)
0–6	n-6	4.4
	n-3	0.5
	Total fats	31.0
7–12	n-6	4.6
	n-3	0.5
	Total fats	30.0

Table 1.10b Daily AIs for fat for children (1+ years), adolescents and adults in Australia and New Zealand

Gender	Age (years)	Amount of fat, by type of fat		
		Linoleic acid (n-6) (g)	Alpha-linolenic acid (n-3) (g)	DHA + EPA + DPA (total long-chain n-3) (mg)
All	1–3	5	0.5	40
	4–8	8	0.8	55
Male	9–13	10	1.0	70
	14–18	12	1.2	125
	19+	13	1.3	160
Female	9–13	8	0.8	70
	14–18	8	0.8	85
	19+	8	0.8	90

Table 1.10c Daily AIs for fat for pregnant and lactating females (14–50 years) in Australia and New Zealand

Stage	Age (years)	Amount of fat, by type of fat		
		Linoleic acid (n-6) (g)	Alpha-linolenic acid (n-3) (g)	DHA + EPA + DPA (total long-chain n-3) (mg)
Pregnancy	14–18	10	1.0	110
	19–50	10	1.0	115
Lactation	14–18	12	1.2	140
	19–50	12	1.2	145

Note: AI: adequate intake; DHA: docosahexaenoic acid; DPA: docosapentaenoic acid; EPA: eicosapentaenoic acid.

Source: NHMRC & Ministry of Health (2006).

Table 1.11 Daily AIs for carbohydrate for infants (0–12 months) in Australia and New Zealand

Age (months)	Amount of carbohydrate (g)
0–6	60
7–12	95

Note: AI: adequate intake.

Source: NHMRC & Ministry of Health (2006).

Table 1.12a Daily AIs for fibre for all ages in Australia and New Zealand

Gender	Age (years)	Amount of fibre (g)
All	1–3	14
	4–8	18
Male	9–13	24
	14–18	28
	19+	30
Female	9–13	20
	14–18	22
	19+	25

Table 1.12b Daily AIs for fibre for pregnant and lactating females (14–51 years) in Australia and New Zealand

Stage	Age (years)	Amount of fibre (g)
Pregnancy (2nd and 3rd trimesters)	14–18	25
	19–30	28
	31–51	28
Lactation	14–18	27
	19–30	30
	31–51	30

Note: AI: adequate intake.

Source: NHMRC & Ministry of Health (2006).

Table 1.13a Daily AIs for water for infants (0–12 months) in Australia and New Zealand

Age (months)	Amount of water (L)
0–6	0.7[a]
7–12	0.8[b]

Table 1.13b Daily AIs for water for all ages in Australia and New Zealand

Gender	Age (years)	Litres (cups) of water from fluids[c]	Total amount of water[d] (L)
All	1–3	1.0 (~4)	1.4
	4–8	1.2 (~5)	1.6
Male	9–13	1.6 (~6)	2.2
	14–18	1.9 (7–8)	2.7
	19+	2.6 (~10)	3.4
Female	9–13	1.4 (5–6)	1.9
	14–18	1.6 (~6)	2.2
	19+	2.1 (~8)	2.8

Table 1.13c Daily AIs for water for pregnant and lactating females (14–51 years) in Australia and New Zealand

Stage	Age (years)	Litres (cups) of water from fluids[c]	Total amount of water[d] (L)
Pregnancy (2nd and 3rd trimesters)	14–18	1.8 (~7)	2.4
	19–30	2.3 (~9)	3.1
	31–51	2.3 (~9)	3.1
Lactation	14–18	2.3 (~9)	2.9
	19–30	2.6 (~10)	3.5
	31–51	2.6 (~10)	3.5

a From breast milk or formula.
b From breast milk, formula, plain water, other beverages and food, including 0.6 litres as fluids.
c Including plain water, milk and other drinks.
d From food and fluids.

Note: AI: adequate intake.

Source: NHMRC & Ministry of Health (2006).

Table 1.14a Daily AIs for vitamins and minerals for infants (0–12 months) in Australia and New Zealand

Nutrient	*Micrograms*/milligrams of nutrients, by age (months)	
	0–6	**7–12**
Biotin	*5.0*	*6.0*
Calcium	210.00	270.00
Choline	125.00	150.00
Chromium	*0.2*	*5.5*
Copper	0.20	0.22
Fluoride	0.01	0.50
Total folate	*65.00*	*80.00*
Iodine	*90.0*	*110.0*
Iron	0.20	–[a]
Magnesium	30.00	75.00
Manganese	*3*	*600*
Molybdenum	*2.0*	*3.0*
Niacin equivalents	6.00	8.00
Pantothenic acid	1.70	2.20
Phosphorus	100.00	275.00
Potassium	400.00	700.00
Riboflavin	0.30	0.40
Selenium	*12.0*	*15.0*
Sodium	120.00	170.00
Thiamin	0.20	0.30
Total vitamin A equivalents	*250.0*	*430.0*
Vitamin B6	0.1	0.3
Vitamin B12	*0.4*	*0.5*
Vitamin C	25.00	30.00
Vitamin D	*5.0*	*5.0*
Vitamin E	4.00	5.00
Vitamin K	*2.0*	*2.5*
Zinc	2.00	–[b]

Table 1.14b Daily RDIs (AIs) for vitamins and minerals for children and adolescents (1–18 years) in Australia and New Zealand

Nutrient	*Micrograms*/milligrams of nutrients, by gender and age (years)					
	Female and male		**Female**		**Male**	
	1–3	**4–8**	**9–13**	**14–18**	**9–13**	**14–18**
Biotin	*(8.0)*	*(12.0)*	*(20.0)*	*(25.0)*	*(20.0)*	*(30.0)*
Calcium[c]	500.0	700.0	1000.0	1300.0	1000.0	1300.0
Choline	(200.0)	(250.0)	(375.0)	(400.0)	(375.0)	(550.0)
Chromium	*(11.0)*	*(15.0)*	*(21.0)*	*(25.0)*	*(25.0)*	*(35.0)*
Copper	(0.7)	(1.0)	(1.1)	(1.1)	(1.3)	(1.5)
Fluoride	(0.7)	(1.0)	(2.0)	(3.0)	(2.0)	(3.0)
Total folate	*150.0*	*200.0*	*300.0*	*400.0*	*300.0*	*400.0*
Iodine	*90.0*	*90.0*	*120.0*	*150.0*	*120.0*	*150.0*
Iron	9.0	10.0	8.0	15.0	8.0	11.0
Magnesium	80.0	130.0	240.0	360.0	240.0	410.0
Manganese	(2.0)	(2.5)	(2.5)	(3.0)	(3.0)	(3.5)
Molybdenum	*17.0*	*22.0*	*34.0*	*43.0*	*34.0*	*43.0*
Niacin equivalents	6.0	8.0	12.0	14.0	12.0	16.0

Nutrient	***Micrograms*/milligrams of nutrients, by gender and age (years)**					
	Female and male		**Female**		**Male**	
	1–3	**4–8**	**9–13**	**14–18**	**9–13**	**14–18**
Pantothenic acid	(3.5)	(4.0)	(4.0)	(4.0)	(5.0)	(6.0)
Phosphorus	460.0	500.0	1250.0	1250.0	1250.0	1250.0
Potassium	(2000)	(2300)	(2500)	(2600)	(3000)	(3600)
Riboflavin	0.5	0.6	0.9	1.1	0.9	1.3
Selenium	*25.0*	*330.0*	*50.0*	*60.0*	*50.0*	*70.0*
Sodium	(200.0–400.0)	(300.0–600.0)	(400.0–800.0)	(460.0–920.0)	(400.0–800.0)	(460.0–920.0)
Thiamin	0.5	0.6	0.9	1.1	0.9	1.2
Total vitamin A equivalents	*300.0*	*400.0*	*600.0*	*700.0*	*600.0*	*900.0*
Vitamin B6	0.5	0.6	1.0	1.2	1.0	1.3
Vitamin B12	*0.9*	*1.2*	*1.8*	*2.4*	*1.8*	*2.4*
Vitamin C	35.0	35.0	40.0	40.0	40.0	40.0
Vitamin D	*(5.0)*	*(5.0)*	*(5.0)*	*(5.0)*	*(5.0)*	*(5.0)*
Vitamin E	(5.0)	(6.0)	(8.0)	(8.0)	(9.0)	(10.0)
Vitamin K	*(25.0)*	*(35.0)*	*(45.0)*	*(55.0)*	*(45.0)*	*(55.0)*
Zinc	3.0	4.0	6.0	7.0	6.0	13.0

Table 1.14c Daily RDIs (AIs) for vitamins and minerals for adults (19+ years) in Australia and New Zealand

Nutrients	***Micrograms*/milligrams of nutrients, by gender and age (years)**							
	Female				**Male**			
	19–30	**31–50**	**51–70**	**70+**	**19–30**	**31–50**	**51–70**	**70+**
Biotin	*(25.0)*	*(25.0)*	*(25.0)*	*(25.0)*	*(30.0)*	*(30.0)*	*(30.0)*	*(30.0)*
Calcium	1000.0	1000.0	1300.0	1300.0	1000.0	1000.0	1000.0	1300.0
Choline	(425.0)	(425.0)	(425.0)	(425.0)	(550.0)	(550.0)	(550.0)	(550.0)
Chromium	*(25.0)*	*(25.0)*	*(25.0)*	*(25.0)*	*(35.0)*	*(35.0)*	*(3.05)*	*(35.0)*
Copper	(1.2)	(1.2)	(1.2)	(1.2)	(1.7)	(1.7)	(1.7)	(1.7)
Fluoride	(3.0)	(3.0)	(3.0)	(3.0)	(4.0)	(4.0)	(4.0)	(4.0)
Total folate	*400.0*	*400.0*	*400.0*	*400.0*	*400.0*	*400.0*	*400.0*	*400.0*
Iodine	*150.0*	*150.0*	*150.0*	*150.0*	*150.0*	*150.0*	*150.0*	*150.0*
Iron	18.0	18.0	8.0	8.0	8.0	8.0	8.0	8.0
Magnesium	310.0	320.0	320.0	320.0	400.0	420.0	420.0	420.0
Manganese	(5.0)	(5.0)	(5.0)	(5.0)	(5.5)	(5.5)	(5.5)	(5.5)
Molybdenum	*45.0*	*45.0*	*45.0*	*45.0*	*45.0*	*45.0*	*45.0*	*45.0*
Niacin equivalents	14.0	14.0	14.0	14.0	16.0	16.0	16.0	16.0
Pantothenic acid	(4.0)	(4.0)	(4.0)	(4.0)	(6.0)	(6.0)	(6.0)	(6.0)
Phosphorus	1000.0	1000.0	1000.0	1000.0	1000.0	1000.0	1000.0	1000.0
Potassium	(2800.0)	(2800.0)	(2800.0)	(2800.0)	(3800.0)	(3800.0)	(3800.0)	(3800.0)
Riboflavin	1.1	1.1	1.1	1.3	1.3	1.3	1.3	1.6
Selenium	*60.0*	*60.0*	*60.0*	*60.0*	*70.0*	*70.0*	*70.0*	*70.0*
Sodium	(460.0–920.0)	(460.0–920.0)	(460.0–920.0)	(460.0–920.0)	(460.0–920.0)	(460.0–920.0)	(460.0–920.0)	(460.0–920.0)
Thiamin	1.1	1.1	1.1	1.1	1.2	1.2	1.2	1.2
Total vitamin A equivalents	*700.0*	*700.0*	*700.0*	*700.0*	*900.0*	*900.0*	*900.0*	*900.0*
Vitamin B6	1.3	1.5	1.5	1.5	1.3	1.3	1.7	1.7

continues

Table 1.14c Daily RDIs (AIs) for vitamins and minerals for adults (19+ years) in Australia and New Zealand *continued*

Nutrients	*Micrograms*/milligrams of nutrients, by gender and age (years)							
	Female				Male			
	19–30	31–50	51–70	70+	19–30	31–50	51–70	70+
Vitamin B12	*2.4*	*2.4*	*2.4*	*2.4*	*2.4*	*2.4*	*2.4*	*2.4*
Vitamin C	45.0	45.0	45.0	45.0	45.0	45.0	45.0	45.0
Vitamin D	*(5.0)*	*(5.0)*	*(10.0)*	*(15.0)*	*(5.0)*	*(5.0)*	*(10.0)*	*(15.0)*
Vitamin E	(7.0)	(7.0)	(7.0)	(7.0)	(10.0)	(10.0)	(10.0)	(10.0)
Vitamin K	*(60.0)*	*(60.0)*	*(60.0)*	*(60.0)*	*(70.0)*	*(70.0)*	*(70.0)*	*(70.0)*
Zinc	14.0	14.0	14.0	14.0	8.0	8.0	8.0	8.0

Table 1.14d RDI and (AI) for vitamins and minerals for pregnant and lactating females (14–50 years) in Australia and New Zealand

Nutrient	*Micrograms*/milligrams of nutrients, by stage and age (years)					
	Pregnancy			Lactation		
	14–18	19–30	31–50	14–18	19–30	31–50
Biotin	*(30.0)*	*(30.0)*	*(30.0)*	*(35.0)*	*(35.0)*	*(35.0)*
Calcium	1300.0	1000.0	1000.0	1300.0	1000.0	1000.0
Choline	(415.0)	(440.0)	(440.0)	(525.0)	(550.0)	(550.0)
Chromium	*(30.0)*	*(30.0)*	*(30.0)*	*(45.0)*	*(45.0)*	*(45.0)*
Copper	(1.2)	(1.3)	(1.3)	(1.4)	(1.5)	(1.5)
Fluoride	(3.0)	(3.0)	(3.0)	(3.0)	(3.0)	(3.0)
Total folate	*600.0*	*600.0*	*600.0*	*500.0*	*500.0*	*500.0*
Iodine	*220.0*	*220.0*	*220.0*	*270.0*	*270.0*	*270.0*
Iron	27.0	27.0	27.0	10.0	9.0	9.0
Magnesium	400.0	350.0	360.0	360.0	310.0	320.0
Manganese	(5.0)	(5.0)	(5.0)	(5.0)	(5.0)	(5.0)
Molybdenum	*50.0*	*50.0*	*50.0*	*50.0*	*50.0*	*50.0*
Niacin equivalents	18.0	18.0	18.0	17.0	17.0	17.0
Pantothenic acid	(5.0)	(5.0)	(5.0)	(6.0)	(6.0)	(6.0)
Phosphorus	1250.0	1000.0	1000.0	1250.0	1000.0	1000.0
Potassium	(2800.0)	(2800.0)	(2800.0)	(3200.0)	(3200.0)	(3200.0)
Riboflavin	1.4	1.4	1.4	1.6	1.6	1.6
Selenium	*65.0*	*65.0*	*65.0*	*75.0*	*75.0*	*75.0*
Sodium	(460.0–920.0)	(460.0–920.0)	(460.0–920.0)	(460.0–920.0)	(460.0–920.0)	(460.0–920.0)
Thiamin	1.4	1.4	1.4	1.4	1.4	1.4
Total vitamin A equivalents	*700.0*	*800.0*	*800.0*	*1100.0*	*1100.0*	*1100.0*
Vitamin B6	1.9	1.9	1.9	2.0	2.0	2.0
Vitamin B12	*2.6*	*2.6*	*2.6*	*2.8*	*2.8*	*2.8*
Vitamin C	55.0	60.0	60.0	80.0	85.0	85.0
Vitamin D	*(5.0)*	*(5.0)*	*(5.0)*	*(5.0)*	*(5.0)*	*(5.0)*
Vitamin E	(8.0)	(7.0)	(7.0)	(12.0)	(11.0)	(11.0)
Vitamin K	*(60.0)*	*(60.0)*	*(60.0)*	*(60.0)*	*(60.0)*	*(60.0)*
Zinc	10.0	11.0	11.0	11.0	12.0	12.0

a No AIs for iron, seven to twelve months. RDI: 11 mg/day; EAR: 7 mg/day.

b No AIs for zinc, seven to twelve months. RDI: 3.0 mg/day; EAR: 2.5 mg/day.

c Oldest two age groups for calcium are nine to eleven years and twelve to thirteen years.

Note: AI: adequate intake (shown in brackets); EAR: estimated average requirement; RDI: recommended dietary intake.

Source: NHMRC & Ministry of Health (2006).

NUTRIENT INTAKES AND FOOD SOURCES FOR ADULTS

Both Australia and New Zealand conduct national surveys to gather information on food and thus intakes of their populations. Both countries used the 24-hour diet recall method to collect information on all foods and drinks consumed in a 24-hour period. Until as recently as 2014, data describing the consumption of foods and nutrients for adults at a national level in Australia was limited to the 1995 National Nutrition Survey results. The 2011–12 Australian Health Survey (ABS 2014) has been completed and is the first national nutrition survey of adults (and children over two years old) conducted in over fifteen years. There are three components to the survey: the National Health Survey, the National Nutrition and Physical Activity Survey and the National Health Measures Survey with data collected from 12,000 participants across Australia, including Aboriginals and Torres Strait Islanders. The results are being released by the Australian Bureau of Statistics progressively. The most recent national report for New Zealand is the 2008–09 New Zealand Adult Nutrition Survey, which collected information from 4721 adults aged fifteen years and over (University of Otago & Ministry of Health 2011). Table 1.15 describes energy, macronutrient and micronutrient intakes for Australian and New Zealand adults.

Nutrient intakes and food sources for infants, children and adolescents have been included within individual chapters in Part II.

Australia and New Zealand report the proportion of nutrients from food groups in their respective surveys. The selection of nutrients varies between the countries and Table 1.16 describes the proportion of protein, total, saturated, monounsaturated and poly-unsaturated fats, carbohydrate and fibre from foods for both Australian and New Zealand adults aged 19–30, 31–50, 51–70 and over 70 years. Table 1.17a and b describe the proportion of vitamin B12, total folate, calcium, iodine, iron, sodium and zinc from foods for Australians aged 19–30, 31–50, 51–70 and over 70 years and Tables 1.17c and d describe the proportion of vitamin B12, calcium, iron and zinc from foods for New Zealanders aged 19–30, 31–50, 51–70 and over 70 years. Note that the New Zealand survey did not report on total folate, iodine and sodium.

Australia and New Zealand categorise foods differently for their respective national surveys. The major food groups and examples for Australia are listed in Table 1.18a and for New Zealand in Table 1.18b.

GLYCAEMIC INDEX

The glycaemic index (GI) is a ranking of carbohydrate in food on a scale from 0 to 100 according to the extent by which foods raise blood sugar levels after eating in comparison with direct consumption of glucose. Both the food and test glucose must contain the same amount of carbohydrate. Foods with a high GI (generally regarded as having a GI score of 70 or more) are rapidly digested and result in high glucose levels, leading to high insulin levels. Low-GI foods (with a score of 55 or less) are digested and absorbed at a slower rate and therefore produce gradual rises in glucose and insulin levels. The GI of a food is influenced by:

- the type of starch (amylopectin has a higher GI than amylose due to the molecule being more open to digestion)
- the amount of processing (e.g., wholegrain bread has a lower GI than wholemeal)
- the sugar type (lactose and fructose have a lower GI than glucose)
- the type of fibre (viscous soluble fibres slow enzyme activity and have low GI)
- cooking (which usually speeds up digestion leading to higher GI)
- the fat, acid and protein content (which slows gastric emptying and digestion and lowers GI).

Regular meals throughout the day and use of low-GI foods can help ensure a constant supply of glucose to the brain. Examples of high-, medium- and low-GI foods are shown in Table 1.19. The GI of a mixed meal can be calculated by finding the weighted average of the GI of the carbohydrate foods within the meal.

Glycaemic load

The amount of carbohydrate in foods will also influence blood glucose levels. High intake of a low-GI food can still result in high blood glucose, whereas a small amount of a high-GI food may not. The glycaemic load takes these issues into account, as shown in the following equation, where the GI and the quantity of carbohydrate are considered:

$$\text{Glycaemic load} = \frac{(\text{GI} \times \text{amount of available carbohydrate})}{100}$$

It is best to use glycaemic load rather than GI to predict the effect that a carbohydrate food will have on blood glucose levels because people mostly eat mixed meals and varying proportions rather than standard amounts.

Table 1.15a Energy, macronutrient and micronutrient intakes and percentage of energy (kJ) from macronutrients for adults (19–30 and 31–50 years) in Australia and New Zealand

	19–30 years				31–50 years			
	Australia		New Zealand		Australia		New Zealand	
Nutrient	**Males[a]**	**Females[a]**	**Males[b]**	**Females[b]**	**Males[a]**	**Females[a]**	**Males[b]**	**Females[b]**
Energy[c]	11,004	7863	11,817	8245	10,220	7540	11,376	7821
	Grams/*milligrams*/(% energy) of nutrients							
Protein	117.0 (18.5)	78.1 (17.7)	111.0 (15.8)	72.0 (15.4)	107.0 (17.8)	79.7 (18.0)	111.0 (16.7)	77.0 (17.0)
Total fat[d]	92.5 (30.5)	69.9 (31.9)	102.0 (33.6)	71.0 (33.0)	86.3 (30.3)	65.3 (31.2)	104.0 (34.3)	74.0 (34.5)
Saturated fat	34.7 (11.4)	26.7 (12.0)	40.7 (13.1)	27.9 (12.9)	32.5 (11.5)	24.4 (11.6)	40.3 (13.3)	28.8 (13.6)
Monounsaturated fat	36.0 (11.9)	26.4 (12.0)	38.5 (13.0)	26.1 (13.0)	33.5 (11.8)	25.0 (12.0)	38.3 (13.0)	26.5 (13.0)
Polyunsaturated fat[d]	14.1 (4.7)	11.0 (5.1)	13.1 (4.0)	9.3 (4.0)	13.2 (4.7)	10.3 (5.0)	14.5 (5.0)	10.3 (5.0)
Linoleic acid	11.8	9.3	–	–	11.0	8.5	–	–
Alpha-linolenic acid	1.6	1.3	–	–	1.6	1.3	–	–
Total long chain omega-3 (n-6) fatty acids	*324.2*	*177.7*	–	–	*256.8*	*266.5*	–	–
Carbohydrate[c]	297.2 (45)	213.7 (45.5)	327.0 (47.1)	239.0 (49.3)	263.9 (43.9)	196.5 (44.3)	299.0 (44.8)	209.0 (45.5)
Total sugars	131.0 (19.1)	99.0 (20.5)	140.0 (28.8)	120.0 (22.9)	119.4 (18.9)	91.1 (19.7)	129.0 (16.0)	94.0 (14.8)
Starch	159.0	112.4	–	–	137.0	103.0	–	–
Dietary fibre	24.4	20.3	23.1	17.0	24.9	20.7	23.4	17.9
Alcohol[e]	14.4	7.0	19.3	11.9	19.7	10.3	20.6	10.7
Vitamins/minerals								
Calcium	954.3	765.3	967.0	704.0	910.8	758.1	992.0	810.0
Total folate	*546.4*	*424.5*	–	–	*535.5*	*424.2*	–	–
Iodine	212.5	152.6	–	–	198.9	153.2	–	–
Iron	13.5	9.7	13.9	10.2	12.7	9.6	13.8	10.2
Magnesium	390.6	291.9	–	–	392.6	308.8	–	–
Niacin equivalents	54.4	35.5	47.4	31.2	49.3	36.0	45.8	31.7
Phosphorus	1774.5	1282.4	–	–	1703.3	1301.6	–	–
Potassium	3279.9	2509.7	3426.0	2643.0	3305.8	2659.6	3724.0	2869.0
Riboflavin	2.4	1.8	2.4	1.6	2.2	1.7	2.3	1.7
Selenium	118.4	76.5	66.3	46.4	104.8	80.8	78.0	51.9
Sodium[f]	3120.2	2303.4	–	–	2915.4	2153.9	–	–
Thiamin	1.9	1.4	1.8	1.1	1.7	1.3	1.6	1.1
Total vitamin A equivalents	*866.1*	*748.8*	*855.0*	*654.0*	*826.3*	*797.4*	*895.0*	*735.0*
Vitamin B6	2.2	1.4	2.9	2.1	1.8	1.3	2.4	1.6
Vitamin B12	*5.9*	*3.7*	*5.2*	*3.0*	*5.3*	*3.8*	*5.0*	*3.3*
Vitamin C	116.8	96.4	94.0	107.0	110.6	93.7	90.0	97.0
Vitamin E	12.4	10.0	11.7	8.8	11.7	9.9	12.5	9.5
Zinc	13.5	9.2	14.4	8.9	12.9	9.4	13.7	9.7

a Australia reports mean intakes.
b New Zealand reports median intakes except for alcohol which is reported as mean.
c Energy includes energy from dietary fibre (Australian data only).
d Components may not sum to total (Australian data only).
e Represents pure alcohol (ethanol) (Australian data only).
f Includes sodium naturally present in foods as well sodium added during processing, but excludes the 'discretionary salt' added by consumers in home-prepared foods or 'at the table' (Australian data only).

Note: – denotes data not reported in the 2008–09 New Zealand Adult Nutrition Survey.

Source: ABS (2014); University of Otago & Ministry of Health (2011).

Table 1.15b Energy, macronutrient and micronutrient intakes and percentage of energy (kJ) from macronutrients for adults (51–70 years and 71 years and over) in Australia and New Zealand

	51–70 years				71 years and over			
	Australia		New Zealand		Australia		New Zealand	
Nutrient	**Males[a]**	**Females[a]**	**Males[b]**	**Females[b]**	**Males[a]**	**Females[a]**	**Males[b]**	**Females[b]**
Energy[c]	9345	7268	9158	7071	8174	6570	7926	6014
	Grams/*milligrams*/(% energy) of nutrients							
Protein	97.5 (18.4)	77.9 (18.9)	89 (16.8)	69 (16.7)	82.6 (17.7)	71.5 (19.0)	78 (16.4)	60 (17.0)
Total fat[d]	78.6 (30.5)	62.0 (30.9)	81 (33.3)	64 (34.1)	66.9 (29.5)	55.6 (30.8)	67 (31.8)	51 (32.1)
Saturated fat	29.4 (11.3)	22.5 (11.2)	31.0 (12.9)	23.2 (12.6)	26.6 (11.7)	21.3 (11.7)	25.2 (12.1)	19.5 (12.3)
Monounsaturated fat	30.5 (11.8)	24.0 (11.9)	29.5 (12.0)	23.5 (13.0)	24.5 (10.8)	20.2 (11.3)	24.0 (12.0)	18.0 (11.0)
Polyunsaturated fat[d]	11.8 (4.6)	9.9 (5.0)	11.5 (5.0)	9.6 (5.0)	9.6	8.8 (4.9)	9.9 (5.0)	7.8 (5.0)
Linoleic acid	9.7	8.1	–	–	7.8	7.1	–	–
Alpha-linolenic acid	1.5	1.3	–	–	1.3	1.2	–	–
Total long chain omega-3 (n-6) fatty acids	*352.8*	*323.4*	–	–	*273.1*	*252.3*	–	–
Carbohydrate[c]	233.8 (41.7)	183.6 (42.2)	243.0 (45.5)	192.0 (46.2)	219.4 (44.6)	172.3 (43.5)	223.0 (47.9)	173.0 (48.8)
Total sugars	101.0 (17.3)	87.0 (19.2)	103.0 (11.6)	91.0 (12.3)	101.8 (19.8)	84.4 (20.3)	100.0 (6.9)	82.0 (8.0)
Starch	125.2	94.2	–	–	112.4	85.7	–	–
Dietary fibre	24.8	22.2	20.8	18.1	25.1	21.0	20.4	17.1
Alcohol[e]	22.6	13.3	19.3	10.1	14.2	8.0	13.4	6.0
Vitamins/minerals								
Calcium	780.8	740.7	828.0	737.0	726.4	674.2	743.0	676.0
Total folate	*520.7*	*438.7*	–	–	*526.5*	*437.0*	–	–
Iodine	181.1	148.1	–	–	171.5	146.5	–	–
Iron	12.2	9.9	12.8	9.9	11.6	9.2	11.4	8.9
Magnesium	364.4	313.3	–	–	315.7	270.7	–	–
Niacin equivalents	44.4	34.6	36.6	27.7	37.5	31.6	31.4	23.3
Phosphorus	1577.0	1294.7	–	–	1392.7	1212.9	–	–
Potassium	3144.2	2699.9	3319.0	2878.0	2889.9	2508.6	3138.0	2567.0
Riboflavin	1.9	1.6	2.0	1.7	1.8	1.6	1.8	1.5
Selenium	97.4	81.5	61.0	47.0	80.0	71.4	52.0	39.5
Sodium[f]	2509.8	1972.1	–	–	2216.7	1772.5	–	–
Thiamin	1.7	1.3	1.5	1.2	1.7	1.3	1.4	1.1
Total vitamin A equivalents	*933.6*	*907.3*	*825.0*	*828.0*	*957.5*	*853.6*	*851.0*	*768.0*
Vitamin B6	1.5	1.3	1.7	1.4	1.3	1.1	1.6	1.3
Vitamin B12	4.9	4.0	4.5	3.4	4.4	3.8	4.2	2.7
Vitamin C	107.1	97.3	90.0	97.0	100.0	88.2	96.0	89.0
Vitamin E	10.4	10.0	10.9	9.7	9.3	8.3	10.2	8.5
Zinc	12.1	9.8	11.8	8.9	10.5	9.0	9.7	7.6

a Australia reports mean intakes.

b New Zealand reports median intakes except for alcohol which is reported as mean.

c Energy includes energy from dietary fibre (Australian data only).

d Components may not sum to total (Australian data only).

e Represents pure alcohol (ethanol) (Australian data only).

f Includes sodium naturally present in foods as well sodium added during processing, but excludes the 'discretionary salt' added by consumers in home-prepared foods or 'at the table' (Australian data only).

Note: – denotes data not reported in the 2008–09 New Zealand Adult Nutrition Survey.

Source: ABS (2014); University of Otago & Ministry of Health (2011).

Table 1.16a Proportion of protein, total, saturated, monounsaturated and polyunsaturated fats, carbohydrate and fibre from foods for adult males in Australia

Major food groups	Protein				Total fat				Saturated fat			
	Age ranges (years)											
	19–30	31–50	51–70	71+	19–30	31–50	51–70	71+	19–30	31–50	51–70	71+
Non-alcoholic beverages	2.1	3.1	2.5	2.1	1.6	2.5	1.8	1.6	2.3	3.6	2.4	1.7
Cereals and cereal products	12.0	12.8	15.2	18.3	6.0	6.5	7.3	8.2	4.3	4.0	4.4	3.9
Cereal-based products and dishes	19.0	16.6	11.8	10.4	23.9	22.0	17.9	19.3	25.3	24.0	20.3	23.2
Fats and oils	0.0	0.0	0.1	0.1	3.1	4.6	7.5	10.3	3.5	4.5	7.3	10.6
Fish and seafood products and dishes	5.8	5.2	6.9	6.7	3.4	3.2	4.3	4.5	2.2	1.8	2.5	2.3
Fruit products and dishes	0.8	1.0	1.2	1.8	0.7	0.5	0.6	0.6	0.4	0.2	0.2	0.2
Egg products and dishes	1.9	2.1	2.4	2.6	2.3	2.6	2.7	2.8	1.8	2.1	2.1	1.9
Meat, poultry and game products and dishes	35.4	35.3	37.4	33.6	23.9	22.6	23.8	20.0	20.9	20.6	22.5	18.0
Milk products and dishes	10.7	10.2	10.5	13.0	14.2	12.7	13.7	16.5	24.3	22.1	24.0	27.1
Dairy and meat substitutes	0.2	0.5	0.3	0.4	0.2	0.5	0.3	0.3	0.1	0.2	0.1	0.1
Soup	0.8	1.4	1.5	2.2	0.6	1.0	1.2	1.7	0.5	0.7	0.9	1.2
Seed and nut products and dishes	1.0	1.1	1.7	1.0	3.3	4.0	6.0	3.7	1.3	1.8	2.2	1.7
Sauces, dips and condiments	0.5	0.4	0.3	0.2	2.8	3.5	2.6	2.0	1.8	1.9	1.7	1.2
Vegetable products and dishes	3.9	4.9	4.9	4.9	8.2	7.7	6.2	5.2	5.6	5.6	4.5	3.1
Legume and pulse products and dishes	0.7	0.5	0.5	0.7	0.1	0.1	0.2	0.1	0.1	0.0	0.1	0.0
Snack foods	0.6	0.5	0.3	0.1	2.7	2.3	1.1	0.9	2.2	1.8	0.8	0.6
Sugar products and dishes	0.1	0.1	0.0	0.1	0.3	0.2	0.0	0.0	0.3	0.3	0.1	0.0
Confectionery and cereal/nut/fruit/seed bars	0.8	0.9	0.6	0.6	2.1	3.1	2.5	2.1	2.6	4.3	4.0	3.1
Alcoholic beverages	0.7	1.0	1.2	0.8	0.1	0.1	0.0	0.1	0.2	0.1	0.0	0.1
Special dietary foods	2.8	2.3	0.5	0.2	0.4	0.3	0.1	0.1	0.5	0.3	0.1	0.1
Miscellaneous	0.1	0.2	0.2	0.2	0.0	0.0	0.0	0.0	0.0	0.0	0.0	0.0

Source: ABS (2014); University of Otago & Ministry of Health (2011).

Monounsaturated fat				Polyunsaturated fat				Carbohydrate				Fibre			
Age ranges (years)															
19–30	31–50	51–70	71+	19–30	31–50	51–70	71+	19–30	31–50	51–70	71+	19–30	31–50	51–70	71+
1.0	1.5	1.0	0.8	0.9	1.0	0.5	0.5	16.1	12.8	8.7	6.1	2.9	3.7	3.6	4.1
5.0	5.4	5.9	7.2	11.7	14.2	17.0	21.9	27.1	27.3	32.7	31.7	28.8	29.1	33.2	34.3
22.6	20.8	17.0	17.7	23.7	20.7	15.3	14.4	20.2	20.5	17.5	17.4	18.8	16.4	12.3	9.4
2.6	4.4	7.6	9.7	3.2	5.3	8.4	11.3	0.0	0.0	0.0	0.0	0.0	0.0	0.0	0.0
3.3	3.4	4.7	5.1	6.4	6.6	7.8	9.6	0.6	0.6	0.6	0.7	0.8	0.6	0.6	0.4
0.5	0.2	0.3	0.2	1.2	0.4	0.7	0.3	4.8	7.1	8.4	11.1	10.4	13.3	14.0	17.4
2.4	2.7	2.9	3.3	2.2	2.5	2.7	3.1	0.1	0.1	0.1	0.1	0.2	0.2	0.2	0.3
28.5	26.8	27.8	24.8	20.4	17.6	17.4	14.3	3.4	3.1	2.6	2.1	6.4	6.2	5.2	4.4
9.3	8.2	8.7	11.3	3.4	2.6	2.9	3.8	7.8	7.0	7.7	9.5	1.0	0.7	0.4	0.3
0.1	0.4	0.2	0.2	0.7	1.4	0.9	0.8	0.1	0.2	0.2	0.2	0.3	0.5	0.3	0.3
0.8	1.3	1.5	2.1	0.6	1.1	1.3	1.7	0.4	0.7	1.1	1.8	1.4	2.3	3.1	4.6
5.1	5.9	9.0	5.6	4.3	5.2	9.4	5.5	0.3	0.3	0.4	0.2	1.8	1.9	2.9	2.1
2.8	3.8	2.6	2.4	5.5	7.3	5.3	3.9	1.5	0.8	0.9	0.5	1.9	1.0	0.8	0.3
9.4	8.5	6.9	6.2	11.4	10.3	8.0	7.1	5.7	6.0	6.7	6.7	18.6	18.7	19.6	18.1
0.1	0.1	0.2	0.0	0.3	0.3	0.3	0.2	0.5	0.4	0.4	0.6	2.6	1.7	1.6	2.5
3.6	3.3	1.6	1.3	2.1	1.6	0.9	0.7	1.7	1.4	0.9	0.4	1.6	1.1	0.6	0.3
0.3	0.1	0.0	0.0	0.2	0.1	0.0	0.0	2.6	4.7	5.0	6.4	0.2	0.1	0.2	0.3
2.1	2.9	2.0	1.9	1.4	1.5	0.9	0.7	3.0	3.1	2.7	2.4	1.6	1.7	0.9	0.6
0.1	0.0	0.0	0.0	0.0	0.0	0.0	0.0	3.5	3.4	3.2	1.8	0.0	0.0	0.1	0.1
0.4	0.3	0.1	0.1	0.2	0.2	0.1	0.0	0.7	0.4	0.3	0.3	0.4	0.4	0.1	0.0
0.0	0.0	0.0	0.0	0.0	0.0	0.1	0.0	0.0	0.1	0.1	0.1	0.3	0.3	0.3	0.2

Table 1.16b Proportion of protein, total, saturated, monounsaturated and polyunsaturated fats, carbohydrate and fibre from foods for adult females in Australia

Major food groups	Protein				Total fat				Saturated fat			
	Age ranges (years)											
	19–30	31–50	51–70	71+	19–30	31–50	51–70	71+	19–30	31–50	51–70	71+
Non-alcoholic beverages	3.4	4.2	3.4	2.9	2.0	2.8	2.5	2.5	2.7	3.6	3.1	2.8
Cereals and cereal products	12.3	12.9	14.1	16.2	5.8	6.9	7.7	8.3	3.6	4.4	4.6	4.9
Cereal-based products and dishes	21.2	15.5	11.5	8.7	26.0	21.3	18.1	17.2	27.9	23.3	20.9	20.4
Fats and oils	0.0	0.0	0.1	0.1	4.6	4.8	6.1	8.8	4.9	5.1	6.4	8.3
Fish and seafood products and dishes	4.4	6.7	8.3	8.0	2.2	4.0	5.1	4.3	1.2	2.3	3.2	2.2
Fruit products and dishes	1.1	1.4	1.8	2.0	0.7	0.7	1.0	0.7	0.2	0.2	0.3	0.2
Egg products and dishes	2.1	2.1	2.1	2.6	2.2	2.2	2.4	3.5	2.0	1.8	2.1	3.4
Meat, poultry and game products and dishes	29.5	31.9	32.4	34.3	17.8	18.3	18.3	18.7	15.8	17.0	17.4	17.1
Milk products and dishes	11.6	11.7	12.3	13.7	12.3	13.3	12.7	16.4	21.0	23.1	22.9	28.2
Dairy and meat substitutes	0.9	0.5	0.7	0.5	0.7	0.5	0.7	0.4	0.3	0.2	0.3	0.2
Soup	1.8	1.6	2.0	1.9	1.1	1.3	1.8	1.8	0.9	0.9	1.3	1.3
Seed and nut products and dishes	1.1	1.3	1.7	1.0	3.4	4.3	6.3	3.7	1.2	1.6	2.2	1.1
Sauces, dips and condiments	0.4	0.5	0.4	0.3	3.7	4.4	4.3	3.4	1.9	2.4	2.8	1.7
Vegetable products and dishes	6.5	6.2	6.0	5.8	10.5	9.2	7.8	6.8	7.5	6.7	5.3	4.0
Legume and pulse products and dishes	0.6	0.6	0.6	0.5	0.2	0.2	0.2	0.1	0.1	0.1	0.1	0.0
Snack foods	0.6	0.4	0.2	0.3	2.4	2.1	1.1	0.9	2.2	1.8	0.8	0.9
Sugar products and dishes	0.1	0.0	0.1	0.1	0.1	0.1	0.1	0.0	0.1	0.1	0.2	0.1
Confectionery and cereal/nut/fruit/seed bars	1.2	1.1	0.9	0.5	4.1	3.5	3.6	2.0	6.4	5.2	5.5	2.8
Alcoholic beverages	0.2	0.3	0.4	0.3	0.0	0.0	0.2	0.2	0.1	0.0	0.3	0.3
Special dietary foods	0.7	1.0	1.0	0.3	0.1	0.1	0.2	0.1	0.1	0.1	0.2	0.1
Miscellaneous	0.3	0.2	0.2	0.2	0.0	0.0	0.0	0.1	0.0	0.0	0.0	0.1

Source: ABS (2014); University of Otago & Ministry of Health (2011).

Monounsaturated fat				Polyunsaturated fat				Carbohydrate				Fibre			
Age ranges (years)															
19–30	31–50	51–70	71+	19–30	31–50	51–70	71+	19–30	31–50	51–70	71+	19–30	31–50	51–70	71+
1.2	1.7	1.3	1.4	1.1	1.6	0.8	0.8	13.7	10.4	7.5	7.2	4.3	4.3	3.9	3.8
4.9	5.9	6.5	7.2	12.3	14.5	17.0	18.9	24.5	27.3	29.3	30.1	23.7	26.0	28.0	32.2
24.9	20.5	16.9	15.9	24.1	19.1	15.3	13.9	23.4	21.0	18.9	17.4	18.7	15.0	11.4	9.0
4.2	4.7	5.9	8.7	4.7	4.6	6.5	10.9	0.0	0.0	0.0	0.0	0.0	0.0	0.0	0.0
2.4	4.3	5.4	5.0	4.2	7.3	8.7	7.7	0.5	0.8	0.9	0.8	0.6	0.7	0.8	0.5
0.4	0.4	0.6	0.3	1.0	0.7	1.3	0.4	6.5	8.8	11.5	12.7	12.6	15.2	17.3	18.7
2.3	2.4	2.5	3.5	1.9	2.0	2.2	2.8	0.2	0.1	0.1	0.2	0.3	0.3	0.3	0.4
21.7	21.5	21.4	23.2	14.1	14.0	13.3	13.0	2.9	2.8	2.9	2.0	4.8	5.1	4.3	3.8
8.0	8.8	8.1	11.0	2.4	2.9	2.6	3.3	7.6	7.8	8.0	9.3	0.9	0.7	0.6	0.3
0.6	0.4	0.5	0.3	2.2	1.3	2.1	1.4	0.5	0.4	0.5	0.2	0.7	0.5	0.6	0.3
1.3	1.6	2.3	2.4	1.1	1.3	1.7	1.7	1.1	1.1	1.5	1.3	3.0	3.0	4.4	4.3
4.8	6.1	9.4	5.5	6.1	7.6	9.5	7.2	0.2	0.3	0.4	0.2	2.2	2.3	3.0	1.7
4.1	5.0	4.6	3.8	7.7	8.5	7.8	7.5	1.2	1.2	1.0	1.1	0.9	1.1	0.8	0.6
12.3	10.7	9.4	8.4	13.0	11.1	8.6	8.9	6.8	6.6	6.4	7.7	22.2	21.1	20.4	21.1
0.2	0.3	0.3	0.0	0.4	0.5	0.5	0.2	0.5	0.5	0.6	0.4	1.8	1.7	1.6	1.7
2.8	2.7	1.5	1.2	1.9	1.5	0.6	0.5	1.4	1.2	0.5	0.4	1.3	0.9	0.3	0.2
0.2	0.1	0.1	0.0	0.2	0.1	0.0	0.0	3.2	4.2	4.2	5.4	0.1	0.2	0.2	0.3
3.4	3.1	3.1	1.9	1.4	1.4	1.3	0.7	4.3	4.0	3.9	2.2	1.6	1.4	1.3	0.5
0.0	0.0	0.1	0.1	0.0	0.0	0.0	0.0	1.4	1.3	1.2	0.8	0.0	0.0	0.0	0.0
0.1	0.2	0.2	0.2	0.0	0.1	0.1	0.1	0.2	0.3	0.5	0.3	0.1	0.2	0.2	0.1
0.0	0.0	0.0	0.0	0.0	0.0	0.0	0.2	0.1	0.1	0.1	0.1	0.4	0.3	0.3	0.3

Table 1.16c Proportion of protein, total, saturated, monounsaturated and polyunsaturated fats, carbohydrate and fibre from foods for adult males in New Zealand

Major food groups	Protein				Total fat				Saturated fat			
	Age ranges (years)											
	19–30	31–50	51–70	71+	19–30	31–50	51–70	71+	19–30	31–50	51–70	71+
Alcoholic beverages	1.2	1.0	1.0	0.9	0.1	0.1	0.1	0.2	0.0	0.0	0.0	0.1
Beef and veal	6.5	8.6	8.6	10.1	4.6	5.3	5.2	6.1	4.5	5.3	5.2	6.6
Biscuits	0.5	0.9	1.0	1.4	1.2	2.5	2.7	4.5	1.6	3.3	3.7	6.3
Bread	10.0	11.7	11.5	14.3	4.5	4.9	4.4	5.5	2.8	2.8	2.1	2.5
Bread-based dishes	12.9	6.7	5.4	2.1	11.4	6.8	6.0	2.1	11.9	7.0	6.1	2.1
Breakfast cereals	2.2	2.4	4.0	3.5	0.9	1.6	3.0	2.1	0.5	1.3	2.4	1.5
Butter and margarine	0.0	0.1	0.1	0.1	6.3	9.0	11.9	15.8	5.0	8.4	10.7	14.6
Cakes and muffins	1.4	1.6	1.7	1.6	2.9	4.3	4.4	4.0	3.4	4.8	5.4	4.9
Cheese	3.3	3.1	2.4	2.3	4.1	4.4	3.4	3.2	6.4	6.9	5.4	5.1
Dairy products	1.3	1.6	1.7	1.6	2.0	2.7	4.1	4.0	2.9	3.9	5.8	6.1
Dietary supplements	0.8	0.5	0.1	0.0	0.6	0.2	0.1	0.0	0.4	0.2	0.0	0.0
Eggs and egg dishes	2.1	2.8	3.3	3.7	1.9	2.7	3.0	3.5	1.3	1.9	2.3	3.0
Fats and oils	0.0	0.0	0.0	0.0	1.6	0.8	0.3	0.9	0.8	0.3	0.1	0.4
Fish and seafood	3.6	6.5	6.1	7.0	2.3	4.3	4.0	4.5	1.6.	2.8	3.0	3.2
Fruit	1.1	1.2	1.5	2.1	1.7	1.9	1.5	2.5	0.9	1.1	0.8	1.4
Grains and pasta	9.4	6.8	6.5	4.7	6.1	3.7	3.8	3.0	5.7	3.2	3.2	3.0
Lamb and mutton	1.7	1.7	2.6	1.9	1.9	1.8	3.0	1.9	2.1	2.0	3.4	2.1
Milk	6.1	7.7	9.7	10.8	4.5	4.9	5.0	5.5	7.2	7.8	7.7	8.6
Non-alcoholic beverages	2.3	2.1	2.2	1.3	1.5	1.3	1.1	0.5	2.1	1.9	1.5	0.8
Nuts and seeds	0.7	1.1	1.4	1.0	1.6	2.3	2.4	2.2	1.1	1.4	1.8	1.2
Other meat	0.2	0.5	0.9	0.8	0.2	0.2	0.6	0.6	0.2	0.1	0.7	0.6
Pies and pasties	4.6	3.2	2.1	1.8	6.1	4.3	2.8	2.4	7.0	4.9	3.2	2.7
Pork	5.9	4.1	6.0	6.0	3.3	2.3	3.5	3.2	3.3	2.3	3.7	3.3
Potatoes, kumara and taro	2.9	3.0	3.2	3.4	7.6	6.5	5.5	4.3	6.5	7.5	4.8	3.8
Poultry	9.1	10.1	7.7	5.1	6.7	6.6	5.1	2.9	6.0	5.7	4.6	2.4
Puddings and desserts	0.1	0.5	0.5	1.4	0.5	1.0	1.2	2.5	0.6	1.1	1.3	3.4
Sausages and processed meats	3.9	3.5	2.9	3.1	4.7	4.6	3.8	3.9	5.1	4.9	4.1	4.4
Savoury sauces and condiments	0.7	0.7	0.8	0.7	1.3	1.8	1.7	1.1	0.8	1.3	1.3	0.9
Snack bars	0.3	0.6	0.3	0.1	0.7	1.2	0.5	0.2	0.9	1.3	0.5	0.2
Snack foods	0.5	0.3	0.1	0.0	1.2	0.6	0.2	0.1	1.3	0.6	0.2	0.1
Soups and stocks	0.9	0.6	0.8	1.7	0.6	0.3	0.8	1.5	0.7	0.3	0.8	1.6
Sugar and sweets	1.2	0.8	0.4	0.3	3.2	2.0	1.1	1.0	3.8	2.6	1.5	1.3
Vegetables	2.8	3.7	3.7	5.0	2.5	2.9	3.9	4.5	1.5	1.9	2.4	2.9

Source: ABS (2014); University of Otago & Ministry of Health (2011).

Monounsaturated fat				Polyunsaturated fat				Carbohydrate				Fibre			
Age ranges (years)															
19–30	**31–50**	**51–70**	**71+**	**19–30**	**31–50**	**51–70**	**71+**	**19–30**	**31–50**	**51–70**	**71+**	**19–30**	**31–50**	**51–70**	**71+**
0.0	0.0	0.0	0.0	0.0	0.0	0.0	0.0	3.8	3.0	2.5	2.0	0.1	0.0	0.3	0.0
5.5	6.3	6.3	7.7	3.1	3.3	3.2	2.9	0.8	0.8	0.5	0.6	1.9	1.2	0.8	1.1
1.0	2.2	2.3	3.9	0.8	1.9	2.0	2.8	1.1	3.1	3.2	4.6	0.7	1.6	1.9	2.1
3.4	3.8	3.3	4.2	8.1	9.5	8.8	10.6	14.4	18.9	18.7	21.1	16.2	19.3	18.6	20.2
11.6	6.8	6.0	2.2	11.5	6.8	6.6	2.4	7.4	4.5	4.3	1.5	8.9	5.2	5.1	1.6
0.7	1.4	2.4	1.7	2.2	3.3	5.5	4.4	3.9	4.2	7.3	6.2	6.1	6.7	10.3	8.4
7.0	9.3	12.8	16.8	10.2	11.7	15.2	19.5	0.0	0.0	0.0	0.0	0.0	0.0	0.0	0.0
2.5	4.0	3.9	3.8	2.4	4.4	4.1	3.9	2.7	4.6	4.7	4.4	1.6	2.2	2.9	2.0
2.5	2.9	2.4	2.1	1.2	1.1	0.9	0.8	0.0	0.0	0.0	0.0	0.0	0.0	0.0	0.0
1.4	2.0	3.2	3.1	0.5	0.9	1.3	1.0	1.8	2.0	1.7	2.0	0.8	0.5	0.3	0.5
0.4	0.1	0.1	0.0	1.1	0.3	0.1	1.0	0.6	0.1	0.1	0.0	0.3	0.1	0.0	0.0
2.2	3.2	3.4	4.0	1.8	2.7	2.5	2.7	0.0	0.1	0.1	0.2	0.0	0.1	0.0	0.1
2.3	1.1	0.4	1.2	2.0	1.1	0.5	1.4	0.0	0.0	0.0	0.0	0.0	0.0	0.0	0.0
2.5	5.1	4.7	5.3	3.3	6.2	5.2	5.7	0.9	1.2	1.1	1.0	0.5	0.7	0.6	0.4
1.7	1.9	1.4	2.4	2.8	2.6	2.3	3.2	5.7	6.8	8.3	11.7	8.3	8.8	10.2	13.7
6.0	3.8	3.6	2.1	6.9	4.7	4.9	4.9	12.1	9.8	9.1	6.2	11.1	8.6	8.0	6.9
1.9	1.9	3.1	2.0	1.0	0.9	1.6	0.8	0.2	0.0	0.1	0.1	0.3	0.1	0.3	0.2
3.0	3.5	3.6	4.0	1.3	1.5	1.6	1.6	3.0	4.1	4.5	4.9	0.1	0.1	0.1	0.1
0.9	0.9	0.8	0.4	0.8	0.6	0.9	0.4	14.1	8.0	6.2	3.7	3.3	4.0	4.2	2.6
1.9	2.8	2.6	2.8	2.1	3.3	3.2	3.3	0.2	0.3	0.5	0.3	0.7	1.1	1.4	1.0
0.3	0.2	0.6	0.6	0.2	0.2	0.5	0.6	0.1	0.0	0.0	0.1	0.0	0.0	0.0	0.0
5.9	4.3	2.8	2.4	3.6	2.6	1.6	1.3	3.9	2.5	1.4	1.2	4.4	2.6	1.4	1.3
3.7	2.7	4.1	3.7	2.8	1.9	2.7	2.5	0.5	0.1	0.3	0.2	0.6	0.1	0.3	0.2
8.9	7.1	6.0	4.5	7.7	6.2	6.5	5.3	7.8	8.0	9.2	9.4	12.7	12.2	11.6	11.2
8.0	8.3	6.4	3.8	7.3	6.6	4.1	2.8	1.1	0.8	0.5	0.3	1.7	1.6	0.7	0.3
0.4	0.9	1.1	2.4	0.3	0.7	1.0	1.5	0.4	1.0	1.3	2.4	0.2	0.4	0.8	0.9
5.0	5.0	4.2	4.3	3.2	2.7	2.0	2.0	0.9	0.8	0.6	0.6	2.6	2.1	1.4	1.4
1.4	1.9	1.6	1.1	2.3	3.0	2.5	2.0	1.4	1.3	1.2	1.2	1.4	1.3	0.7	0.6
0.5	1.2	0.5	0.2	0.9	1.8	0.8	0.3	0.7	1.3	0.5	0.2	0.5	1.3	0.4	0.2
1.1	0.6	0.3	0.1	1.6	0.7	0.2	0.1	0.9	0.6	0.2	0.1	1.4	0.7	0.4	0.1
0.5	0.3	0.9	1.4	0.4	0.4	0.6	1.4	0.7	0.4	0.6	1.3	1.6	0.9	0.7	2.3
3.0	1.6	0.9	0.8	2.6	1.4	0.6	0.7	6.5	8.0	7.3	8.3	1.4	1.0	0.6	0.6
2.7	3.0	4.5	5.0	4.2	5.0	6.6	7.2	2.5	3.5	3.8	4.3	10.7	15.2	15.8	19.8

Table 1.16d Proportion of protein, total, saturated, monounsaturated and polyunsaturated fats, carbohydrate and fibre from foods for adult females in New Zealand

Major food groups	Protein				Total fat				Saturated fat			
	Age ranges (years)											
	19–30	31–50	51–70	71+	19–30	31–50	51–70	71+	19–30	31–50	51–70	71+
Alcoholic beverages	0.2	0.6	0.3	0.3	0.1	0.1	0.1	0.2	0.7	0.2	0.0	0.0
Beef and veal	4.6	7.7	8.5	9.3	4.6	5.3	5.2	6.1	3.1	5.1	5.6	6.1
Biscuits	1.0	1.4	1.3	1.5	1.2	2.5	2.7	4.5	3.7	4.4	4.4	5.9
Bread	9.0	10.4	11.2	14.2	4.5	4.9	4.4.	5.5	1.8	1.9	2.1	2.5
Bread-based dishes	7.7	5.8	2.6	1.9	11.4	6.8	6.0	2.1	6.1	5.7	3.1	2.1
Breakfast cereals	1.9	2.6	3.2	3.5	0.9	1.6	3.0	2.1	0.7	1.4	1.3	1.6
Butter and margarine	0.0	0.0	0.1	0.1	6.3	9.0	11.9	15.8	6.3	7.8	9.6	13.3
Cakes and muffins	2.2	1.6	2.0	2.6	2.9	4.3	4.4	4.0	5.4	4.7	6.8	6.6
Cheese	2.0	4.1	3.2	3.1	4.1	4.4	3.4	3.2	4.5	8.1	6.4	6.8
Dairy products	2.6	1.8	2.8	2.8	2.0	2.7	4.1	4.0	5.0	4.0	5.7	6.1
Dietary supplements	0.7	0.4	0.1	0.3	0.6	0.2	0.1	0.0	0.4	0.3	0.1	0.2
Eggs and egg dishes	3.9	2.3	3.0	3.9	1.9	2.7	3.0	3.5	3.0	1.8	2.6	3.4
Fats and oils	0.0	0.0	0.0	0.0	1.6	0.8	0.3	0.9	0.8	0.4	0.5	0.3
Fish and seafood	5.2	6.4	8.1	6.0	2.3	4.3	4.0	4.5	2.3	2.9	3.4	3.0
Fruit	1.5	1.8	2.5	2.6	1.7	1.9	1.5	2.5	1.1	1.5	2.3	1.8
Grains and pasta	9.7	6.7	4.4	4.2	6.1	3.7	3.8	3.2	5.4	3.9	1.6	2.1
Lamb and mutton	2.4	1.4	2.9	2.5	1.9	1.8	3.0	1.9	2.6	1.4	3.3	2.4
Milk	7.7	9.9	9.8	11.5	4.5	4.9	5.0	5.5	6.9	8.0	7.0	9.0
Non-alcoholic beverages	2.6	3.3	3.1	1.9	1.5	1.3	1.1	0.5	1.9	2.0	1.7	1.1
Nuts and seeds	0.6	1.5	1.5	0.7	1.6	2.3	2.4	2.2	1.0	2.4	2.3	1.0
Other meat	0.2	0.3	0.4	0.5	0.2	0.2	0.6	0.6	0.1	0.1	0.4	0.4
Pies and pasties	3.0	1.9	1.7	1.5	6.1	4.3	2.8	2.4	4.4	3.1	2.8	2.4
Pork	3.3	3.9	4.2	4.3	3.3	2.3	3.5	3.2	1.7	2.1	2.4	2.5
Potatoes, kumara and taro	4.2	3.1	2.6	2.9	7.6	6.5	5.5	4.3	7.7	5.6	4.2	2.8
Poultry	11.5	9.3	7.0	6.0	6.7	6.6	5.1	2.9	6.7	5.0	3.9	3.5
Puddings and desserts	0.5	0.4	0.6	0.8	0.5	1.0	1.2	2.5	1.5	1.0	1.5	1.9
Sausages and processed meats	3.0	2.9	2.8	2.7	4.7	4.6	3.8	3.9	4.5	4.2	4.0	3.8
Savoury sauces and condiments	0.8	0.9	0.8	0.8	1.3	1.8	1.7	1.1	1.6	1.7	1.5	1.1
Snack bars	0.3	0.4	0.4	0.2	0.7	1.2	0.5	0.2	0.5	0.7	1.0	0.4
Snack foods	0.6	0.4	0.2	0.0	1.2	0.6	0.2	0.1	2.3	0.9	0.4	0.0
Soups and stocks	1.0	1.2	1.7	1.7	0.6	0.3	0.8	1.5	0.9	1.0	1.2	1.4
Sugar and sweets	1.3	1.1	0.7	0.4	3.2	2.0	1.1	1.0	3.1	4.2	2.5	1.3
Vegetables	4.7	4.5	6.3	5.5	2.5	2.9	3.9	4.5	2.2	2.5	4.4	3.1

Source: ABS (2014); University of Otago & Ministry of Health (2011).

Monounsaturated fat				Polyunsaturated fat				Carbohydrate				Fibre			
Age ranges (years)															
19–30	**31–50**	**51–70**	**71+**	**19–30**	**31–50**	**51–70**	**71+**	**19–30**	**31–50**	**51–70**	**71+**	**19–30**	**31–50**	**51–70**	**71+**
0.2	0.2	0.0	0.0	0.1	0.0	0.0	0.0	3.7	2.2	1.2	0.7	0.0	0.0	0.0	0.0
3.8	5.7	5.8	6.9	1.5	2.5	2.2	2.6	0.3	0.4	0.5	0.4	0.5	0.5	1.0	0.6
2.4	3.0	2.5	3.7	2.4	2.6	2.1	2.8	2.9	3.9	3.9	4.4	1.5	2.3	2.3	2.2
2.5	2.9	3.0	4.5	7.0	8.0	8.2	11.3	13.0	17	17.1	21.7	14.4	16.3	15.3	19.8
6.7	5.8	2.9	2.2	7.0	6.1	3.0	2.4	4.1	3.9	2.0	1.5	5.4	4.4	2.3	1.6
0.9	2.0	1.8	2.1	2.7	4.4	4.8	5.3	3.8	5.1	6.1	6.4	4.7	6.7	7.8	8.1
6.8	8.4	10.8	16.1	8.8	10.3	13.1	19.2	0.0	0.0	0.0	0.0	0.0	0.0	0.0	0.0
4.7	3.2	4.6	5.0	4.5	3.0	4.3	5.1	4.4	4.2	5.6	5.9	3.1	1.8	2.7	2.7
2.1	3.4	2.8	3.1	0.8	1.3	1.1	1.0	0.0	0.0	0.0	0.0	0.0	0.0	0.0	0.0
2.6	2.2	3.0	3.2	1.1.	1.0	1.0	1.0	2.9	1.8	2.5	2.6	0.8	0.4	0.6	0.6
0.3	0.2	0.1	0.4	0.1	0.0	0.0	0.5	0.3	0.2	0.1	0.3	0.0	0.0	0.0	0.1
4.0	2.5	3.5	4.2	2.9	1.9	2.5	2.9	0.2	0.1	0.1	0.1	0.1	0.0	0.1	0.2
2.5	1.3	1.9	1.1	2.3	1.5	1.7	0.8	0.0	0.0	0.0	0.0	0.0	0.0	0.0	0.0
3.3	4.8	5.2	4.6	4.5	6.0	5.6	5.2	0.9	0.8	1.1	0.8	0.6	0.4	0.5	0.5
1.5	3.1	4.1	3.0	3.1	3.8	4.9	4.5	7.8	9.4	12.9	14.5	11.8	12.1	15.4	17.7
5.7	4.3	2.0	2.2	6.5	4.9	3.4	3.6	11.3	9.6	7.5	5.3	10.6	8.1	6.2	5.3
2.6	1.2	3.1	2.3	1.4	0.5	1.2	0.7	0.1	0.0	0.2	0.0	0.2	0.1	0.3	0.1
3.2	4.4	3.5	4.4	1.8	2.4	1.9	1.8	3.3	5.1	4.6	5.1	0.3	0.2	0.2	0.1
1.0	1.1	0.8	0.5	0.9	1.1	0.8	0.4	12.0	7.7	6.6	4.3	3.3	5.1	3.3	2.7
1.9	3.9	4.4	2.2	2.1	4.7	5.4	2.9	0.2	0.5	0.4	0.1	0.9	1.9	1.7	0.7
0.1	0.1	0.4	0.5	0.1	0.1	0.3	0.4	0.0	0.0	0.0	0.1	0.0	0.0	0.0	0.0
4.0	2.7	2.3	2.0	2.6	1.7	1.5	1.1	2.4	1.5	1.3	1.0	2.4	1.7	1.3	0.9
2.1	2.3	2.9	3.0	1.5	1.4	1.8	1.7	0.2	0.1	0.1	0.1	0.3	0.1	0.1	0.1
9.2	6.9	4.7	3.6	8.2	6.6	4.4	4.3	8.7	8.0	7.9	8.5	13.7	10.8	9.1	9.4
9.5	7.2	5.3	5.0	6.9	5.7	3.6	3.3	0.6	0.8	0.4	0.2	1.1	1.2	0.5	0.3
0.9	0.6	1.0	1.4	0.8	0.5	0.7	1.0	1.3	1.0	1.4	1.5	0.6	0.5	0.7	0.6
4.2	4.2	3.9	3.9	2.1	2.2	2.0	1.8	0.5	0.7	0.6	0.5	1.3	2.0	1.5	1.3
2.5	2.7	2.4	1.4	4.4	4.3	4.3	2.3	1.3	1.5	1.2	1.1	1.3	0.9	0.7	0.7
0.5	0.9	0.6	0.3	0.8	1.4	1.0	0.4	0.4	0.7	1.0	0.3	0.5	0.6	0.8	0.3
2.1	1.0	0.4	0.0	2.3	1.2	0.4	0.0	1.3	0.8	0.5	0.0	1.3	1.0	0.6	0.0
0.8	1.0	1.4	1.2	1.4	0.9	1.3	1.1	0.9	1.0	1.4	1.9	1.7	1.8	2.4	2.7
2.1	2.8	1.7	1.0	1.4	1.6	1.4	0.5	7.6	7.7	5.9	5.5	1.6	1.4	0.8	0.5
3.4	4.2	7.2	5.0	5.6	6.6	9.7	7.8	3.8	4.6	5.8	4.9	16.1	17.5	21.9	20.4

Table 1.17a Proportion of vitamin B12, total folate, calcium, iodine, iron, sodium and zinc from foods for adult males in Australia

Major food groups	Vitamin B12				Total folate				Calcium			
	Age ranges (years)											
	19–30	31–50	51–70	71+	19–30	31–50	51–70	71+	19–30	31–50	51–70	71+
Non-alcoholic beverages	9.9	8.8	4.4	2.4	7.6	7.9	8.3	7.8	11.3	13.9	10.6	7.5
Cereals and cereal products	2.9	3.1	3.6	3.7	44.9	46.4	51.0	55.8	10.6	11.8	14.7	15.8
Cereal-based products and dishes	16.1	14.9	9.8	8.0	14.7	12.1	7.3	5.8	16.7	13.4	9.4	7.7
Fats and oils	0.0	0.0	0.0	0.0	0.0	0.0	0.1	0.1	0.0	0.1	0.1	0.1
Fish and seafood products and dishes	10.2	6.8	11.8	12.9	0.8	0.6	0.6	0.6	1.6	1.5	2.0	2.8
Fruit products and dishes	0.0	0.0	0.0	0.0	3.1	3.4	3.5	4.4	1.4	1.8	2.2	3.1
Egg products and dishes	3.6	4.0	4.5	4.5	2.1	2.0	2.2	1.9	1.1	1.1	1.5	1.3
Meat, poultry and game products and dishes	26.7	31.4	37.6	35.9	5.9	4.7	4.5	3.5	3.4	3.1	3.2	2.7
Milk products and dishes	25.6	25.4	24.7	29.4								
Dairy and meat substitutes	0.3	0.7	0.5	1.2								
Soup	0.5	0.9	0.7	0.6								
Seed and nut products and dishes	0.0	0.0	0.0	0.0								
Sauces, dips and condiments	0.2	0.2	0.2	0.2								
Vegetable products and dishes	0.8	1.3	1.2	0.6	6.5	8.0	8.5	7.0	4.4	5.4	6.1	5.5
Legume and pulse products and dishes	0.0	0.0	0.0	0.0	0.9	0.6	0.6	0.9	0.4	0.3	0.4	0.6
Snack foods	0.0	0.1	0.0	0.0	0.6	0.5	0.2	0.1	0.5	0.5	0.2	0.1
Sugar products and dishes	0.0	0.0	0.0	0.0	0.0	0.1	0.1	0.1	0.1	0.2	0.2	0.3
Confectionery and cereal/nut/fruit/seed bars	0.1	0.2	0.2	0.1	0.5	0.5	0.4	0.2	1.2	1.8	1.6	1.1
Alcoholic beverages	0.0	0.0	0.0	0.0	0.1	0.2	0.3	0.3	1.3	1.8	2.4	1.4
Special dietary foods	3.0	2.1	0.6	0.5	0.9	1.1	0.4	0.4	3.3	2.4	1.1	0.6
Miscellaneous	0.0	0.0	0.2	0.0	2.8	4.5	4.6	3.3	0.1	0.1	0.1	0.1

Source: ABS (2014); University of Otago & Ministry of Health (2011).

Iodine				Iron				Sodium				Zinc			
Age ranges (years)															
19–30	**31–50**	**51–70**	**71+**	**19–30**	**31–50**	**51–70**	**71+**	**19–30**	**31–50**	**51–70**	**71+**	**19–30**	**31–50**	**51–70**	**71+**
14.0	15.8	12.9	9.5	3.7	3.9	4.0	3.9	4.6	4.4	3.5	2.8	2.0	2.9	2.3	1.7
22.8	27.1	32.3	36.7	31.4	29.4	32.7	38.6	15.8	16.6	19.8	24.0	17.5	16.1	18.6	22.7
15.9	13.1	8.5	7.4	18.5	16.4	12.1	10.2	27.9	25.2	20.2	17.9	18.5	15.3	10.7	8.5
0.1	0.1	0.2	0.3	0.0	0.0	0.0	0.0	0.7	0.9	1.5	2.4	0.0	0.0	0.0	0.0
4.6	3.7	6.3	4.9	2.6	1.9	2.4	2.1	3.3	3.6	3.7	4.4	2.5	1.8	3.1	2.4
0.3	0.4	0.5	0.8	2.6	3.2	3.7	4.9	0.1	0.1	0.2	0.3	1.3	1.5	1.8	2.7
4.5	4.5	5.4	5.0	2.4	2.4	2.8	2.5	0.9	1.1	1.3	1.5	1.5	1.6	1.8	1.9
4.2	3.3	3.0	2.3	18.0	19.7	19.5	16.5	19.0	20.9	21.9	18.6	33.0	35.9	37.2	33.7
26.5	24.8	24.7	28.9	1.6	1.2	1.2	1.6	6.8	6.7	7.4	8.5	10.3	9.5	9.4	11.4
0.0	0.2	0.1	0.2	0.4	0.9	0.5	0.4	0.1	0.3	0.2	0.3	0.2	0.5	0.3	0.3
0.3	0.7	0.9	0.8	0.8	1.7	2.2	3.0	2.2	3.6	5.9	8.2	0.6	1.5	1.4	2.5
0.1	0.1	0.1	0.1	1.2	1.5	2.2	1.2	0.5	0.4	0.6	0.4	1.4	1.7	2.5	1.6
0.6	0.5	0.4	0.2	1.7	1.1	0.8	0.3	8.3	5.9	4.9	2.4	0.5	0.4	0.3	0.1
1.9	2.1	2.1	1.5	8.7	9.3	9.7	9.2	4.1	4.7	3.7	3.4	5.5	6.4	7.0	6.8
0.0	0.1	0.1	0.1	1.5	0.9	0.9	1.1	0.6	0.7	0.9	1.7	0.7	0.4	0.4	0.6
0.2	0.1	0.1	0.0	1.2	0.9	0.4	0.3	2.4	1.6	0.8	0.5	0.9	0.6	0.4	0.3
0.1	0.1	0.1	0.1	0.2	0.3	0.3	0.3	0.1	0.1	0.1	0.2	0.5	0.5	0.5	0.8
0.3	0.5	0.5	0.3	1.5	1.8	1.3	1.5	0.5	0.5	0.4	0.4	0.9	1.0	0.8	0.7
1.1	1.1	1.2	0.7	0.6	1.5	2.1	1.7	0.9	1.2	1.6	1.0	0.1	0.4	0.7	0.6
2.5	1.6	0.7	0.4	1.0	1.4	0.6	0.5	0.6	0.4	0.2	0.1	1.8	1.5	0.6	0.5
0.0	0.0	0.1	0.0	0.3	0.5	0.5	0.3	0.7	1.0	1.3	1.1	0.2	0.3	0.3	0.3

Table 1.17b Proportion of vitamin B12, total folate, calcium, iodine, iron, sodium and zinc from foods for adult females in Australia

Major food groups	Vitamin B12				Total folate				Calcium			
	Age ranges (years)											
	19–30	31–50	51–70	71+	19–30	31–50	51–70	71+	19–30	31–50	51–70	71+
Non-alcoholic beverages	10.8	9.8	6.3	4.4	9.2	9.7	10.4	10.4	13.7	15.8	12.3	10.0
Cereals and cereal products	3.3	3.5	3.9	4.4	39.4	42.9	43.3	49.4	10.7	11.2	12.4	15.7
Cereal-based products and dishes	16.9	13.3	8.8	6.9	14.3	9.9	6.7	4.5	16.8	11.2	8.2	6.4
Fats and oils	0.0	0.0	0.0	0.0	0.0	0.0	0.0	0.1	0.1	0.1	0.1	0.1
Fish and seafood products and dishes	5.5	10.3	12.8	12.2	0.5	0.8	0.9	1.1	0.9	2.2	2.6	2.3
Fruit products and dishes	0.0	0.0	0.0	0.0	3.8	4.6	5.4	5.1	1.7	2.1	2.8	2.9
Egg products and dishes	3.8	4.0	4.1	4.6	1.9	2.0	1.9	2.1	1.2	1.2	1.1	1.3
Meat, poultry and game products and dishes	24.0	25.9	31.8	32.3	3.6	4.1	4.3	3.4	2.3	2.6	2.6	2.2
Milk products and dishes	28.9	27.6	26.4	31.2	6.2	6.1	5.8	6.8	36.5	37.9	40.1	45.6
Dairy and meat substitutes	1.9	0.6	1.2	1.1	0.5	0.3	0.4	0.2	2.3	1.4	2.1	1.4
Soup	1.1	1.0	1.1	0.6	0.7	0.9	1.4	0.9	1.0	1.0	1.4	1.4
Seed and nut products and dishes	0.0	0.0	0.0	0.0	0.6	0.7	0.8	0.4	0.7	0.9	1.2	0.6
Sauces, dips and condiments	0.3	0.3	0.3	0.3	0.4	0.5	0.4	0.2	0.6	0.7	0.8	0.5
Vegetable products and dishes	2.0	2.2	1.6	1.0	9.1	10.2	11.3	9.3	5.9	6.5	6.6	5.6
Legume and pulse products and dishes	0.0	0.0	0.0	0.0	0.7	0.6	0.6	0.7	0.4	0.4	0.4	0.4
Snack foods	0.2	0.1	0.0	0.2	0.4	0.4	0.2	0.1	0.8	0.5	0.2	0.6
Sugar products and dishes	0.0	0.0	0.0	0.0	0.0	0.0	0.1	0.1	0.3	0.2	0.1	0.3
Confectionery and cereal/nut/fruit/seed bars	0.4	0.3	0.2	0.1	0.9	0.6	0.5	0.2	2.4	1.9	2.1	1.0
Alcoholic beverages	0.0	0.0	0.0	0.0	0.1	0.1	0.1	0.1	0.6	0.9	1.2	0.7
Special dietary foods	0.7	0.9	1.3	0.5	0.5	0.6	1.2	0.4	0.9	1.1	1.7	0.7
Miscellaneous	0.2	0.1	0.0	0.2	7.2	4.9	4.2	4.5	0.2	0.2	0.2	0.2

Source: ABS (2014); University of Otago & Ministry of Health (2011).

Iodine				Iron				Sodium				Zinc			
Age ranges (years)															
19–30	31–50	51–70	71+	19–30	31–50	51–70	71+	19–30	31–50	51–70	71+	19–30	31–50	51–70	71+
16.3	17.9	16.1	12.9	7.2	5.3	4.2	4.5	4.3	4.9	4.1	3.5	3.2	3.8	2.9	2.4
24.4	26.5	29.1	33.4	25.1	26.8	28.3	35.5	15.1	16.5	18.6	22.0	15.3	15.6	16.3	19.8
16.5	11.5	8.3	5.9	19.3	15.4	11.7	9.6	29.4	22.6	19.8	15.6	19.1	14.5	9.8	7.0
0.1	0.1	0.2	0.2	0.0	0.0	0.0	0.0	0.8	1.0	1.3	1.9	0.0	0.0	0.0	0.0
2.7	4.6	6.2	5.2	1.5	3.1	3.1	2.4	2.4	4.3	5.2	4.9	1.4	3.1	5.4	4.8
0.4	0.5	0.7	0.8	3.4	4.3	5.1	5.3	0.1	0.2	0.2	0.2	1.8	2.2	2.6	2.8
4.0	4.3	4.4	4.9	2.3	2.5	2.4	2.8	1.1	1.2	1.1	1.8	1.7	1.7	1.6	2.0
2.7	2.7	2.8	2.2	14.7	15.9	16.7	16.8	15.3	16.5	16.1	19.2	27.8	30.2	31.7	33.8
26.3	25.1	25.6	29.9	1.4	1.5	1.4	1.4	6.8	7.9	9.2	9.5	11.0	11.2	11.0	12.2
0.3	0.2	0.2	0.1	1.3	0.8	1.0	0.6	0.5	0.4	0.5	0.2	0.7	0.5	0.5	0.3
0.7	1.0	0.8	0.7	2.1	2.2	3.2	3.0	5.8	5.4	7.1	6.9	1.7	1.5	2.1	1.7
0.1	0.1	0.1	0.1	1.7	2.0	2.5	1.3	0.3	0.4	0.5	0.2	2.0	2.1	2.7	1.5
0.6	0.6	0.6	0.4	1.1	1.3	1.0	0.8	6.9	8.0	5.9	5.1	0.5	0.5	0.5	0.3
2.6	2.5	2.3	1.8	11.9	11.7	11.5	11.0	5.9	5.5	5.2	3.7	8.7	8.3	8.2	8.0
0.1	0.1	0.1	0.1	1.2	1.1	1.1	0.8	0.6	0.8	1.0	1.3	0.6	0.6	0.5	0.4
0.2	0.2	0.1	0.1	1.0	0.7	0.3	0.2	1.8	1.7	0.7	0.7	0.9	0.7	0.3	0.3
0.1	0.1	0.1	0.1	0.3	0.3	0.2	0.3	0.1	0.1	0.1	0.1	0.6	0.5	0.3	0.7
0.9	0.7	0.7	0.3	2.1	2.0	2.3	1.2	0.7	0.6	0.5	0.3	1.4	1.1	1.1	0.6
0.5	0.5	0.6	0.3	0.8	1.4	1.7	1.0	0.5	0.8	1.2	0.7	0.3	0.6	0.9	0.5
0.7	0.9	1.2	0.5	0.8	0.9	1.6	0.4	0.2	0.3	0.4	0.1	0.8	1.0	1.3	0.5
0.1	0.0	0.1	0.1	0.8	0.7	0.6	0.8	1.3	1.2	1.2	1.9	0.5	0.3	0.3	0.3

Table 1.17c Proportion of vitamin B12, calcium, iodine, iron and zinc from foods for adult males in New Zealand

Major food groups	Vitamin B12				Calcium				Iron				Zinc			
	Age ranges (years)															
	19–30	31–50	51–70	71+	19–30	31–50	51–70	71+	19–30	31–50	51–70	71+	19–30	31–50	51–70	71+
Alcoholic beverages	0.0	0.0	0.1	0.1	2.0	1.8	1.9	1.6	0.5	1.0	0.8	NA	0.4	0.4	0.3	0.4
Beef and veal	8.8	12.0	11.2	14.1	0.8	1.0	0.8	0.7	5.5	8.0	7.3	7.9	8.4	11.0	10.7	12.5
Biscuits	0.0	0.3	0.1	0.3	0.4	0.9	0.7	1.0	0.8	1.5	1.5	1.8	0.5	1.0	1.1	1.4
Bread	1.1	0.2	0.1	0.2	8.8	10.9	11.1	12.4	10.6	12.7	11.9	14.0	8.0	10.1	10.1	12.8
Bread-based dishes	12.5	6.6	4.4	2.2	9.6	5.5	4.4	1.3	10.6	5.7	4.6	1.7	13.0	7.2	5.3	2.1
Breakfast cereals	0.0	0.0	0.1	0.2	1.5	1.6	2.8	2.0	8.7	8.7	14.3	13.2	1.9	2.8	5.1	4.2
Butter and margarine	0.3	1.0	0.8	1.1	0.1	0.2	0.2	0.2	0.0	0.2	0.1	0.1	0.0	0.1	0.1	0.1
Cakes and muffins	1.0	1.7	1.7	1.6	1.7	2.4	2.7	2.0	1.9	2.4	2.5	2.3	1.3	1.5	1.6	1.5
Cheese	5.3	4.7	3.5	3.4	9.2	8.8	6.4	6.2	0.2	0.3	0.1	0.2	3.6	3.4	2.5	2.4
Dairy products	2.6	3.3	4.1	3.8	4.3	5.1	5.0	4.8	0.1	0.2	0.2	0.2	1.1	1.3	1.4	1.4
Dietary supplements	1.0	0.3	0.1	0.1	1.9	0.5	0.1	0.0	0.4	0.2	0.1	0.0	0.6	0.1	0.1	0.0
Eggs and egg dishes	4.4	6.0	7.0	8.0	1.1	1.6	1.9	1.9	2.1	3.1	3.2	3.3	1.4	1.9	2.3	2.8
Fats and oils	0.0	0.0	0.0	0.0	0.0	0.0	0.0	0.0	0.0	0.0	0.0	0.0	0.0	0.0	0.0	0.0
Fish and seafood	5.0	9.5	10.5	10.1	1.2	2.0	2.8	2.5	1.9	3.7	4.4	3.5	1.5	2.5	2.9	2.9
Fruit	0.0	0.0	0.0	0.0	1.6	1.8	2.1	3.4	3.1	3.0	3.7	5.3	1.4	1.4	2.0	2.9
Grains and pasta	6.9	3.8	3.4	1.5	6.0	2.9	3.4	2.5	9.3	7.3	6.6	7.1	13.0	9.5	8.6	5.3
Lamb and mutton	2.8	2.7	3.7	3.0	0.4	0.2	0.4	0.2	1.5	1.4	2.1	1.3	2.1	2.2	2.9	2.3
Milk	15.1	19.4	22.7	26.1	21.2	26.1	30.0	34.0	0.5	0.6	0.5	0.5	5.1	6.7	8.1	9.3
Non-alcoholic beverages	10.1	5.2	3.4	1.2	11.2	9.3	7.5	5.0	4.7	4.3	4.8	0.8	3.4	3.0	3.3	2.9
Nuts and seeds	0.0	0.0	0.0	0.0	0.3	0.5	0.6	0.3	0.6	1.0	1.3	0.7	0.7	1.3	2.0	1.2
Other meat	0.3	0.8	2.9	2.4	0.0	0.0	0.1	0.1	0.7	0.5	1.4	1.6	0.2	0.6	1.0	0.9
Pies and pasties	5.4	3.6	2.7	2.3	3.0	2.2	1.2	1.0	4.5	2.9	1.8	1.4	5.5	3.5	2.1	1.9
Pork	3.2	2.4	3.7	5.1	0.7	0.4	0.7	0.6	3.4	2.3	3.5	3.0	5.6	4.2	6.2	6.4
Potatoes, kumara and taro	0.2	0.5	0.5	0.4	1.5	2.0	2.0	1.7	6.0	6.0	5.7	5.8	4.0	4.3	4.4	4.4
Poultry	5.6	6.7	6.2	3.7	1.7	1.5	1.0	0.6	4.7	4.9	3.4	1.9	5.2	6.4	4.8	3.2
Puddings and desserts	0.2	0.7	0.6	1.8	0.2	0.9	0.6	2.3	0.1	0.4	0.6	0.8	0.1	0.4	0.4	1.1
Sausages and processed meats	3.7	3.8	3.6	3.2	0.9	0.7	0.5	0.4	3.2	2.7	1.9	2.0	4.0	3.8	2.8	3.2
Savoury sauces and condiments	1.4	2.5	1.6	2.4	1.1	1.1	1.4	1.2	2.8	3.2	2.3	2.8	1.0	0.9	1.1	1.0
Snack bars	0.0	0.3	0.0	0.0	0.3	0.6	0.2	0.1	0.6	1.5	0.4	0.2	0.5	1.0	0.3	0.2
Snack foods	0.1	0.0	0.0	0.0	0.4	0.2	0.1	0.0	0.5	0.3	0.1	0.1	0.6	0.4	0.2	0.0
Soups and stocks	1.6	0.3	0.9	1.1	0.8	0.5	0.6	1.7	1.2	1.0	0.9	2.4	1.0	0.5	0.7	1.6
Sugar and sweets	0.5	0.4	0.1	0.1	2.4	1.9	1.3	1.2	2.6	1.9	1.4	1.4	1.4	1.1	0.8	0.5
Vegetables	0.8	0.6	0.2	0.4	3.7	5.2	5.7	7.0	5.3	7.0	6.6	8.6	3.5	5.2	4.8	6.9

Source: University of Otago & Ministry of Health (2011).

Table 1.17d Proportion of vitamin B12, calcium, iodine, iron and zinc from foods for adult females in New Zealand

Major food groups	Vitamin B12				Calcium				Iron				Zinc			
	Age ranges (years)															
	19–30	31–50	51–70	71+	19–30	31–50	51–70	71+	19–30	31–50	51–70	71+	19–30	31–50	51–70	71+
Alcoholic beverages	0.0	0.0	0.0	0.0	1.9	1.2	1.0	0.6	2.1	1.3	1.4	0.8	0.9	0.5	0.5	0.3
Beef and veal	5.5	11.1	11.8	3.8	0.4	0.6	0.6	0.6	4.1	7.0	7.0	7.1	5.9	9.8	10.5	11.4
Biscuits	0.3	0.3	0.3	0.5	0.8	0.9	0.8	0.8	1.5	1.8	1.7	1.9	1.1	1.6	1.4	1.4
Bread	0.2	0.2	0.2	0.1	8.3	8.6	9.4	11.8	9.2	11.2	11.2	14.4	7.6	9.8	9.6	12.8
Bread-based dishes	6.9	5.7	2.5	1.3	5.3	4.2	2.0	1.4	5.8	4.5	2.1	1.4	7.4	5.5	2.7	1.8
Breakfast cereals	0.0	0.0	0.1	0.2	1.6	1.9	1.9	2.6	7.9	9.7	11.2	12.7	2.6	3.9	4.0	4.2
Butter and margarine	0.7	0.8	0.8	1.0	0.1	0.1	0.2	0.2	0.1	0.1	0.1	0.1	0.1	0.1	0.1	0.1
Cakes and muffins	2.0	1.4	1.8	2.4	2.5	1.9	2.2	2.8	3.1	2.3	3.1	3.4	2.4	1.5	1.9	2.3
Cheese	3.2	6.9	4.3	4.5	4.9	9.5	7.4	6.7	0.1	0.4	0.2	0.2	2.1	4.5	3.2	3.0
Dairy products	5.5	3.5	5.3	5.7	7.4	5.0	7.5	7.8	0.5	0.2	0.2	0.2	2.4	1.5	2.5	2.3
Dietary supplements	0.4	0.4	0.1	0.6	0.8	0.6	0.1	0.5	0.4	0.3	0.1	0.5	0.3	0.3	0.1	0.5
Eggs and egg dishes	9.7	5.2	6.4	8.1	2.3	1.2	1.4	1.7	3.9	2.4	2.9	3.6	2.8	1.7	2.2	3.0
Fats and oils	0.0	0.0	0.0	0.0	0.0	0.0	0.0	0.0	0.0	0.0	0.0	0.0	0.0	0.0	0.0	0.0
Fish and seafood	7.6	11.4	13.2	9.4	1.5	1.6	2.6	1.8	3.3	3.3	3.8	3.2	2.3	2.6	3.3	2.6
Fruit	0.0	0.0	0.0	0.0	2.6	2.6	3.3	3.7	4.1	5.1	6.4	6.9	2.0	2.5	3.4	3.5
Grains and pasta	6.6	4.2	1.7	2.3	5.4	3.4	1.6	2.4	8.6	6.6	4.9	5.6	12.9	9.2	6.5	5.1
Lamb and mutton	3.4	2.4	4.4	3.6	0.4	0.2	0.4	0.2	1.4	1.0	2.1	1.5	2.7	1.7	3.5	2.9
Milk	17.8	22.1	22.9	25.5	21.8	28.0	29.2	31.7	0.9	0.8	0.7	0.6	6.6	8.1	8.5	9.7
Non-alcoholic beverages	8.4	5.6	4.0	2.3	11.5	11.2	10.5	6.5	6.3	6.4	5.3	5.2	4.3	4.4	4.2	4.0
Nuts and seeds	0.0	0.0	0.0	0.0	0.5	0.6	0.7	0.3	0.8	1.7	1.9	0.7	0.7	2.0	2.0	0.8
Other meat	1.0	0.5	1.3	1.9	0.0	0.0	0.0	0.2	0.3	0.3	0.7	0.8	0.2	0.4	0.6	0.7
Pies and pasties	2.9	1.9	1.8	1.7	1.9	1.2	1.1	1.2	2.7	1.9	1.6	1.1	3.3	2.0	1.8	1.3
Pork	1.7	2.8	2.6	3.1	0.4	0.4	0.3	0.3	2.0	2.3	2.2	2.1	3.9	3.8	4.1	4.3
Potatoes, kumara and taro	0.4	0.6	0.2	0.2	2.2	1.7	1.2	1.3	7.2	5.7	4.9	5.0	5.2	4.0	3.4	3.8
Poultry	7.9	5.3	5.1	4.5	1.4	1.2	0.7	0.6	4.9	4.0	2.8	2.2	6.9	5.4	4.3	3.9
Puddings and desserts	0.6	0.4	0.8	1.1	1.1	0.7	0.9	1.1	0.5	0.5	0.5	0.6	0.5	0.4	0.5	0.7
Sausages and processed meats	3.4	2.6	3.0	2.8	0.9	0.6	0.6	0.4	2.0	2.3	1.9	1.8	3.1	2.9	2.7	2.8
Savoury sauces and condiments	1.7	2.5	3.3	2.4	1.7	1.3	1.1	1.0	3.5	3.1	3.2	2.8	1.0	1.1	1.1	1.1
Snack bars	0.1	0.0	0.3	0.0	0.3	0.2	0.5	0.1	0.6	0.7	0.9	0.3	0.4	0.5	0.6	0.2
Snack foods	0.1	0.1	0.0	0.0	0.6	0.4	0.1	0.0	0.6	0.5	0.3	0.0	0.8	0.5	0.3	0.0
Soups and stocks	0.5	1.0	0.9	0.5	0.7	1.1	1.3	1.4	1.4	1.7	2.5	2.5	0.8	1.1	1.6	1.7
Sugar and sweets	0.3	0.4	0.3	0.1	2.5	2.3	1.2	1.0	2.6	2.5	1.7	1.3	1.6	1.6	0.9	0.7
Vegetables	0.8	0.7	0.5	0.5	6.3	5.7	8.3	7.4	7.6	8.6	10.6	9.4	5.3	6.0	8.2	7.2

Source: University of Otago & Ministry of Health (2011).

Table 1.18a Examples of foods in major food groups: 2011–12 Australian Health Survey

Major food group	Examples
Non-alcoholic beverages	Tea, coffee, fruit juice, cordial, soft drink, water, electrolyte drink
Cereals and cereal products	Bread, rice, noodles, pasta, breakfast cereal
Cereal-based products and dishes	Sweet biscuits, savoury biscuits, cake, sweet pastry, savoury pastry, pizza, sandwiches, burgers
Fats and oils	Butter, margarine, oils
Fish and seafood products and dishes	Fish, prawns, canned tuna, fish with pasta, fish with rice
Fruit products and dishes	Apples, pears, berries, oranges, peaches, bananas, melons, dried fruit, banana split, apple crumble
Egg products and dishes	Eggs, omelette, souffle, frittata
Meat, poultry and game products and dishes	Beef, lamb, pork, veal, kangaroo, chicken, ham, dried meats, sausages, casseroles, curries
Milk products and dishes	Milk, yoghurt, cream, cheese, ice cream, dairy desserts, cheesecake
Dairy and meat substitutes	Soy beverages, almond milk, tofu, quorn, tofu stir-fry
Soup	Canned soup, homemade soup, dried soup mix
Seed and nut products and dishes	Peanuts, peanut butter, pumpkin seeds, coconut milk
Savoury sauces and condiments	Tomato sauce, chutney, salad dressings, mayonnaise, vinegar, dips
Vegetable products and dishes	Potatoes, carrots, beans, tomato, corn, salads, potato bake
Legume and pulse products and dishes	Lentils, soy beans, chickpeas, kidney beans, falafel, dhal
Snack foods	Potato crisps, popcorn, corn chips, rice crisps, pretzels
Sugar products and dishes	Sugar, honey, jam, icing sugar, apple sauce, meringue
Confectionery and cereal/nut/fruit/seed bars	Chocolate, muesli bars, fruit bars, lollies, chewing gum
Alcoholic beverages	Beer, wine, spirits, cocktails, liqueurs
Special dietary foods	Liquid and powdered meal replacements, protein drinks and powders, oral supplement powder and beverages (excluding electrolyte drinks)
Miscellaneous	Yeast, salt, intense sweeteners, herbs, stock, essences, gelatine, spreadable yeast extract
Infant formulae and foods	Toddler formula, rusks, infant cereals, infant fruit, infant custards, infant fruit juices
Dietary supplements	Vitamins and mineral supplements, fish oil supplements, fibre supplements
Reptiles, amphibians and insects	Crocodile, turtle, goanna

Source: ABS (2014).

Table 1.18b Examples of foods in major food groups: 2008–09 New Zealand Adult Nutrition Survey

Major food group	Examples
Grains and pasta	Rice (boiled, fried, risotto, sushi, salad), flour, pasta/noodles, bran, cereal-based products and dishes (pasta and sauce, lasagne, pasta salad, noodle soup, chow mein)
Bread	All types of bread (rolls, pita, foccacia, garlic), bagels, crumpets, sweet buns
Breakfast cereals	All types (muesli, wheat biscuits, porridge, puffed/flaked/extruded cereals)
Biscuits[a]	Sweet biscuits (plain, chocolate coated, fruit filled, cream filled), crackers
Cakes and muffins[a]	All cakes and muffins, slices, scones, pancakes, doughnuts, pastry
Bread-based dishes	Sandwiches, filled rolls, hamburgers, hotdogs, pizza, nachos, doner kebabs, wontons, spring rolls, stuffings
Puddings and desserts	Milk puddings, cheesecake, fruit crumbles, mousse, steamed sponges, sweet pies, pavlova, meringues
Milk	All milk (cow, soy, rice, goat and flavoured milk), milkshakes, milk powder
Dairy products	Cream, sour cream, yoghurt, dairy food, ice cream, dairy-based dips

Major food group	Examples
Cheese	Cheddar, Edam, specialty (blue, brie, feta, etc.), ricotta, cream cheese, cottage cheese, processed cheese
Butter and margarine	Butter, margarine, butter/margarine blends, reduced-fat spreads
Fats and oils	Canola, olive, sunflower and vegetable oils, dripping, lard
Eggs and egg dishes	Poached, boiled, scrambled and fried eggs, omelettes, self-crusting quiches, egg stir-fries
Beef and veal	All muscle meats (steak, mince, corned beef, roast, schnitzel, etc.), stews, stir-fries
Lamb and mutton	All muscle meats (chops, roast, mince, etc.), stews, stir-fries, curries
Pork	All muscle meats (roast, chop, steak, schnitzel, etc.), bacon, ham, stews, stir-fries
Poultry	All chicken, duck, turkey and muttonbird muscle meats and processed meat, stews and stir-fries
Other meat	Venison, rabbit, goat, liver (lambs fry), pâté (liver), haggis
Sausages and processed meats	Sausages, luncheon meat, frankfurters, saveloys/cheerios, salami, meatloaf and patties
Pies and pasties	All pies including potato top, pasties, savouries, sausage rolls, quiche with pastry
Fish and seafood	All fish (fresh, frozen, smoked, canned, battered, fingers, etc.), shellfish, squid, crab, fish/seafood dishes (pies, casseroles and fritters), fish/seafood products
Vegetables	All vegetables (fresh, frozen, canned) including mixes, coleslaw, tomatoes, green salads, legumes and pulses, legume products and dishes (baked beans, hummus, tofu), vegetable dishes
Potatoes, kumara and taro	Mashed, boiled, baked potatoes and kumara, hot chips, crisps, hash browns, wedges, potato dishes (stuffed, scalloped potatoes), taro roots and stalks
Snack foods	Corn chips, popcorn, extruded snacks (burger rings etc.), grain crisps
Fruit	All fruit, fresh, canned, cooked and dried
Nuts and seeds	Peanuts, almonds, sesame seeds, peanut butter, chocolate/nut spreads, coconut (including milk and cream), nut-based dips (pestos)
Sugar and sweets	Sugars, syrups, confectionery, chocolate, jam, honey, jelly, sweet toppings and icing, ice-blocks, artificial sweeteners
Soups and stocks	All instant and homemade soups (excluding noodle soups), stocks and stock powder
Savoury sauces and condiments	Gravy, tomato- and cream-based sauces, soy, tomato and other sauces, cheese sauces, mayonnaise, oil and vinegar dressings, chutney, Marmite
Non-alcoholic beverages	All teas, coffee and substitutes, hot chocolate drinks, juices, cordial, soft drinks, water, powdered drinks, sports and energy drinks
Alcoholic beverages	Wine, beer, spirits, liqueurs and cocktails, ready-to-drink alcoholic sodas (RTDs)
Supplements providing energy[a]	Meal replacements, protein supplements (powders and bars)
Snack bars[a]	Muesli bars, wholemeal fruit bars, puffed cereal bars, nut and seed bars

a Comparable with 2002 National Children's Nutrition Survey but not comparable with 1997 National Nutrition Survey.

Note: Some foods may not be assigned to the same food groups as in the 1997 National Nutrition Survey so care should be taken when making direct comparisons. For example, muesli bars were assigned to biscuits in the 1997 National Nutrition Survey, but to snack bars in the 2008–09 New Zealand Adult Nutrition Survey.

Source: University of Otago & Ministry of Health (2011).

Table 1.19 Examples of low-, moderate- and high-GI foods

Low GI (less than 55)	Moderate GI (55–69)	High GI (70 or greater)
Apple, apricot, banana, orange, peach, pear, pineapple, carrot, sweet corn, butternut pumpkin, barley, pasta, milk, soy milk, chocolate, honey, yoghurt	Arborio rice, basmati rice, Doongara rice, porridge, couscous, polenta, potato, buckwheat, raisins, sucrose, golden syrup, beetroot, cherries	Glucose, potato, long-grain rice, white bread, wholemeal bread, rye bread, cornflakes, instant porridge, corn thins, rice cakes, watermelon

Note: Differences in GI may occur across different varieties of fruits and vegetables and grains. Those listed in the table are representative and details of specific cultivars and/or processed varieties should be sought.

Source: University of Sydney (2014).

Glycaemic index and sports performance

Using GI principles can optimise the availability of carbohydrate during different types of physical activities. Sports nutrition guidelines include recommendations for carbohydrate intake before, during and after exercise:

- *Pre-exercise*—Some studies have shown that a person can benefit from consuming a low-GI meal prior to endurance exercise, as glucose is released slowly into the bloodstream and blood glucose levels are therefore likely to be sustained.
- *During exercise*—During prolonged exercise, consumption of moderate- or high-GI foods and/or drinks can maintain adequate blood glucose levels and promote carbohydrate as a fuel.
- *Post-exercise*—Consumption of high-GI foods and/or drinks after exercising can promote rapid glycogen repletion in the muscles and is therefore often encouraged.

Specific individualised recommendations should be made by a dietitian with experience in sports nutrition, considering the type, frequency, intensity and duration of activity and the person's goals.

LABEL READING

Packaged food must include the following information:

- name of food
- lot identification
- name and address of supplier
- list of ingredients:
  - *Statement of ingredients*—All ingredients including food additives must be listed in a statement of ingredients. The ingredients are listed in descending order by weight, including added water; i.e., the first ingredient listed is present in the greatest amount, and the last ingredient listed is present in the smallest amount.
  - *Compound ingredients*—All ingredients which make up a compound ingredient must be declared in the ingredients list (in brackets after the named compound ingredient), except when the compound ingredient is used in amounts of less than 5 per cent of the final food. However, if an ingredient that makes up a compound ingredient is a known allergen it must be declared regardless of how much is used.
  - *Characterising ingredients*—All ingredients that are mentioned in the name of a food or are emphasised on the label of a food in words, pictures or graphics are called characterising ingredients and must be listed as a percentage of the final food. For example, a 'tomato and mushroom pasta sauce' would need to list its ingredients as 'tomatoes (85 per cent), mushroom (4.5 per cent), sugar, salt', and so on. Foods such as white bread or cheese may have no characterising ingredients or components.
- list of certain substances when present in the food as an ingredient, a food additive, a processing aid or a component of any of these, declaring:
  - added sulfites in concentrations of 10 milligrams per kilogram or more
  - cereals containing gluten and cereal products—namely, wheat, rye, barley, oats and spelt and their hybridised strains (except when present in beer and spirits)
  - crustaceans and crustacean products
  - eggs and egg products
  - fish and fish products (except for isinglass derived from swim bladders and used as a clarifying agent in beer and wine)
  - milk and milk products
  - peanuts and peanut products
  - sesame seeds and sesame seed products
  - soybeans and soybean products
  - tree nuts and tree nut products (other than coconut from the fruit of the palm *Cocos nucifera*).
- directions for use and storage when incorrect handling and/or use of the food could adversely affect consumer health or safety
- a **use by date** or **best before date** for packaged foods unless the food is safe for human consumption for a period of greater than two years.

Legibility

All information on labels must be legible, prominent, distinct from the background and written in English.

Use by date is the date after which the food contained within the package (stored as recommended on the label) should not be consumed for health or safety reasons. It is illegal to sell food after a use by date.

Best before date is the date by which the food contained within the package (stored as recommended on the label) will still be safe for consumption and will retain any specific qualities for which expressed or implied claims have been made.

The size of the type in warning statements must be at least 3 millimetres high, except on small packages.

Country of origin

All packaged food must state where the food comes from—that is, its country of origin. Additionally, a separate statement that identifies where the food was produced, made or packaged must be listed. If the food is a mix of foods from different countries, it must be listed with the major source first; for example, 'Made from local and imported foods' indicates 50 per cent or more local produce.

Nutrition information panels

Nutrition information panels are required for all packaged foods except those in small packages (100 square centimetres or less), those with no significant nutritional value (for example, herbs, spices, vinegar, gelatine, salt, mineral water, tea and coffee), single-ingredient foods (for example, fruit, vegetables and meat) and alcoholic beverages. Foods sold unpackaged or made and packaged at the point of sale are also exempt. An example of a nutrition information panel is shown in Figure 1.2. Mandatory information displayed in nutrition information panels includes:

- number of servings of the food in the package
- average quantity of the food in a serving (millilitres or grams)
- unit quantity of the food
- average energy content (kilojoules), protein (grams), fat (total and saturated fat in grams), total carbohydrate (grams), sugars (grams), sodium (milligrams)
- name and average quantity of any biologically active substance mentioned in a nutrition claim. For example, if a product claims it is 'high fibre' then fibre content must be listed in the nutrition information panel, and if a product claims it is 'gluten free' then gluten content must be listed in the panel. There are conditions that must be met in order for a nutrition claim regarding a food to be permitted. For a full list of these, see Food Standards Code—Standard 1.2.7—Nutrition, Health Related Claims (FSANZ 2013: 1.2.7).

A percentage of daily intake information regarding nutrients may be included (voluntarily) in the nutrition information panel. If used, it must include the statement 'Percentage daily intakes are based on an average adult diet of 8700 kilojoules. Your daily intakes may be higher or lower depending upon your energy needs.' The percentage daily intakes are calculated using the following reference values:

Energy	8700 kilojoules
Protein	50 grams
Fat	70 grams
Saturated fatty acids	24 grams
Carbohydrate	310 grams
Sodium	2300 milligrams
Sugars	90 grams
Dietary fibre (if included)	30 grams.

NUTRITION INFORMATION Servings per package: (insert number of servings) Serving size: g (or mL or other units as appropriate)		
	Quantity per serving	Quantity per 100 g (or 100 mL)
Energy	kJ (Cal)	kJ (Cal)
Protein	g	g
Fat, total	g	g
—saturated	g	g
Carbohydrate	g	g
—sugars	g	g
Sodium	mg (mmol)	mg (mmol)
(insert any other nutrient or biologically active substance to be declared)	g, mg, mg (or other units as appropriate)	g, mg, mg (or other units as appropriate)

Figure 1.2 Mandatory information required in a nutrition information panel

Source: FSANZ (2013: 1.2.8).

FOOD SAFETY

Food safety encompasses principles and practices to help prevent food-borne illnesses (also called food poisoning). It is estimated that approximately 1.8 million people die each year from diseases caused by pathogens from food or water sources, the majority occurring in developing countries. However, food poisoning affects developed countries as well, and is a significant cause of illness; there are approximately 5.4 million cases of food poisoning reported annually in Australia (120 deaths). In New Zealand there are approximately 1.4 million cases of food-borne illness and seventeen deaths annually.

Food poisoning can be caused by foods that are contaminated with pathogenic microorganisms—for example, bacteria (such as *Campylobacter jejuni*, *Listeria monocytogenes*, *Escherichia coli* and *Salmonella typhimurium*), viruses (including rotavirus, enterovirus and hepatitis A), prions, parasites, environmental contaminants (such as persistent organic pollutants and heavy metals), naturally occurring substances within foods (such as ciguatera and cyanide in cassava), or toxic substances produced by microorganisms called enterotoxins (such as *Clostridium botulinum*, *Staphylococcus aureus*, aflotoxins and mycotoxins).

Variation can occur in how severely people are affected after eating affected foods; common food poisoning symptoms include nausea, vomiting, diarrhoea (with or without blood), fever and headache. Some food-borne illnesses can cause death, including the botulism-causing agent, *C. botulinum*. The people most vulnerable to food-borne illnesses include those who are immune-compromised (due to medications or diseases such as AIDS or cancer), older people, pregnant women, and young children and infants.

High-risk foods include eggs, pasta, rice, poultry, seafood and meats (including beef, lamb, pork, venison and kangaroo)—especially when raw or undercooked—unpasteurised milk and cheese (particularly soft cheeses), fruit and vegetables (especially unwashed).

Foods can naturally contain low levels of harmful bacteria; in very small quantities, such bacteria is usually not harmful. However, under certain conditions, bacteria numbers can rapidly increase, making the food toxic. The following safe food handling tips can help to minimise the risk of food-borne illnesses:

- Wash hands properly (with clean water) before handling food, and during handling, especially when touching raw meat, visiting the toilet, touching animals, changing baby nappies or blowing nose.
- If using disposable gloves, use them for one task only. Put on new gloves for each new task.
- Use clean surfaces, utensils and cutting boards before, during and after food preparation.
- Wash fruit and vegetables with clean water before cooking or eating.
- Do not handle food if unwell with food poisoning. People with food poisoning should wait 48 hours after symptoms have ceased before handling food again.
- Keep raw and cooked food separate:
  - Use separate cutting boards for different foods: raw meat and poultry, raw fruit/vegetables, cheeses, etc.
  - Clean cutting boards and utensils well with hot water and detergent when switching between raw and cooked foods.
  - Do not allow meat/poultry juices to come into contact with foods to be eaten raw.
  - Store raw and cooked foods on different shelves in the refrigerator—raw meats should be kept covered on the bottom shelf or in a dedicated meat drawer. Fruits and vegetables should be kept in a crisper.
- Store foods well covered to ensure they do not come into contact with each other in the refrigerator and/or freezer.
- Defrost frozen foods in the refrigerator—never at room temperature.
- Do not refreeze thawed foods.
- Cooked food should be kept below 5°C or above 60°C to prevent bacterial growth. If refrigerating, cool as quickly as possible (spread out on large trays and cover), then place in the refrigerator.
- Do not keep cooked food at room temperature for more than 2 hours.
- When reheating foods, ensure they are heated to a steaming hot temperature (at least 70°C) as soon as possible, and within two hours of being taken from the refrigerator.
- Do not reheat leftovers more than once.
- When serving food, keep separate serving utensils for different foods. This is especially important in buffet-style food serveries. Keep utensil handles out of foods.
- Do not cough or sneeze over served foods.
- Check the refrigerator temperature to ensure it is less than 5°C. Also check use by dates and discard any foods past the expiry date. (WHO 2009)

Commercial production of food is regulated in Australia and New Zealand to ensure best practices are followed to protect consumers. Hazard Analysis Critical Control Point (HACCP) is enforced in Australia and the Ministry for Primary Industries legislates for safe and suitable food in New Zealand. These regulations enforce strict food safety practices and documentation; such practices have enabled faster recognition of food hazard risks so that products can be discarded before reaching the consumer. Although the risk to the consumer is minimised, some commercially prepared food products can still cause food-borne illnesses, and such products, when identified, need to be recalled. FSANZ regulates food recalls in Australia and New Zealand (FSANZ 2014).

REFERENCES

ABS (Australian Bureau of Statistics) 2013. *Australian Health Survey: Updated Results, 2011–2012.* Cat. no. 4364.0.55.003. Commonwealth of Australia. <www.abs.gov.au/ausstats/abs@.nsf/Lookup/33C64022ABB5ECD5CA257B8200179437?opendocument>, accessed 14 November 2014.

——2014. *Australian Health Survey: Nutrition First Results—Food and Nutrients, 2011–12.* Cat. no. 4364.0.55.007. Commonwealth of Australia. <www.abs.gov.au/AUSSTATS/abs@.nsf/DetailsPage/4364.0.55.0072011-12?OpenDocument>, accessed 23 November 2014.

Department of Health 2010. *How to Measure Yourself.* Commonwealth of Australia. <www.measureup.gov.au/internet/abhi/publishing.nsf/Content/How+do+I+measure+myself-lp>, accessed 18 September 2014.

FNB:IOM (Food and Nutrition Board: Institute of Medicine) 2002. *Dietary Reference Intakes for Energy, Carbohydrates, Fiber, Fat, Fatty Acids, Cholesterol, Protein and Amino Acids.* National Academy Press, Washington, DC.

FSANZ (Food Standards Australia New Zealand) 2013. *Food Standards Code.* Commonwealth of Australia. <www.foodstandards.gov.au/code/Pages/default.aspx>, accessed 16 March 2014.

——2014. *Food Safety and Recalls.* <www.foodstandards.gov.au/consumer/safety/Pages/default.aspx>, accessed 15 November 2014.

Kuczmarski, R.J., Ogden, C.L., Grummer-Strawn, L.M., Flegal, K.M. et al. (2000). *CDC Growth Charts: United States; Advance data from vital and health statistics* 314: 1–28. National Center for Health Statistics.

Ministry of Health 2003. *Food and Nutrition Guidelines for Healthy Adults: A background paper.* New Zealand Government. <www.health.govt.nz/system/files/documents/publications/foodandnutritionguidelines-adults.pdf>, accessed 16 March 2014.

——2006. *Food and Nutrition Guidelines for Healthy Pregnant and Breastfeeding Women: A background paper.* New Zealand Government. <www.health.govt.nz/system/files/documents/publications/food-and-nutrition-guidelines-preg-and-bfeed.pdf>, accessed 18 September 2014.

——2008. *Food and Nutrition Guidelines for Healthy Infants and Toddlers (Aged 0–2): A background paper* 4th edn, partially revised December 2012. New Zealand Government. <www.health.govt.nz/system/files/documents/publications/food-and-nutrition-guidelines-healthy-infants-and-toddlers-revised-dec12.pdf>, accessed 16 March 2014.

——2012a. *Food and Nutrition Guidelines for Healthy Children and Young People (Aged 2–18 years): A background paper.* New Zealand Government. <www.health.govt.nz/system/files/documents/publications/food-and-nutrition-guidelines-for-healthy-children-and-young-people-aged-2-18-years.pdf>, accessed 16 March 2014.

——2012b. *Eating for Healthy Children: 2–12 years.* New Zealand Government. <www.healthed.govt.nz/system/files/resource-files/HE1302%20Eating-for-healthy-children.pdf>, accessed 14 November 2014.

——2013a. *Food and Nutrition Guidelines for Healthy Older People: A background paper.* New Zealand Government. <www.health.govt.nz/system/files/documents/publications/food-and-nutrition-guidelines-healthy-older-people-background-paper-jan2013.pdf>, accessed 16 March 2014.

——2013b. *Eating for Healthy Adults*. New Zealand Government. <www.healthed.govt.nz/system/files/resource-files/HE1518_2.pdf>, accessed 18 September 2014.

——2013c. *New Zealand Health Survey: Annual update of key findings 2012/13*. New Zealand Government. <www.health.govt.nz/publication/new-zealand-health-survey-annual-update-key-findings-2012-13> accessed 14 November 2014.

Monash Obesity and Diabetes Institute 2013. Obesity in Australia (website). <www.modi.monash.edu.au/obesity-facts-figures/obesity-in-australia/>, accessed 17 November 2014.

NHMRC (National Health and Medical Research Council) 2013. *Eat for Health: Australian Dietary Guidelines; Providing the scientific evidence for healthier Australian diets*. Commonwealth of Australia. <www.eatforhealth.gov.au/sites/default/files/files/the_guidelines/n55_australian_dietary_guidelines.pdf>, accessed 16 March 2014.

——n.d. *Serve Sizes*. Commonwealth of Australia. <www.eatforhealth.gov.au/food-essentials/how-much-do-we-need-each-day/serve-sizes>, accessed 11 October 2014.

NHMRC (National Health and Medical Research Council) & Ministry of Health 2006. *Nutrient Reference Values for Australia and New Zealand: Including recommended dietary intakes*. Commonwealth of Australia & New Zealand Government. <www.nhmrc.gov.au/_files_nhmrc/publications/attachments/n35.pdf>, accessed 16 March 2014.

Schofield, W.N. 1985. Predicting basal metabolic rate, new standards and review of previous work. *Human Nutrition. Clinical Nutrition* 39: 5–41.

University of Otago & Ministry of Health 2011. *A Focus on Nutrition: Key findings of the 2008/09 New Zealand Adult Nutrition Survey*. New Zealand Government. <www.health.govt.nz/system/files/documents/publications/a-focus-on-nutrition-v2.pdf>, accessed 23 November 2014.

University of Sydney 2014. Glycemic Index (website). <www.glycemicindex.com/>, accessed 30 March 2014.

WHO (World Health Organization) 1998. *Obesity: Preventing and managing the global epidemic. Report of a World Health Organization consultation on obesity*. <www.who.int/nutrition/publications/obesity/WHO_TRS_894/en/>, accessed 16 March 2014.

——2009. *Prevention of Foodborne Disease: The five keys to safer food*. <www.who.int/topics/food_safety/flyer_keys_en.pdf>, accessed 15 November 2014.

——n.d. BMI Classification (website). <http://apps.who.int/bmi/index.jsp?introPage=intro_3.html>, accessed 14 November 2014.

2

Cultures, beliefs and food habits

Sharon Croxford and Catherine Itsiopoulos

LEARNING OUTCOMES

Upon completion of this chapter you will be able to:

- define the term 'food habits' and describe aspects of Australian and New Zealand food habits
- describe the immigration patterns in both Australia and New Zealand
- outline the characteristics of key culinary influences in Australia and New Zealand and their respective culinary practices
- list food restrictions and describe food-related rituals associated with religious beliefs
- know where to go for additional information about cultures and their beliefs and food habits.

Australia and New Zealand are among the world's most multicultural countries. With multiculturalism comes a vibrant and varied food culture. This chapter introduces food habits and then looks into the history of immigration and recent immigration patterns in both countries and how changes in ethnic demographics (with related food beliefs and practices) have changed their food culture landscapes. Religions and religious beliefs also impact on culinary cultures, and the last section of the chapter reviews the main religious groups in each country and the links between food and drink and religious practices.

FOOD HABITS IN AUSTRALIA AND NEW ZEALAND

Food habits in its broadest sense describes a range of behaviours related not just to eating and drinking patterns but also to food and drink procurement, storage, preparation and cooking, the times and spaces in which these things happen and the people involved. Food habits with respect to current food intakes at various life stages are presented in Chapter 1 for adults, and in the chapters that follow for infants, children and adolescents.

Shopping, preparing and cooking food

Before the arrival of the supermarket, shopping for ingredients involved visits to a number of specialist stores: the baker for bread, the greengrocer for fruit and vegetables, and the grocer for dry goods. Milk was delivered to many homes early in the morning, with a horse and cart used until the 1970s. Bread was also sometimes delivered, and mobile fruit and vegetable vans visited some suburbs.

Supermarkets have afforded most shoppers the ability to buy everything in one shop. Through clever marketing, superstores have influenced purchasing habits. While the sense of seasonality has diminished, as products and produce are available all year round, the convenience of being able to shop quickly and purchase foods at competitive prices appeals to consumers.

At the other end of the spectrum, farmers' markets have become popular again in both Australia and New Zealand with their appeal to consumers interested in fresh, local produce. The renewed interest in supporting farmers and small producers has drawn people back out of the supermarkets. Both Australia and New Zealand have associations to support the growing farmers' market business.

Foods purchased for preparation and cooking at home have seen a dramatic shift in the last 30 years. Olive oil once came in a small bottle from the pharmacy shelf and now has a large section of a supermarket aisle devoted to it. Tomatoes come in all shapes and sizes, with semi-sundried tomatoes a common purchase. Cheese was once limited to bland, processed varieties but nowadays comes in too many shapes, sizes, flavours and packages to count. Foods that were exotic in the 1980s are common fare today—a demonstration of how rapidly the culinary landscape changes and how quickly the Australian and New Zealand cultures adapt.

The place of food preparation and cooking has changed enormously since colonisation in both Australia and New Zealand. With the limited foods available for creating meals, cooking required preparing staples for sustenance rather than enjoyment. As the colonies developed, with more food produced and better transport, the range of meals prepared grew, but it was not until the late 1970s that food and eating, particularly in Australia, took on a more fashionable status (Bannerman 2011). Bannerman suggests that food and eating have become associated with the concepts of 'convenience', 'gourmet' and 'organic'. This shift in the place and different roles that food and eating play has been seen in many countries. Data on food habits (see Chapter 1 for adults and Chapters 7–9 for infants, children and adolescents) show the growing place that convenience and fast foods have within the current Australian and New Zealander diet, the worldwide popularity of television shows focused on food and cooking provides an indication of how people want to connect with the gourmet aspect, and the place of organic foods, while tentative and too expensive for many, is firmly established. The elevation of food and eating within our cultures, along with the broader range of available foods and multiculturalism, means that while people shop and cook for many reasons, one of these is for pleasure.

Meal structure

Food habits evolve over time within any given culture. It is relatively usual for Australians and New Zealanders to eat three meals per day: breakfast followed by lunch, dinner or midday meal then supper, dinner, tea or evening meal. The title of the meal depends on cultural background. 'What are you having for tea?' is bound to confuse many today; for some 'tea' means a cup of tea, while for others it means the evening meal.

Breakfast is usually eaten before the work or school day begins, lunch time breaks up the day, and the last meal of the day follows the finish of work. Morning and afternoon teas might be taken between meals, and supper, taken between the last main meal and bed time, might be included. Morning tea historically meant the drink tea, often alongside a light snack; however, these days it signifies a mid-morning break at which coffee is the drink of choice for many.

The aisles of the supermarket suggest that dry cereal with milk is a common breakfast choice, alongside toast with a range of spreads. Lunches include filled sandwiches or bread rolls, wraps, pastries, variations on evening meals or sushi, *börek* or rice-paper wraps. Evening meals are typically hot and might include traditional meat and vegetables, pasta, pizza, noodle or rice dishes, curries, stews, braises, salads and barbecues. Often, individual meals are served, and people eat from their own plates. Dessert, if eaten, is likely to be served after the main course and is commonly associated with foods that are sweet in taste.

Families tend to eat together for the first and last meals of the day on weekdays and more so on weekends, although the changing demographics of the family and growth in hours spent at work for both mothers and fathers have made this increasingly difficult. Family meals have seen a decline in many societies, but especially in Western countries with more children; families regularly eat while watching television rather than sitting around a table and talking, discussing the day's events over food.

Social eating

The style of social eating is constantly evolving. Gatherings that involve food might invite participants to 'bring a plate', also called a pot-luck dinner, where everyone contributes food to share with all those present. Progressive dinners, where a group start with entrée at one person's house, main at another, desserts and more at others, seem a relic of the 1970s; however, a new style of progressive dinner sees groups of diners move from restaurant to restaurant, sampling dishes and courses as they go.

This change to dining in restaurants perhaps signifies the most profound change in social eating habits of recent times: eating away from the home. The total share of Australian households' food dollars spent on meals away from home has steadily increased from 21.6 per cent in 1984 to 30.8 per cent in 2009–10. Interestingly, despite the global financial crisis in 2008–09, Australians still ate out but chose cheaper dining spaces; cafes, restaurants and specialised food outlets saw a decline in business, with takeaway, supermarket and grocery shopping increasing (Ting 2013). Despite the downturn of 2008–09, cafe (and coffee) culture is worth noting as a strong component in the food scene in both Australia and New Zealand. Street vendors, or food trucks, have become more popular on city and suburban streets, and perhaps this is part of a continued response to cheaper alternatives for prepared food away from home for still-cautious spenders. Street vendors also deliver a quicker service with variety of choice for time-poor workers—for example, lunches for office workers.

CULTURAL GROUPS IN AUSTRALIA AND NEW ZEALAND

Both Australia and New Zealand are multicultural countries. Multiculturalism as a policy platform relies on building a society that is ethnically, culturally and linguistically diverse and in which all members accept and respect this diversity. Prior to the arrival of the British, Aboriginal people were culturally and linguistically diverse, yet in the 200 and more years since then this diversity has been lost; from 250 language groupings and 600–700 dialects to less than twenty spoken widely today (Obata & Lee 2010). From being the country's only inhabitants, Aboriginals and Torres Strait Islanders now make up 2.5 per cent of the Australian population. Most of the remainder of the population was born in Australia, with 27 per cent born overseas (ABS 2012).

Australia's policy regarding immigration has changed markedly over recent decades. For much of the period following Federation, non-Europeans were excluded from immigrating to Australia, a principle known as the White Australia policy. The view that only European settlers were welcome was challenged following the Second World War; however, it was not until the mid-1960s to 1970s that the policy was fully dismantled. Non-European arrivals increased from 746 in 1966 to 2696 in 1971, with part-European arrivals quadrupling to 6054 in the same period (DIBP 2008).

The current policy in Australia allows that anyone can apply to migrate providing they meet criteria set by law. Australia's immigration program is designed to meet its economic and social needs and has a range of categories that limits the total number of migrant applications that will be accepted in any one year. In 2013–14 the program had 190,000 places available, and all were predicted to be taken (DIBP 2013).

The migrant profile of Australia is changing. Table 2.1 shows that of the 28 per cent of residents in Australia who are migrants, almost a quarter are from the United Kingdom, with more than half of migrants being represented by the top ten countries in 2010. Both Italy and Poland saw a decline in numbers between 2006 and 2010, while South Korea and India saw their population groups double. China, Japan and Thailand have also seen their communities within Australia grow substantially.

The 2011 census has added some more recent data and has shown that from 2001 to 2011, the proportion of total overseas-born migrants from the top five countries has changed. Those from the United Kingdom and Italy have seen a decline, while India, China and New Zealand have increased. Italy, Greece and Poland saw the largest decreases. (Please note that in Table 2.1 there is a net increase from 2006 to 2010, but overall a decline from 2001 to 2011; ABS 2012.) These declines are related to deaths of longer-standing migrants where new migration has not filled the gap. Figure 2.1 shows the top ten countries of birth for migrants in Australia.

Nearly 50 per cent of Indian migrants are new settlers, with 35 per cent of the Chinese population in Australia arriving since 2007. Seven out of ten countries of the top ten birthplaces for recent arrivals are Asian: India, China, the Philippines, Malaysia, South Korea, Vietnam and Sri Lanka. The median age for these arrivals is 27 years, substantially younger than the median 50 years for longer-standing migrants (ABS 2012).

East Polynesians were the first people to settle in New Zealand, hundreds of years before James Cook lay claim to the fertile islands for Britain. Europeans

Table 2.1 Numbers of migrants taking up residence in Australia, by country of birth (2006 and 2010)

Country of birth[a]	Estimated resident population		Change (%)	Proportion of Australian population[b] (2010)
	2006	2010		
United Kingdom	1,153,264	1,192,878	+3.0	5.3
New Zealand	476,719	544,171	+14.1	2.4
China[c]	203,143	379,776	+87.0	1.7
India	153,579	340,604	+121.8	1.5
Italy	220,469	216,303	−1.9	1.0
Vietnam	180,352	210,803	+16.9	1.0
Philippines	135,619	177,389	+30.8	0.8
South Africa	118,816	155,692	+31.0	0.7
Malaysia	103,947	135,607	+30.4	0.6
Germany	114,921	128,558	+11.9	0.6
Greece	125,849	127,195	+1.1	0.6
South Korea	49,141	100,255	+104.0	0.4
Sri Lanka	70,908	92,243	+30.1	0.4
Lebanon	86,599	90,395	+4.4	0.4
Hong Kong	76,303	90,295	+18.3	0.4
Netherlands	86,950	88,609	+1.9	0.4
United States	64,832	83,996	+29.5	0.4
Indonesia	67,952	73,527	+8.2	0.3
Ireland	57,338	72,378	+26.2	0.3
Croatia	56,540	68,319	+20.8	0.3
Fiji	58,815	62,778	+6.7	0.3
Singapore	49,819	58,903	+18.2	0.3
Poland	59,221	58,447	−1.3	0.3
Thailand	32,747	53,393	+63.0	0.2
Japan	29,469	52,111	+76.8	0.2

a Lists the 25 countries from which Australia received the highest numbers of migrants.
b *N* = 22,328,847.
c Excludes Special Administrative Regions and Taiwan.

Source: ABS (2007, 2011).

arrived from Australia after 1788 but were also attracted to whaling and sealing in the surrounding ocean. The Treaty of Waitangi of 1840, between 240 Māori chiefs and the British Crown's representatives, opened the gates for British and Irish migrants.

New Zealand's immigration policies followed a similar pattern to Australia's prior to the Second World War, and it was not until 1987 that a new Immigration Act came into law and ended the pro-European bias. Contribution to society through skills and personal attributes is now considered for migrant applications to New Zealand.

While Australia's population has seen continual growth, New Zealand has experienced some small dips in its total population through increased permanent departures, despite immigration targets. The 2013 census in New Zealand showed 4,242,048 residents. Twenty-four per cent were born overseas. The top seven overseas birthplaces in descending order were Asia (7 per cent), the United Kingdom and Ireland (6), the Pacific Islands (4), the Middle East and Africa (2), Europe (2), Australia (1) and North America (1). Ethnic groups that New Zealanders identify with are European (74 per cent), Māori (15), Asian (12), Pacific peoples (7) and Middle Eastern, Latin American and African (1) (people may identify with more than one group). The Asian and Middle Eastern, Latin American and African ethnic groups have increased by one-third since 2006 (Statistics New Zealand 2013a).

These data tell us that the mosaic of cultures within Australia and New Zealand is constantly changing, and thus the associated food cultures are

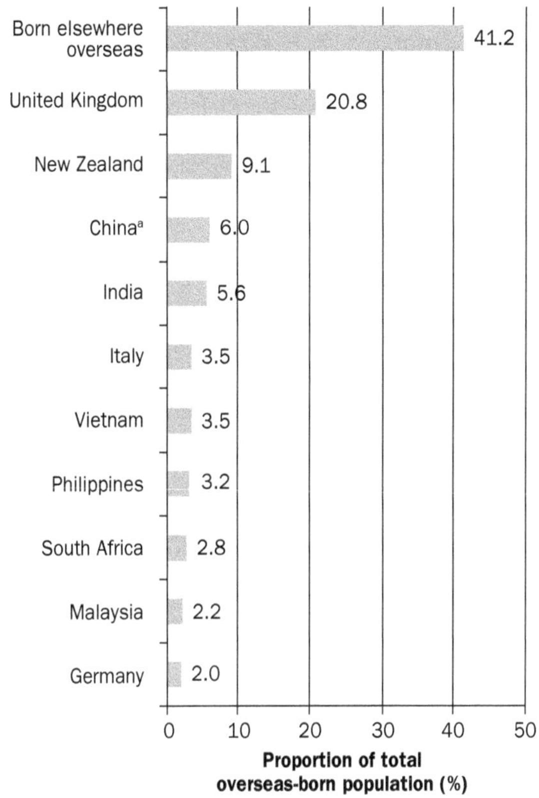

Figure 2.1 Ten countries of birth from which Australia received the most migrants, as proportions of the total overseas-born population (2011)

a Excludes Special Administrative Regions and Taiwan.

Source: ABS (2012).

never-ending works in progress. Migrant groups, while often adopting or acculturating to some of their new country's food habits, will typically bring their food cultures with them (see Chapter 20). New migrants may struggle in the early years, and food insecurity through lack of access to familiar foods may be a real issue. However, history shows that as policies around immigration have changed and communities have grown, infrastructure that supports those communities has also developed. Food stores selling country-specific and sometimes regional-specific items have opened; restaurants and other shops selling prepared food have begun to cater for people from a more diverse society. The Australian and New Zealand palates have become richer and more sophisticated through immigration and multiculturalism.

INFLUENCES OF MULTICULTURAL FOOD HABITS IN AUSTRALIA AND NEW ZEALAND

Some of the key migrant culinary cultures that have influenced Australian and New Zealand cuisine include those listed below.

Aboriginal and Torres Strait Islander

It has only been in recent years that traditional foods from Aboriginal and Torres Strait Islander food cultures have made a mark on Australian non-Indigenous cuisine. Bush food or bush-tucker is known for its nutritive as well as medicinal properties. The early colonists, while using bush foods to supplement their diet, preferred foods from England. Australia's first cookbook, *The English and Australian Cookery Book* (1864) by Tasmanian Edward Abbott contained recipes with kangaroo meat but it was not for almost another two centuries, and the publication of several books focusing on the plant base of Indigenous bush foods, that a broader audience learned of native foods. Mark Olive is Australia's most well-known Indigenous chef and has introduced viewers to bush foods through the television series *The Outback Café*. However, most Australians have limited real experience of traditional foods. Kangaroo is available for consumption across Australia, and a niche market for dried foods, especially condiments, is developing. See Chapter 14 for more information.

American

While migrants from the United States make up a relatively small proportion of the Australian and New Zealand populations, the pervasive nature of American culture through television, cinema and the internet, including social media, has had a substantial influence on food culture. Hamburgers, French fries and other fast foods, frozen foods and especially frozen meals, soft drinks and corn flakes are just some of the foods that have been incorporated into Australian and New Zealand cuisines. The shopping experience also changed through American influence, with larger shopping centres and supermarkets taking over from smaller traders in much of suburbia.

American eating habits largely reflect British food culture, with emphasis on **family-style service** in which dishes are passed around so everyone can take a portion. One major difference at the table is how Americans use cutlery. They cut up food using their dominant hand, then replace their knife with

Family-style service involves foods for a meal served in larger dishes from which individuals help themselves to portions as required or as etiquette dictates.

their fork to scoop food into their mouth. The English and European style, also adopted in Australia and New Zealand, is to use the knife in the dominant hand and the fork in the other, and use both at once to push food onto the fork.

British

British food was the main influence in Australian cuisine following colonisation and in New Zealand following the arrival of the pioneers in the 1800s. The foods were basic, with potatoes, flour, milk and meat being staples along with tea, sugar and salt, the last of which was used particularly for preserving foods in the absence of refrigeration. In Australia, once the wide-open plains of the broader countryside were uncovered, grazing was possible, and thus meat and wheat became mainstays in the diet.

Meat and three vegetables, one of which was potato, was and still remains a typical evening meal in many Australian households. Lamb chops, sausages and Sunday roasts of beef or mutton with all the trimmings, heavy desserts such as puddings with custard and cream, a range of buns and cakes, battered fish and chips, porridge, toast and tea have left their mark on both Australian and New Zealand cuisine. The modern British meal pattern includes breakfast, lunch and evening meal. Breakfast is lighter than in previous times, but a cooked breakfast is often consumed on weekends. Lunch is moderate, with the evening meal forming the main meal of the day and often shared by families. Meat, fish or poultry typically make up the main component of the meal, with a starchy and some vegetable sides. Meals are usually served for individuals on single plates; however, for larger occasions a family-style service may be used. Courses are usually served sequentially, with entrée followed by the main course and then dessert.

Chinese

The Chinese arrived in large numbers in Australia and headed to the goldfields in the 1850s and in New Zealand in the 1860s. While some returned home, many remained, and their descendants have become part of the fabric of both countries. Chinatowns are established in cities and rural towns across Australia, and in New Zealand particular streets focus on trade for Asian businesses. As dining out became a more popular and feasible option for families, many would head to a Chinese restaurant. The award-winning Flower Drum restaurant opened in Melbourne in 1975 and showcased the finest Cantonese cuisine. It is this cuisine that many people understand as Chinese. Meats in oyster, plum, black bean and sweet and sour sauce, fried and steamed rice and noodles are not just eaten out but cooked in many households. Popular brunch options include dim sum—steamed and fried filled dumplings, buns, rolls and cakes taken with tea.

Chinese people tend to eat three meals per day plus a number of snacks. While snacks may be purchased from street vendors, main meals are eaten in a shared fashion. Foods are placed in a central space, and diners select from the communal dish, placing their selection in a bowl or on a small plate at their place. Food selection may be based on the need to balance Yin and Yang; that is, if there is an imbalance in body functions, foods with either Yin or Yang characteristics can be eaten to attain equilibrium. Additionally, the concept of balancing *fan* and *ts'ai* foods is important. *Fan* is the staple of the meal, a cereal grain or foods made from cereals, and *ts'ai* is the meat and vegetables, legumes, fruits, nuts and seasonings of the meal.

Greek and Italian

Migrants from southern Europe, mainly from Greece and Italy, were among the largest groups arriving from overseas from the early 1950s to the 1970s. In 2001, Greek and Italian migrants formed the third-largest overseas-born group in Australia, after groups from the United Kingdom and New Zealand. Furthermore, Italian and Greek were the second and third (respectively) most common languages spoken at home in Australia (Chinese being the first). (Greek and Italian migrant populations make up much smaller numbers in New Zealand.) This signifies retention of the languages in the following generations, and with language comes the retention of culture, especially food culture.

The traditional cuisines of Greece and Italy have many similarities, such as olive oil being the principal fat used in cooking and on salads, plentiful leafy vegetables and fresh fruits, and legumes used in **vegetarian** dishes.

Vegetarian diets omit meat, poultry and seafood. Lacto-ovo vegetarians avoid flesh foods but consume eggs and dairy foods. Pescetarians consume fish but no meat or poultry.

There are, however, some important differences, which are described below.

The typical meal pattern of the traditional Greek diet consists of three meals per day, the largest in the middle of the day (followed by a rest or sleep period), and snacks of fresh fruit, yoghurt and nuts. A typical breakfast includes wholegrain sourdough-style bread with tomato, cucumber, cheese, olives or a boiled egg or semolina-type porridge. Many Greeks, however, typically have only a short black Greek or Turkish coffee with sugar.

Lunch is usually the main meal of the day and consists of foods such as baked fish with boiled *horta*, a lamb shoulder with baked potatoes or casserole dishes with meat and vegetables in a tomato and olive oil base. Dinner is usually a lighter meal, such as lentil soup or spinach and rice casserole with a small piece of bread. All meals are usually served with at least one or two salads and feta cheese and olives on the side. Bread is a staple and eaten with most meals. Greeks always drink water with meals and like a glass (or two) of wine (retsina or a homemade red).

During religious festivals Greeks usually follow an Eastern Orthodox religious fast, which involves avoiding all animal products and following a **vegan** diet for periods of up to 40 days. During fasting days legume-based dishes are consumed as a substitute for meat. During celebrations Greeks eat festive foods such as a whole lamb cooked on a spit, with many salads and other cooked dishes as accompaniments. Celebratory meals are usually followed by traditional sweets such as buttery shortbreads crescents called *kourabiedes* or syrupy pastries including baklava and *galaktoboureko*.

Greek-born migrants to Australia have made some changes to their diet—namely, increasing meat and dairy food intake. Many arrived in Australia from poverty-stricken areas in Greece where meat was scarce. Here, they found meat (especially lamb) was abundant and affordable and they enjoyed large quantities regularly. They also consumed more cheese and butter, margarine and white bread.

Greek migrants experienced many difficulties in sourcing traditional ingredients such as olive oil, herbs and spices and leafy vegetables. Today, with an abundance of traditional foods and ingredients at hand, they have no difficulty in following a more traditional diet. In fact, data show that elderly Greek migrants have returned to a traditional plant-based diet, eating smaller portions of meat, more vegetarian legume-based dishes and vegetables that are home grown. The retention of the traditional pattern by this generation is likely to be contributing to the health benefits associated with the Mediterranean diet (see box and Chapter 20).

Vegan diets omit all animal-derived foods and products, including meat, poultry, seafood, eggs and dairy.

The Mediterranean diet

The Greek diet is famously known as the archetypal Mediterranean diet—that is, cardio-protective and diabetes preventive. It was first identified by researcher Ancel Keys (1986), who conducted a large project called the Seven Countries Study in the 1950s. In follow-up research he showed that the Cretan diet was linked to low mortality from heart disease and all causes. Since this study, the traditional Cretan diet has been studied extensively and shown to promote longevity as well as preventing heart disease, diabetes, some cancers and neurodegenerative conditions such as Alzheimer's and Parkinson's diseases.

The health benefits of the diet have been attributed in part to the high content of plant-derived bioactive phytochemicals (such as carotenoids, flavonoids and polyphenols) from fruits, vegetables, olive oil and wine, which could play a significant role in reducing chronic disease, particularly cardiovascular disease (CVD) (Kapiszewska et al. 2005; Owen et al. 2000; Serra-Majem et al. 2003; Simopoulos 2001; Visioli et al. 2005; Visioli & Galli 2001).

The Mediterranean diet is predominately plant based and is characterised by:

- an abundance and wide variety of plant foods (fruits and vegetables, particularly wild edible leafy greens, called *horta*, wholegrain cereals, nuts and legumes)
- olive oil, as the principal source of fat and minimal use of other added fats
- low intake of red meat
- moderate to low intake of dairy foods (mostly as cheese and yoghurt)
- moderate to high intake of fish
- moderate intake of wine, normally with meals.

(Kromhout et al. 1989; Simopoulos 2001; Willett et al. 1995)

Little is known, however, about the second and third generations, though it is thought that they are probably not following a traditional diet but consuming a wide variety of foods from different cultures.

The typical Italian meal pattern consists of three meals with a snack in the afternoon. Breakfast is usually an espresso coffee with hot milk and a sweet pastry, biscuit or small bread roll with cheese and a slice of cured meat. Lunch is normally the main meal of the day and is followed by a rest period, as for the Greeks. The Italians, however, serve multiple courses in a main meal, such as the first plate (*primo piatto*), usually a pasta or rice dish (risotto), followed by a second dish (*secondo piatto*), a braised meat or fish dish with vegetables or salad, then fruit and hard cheese (parmesan or provolone). A mid-afternoon snack is usually eaten—a sweet pastry or biscuit with an espresso (strong black) coffee, fruit or nuts. The evening meal is usually light and includes a soup (minestrone) or cold meats and cheese with bread, or leftovers from lunch. Wine (usually homemade red) is often enjoyed with meals. Popular quick Italian meals include the well-known pizza.

Indian

Just as Chinatown and Little Italy can been found in our major cities, so can Little India. In areas where migrants have settled, shops and restaurants have sprung up to support local ethnic groups. Victoria, particularly Melbourne, has the largest number of Indian migrants in Australia, and Auckland has the largest number in New Zealand. The culinary concept of curry, while in a vastly inferior form, was introduced over 150 years ago to households down under by the Briton Joseph Keen, who had settled in Tasmania. Curries were introduced to Britain from India during British colonial rule. Following Australian and New Zealand travellers' experiences in different regions of India and increased Indian immigration, Australians' and New Zealanders' understanding of curries and the range available increased. While perhaps the form is still anglicised, we know of korma, tikka masala, madras, rogan josh and vindaloo curries, butter chicken, dahl and the range of flatbreads, including naan, chapatti, roti and paratha. Worcestershire sauce, while an English invention, was concocted with Indian spices and an understanding of Indian cuisine. Mulligatawny and kedgeree are examples of early fusion cuisine using English and Indian traditions.

In India breakfast is eaten in the morning, and another meal consisting of small snacks and foods similar to those eaten at breakfast is eaten later in the afternoon. The main meal is eaten in the evening, and while a number of foods may be served, they are not necessarily served sequentially. Savoury and sweet dishes arrive at the same time.

Traditional Indian cuisine reflects several ancient and religious influences. Vegetarianism is associated with Hinduism, according to which cattle are sacred. Some meat-eating practices show influences of the Mughal Empire. Ancient Vedic philosophy categorises food as *sattvic* (pure, uncontaminated), *rajasic* (stimulant) or *tamasic* (sedative). Meat, fish, onion, garlic and egg-plant are some of the foods considered *tamasic*; they are avoided by those following **Ayurvedic** principles, as they are said to lead to mental or physical harm.

Japanese

The Japanese tend to consume simple meals at breakfast and lunch followed by a more substantial evening meal. Foods are generally served in separate bowls at one time. The Japanese are a relatively new migrant group to Australia, and their numbers have grown markedly since 1980. Without doubt, sushi and the sushi bar have had the biggest influence of Japanese cuisine on Australia and New Zealand. While known to elite dining circles for many decades, it is only in recent years that sushi has become fast food, especially lunch, for the masses.

Māori

Māori brought the starchy sweet potato and taro to New Zealand when they first arrived from Polynesia. Traditional cooking of food occurs on hot stones in an earth oven called a hangi, with wrapped food placed on the stones and covered with leaves or wet hessian bags. A boil-up of pork bones, *puha* (sow-thistle) and doughboys (dumplings) is a significant cultural dish. See Chapter 15.

Samoan

Samoans make up 50 per cent of Pacific peoples in New Zealand. Samoan cuisine features taro, coconut cream and bananas and other tropical fruits alongside fish, poultry and pork. An *umu* is the Samoan equivalent of the Māori hangi, in which hot volcanic rocks were

Ayurvedic (or Ayurveda) medicine is a traditional Hindu system that incorporates elements of food and diet, herbs and yogic breathing to maintain equilibrium in the body.

traditionally used for cooking foods wrapped in taro leaves.

Thai

Thai cuisine has influenced Australian cuisine partly through Australian travellers bringing home a desire for Thai food. The Thai ethnic group in Australia is relatively small, although rapid growth began in the 1980s and has continued strongly. If eating at Chinese restaurants was the thing to do in the 1980s, Thai became the fashion in the late 1990s and 2000s. Green and red curries (based on the chilli used), tom yum, pad thai and *yum nua* are not just favourite restaurant dishes but commonly cooked at home, sometimes from scratch but often from pre-prepared sauces available at any supermarket.

Thais eat three meals a day plus snacks, and the volume of street vendors in Thailand supplying easy snacks testifies to the latter. Breakfast may be a curry but is changing increasingly to a lighter option. Meals are typically shared with family and friends, especially the evening meal. There are no courses, and the foods are served all at once.

Turkish

Turkish migrants first came to Australia in significant numbers following the assisted passage agreement in 1967. Family reunions then increased their communities, with almost 60,000 Turkish speakers resident in Australia according to the 2011 census (SBS n.d.). New Zealand has substantially fewer Turkish migrants. Turkish delight and kebabs are what many people would associate with Turkish cuisine; however, the gozleme, a flat unleavened bread typically filled with feta cheese, meat or potato, has become a popular market food in recent years. Turkish cuisine has many similarities to other eastern Mediterranean and Middle Eastern cuisines.

A Turkish breakfast consists of bread, cheese (with the type depending on the region), tomato, cucumber, olives, spicy pepper paste (again depending on the region), maybe some butter, jam and lots of tea. Tea is served black and taken without milk. Lunch and the evening meal are similar to each other, with rice or pasta as the main cereal and a range of meat and vegetable dishes. Street vendors and small shops sell a range of snacks, including *simit* (a bread ring coated in sesame), *börek*, freshly peeled cucumber and chargrilled corn.

Families eat together at breakfast and in the evening, with lunch often provided at work. Family service is typical, with a table filled with dishes and the cook or head of the table offering portions to each diner. Meals are eaten at high tables with chairs or at low tables with people seated on cushions on the floor, typically depending on socio-economic status.

Vietnamese

Refugees from the Vietnam War reached Australia and New Zealand, many by boat, towards the end of the war, and this formed the greatest increase seen in the number of Vietnamese people settling in both countries. With these settlers came pho (pronounced 'far'), a popular street food in Vietnam. However, in recent years rice-paper wraps have become a mainstay in many cafes offering lunch on the run.

In Vietnam, a balance of heating and cooling foods following the principles of Yin and Yang is important in a meal. Vietnamese cuisine balances five elements, which correspond with wood, fire, earth, metal and water. When preparing a meal, a Vietnamese cook will include five tastes gained from different spices (sour, bitter, sweet, spicy, salty), five colours (green, red, yellow, white, black) and five nutrients (carbohydrates, fat, protein, minerals, water). Meals will also be designed to appeal to the five senses, so they will look, taste, feel, smell and sound good when eaten.

Vietnamese meals are generally eaten in the same shared style as Chinese meals, with diners selecting from a range of communal fish, meat or vegetable dishes and placing their portion in an individual bowl of rice. Rice is ubiquitous throughout the country, and fermented fish sauce is the condiment of choice. Breakfast often starts the day and includes pho or other soup-like dishes with noodles or rice. Midday and evening meals are similar to each other.

Both the Vietnamese rice paper roll and the Japanese sushi options have in recent years been included and recommended as 'healthy lunch options' by Australian and New Zealand dietitians and nutritionists. This reflects the inclusion and acceptance of new and different foods and the expansion of food culture into the Australian mainstream that was lacking in post-Second World War Australia.

RELIGION AND ITS IMPACT ON FOOD HABITS

The main religions of the world can be divided into east and west. Eastern religions developed in India and include Hinduism and Buddhism. Western religions originated in the Middle East and include Christianity,

Islam and Judaism. Within each of these religions, relationships with food can be starkly different.

Religious groups in Australia and New Zealand

According to the 2011 Australian census data, 75 per cent of Australians described themselves as having a religious affiliation. Of these, 89 per cent were Christian and 11 per cent non-Christian (ABS 2012) (see Figure 2.2). Results from the New Zealand 2013 census showed that 45.6 per cent of those who answered the questions related to religion affiliated with a Christian denomination (including Māori Christian), 6 per cent with a non-Christian religion and 38.5 per cent with no religion (Statistics New Zealand n.d.).

In New Zealand an increase in people reporting an affiliation with a non-Christian religion between 2001 and 2013 has been attributed to the arrival of Asian migrants who follow Hinduism and Buddhism and of people born in southern Asia and the Middle East who follow Islam. In 2013 the four largest Christian denominations followed in New Zealand were Catholic, Anglican, Presbyterian (including Congregational and Reformed) and Methodist. Of these, Anglicanism, Presbyterianism and Methodism saw a decrease in followers, although the number of Catholic followers has increased since 2001. Other Christian denominations that have grown in New Zealand are Orthodox, Evangelical and Seventh Day Adventist Christian denominations (Statistics New Zealand 2013b).

Between 2001 and 2011 there was a primary shift in religious demographics in Australia. The changes, as in New Zealand, can be attributed largely to immigration. The main Christian denominations, with the exception of Catholicism, experienced reductions in both absolute and relative numbers. Conversely, Hinduism, Islam and Buddhism witnessed substantial growth in the same ten years. Figure 2.3 shows the

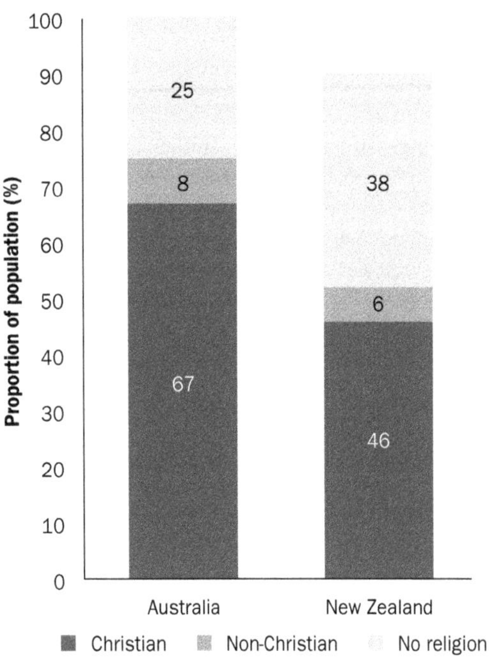

Figure 2.2 Religious affiliation in Australia (2011) and New Zealand (2013)

Note: For New Zealand, 10 per cent of the population did not answer the question.

Source: ABS (2012); New Zealand Statistics (n.d.).

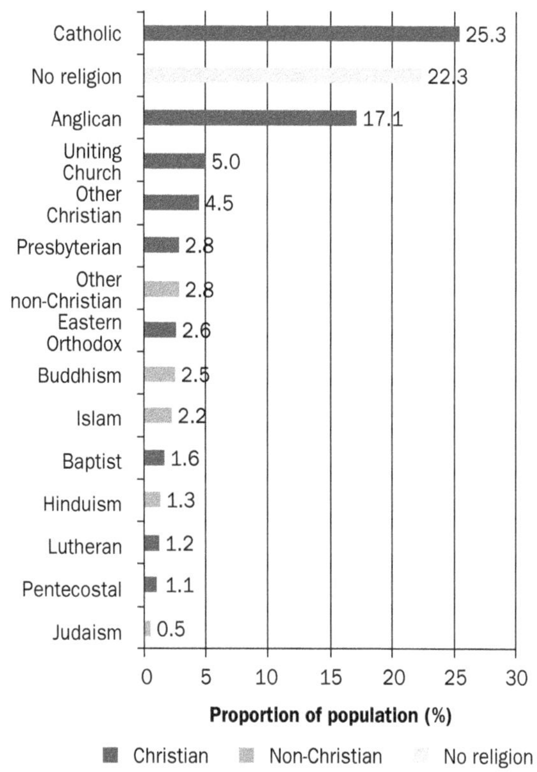

Figure 2.3 Self-described religious affiliations in Australia (2011)

Note: 'Other Christian' denominations include, but are not limited to, Brethren, Methodist, Charismatic, Religious Society of Friends (Quaker), Millerite and Mormon. 'Other non-Christian' religions include Daoism, Sikhism, Jainism, Zoroastrianism, Spiritualism and New Age religions including Rastafarianism.

Source: ABS (2012).

distribution of different religions and no religion within the Australian population.

Main religions and associated food habits

Not all religions include food or drink restrictions or laws as part of their doctrine. For the religions that do, the lists of proscribed and permitted foods and food practices vary widely. Religions that commonly observe rules pertaining to food and food practices include Buddhism, Catholicism, Eastern Orthodox, Hinduism, Islam, Judaism, Mormonism, Rastafarianism and Seventh-day Adventists. See Table 2.2 for a general overview of the feasting and fasting practices of these religions. The customs listed below reflect those that are well known; however, not every person who affiliates with a religion observes each of the customs. It is always good practice to ascertain which food-related traditions are followed when working with people from culturally and ethnically diverse backgrounds.

Religious symbolism in foods

Many everyday foods have their origins in religion. In Christianity, pancakes are associated with Shrove Tuesday and signify the beginning of fasting during Lent, hot cross buns, eaten at Easter, show the cross sign and were traditionally eaten to break the Lenten fast on Good Friday, Easter eggs symbolise new life, the **simnel cake** of Ireland and the United Kingdom has eleven marzipan balls arranged around the top to represent Jesus' twelve apostles minus Judas, and pretzels show a child's arms folded in prayer. Helva, often made from semolina, is prepared and eaten on the anniversary and the seventh and fortieth days after the death of a Muslim in some countries. Matzah, an unleavened dry bread, is traditionally Passover food for the Jewish faith, while the braided bread challah is blessed before the Sabbath meal. While some of these foods make only fleeting appearances during the year, many are more readily available and have become integrated into regular food culture.

Simnel cake A light fruit cake with a marzipan layer in the middle and on top with eleven (sometimes twelve) marzipan balls at equal distances around the rim symbolising the twelve apostles minus Judas.

Fasting and abstinence

Fasting practices vary, with some religions advocating many fast days and others few. Not all those who identify themselves as following a religion will observe all fasting rules—for example, one Muslim may not fast every day during Ramadan yet another will, and not all Eastern Orthodox followers adhere to all fast days. Fasting is a method used for people to grow closer to God, for purification, commemorative mourning or gratitude or to help achieve atonement in some faiths. A list of the fasting practices of some religions follows.

Table 2.2 Food regulations, fasting and major feast days or festivals by religion

Religion	Food regulations (or taboo foods)	Fasting	Major feast days or festivals involving food
Buddhism	Restrictions vary depending on tradition and culture	Yes	None
Catholicism	Meat on Friday	Yes	Christmas, Easter
Eastern Orthodox	Meat, poultry, dairy products, eggs, olive oil, wine on fast days	Yes	Christmas, Lent, Easter
Hinduism	Beef avoided; other restrictions vary	Yes	Diwali (Festival of Lights), Holi (Festival of Colour), Navratri
Islam	Non-halal (haram) foods, especially pork and pork products; alcohol	Yes	Eid al-Fitr (end of Ramadan), Eid al-Adha (Feast of the Sacrifice)
Judaism	Non-kosher foods, especially pork and shellfish; meat and milk not eaten together	Yes	Rosh Hashanah (Jewish New Year), Sukkot (Feast of the Tabernacles), Hanukkah (Festival of Lights), Passover
Mormonism	Tea, coffee, other caffeine, alcohol	Yes	Christmas, Easter
Protestantism	None	No	Christmas, Easter
Seventh-day Adventists	Pork, shellfish, coffee, tea, alcohol	Yes	None
Rastafarianism	All meat, shellfish, scavengers, fish greater than 30 cm long, alcohol, salt, milk, coffee	No	Christmas

Fasts in other religious groups are observed at the discretion of local organisations and individuals.

Buddhism

Fasting involves abstaining from solid food. Fast days are generally observed by monks, and practices depend on the area but usually involve refraining from eating solid food, with some liquid allowed.

Catholicism

Fasting means abstaining from practices, including reducing food intake. Modern fast days include all Fridays in Lent (which lasts for approximately 40 days, excluding Sundays, prior to Easter), but more importantly Ash Wednesday and Good Friday. Strict Catholics from the age of 18 to 59 years will have one full meal at midday that does not contain meat or poultry or products made from these foods. Two other smaller meals may be consumed on these days. The idea is to reduce one's overall intake. Abstinence on Fridays traditionally involved not eating meat, but may now mean other penitential deeds. Fish is often used as a substitute for meat on these days. Food should not be consumed in the hour before Holy Communion.

Eastern Orthodox

Lent, Apostles', Dormition and Nativity fasts are practised, as well as three other one-day fasts, on the Eve of Theophany, the Beheading of John the Baptist and the Elevation of the Holy Cross. Wednesdays and Fridays are also fasting days, except those in 'fast-free' weeks, which occur three times a year. Dates for fasting depend on whether the Julian or Gregorian calendar is followed. The longest fast (Great Lent) takes place during the 40 days prior to Easter. Strict observers may fast up to 180 days per year.

Meat, poultry, dairy products and eggs are prohibited on fast days, along with olive oil and wine. Fish with a backbone is generally not allowed; however, shellfish is permitted. Young children, pregnant and nursing mothers, the very old and the ill (e.g. those suffering from diabetes) are exempt from fasting.

Hinduism

Fasting may mean abstinence from or restriction of food and drink. Practices vary according to caste, family and individual, and the type of fast also varies. A fast may mean no food at all or abstinence from particular foods or meals, with or without the consumption of liquids. During the festival of Navratri abstinence is observed, with the devout eating once per day and avoiding many foods. Hinduism follows a lunar calendar, so fasting dates change each year.

Islam

Fasting is seen as abstaining from a broader range of practices than just eating and drinking. Ramadan, the ninth month in the Muslim calendar and the most sacred month of the year, requires fasting in which no food or drink is consumed from dawn to dusk. The fast is broken with light foods, often dates, as this was one of the foods that the prophet Mohammed ate to break his fast. Ashura is another day of fasting for Sunni Muslims and commemorates the day on which Noah left the Ark and Moses was saved by Allah.

All adult Muslims are required to fast. Those who are exempt from fasting (pregnant, lactating or menstruating women, women who have given birth within 40 days and travellers) are expected to make up the fast days before the next Ramadan. Other exceptions include pre-pubescent children, those with short-term illnesses especially in which vomiting occurs, people who are seriously ill, people experiencing mental illness and those physically unable to eat.

Judaism

Fasting involves abstaining from all food and drink. Fast days include the holiest day in the Jewish calendar, Yom Kippur (Day of Atonement), and Tishah b'Av, the day when the destruction of the First and Second temples is remembered. Food and drink are not permitted from sundown to sundown. Another four to five fast days per year commemorate other events in Jewish history, with eating and drinking prohibited between sunrise and sundown. Meals before fasts may be as simple as a boiled egg and afterwards consist of light foods. In the tradition of the Ashkenazim (an ethnic group within Judaism), the bride and groom fast immediately before their wedding ceremony. Other communal or individual fasts may also occur.

Post-pubescent Jews are expected to fast. Very ill people and women in childbirth are not required to participate in the two main fasts, and pregnant and lactating women and others who would find the practice detrimental to their health are also exempt from other fasts.

Mormonism

Fasting involves abstaining from food and drink. The first Sunday of the month is typically observed as a fast day on which food and drink are not taken for two consecutive meals. Donations of food or money, a 'fast offering' equivalent to the food not eaten or the money not spent on those meals, are made to those in need through the church. Other fasts are practised at the discretion of local wards, families or individuals.

Protestantism

Fasting is unusual but can be undertaken at the discretion of local church communities and individuals.

Seventh-day Adventists

No particular days of fasting are observed, however the first Sunday of each quarter is a day of prayer and fasting.

Rastafarianism

No fasting is required.

Food restrictions

Meat, eggs, dairy, fish, shellfish, leavened foods, alcohol, coffee and tea are foods and drinks that are restricted to some degree across many religions.

Buddhism

Restrictions of food and drink vary widely with sect and country. The taking of life is forbidden, so Buddhists will not kill animals; however, some may eat the meat. Many are lacto-ovo-vegetarians. Tibetan Buddhists do not eat fish, and Chinese and Vietnamese Buddhists avoid onion, garlic and leek, as these and other pungent spices are believed to increase libido.

Catholicism

Meatless Fridays existed prior to 1966 and some Catholics still follow this rule.

Eastern Orthodox

During non-fasting periods all food is considered pure and there are no restrictions. Strict followers of religious fasting will continue to fast from all animal foods and olive oil and wine on Wednesdays and Fridays.

Hinduism

Hindu society divides people into four classes: Brahmin (priests), Kshatriya (soldiers), Vaisya (farmers and merchants) and Sudra (menial labourers). People of the Brahmin caste do not touch or eat meat, fish or eggs; many also avoid onions and garlic. Dairy products are revered and consumed. People of other castes (except widows, because it is believed these foods may increase libido) may avoid or consume meat, fish or eggs but will not eat them during religious events, marriages, when in mourning, when breaking a religious fast and at some other times. Pregnant women may avoid certain fruits and nuts that are believed to negatively influence the pregnancy. Alcohol and illicit drugs are prohibited, and traditional Hindus will not use them.

Islam

Eating is part of worship. Muslims believe food is eaten for good health and that they should respect the food provided, eat in moderation, share food and meals and wash their hands and mouths before and after eating. The right hand is often used for eating.

Enzymes used for cheese-making must be halal. Animals must be slaughtered according to Islamic food laws. Animals that are chosen for slaughter are free from disease, handled with care and kindness and killed quickly to minimise trauma. Their blood is then drained. Halal is the word used to identify which foods are permissible under Islamic food laws.

Pork and pork-related products, and food from animals that catch prey in their mouths, including four-footed animals and birds of prey, are prohibited. Alcoholic drinks and blood in liquid form are also forbidden.

Halal certification of products signifies food that is produced according to the Islamic dietary laws. While practices differ, some Muslims avoid eating foods that contain animal by-products or additives of animal origin when their source is not halal. Others do not eat food from restaurants or shops where pig meat is served. Some Muslims will also avoid some medications. Forbidden foods are referred to as haram.

Judaism

Jewish dietary rules are referred to as kashrut, and food needs to be kosher, or fit to eat, to be part of a Jewish diet. Kosher symbols on food signify that it meets the Jewish law. Food falls into four basic categories: meat, milk, neutral and non-kosher. Meat and milk (from kosher animals) must not be eaten together; meat must be eaten six hours before milk and milk one hour before meat. Cooking utensils used to prepare milk must be cleaned before being used for other groups. Some people have different utensils and crockery for meat- and milk-based foods. Neutral foods include eggs (from kosher birds), plants, fish and drinks, and all with the exception of fish can be eaten with meat. Some people, however, do not mix milk and fish. Some other Jewish food laws are listed below:

- Animals must be slaughtered following the *shechita* ritual, which has some similarities to Islam in that the blood must be drained away.
- Kosher animals chew the cud and have split hooves—for example, cattle and sheep.
- Kosher birds have a crop, gizzard and extra talon for scratching for food—for example, chickens, turkeys, ducks and geese. Birds of prey are forbidden.
- Shellfish and fish that do not have fins and scales are forbidden.
- Reptiles, amphibians and invertebrates are forbidden.
- Insects are not kosher. Although honey is kosher, bees are not.
- Cheese- and grape-related food production must be supervised by a rabbi.

Mormonism

The religion's laws of health describe many food-related practices, advocating a diet based on cereals and minimal use of meat. Tea, coffee, caffeine, alcohol and tobacco are prohibited.

Rastafarianism

Rastafarians follow a strict I-tal diet, a vegetarian or vegan diet in which food is prepared without salt and often with coconut oil. Meat, especially pork, shellfish or other animals that scavenge, fish greater than 30 centimetres long, alcohol, salt, milk, coffee are avoided.

Seventh-day Adventist

The 'Eight Laws of Health' encourage eating simply and in moderation, not eating between meals, taking both rest and exercise and washing before and after, but not during, meals. Vegetarianism is recommended, and many followers are lacto-ovo-vegetarians. Foods that are deemed 'unclean' in Leviticus are proscribed, including pork and shellfish, as well as coffee, tea, alcohol, tobacco and highly seasoned meals and condiments.

Feasting

Most religions mark their year with festivals, and many of these involve food in some way. Feasting refers not to tables laden with arrays of food but to specific and more symbolic foods.

Buddhism

The most significant festival commemorates the birth, enlightenment and death of the Buddha. There are many other festivals throughout the year, mostly linked to events in the Buddha's life or in the lives of various bodhisattvas (enlightened beings). Buddhist festivals typically involve followers making food offerings to monks and others in need.

Catholicism

Feast days generally celebrate the death of a saint or a special day within the church calendar. Christmas and Easter are the main festivals, with others, including the foods used, depending on the country, with foods dependent on ethnicity. There are many other feast days listed in the Catholic calendar, but these are not related to food practices.

Eastern Orthodox

The most important feast day is Easter Sunday, with many feast occasions surrounding Lent. After midnight on Holy Saturday a traditional offal soup called *magiritsa* is eaten to break the fast. Coloured eggs are important during Easter. They are painted red in Greece and with decorative patterns in Eastern churches and are a symbol of the tomb of Christ, representing renewal of life and victory over death. Christmas is another important feast day, and there are nine others during the year. Traditional meals on Easter Sunday and Christmas Day usually include meat such as lamb or goat cooked on a spit; many look forward to these feasts, which come after long periods of fasting.

Hinduism

There are eighteen major festivals in the Hindu year, including Ram Navami, Dussehra, Pongal and Janmashtami. The five-day Diwali (Festival of Lights) is a very popular event, with sweetmeats a main feature. Holi, the spring festival of fertility and harvest otherwise known as the Festival of Colour, sees people coloured from head to toe in paint powders. Foods eaten at Holi vary, but the emphasis is on fresh and colourful dishes, and gifts of multicoloured sweets are passed around. Navratri celebrates the triumph of good over evil during nine evenings of dancing and feasting.

Food is offered to the gods before eating, especially during religious festivals. Food also plays a role in personal events, including birthdays, marriages and funerals.

Islam

Eid al-Fitr starts at the end of Ramadan, on the first day of the month of Shawwal in the Islamic calendar, and is celebrated in many countries with a three-day festival that involves feasting. Other main feasts include Eid al-Adha, a four-day festival in which animals are sacrificed and shared among family, friends and importantly those in need.

Many events are celebrated in Islam through food and feasting, and include births, deaths, marriages, the circumcision of a boy and the time when a child starts reading the Qur'anic alphabet. The form of these celebrations and the foods prepared for them depends on the country. Other festivals marking significant points in the year are celebrated through other means, for example with prayer or light.

Judaism

There are many Jewish festivals in which foods have symbolic meanings. The Sabbath, the day of rest for Jewish people, starts at sundown on Friday and ends after nightfall on Saturday. Foods for Saturday night are cooked before the Sabbath begins.

Rosh Hashanah, the Jewish New Year, is celebrated with challah and apples dipped in honey. Sour and bitter foods are forbidden. Sukkoth (the Feast of Tabernacles) requires meals to be eaten in a specially prepared hut, and Hanukkah (the Festival of Lights) is celebrated with foods cooked in oil, such as potato pancakes called *latkes*. Purim, the feast held in honour of Queen Esther, includes *hamantaschen*, *kreplach*, Purim challah and other special dishes. Pesach, or Passover, an eight-day festival of spring and freedom, requires that all foods are kosher. The main celebratory meal at Passover is Seder, for which a traditional Passover Seder plate is arranged, with six symbolic food items.

Mormonism

Christmas and Easter are the main festivals. Also, Pioneer Day is celebrated on 24 July, in commemoration of the day Brigham Young, president of the church, and the pioneer Latter-day Saints arrived in Salt Lake Valley, in 1847.

Protestantism

Christmas and Easter are observed with foods often determined by ethnicity and preference, more so than religious convention.

Rastafarianism

Ethiopian Christmas celebrations on 7 January include large vegetarian or vegan feasts. There are many other events related to Haile Selassie I (the Ethiopian emperor who was regarded as god of the black race by Rastafarians) and the development of the religion.

Seventh-day Adventists

Historically Christmas and other religious holidays were not celebrated by Seventh-day Adventists, however today some people celebrate these events.

SUMMARY AND KEY MESSAGES

The food habits of Australians and New Zealanders have evolved considerably since the first British influences. Multiculturalism is one of the main current influences on food habits and continues to shape broader food culture in both countries.

- Immigration policies in both countries have changed over time, and thus the mix of migrants has varied.
- Migrant groups bring with them diverse culinary beliefs and practices, some related to religion, and these influence local eating habits.
- Food and food preparation in some cultures incorporate a range of beliefs about food combinations and avoidances.
- Many religions place restrictions on food and drink, although the degree to which followers adhere to these restrictions varies.

REFERENCES

ABS (Australian Bureau of Statistics) 2007. *Migration, 2005–06*. Cat. no. 3412.0. Commonwealth of Australia. <www.ausstats.abs.gov.au/ausstats/subscriber.nsf/0/E0A79B147EA8E0B5CA2572AC001813E8/$File/34120_2005-06.pdf>, accessed 30 January 2014.

——2011. *Migration, Australia, 2009–10*. Cat. no. 3412.0. Commonwealth of Australia. <www.abs.gov.au/ausstats/abs@.nsf/Products/52F24D6A97BC0A67CA2578B0001197B8?opendocument>, accessed 30 January 2014.

——2012. *Reflecting a Nation: Stories from the 2011 census, 2012–2013*. Cat. no. 2071.0. Commonwealth of Australia. <www.abs.gov.au/ausstats/abs@.nsf/Lookup/2071.0main+features902012-2013>, accessed 30 January 2014.

Bannerman, C. 2011. Making Australian food history. *Australian Humanities Review* 51: 49–63. <http://press.anu.edu.au//wp-content/uploads/2011/12/making-australia-food.pdf>, accessed 10 November 2014.

DIBP (Department of Immigration and Border Protection) 2008. *Abolition of the 'White Australia Policy'*. Fact Sheet 8. Commonwealth of Australia. <www.immi.gov.au/media/fact-sheets/08abolition.htm>, accessed 30 January 2014.

——2013. *Migration Programme Planning Levels*. Fact Sheet 20. Commonwealth of Australia. <www.immi.gov.au/media/fact-sheets/20planning.htm>, accessed 30 January 2014.

Kapiszewska, M., Soltys, E., Visioli, F., Cierniak, A. et al. 2005. The protective ability of the Mediterranean plant extracts against the oxidative DNA damage: The role of the radical oxygen species and the polyphenol content. *Journal of Physiology and Pharmacology* 56 Suppl 1: 183-97.

Keys, A., Menotti, A., Karonen, M.J., Aravanis, C., et al. 1986. The diet and 15-year death rate in the seven countries study. *American Journal of Epidemiology* 124: 903–13.

Kromhout, D., Keys, A., Aravanis, C., Buzina, R. et al. 1989. Food consumption patterns in the 1960s in seven countries. *American Journal of Clinical Nutrition* 49: 889–94.

Obata, K. & Lee, J. 2010. *Languages of Aboriginal and Torres Strait Islander Peoples: A uniquely Australian heritage*. Feature Article 3. ABS, Commonwealth of Australia. <www.abs.gov.au/AUSSTATS/abs@.nsf/Lookup/1301.0Feature+Article42009–10>, accessed 20 February 2014.

Owen, R., Wen, R.W., Giacosa, A., Hull, W.E. et al. 2000. Olive-oil consumption and health: The possible role of antioxidants. *Lancet Oncology* 1: 107–12.

SBS (Special Broadcasting Service) n.d. SBS Census Explorer. <www.sbs.com.au/censusexplorer/>, accessed 10 November 2014.

Serra-Majem, L., De La Cruz, J.N., Ribas, L. & Salleras, L. 2003. Mediterranean diet and health: Is all the secret in olive oil? *Pathophysiology of Haemostasis Thrombology* 33: 461–5.

Simopoulos, A.P. 2001. The Mediterranean diets: What is so special about the diet of Greece? The scientific evidence. *Journal of Nutrition* 131: 3065S–73S.

Statistics New Zealand 2013a. *2013 Census: Major ethnic groups in New Zealand*. New Zealand Government.<www.stats.govt.nz/Census/2013-census/profile-and-summary-reports/infographic-culture-identity.aspx >, accessed 10 November 2014.

——2013b. *2013 Census: Totals by topic*. New Zealand Government. Table 28: Religious affiliation (total responses). <www.stats.govt.nz/Census/2013-census/data-tables/total-by-topic.aspx>, accessed 20 February 2014.

—— n.d. Religious affiliation (total responses) by age group and sex, for the census usually resident population count, 2001, 2006, and 2013 Census (RC, TA, AU). New Zealand Government. <http://nzdotstat.stats.govt.nz/wbos/Index.aspx#>, accessed 10 November 2014.

Ting, I. 2013. *How Australia Eats: The ultimate pie chart*. Good Food. <www.goodfood.com.au/good-food/food-news/how-australia-eats-the-ultimate-pie-chart-20131102-2wstm.html>, accessed 20 February 2014.

Visioli, F., Caruso, D., Grande, S., Bosisio, R. et al. 2005. Virgin Olive Oil Study (VOLOS): Vasoprotective potential of extra virgin olive oil in mildly dyslipidemic patients. *European Journal of Nutrition* 44: 121–7.
Visioli, F. & Galli, C. 2001. Phenolics from olive oil and its waste products: Biological activities in in vitro and in vivo studies. *World Review of Nutrition and Dietetics* 88: 233–7.
Willett, W.C., Sacks, F., Trichopoulou, A., Drescher, G. et al. 1995. Mediterranean diet pyramid: A cultural model for healthy eating. *American Journal of Clinical Nutrition* 61: 1402S–1406S.

ADDITIONAL READING

Bastian, B. 2012. Immigration, multiculturalism and the changing face of Australia. In Bretherton, D. & Balvin, N. (eds). *Peace Psychology in Australia: Dreamings of peace* 55–70. Springer Science+ Business Media. <www2.psy.uq.edu.au/~uqbbast1/Bastian%20Ch4%20Peace%20Psychology%20Australia.pdf>, accessed 4 October 2014.
Huntley, R. 2008. *Eating Between the Lines: Food and equality in Australia*. Black Inc., Melbourne.
Pearson, S. & Kothari, S. 2007. Menus for a multicultural New Zealand. *Continuum: Journal of Media & Cultural Studies* 21: 45–58. <http://dx.doi.org/10.1080/10304310601103950>, accessed 20 November 2014.
Santich, B. 2014. *Bold Palates*. Wakefield Press, Adelaide.

PART II

THE LIFE STAGES

3

Preconception

Regina Belski and Adrienne Forsyth

> **LEARNING OUTCOMES**
>
> Upon completion of this chapter you will be able to:
>
> - understand how weight impacts on fertility in men and women
> - identify the nutritional factors that affect fertility
> - identify lifestyle factors that affect fertility
> - describe common conditions that may impair fertility.

It is well recognised that the nutritional status of both the woman and the man is important when trying to conceive a baby. This chapter explores the nutrition-related aspects of maximising the chances of conception and addresses the nutritional factors that may improve or impair those chances. Other lifestyle considerations are also discussed, as well as common health conditions that may impair fertility.

INFLUENCES ON CONCEPTION

Fertility

Most men and women of reproductive age will at one stage or another consider having children, and it is this stage prior to conception of a child that is termed preconception. The ability to produce children is known as fertility, and the biological capacity to bear children (after one year of unprotected intercourse) is referred to as fecundity; however, in this textbook fertility will be used to represent both for ease of understanding.

Research reports that at least 9 per cent of couples living in the developed world are childless involuntarily, and the rates of infertility in both men and women are growing (Boivin et al. 2007). These rates of infertility are reported to be linked to increases in sedentary lifestyle, obesity and exposure to chemicals and toxins.

A regular menstrual cycle of 22–36 days in length is generally considered to be required for optimal female fertility. Irregular menstrual cycles have been linked to some dietary choices.

For men, sperm quality and quantity are critical to optimal fertility, and anything that affects these will likely have a significant impact on the chances of conception. It takes approximately 72–90 days to

CASE STUDY 3.1 LOW BODY WEIGHT AND FERTILITY

Profile and problem

Ms K. is a 34-year-old female who has been in a relationship for five years. She and her partner have been trying to conceive for twelve months and have been unsuccessful. Ms K. is 166 centimetres tall and weighs 50 kilograms. She has been this weight since high school. She has no known health problems, but her menstrual cycle is irregular and sometimes absent for months at a time.

Assessment

Ms K. works in a high-profile media job and is concerned with maintaining her body image. She exercises for 60–120 minutes per day, running and lifting weights. She consumes a low-carbohydrate diet because she believes that this helps to control her weight and limit her appetite. Ms K. eats four small meals per day and avoids eating after 5 pm.

Treatment

Ms K. has agreed to do the following:

- consume an additional daily serve each of cereals and dairy food—e.g., a bowl of cereal with milk in the evening
- limit physical activity to 60 minutes per day
- take a daily prenatal vitamin supplement including iron, folate and iodine.

Plan

Ms K. will monitor both her weight, with an aim to gain 2 kilograms, and her menstrual cycle, with an aim to regain a regular monthly cycle. She will follow up with her family physician and a fertility specialist if she is unable to conceive within the next six months.

produce mature sperm. Researchers have identified foods and nutrients that can both impair and improve sperm quality and quantity.

Age

The age of both men and women can impact on the chances of conception. Fertility in men appears to decline from the age of 35 years due to a decrease in **testosterone** levels, sperm production and to **DNA** damage. In women, fertility declines in their thirties as the number of **oocytes** declines and the risk of chromosomal abnormalities increases (Sharma et al. 2013).

Weight

For men and women, both excessive and inadequate body fat levels are connected to reduced fertility. This appears to be linked directly to changes in hormone levels and concentrations. For example, a recent study of obese males by Hofny et al. (2010) concluded that serum leptin, a hormone released by fat cells, regulates a link between obesity and male infertility. Obesity in men is also related to reduced testosterone levels and raised **oestrogen** levels, which can impact on sperm production and quality.

It appears that for overweight men and women even a small amount of weight loss may significantly improve chances of conception. Similarly, for underweight men and women even a small amount of weight gain can improve chances of conception. Hence, it is important that both men and women are of a healthy weight to maximise their chances of conception.

Undernutrition

Undernutrition plays a significant role in impairing fertility in women. Whether undernutrition is chronic or acute affects the severity of the impact; acute undernutrition in women who were previously well

Testosterone is a steroid hormone from the androgen group and is needed for normal reproductive and sexual function.

DNA (deoxyribonucleic acid) is the molecule (double helix structure) in body cells which contains genetic information.

Oocytes Precursors to mature female eggs.

Oestrogen is a steroid hormone produced by the ovaries and by the placenta. It is responsible for the female secondary sexual characteristics that develop at puberty. Oestrogen production is elevated during pregnancy, and oestrogens have growth-promoting effects.

nourished appears to have the most severe impact on fertility. Acute undernutrition leads to a significant decline in fertility, but this has been shown to improve with better nutrient intake. Chronic undernutrition, as may be experienced by someone with a consistent undersupply of food and thus low energy intake—for example, in developing countries—is more likely to lead to the birth of small and frail offspring with higher risk of health complications in early life (Van der Spuy 1985).

A low energy intake, low body weight and/or excessive weight loss in women can have a significant impact on the menstrual cycle and more specifically ovulation. **Amenorrhea** is common in women of a low body weight and in those with excessive weight losses, but low body weight is not always associated with amenorrhoea as, although ovarian activity and body fat content are correlated, they are not causally linked. This has been suggested to be due to energy balance/availability being the regulator of reproductive function, not simply body fat levels (ESHRE, Capri Workshop Group 2006). More specifically, it appears to be frequent or severe dieting, nutrient restriction, high energy expenditure or erratic eating patterns, including bingeing and crash dieting, that may have a negative effect on reproductive cycles, even if weight remains within a normal range. There is a critical level of energy and nutrient intake relative to the body's needs that is necessary to maintain reproductive function. Therefore several lifestyle factors need to be considered for optimising ovulation and hence fertility, not just weight. For example, researchers have shown that menstrual cycles can return in some female athletes when energy expenditure (via exercise) is reduced, long before a change in body weight or an increase in body fat occurs (Loucks 2003).

Diet

In both men and women, the source of energy plays a role in fertility. The semen quality of men improves when diets are relatively high in carbohydrate, fibre, fruit and vegetables. For women, better fertility is associated with choosing more vegetable protein than animal protein, unsaturated fats rather than trans fats and a low glycaemic load (see Chapter 1; Sharma et al. 2013). For men, high consumption of soy-based foods has been linked to reduced sperm count (Chavarro et al. 2008), and men experiencing a low sperm count would be advised to avoid excessive soy intake.

> **Amenorrhoea** is the cessation of menstrual periods in women of reproductive age.

Micronutrients

Particular micronutrients appear to have a significant impact on the chances of conception.

Antioxidants

Evidence is mounting that oxidative stress and free radicals may have a negative impact on fertility. Therefore, high intakes of fruits and vegetables containing antioxidants including beta-carotene, vitamin E and vitamin C, as well as folate and zinc, which can protect from oxidative damage, appear to be beneficial for improving the likelihood of conception. In men, selenium, vitamins E and C and beta-carotene supplements have all been linked to improved sperm number, motility or quality (Eskenazi et al. 2005).

Iron

Women's iron status during preconception is important for improving chances of conception and reducing the likelihood of developing complications during early pregnancy. It is advised that women planning to become pregnant have their iron levels tested and correct any deficiency prior to trying to conceive.

Folate

Folate plays an important role in DNA synthesis and hence reproduction. There is limited understanding about the impact of dietary folate and folate status on fertility; however, there is a substantial evidence base to support the importance of adequate folate for neural tube development in early pregnancy. Problems with neural tube development can lead to **spina bifida**, **anencephaly** and **encephalocele**. It is important that folate status of women planning to conceive is adequate

> **Spina bifida** is a neural tube birth defect in which some of the spinal vertebrae do not close properly and leave some nerves in the spinal cord exposed and damaged. This can affect a number of organs and muscles, depending on the point on the spinal cord at which the damage occurs.

> **Anencephaly** is a neural tube birth defect in which the brain and skull are not properly developed. Stillbirth is common, and most babies born with anencephaly die shortly after birth.

Encephalocele is a neural tube birth defect in which the brain and its membranes protrude through openings in the skull. This can be corrected with surgery, but lifelong neurological problems such as seizures, paralysis and developmental delays can occur.

and the NRVs recommend that women of childbearing age consume at least 400 micrograms of folate per day (NHMRC & Ministry of Health 2006). There has been **mandatory folate fortification** of all wheat flour (except organic) for making bread in Australia since September 2009. For women consuming three slices of bread daily, this will provide 120 micrograms of folate. Fortification is voluntary in New Zealand.

It appears that folate may also play a significant role in male fertility. Studies using supplements of folic acid as a single nutrient have shown mixed results on sperm count and quality (either no effect or a positive effect). Higher levels of dietary folate intake have been associated with reduced numbers of chromosomally abnormal sperm (Young et al. 2008), and simultaneous folate and zinc supplementation has been associated with an increase in normal sperm count (Wong et al. 2002). These studies suggest that appropriate folate intake, and possibly supplementation, may be advantageous for men experiencing impaired fertility.

Zinc

Zinc appears to be important for women in reproduction; however, very few studies have been undertaken. Some animal studies suggest that poor zinc status is linked to subfertility, but the few human studies do not support this. As zinc is an important nutrient and also serves an antioxidant function in the body, it is important that adequate levels of zinc are consumed through the diet, as per the nutrient reference values (NRVs; see Chapter 1; NHMRC & Ministry of Health 2006).

Mandatory folate fortification became enforceable in 2009 in Australia, requiring the addition of folic acid to all wheat flour for bread-making, at 2–3 milligrams per kilogram of flour. Public opposition in New Zealand led to abandoning the same policy and fortification is now voluntary. Folate is necessary for healthy development of babies in early pregnancy, and folic acid taken before conception and for the first trimester can prevent many neural tube defects.

For men, zinc plays a critical role in a number of reproductive processes. Zinc status has been linked to the quality, shape and concentration of sperm produced (Yuyan et al. 2008), and there is evidence that zinc supplementation, either independent or paired with other nutrients including folate and vitamins A and C, may improve sperm quality (Wong et al. 2002).

Non-nutritive substances

Caffeine

In women, high intakes of caffeine can increase the time to conception, as well as the risk of miscarriage and stillbirth that are not related to genetic defects. Inability to conceive and miscarriage are more common in women with higher caffeine intakes: consumption of more than 100 milligrams of caffeine per day (equivalent to one mug of instant coffee or two to three cans of cola) has been shown to increase the risk of miscarriage and stillbirth (Cnattingius et al. 2000; Tolstrup et al. 2003).

Alcohol

There is a link between alcohol and infertility, but the amount of alcohol that causes reproductive problems is unclear. In men, alcohol reduces the sperm count, and, in women, it can have more devastating effects, as it negatively impacts on **blastocyst** development and implantation. There is no known safe level of alcohol consumption with regard to female fertility, so it is recommended that women who are trying to conceive abstain from alcohol.

Lifestyle

Physical activity

Physical activity can have differing impacts on fertility, depending on the type, duration, intensity and frequency of the activity, as well as an individual's weight and energy balance status. In both men and women who are overweight or obese, physical activity that results in weight loss improves fertility. In individuals with a healthy weight, excessive exercise that induces a negative energy balance (see Chapter 1) can negatively impact on fertility.

A **blastocyst** is the product of reproduction from five days after fertilisation until implantation in the uterine lining, at which point the product becomes known as an embryo.

In men, moderate physical activity, such as walking for one hour three times per week, has been demonstrated to increase fertility. More frequent vigorous physical activity, such as cycling for more than five hours per week, has been demonstrated to reduce fertility by decreasing sperm count, concentration and mobility (Vaamonde et al. 2009; Wise et al. 2011).

Moderate physical activity may also improve fertility in women, but more frequent vigorous physical activity decreases the chance of conception. Exercise that causes women to feel sweaty and out of breath presents only a slightly increased risk of infertility compared to exercising at an easy level, while exercise to the point of exhaustion presents a risk of infertility 2.3 times greater than exercising at low intensity (Gudmundsdottir et al. 2009).

Stress

Stressful life events can diminish both the quality and quantity of sperm in men. Stress and depression may cause hormonal changes that affect sperm production. Experiencing infertility may also be stressful and exacerbate difficulties with sperm production. In women, stress may reduce fertility via hormonal changes that impact on blood flow in the reproductive organs.

Cigarette smoking

In men, smoking impairs several sperm characteristics, including both quantity and fertilising capacity. Smoking also increases the rate of infertility in women, possibly by disrupting hormone levels, decreasing the number of mature eggs released by the ovaries and disrupting uterine function.

Drug use

Marijuana negatively impacts fertility in both men and women. In men, it reduces testosterone release and sperm production. In women, it causes changes in hormonal regulation.

Cocaine also reduces sperm production in men and may diminish sexual stimulation. In women, cocaine may impact negatively on the function of the ovaries.

Other factors

Other lifestyle factors that may impact on fertility include exposure to air pollution, heavy metals such as lead and mercury, chemicals such as pesticides, and radiation. Use of oral contraceptives may preserve fertility in women and increase their chances of conceiving in the future. Hormonal contraceptives may have nutritional complications, as oral contraceptives, as well as contraceptive injections, implants and patches, may alter cholesterol levels, increase the requirements for vitamins B6 and B12 and cause weight gain.

HEALTH CONDITIONS

Polycystic ovarian syndrome

Polycystic ovarian syndrome (PCOS) occurs when a woman's ovaries produce higher than normal levels of androgen hormones, interfering with the development and release of eggs and resulting in the growth of cysts within the ovaries. Symptoms of PCOS can include abdominal obesity, acne, male-pattern baldness, increased hair growth on the face, chest and back, thick dark patches of skin, skin tags, pelvic pain, irregular menstruation and infertility.

Most conditions that impact on fertility are not related to diet, but PCOS can be managed with changes in eating patterns. It is associated with both abdominal obesity and **insulin resistance**, and changes in dietary and physical activity patterns can improve reproductive function and reduce the risk of developing other lifestyle-related conditions such as diabetes and cardiovascular disease (CVD). These improvements can occur even without substantial weight loss (Farshchi et al. 2007). In women who are overweight or obese, as little as 5 per cent loss of total body weight can significantly improve fertility and reduce other symptoms of PCOS (JHFWH 2011).

Several studies have investigated altering the macronutrient composition of the diet to manage PCOS. It appears that diets with energy restriction result in weight loss, symptom reduction and improved fertility, regardless of their macronutrient composition. It is recommended that women with PCOS follow the Australian Dietary Guidelines (NHMRC 2013) and the acceptable macronutrient distribution range (AMDR) recommended in the NRVs (see Chapter 1).

Insulin resistance is a general term used to describe any defect in insulin action and is divided into abnormalities of insulin sensitivity or insulin responsiveness.

CASE STUDY 3.2 MANAGING POLYCYSTIC OVARIAN SYNDROME FOR FERTILITY

Profile and problem

Ms Y., a 27-year-old woman, was diagnosed with PCOS in her teens. She has carefully monitored her diet to avoid weight gain but has other symptoms such as acne and facial hair. She takes metformin to help control her blood sugar levels. Ms Y. would like to start a family and is concerned that she will have difficulty given her PCOS.

Assessment

Ms Y. weighs 62 kilograms and is 168 centimetres tall. She follows a low-fat diet and can sometimes be quite restrictive with food in an attempt to maintain her weight. Ms Y. consumes adequate serves of all the core food groups and avoids discretionary foods, including fats and oils. Her diet is high in carbohydrate, as she consumes a lot of rice, breakfast cereals and white bread.

Treatment

Ms Y. has agreed to do the following:

- choose wholegrain and low–glycaemic index (GI) breads and cereals
- eat at least one protein food at each meal
- allow for a small amount of healthy fats in the diet; this can be in the form of nuts and seeds if oils are not preferred
- participate in moderate physical activity, such as walking for 30 minutes daily.

Plan

Ms Y. will return to the dietitian for a follow-up to assess her progress with her diet. She will also see her family physician to explore further options to improve management of her PCOS.

Coeliac disease

Coeliac disease occurs when the digestive system reacts to gluten (a protein found in wheat, rye, barley and oats) contained in foods. It damages the bowel and causes gastrointestinal symptoms such as abdominal pain, bloating and diarrhoea, impaired absorption of nutrients and a number of associated symptoms and conditions including amenorrhea and subfertility in both men and women. Even if able to conceive, women with poorly managed coeliac disease have a high risk of miscarriage. The only treatment for coeliac disease is a gluten-free diet for life. Following a gluten-free diet improves fertility as well as reducing abdominal and other associated symptoms.

Related conditions

While endometriosis, primary ovary insufficiency and uterine fibroids are not nutrition-related, they provide additional background to other impacts on fertility and conception, so are briefly mentioned below.

> **Coeliac disease** is an autoimmune disorder of the small intestine that occurs in genetically predisposed people of all ages from middle infancy onward. It is caused by exposure to gluten (a protein in wheat) and similar proteins in other cereals such as barley, rye and oats. Exposure to gluten causes inflammation in the bowel, leading to damage of the epithelium of the bowel.

Endometriosis

Endometriosis occurs when the cells that normally line the inside of the uterus grow outside the uterus. Approximately one-third of women with endometriosis are unable to conceive. The cause of endometriosis-related infertility is unknown. A variety of changes occur that may negatively impact on the ability to conceive. Structural changes may affect the release and movement of eggs, and chemical changes may affect egg and sperm function and the ability of the blastocyst to properly implant.

Primary ovary insufficiency

Women with primary ovary insufficiency stop producing ovarian hormones and releasing mature eggs at a young age. Ovulation may take place irregularly or not at all. Unassisted pregnancies occur in only 5–10 per cent of women with this condition.

Uterine fibroids

Fibroids are non-cancerous tumours. When they occur within the uterus, they may impact negatively on a woman's ability to conceive by changing the position and shape of the cervix and uterus and by blocking the fallopian tubes.

SUMMARY AND KEY MESSAGES

The time leading up to conception provides an opportunity for both men and women to improve their health and nutritional status to provide the best chance of conceiving a healthy child. In general, couples wanting to become parents should focus on maintaining a healthy weight, following national dietary guidelines and paying extra attention to iron, folate and antioxidant status.

- The rate of involuntary infertility is increasing, mainly due to obesogenic, or obesity-promoting, changes in our lifestyle.
- Both underweight and overweight impair fertility in men and women.
- Diets rich in carbohydrate, fibre, fruit and vegetables improve fertility in men.
- Diets higher in vegetable proteins and unsaturated fats and with a low glycaemic load improve fertility in women.
- Adequate nutrient status, including iron, folate, zinc and antioxidants, may improve fertility in both men and women.
- Caffeine, alcohol and other drugs, toxins and heavy metals can impair fertility.
- Some health conditions, such as PCOS and coeliac disease, may reduce fertility. Chances of conception are improved by following special diets, and individuals with these conditions should see a dietitian for further support.

REFERENCES

Boivin, J., Bunting, L., Collins, J.A. & Nygren, K.G. 2007. International estimates of infertility prevalence and treatment-seeking: Potential need and demand for infertility medical care. *Human Reproduction* 22: 1506–12.

Chavarro, J.E., Toth, T.L., Sadio, S.M. & Hauser, R. 2008. Soy food and isoflavone intake in relation to semen quality parameters among men from an infertility clinic. *Human Reproduction* 23: 2584–90.

Cnattingius, S., Signorello, L.B., Anneren, G., Clausson, B.et al. 2000. Caffeine intake and the risk of first-trimester spontaneous abortion. *New England Journal of Medicine* 343: 1839–45.

ESHRE, Capri Workshop Group 2006. Nutrition and reproduction in women. *Human Reproduction Update* 12: 193–207.

Eskenazi, B., Kidd, S.A, Marks, A.R, Sloter, E. et al. 2005. Antioxidant intake is associated with semen quality in healthy men. *Human Reproduction* 20: 1006–12.

Farshchi, H., Rane, A., Love, A. & Kennedy, R.L. 2007. Diet and nutrition in polycystic ovary syndrome (PCOS): Pointers and nutritional management. *Journal of Obstetrics and Gynaecology* 27: 762–73.

Gudmundsdottir, S.L., Flanders, W.D. & Augestad, L.B. 2009. Physical activity and fertility in women: The North-Trondelag Health Study. *Human Reproduction* 24: 3196–204.

Hofny, E.R., Ali, M.E., Abdel-Hafez, H.Z., Kamal, E. el-D. et al. 2010. Semen parameters and hormonal profile in obese fertile and infertile males. *Fertility and Sterility* 94: 581–4.

JHFWH (Jean Hailes Foundation for Women's Health) 2011. *Evidence-based Guideline for the Assessment and Management of Polycystic Ovary Syndrome.* Jean Hailes Foundation for Women's Health on behalf of the PCOS Australian Alliance, Clayton, Vic.

Loucks, A.B. 2003. Energy availability, not body fatness, regulates reproductive function in women. *Exercise and Sport Sciences Reviews* 31: 144–8.

NHMRC (National Health and Medical Research Council) 2013. *Eat for Health: Australian Dietary Guidelines; Providing the scientific evidence for healthier Australian diets.* Commonwealth of Australia. <www.eatforhealth.gov.au/sites/default/files/files/the_guidelines/n55_australian_dietary_guidelines.pdf>, accessed 28 September 2014.

NHMRC & Ministry of Health 2006. *Nutrient Reference Values for Australia and New Zealand: Including recommended dietary intakes*. Commonwealth of Australia & New Zealand Government. <www.nhmrc.gov.au/_files_nhmrc/publications/attachments/n35.pdf>, accessed 26 September 2014.

Sharma, R., Biedenharn, K.R., Fedor, J.M. & Agarwal, A. 2013. Lifestyle factors and reproductive health: Taking control of your fertility. *Reproductive Biology and Endocrinology* 11: 66.

Tolstrup, J.S., Kjaer, S.K., Munk, C., Madsen, L.B. et al. 2003. Does caffeine and alcohol intake before pregnancy predict the occurrence of spontaneous abortion? *Human Reproduction* 18: 2704–10.

Vaamonde, D., Da Silva-Grigoletto, M.E., Garcia-Manso, J.M., Vaamonde-Lemos, R. et al. 2009. Response of semen parameters to three training modalities. *Fertility and Sterility* 92: 1941–6.

Van der Spuy, Z.M. 1985. Nutrition and reproduction. *Clinical Obstetrics and Gynecology* 12: 579–604.

Wise, L.A., Cramer, D.W., Hornstein, M.D., Ashby, R.K. et al. 2011. Physical activity and semen quality among men attending an infertility clinic. *Fertility and Sterility* 95: 1025–30.

Wong, W.Y., Merkus, H.M., Thomas, C.M., Menkveld, R. et al. 2002. Effects of folic acid and zinc sulfate on male factor subfertility: A double-blind, randomized, placebo-controlled trial. *Fertility and Sterility* 77: 491–8.

Young, S., Eskenazi, B., Marchetti, F., Block, G. et al. 2008. The association of folate, zinc and antioxidant intake with sperm aneuploidy in healthy non-smoking men. *Human Reproduction* 23: 1014–22.

Yuyan, L., Junqing, W., Wei, Y., Weijin, Z. et al. 2008. Are serum zinc and copper levels related to semen quality? *Fertility and Sterility* 89: 1008–11.

ADDITIONAL READING

Harvard Medical School 2009. *Conditions That Affect Fertility*. Harvard Health Publications. <www.health.harvard.edu/newsletters/Harvard_Mental_Health_Letter/2009/May/Conditions-That-Affect-Fertility>, accessed 12 September 2014.

National Institute of Child Health and Human Development 2013. *Diseases and Conditions That Influence Fertility*. NICHHD. <www.nichd.nih.gov/health/topics/infertility/conditioninfo/Pages/health-factors.aspx>, accessed 12 September 2014.

TEST YOUR UNDERSTANDING

Note: There is only one correct answer for each question.

1. What is the rate of involuntary childlessness for couples in the developed world?
   a. 3 per cent
   b. 5 per cent
   c. 7 per cent
   d. 9 per cent
   e. 11 per cent
2. Which nutrient plays a role in the prevention of neural tube defects?
   a. Vitamin C
   b. Iron
   c. Folate
   d. Zinc
   e. Vitamin A
3. How much caffeine per day can increase the chance of miscarriage and stillbirth?
   a. 10 milligrams
   b. 50 milligrams
   c. 100 milligrams
   d. 150 milligrams
   e. 200 milligrams
4. Which of the following is a condition characterised by acne, unwanted hair growth, irregular menstruation and infertility?
   a. Endometriosis
   b. PCOS
   c. Primary ovary insufficiency
   d. Uterine fibroids
   e. Spina bifida

5. Which of the following strategies may increase the chance of conception in women with PCOS?
   a. Weight loss
   b. Weight gain
   c. Low-carbohydrate diets
   d. High-protein diets
   e. Low-fat diets
6. Which of the following has been shown to increase the likelihood of irregular menstrual cycles?
   a. High consumption of plant foods
   b. Low consumption of fat
   c. High consumption of fat
   d. a and b only
   e. a and c only
7. Approximately how long does it take sperm to mature?
   a. 2–4 days
   b. 6–10 days
   c. 14–28 days
   d. 42–60 days
   e. 72–90 days
8. Which of the following antioxidants may play a role in fertility?
   a. Vitamin B6
   b. Vitamin E
   c. Selenium
   d. a and b only
   e. b and c only
9. Fertility in men appears to decline from what age?
   a. 30 years
   b. 35 years
   c. 40 years
   d. 45 years
   e. None of the above
10. Which of the following are factors known to reduce fertility in men?
    a. Exposure to heavy metals
    b. High meat consumption
    c. High soy food intake
    d. All of the above
    e. a and c only

STUDY QUESTIONS

1. What is fertility?
2. How does weight impact on ability to conceive?
3. Which foods should men and women choose when trying to conceive?
4. Which foods should be avoided when trying to conceive?
5. Why might women who are trying to conceive need a vitamin or mineral supplement? Which nutrients should it contain?
6. Describe the health conditions that may reduce a woman's ability to conceive.

4

Conception and pregnancy

Jane Coad and Louise Brough

LEARNING OUTCOMES

Upon completion of this chapter you will be able to:

- discuss nutritional factors that affect conception
- outline the importance of nutritional programming and its relationship to maternal nutrition
- describe maternal physiological adaptations to pregnancy
- list changes in energy and macronutrient requirements for healthy pregnancy
- explain altered vitamin and mineral requirements for healthy pregnancy
- identify particular groups of women who are at increased risk of nutritional compromise during pregnancy.

Nutrition before, during and after pregnancy can affect the subsequent health and wellbeing of the mother and her baby. Compromised nutrient intake is associated with increased risk of premature birth, low birthweight (LBW) and **pre-eclampsia**. Infants born of LBW or subject to **nutrient insults** in foetal life may have altered growth in infancy and childhood and are at increased risk of disease in adulthood. The maternal diet has to provide sufficient energy and nutrients for the woman's own requirements and for the needs of her developing foetus as well as allowing maternal stores to be established to buffer any fluctuations in intake and to support nutrient requirements during lactation. Some of the increased nutrient requirements may be met by physiological adaptation, such as increased absorption. However, the increased requirements for some nutrients, such as iron, iodine, folate and vitamin D, are difficult to meet by dietary changes, and supplementation may be required. Advice about food selection to optimise food safety and avoid toxic constituents and food-borne disease is important in pregnancy. The particular nutrient requirements of some groups of pregnant women, including those who are very young, are having multiple births or are obese, are essential to consider. This chapter provides an overview of the changed nutrient requirements in

Pre-eclampsia is a disorder of pregnancy characterised by high maternal blood pressure, protein in the urine and fluid retention leading to swelling of the hands, feet and face.

A **nutrient insult** is another term for an episode of nutrient deprivation affecting the foetus.

pregnancy, with a focus on those nutrients that are more important to support health in the mother and her infant.

MATERNAL PHYSIOLOGICAL CHANGES

A mother's body must substantially adapt to facilitate a successful pregnancy, including adequate foetal growth and the delivery of an optimally sized and healthy infant. The changes are largely orchestrated by hormones produced by the foetus and placenta. Some of these changes have significant impact on maternal metabolism and the provision of nutrients to the placenta and foetus.

Progesterone relaxes the tone of smooth muscle (muscles involved in the involuntary contraction of hollow organs). This affects smooth muscle throughout the body, causing changes in blood vessels, the airways, the gut and the renal system. The cardiovascular system undergoes marked changes. Blood volume increases—plasma volume more so than the other components. The increased blood volume is matched by increased blood being pumped from the heart and accommodated by decreased resistance in the blood vessels.

Changes in the renal system result in enhanced blood flow through the kidneys, with increased blood filtration. The relaxation of the smooth muscle of the gut results in slower **gut motility**. This can influence the rate at which the stomach empties and leads to increased nutrient and water absorption.

Early in pregnancy, the developing embryonic tissue produces hormones that promote maternal fat production and storage and increase blood glucose levels to facilitate placental glucose transfer to the foetus. In later stages of pregnancy, the changes in maternal metabolism result in the woman tending to use more fat for fuel and sparing glucose for the foetus.

Many women experience nausea and vomiting in pregnancy, probably mediated by the raised hormone levels in the first trimester. Some women will feel too nauseous to eat much; others will eat more frequently or change their diet to reduce nausea. Nausea and vomiting usually resolve after the first trimester of pregnancy and is positively associated with a healthy pregnancy. Dietary strategies to reduce nausea and vomiting include eating small frequent meals, avoiding skipping meals, choosing bland foods like crackers or savoury foods, sucking on ice-blocks and drinking clear fluids frequently. Severe forms of nausea and vomiting during pregnancy require medical attention.

> **Progesterone** is a steroid hormone produced by the ovaries and by the placenta. Progesterone literally means 'for gestation'. It is the dominant hormone produced after ovulation and following fertilisation, and is essential to maintain pregnancy.

> **Gut motility** is the movements of the stomach muscles that push food through the stomach and into the small intestine.

Although it is accepted that the time to conception is longer for both underweight and overweight women, there is no consensus about the optimal body mass index (BMI) for fertility and successful pregnancy outcome. Women with BMI less than 18.5 tend not to be as fertile or to have such good outcomes for pregnancy as heavier women. Women who are obese are also less fertile, and obesity is associated with other problems in pregnancy.

Optimal maternal and foetal outcomes are associated with an infant birthweight of 3.01–3.60 kilograms. LBW infants (born below 2.5 kilograms) have increased morbidity and mortality and an increased risk of chronic disease in adulthood. Prematurity (born before 37 weeks' gestation) and LBW are the greatest risk factors for infant mortality. Problems with subsequent growth and brain development are also associated with LBW. Babies who are large for gestational age or macrosomic (born above 4.5 kilograms) have an increased risk of obstetric complications, such as an instrumental (forceps) delivery and birth trauma, experience higher rates of infant morbidity and mortality and are more likely to become obese.

Weight gain in pregnancy should allow for optimal foetal growth and the laying down of maternal fat stores which could be used to sustain the mother in pregnancy and lactation. For women who have a pre-pregnancy weight in the healthy range, a gestational weight gain of 10–14 kilograms is associated with a reduced risk of having an LBW infant and with the lowest risk of complications in pregnancy and labour. However, gestational weight gain can be extremely variable; fat stores deposited in pregnancy are variable, and water retention, particularly that associated with oedema, can range broadly.

Women who have low pre-pregnancy weight or have multiple foetuses require a higher weight gain in pregnancy for optimal outcome. Women who have low gestational weight gain have a higher risk of having a LBW infant. Excessive weight gain in pregnancy increases the chances that the mother will remain overweight or obese after the birth. The general advice is that pre-pregnancy bodyweight should be considered; heavier women do not need to gain as much weight in pregnancy, and lighter women should aim to gain more. It is preferable for women to make dietary changes to achieve a healthy weight prior to conception. Low socio-economic status, genetic factors and ethnicity are also associated with the likelihood of having an LBW infant.

Weight gained in the first trimester is difficult to assess, because antenatal care and regular weighing usually start well into the first trimester and data about pre-pregnancy weight are skewed, suggesting that when women are asked about their pre-pregnancy weight, they tend to report the weight they wished they had been! The few studies that have assessed weight gain in the first trimester suggest that only about 5 per cent of total gestational weight gain occurs in this trimester. In the second and third trimesters weight gain is fairly even, with 400–500 grams being gained per week. First trimester weight gain, although small, is important. In the second trimester, maternal fat deposition is high, and in the third trimester foetal growth is at its highest.

NUTRIENT REQUIREMENTS AND INTAKES

Human reproduction requires adequate nutrition. When there is a limited supply of nutrients available, the body will suspend reproductive functioning to increase the chances of a healthy pregnancy later and the survival of the foetus and infant. As discussed in Chapter 3, low energy intake, low body weight and weight loss impact on menstrual cycles and ovarian function. Women who consume inadequate energy to maintain reproductive function may experience amenorrhoea. This impacts on fertility and is also a concern because chronic low levels of oestrogen may be associated with lowered bone mineral density and predisposition to osteoporosis.

Both high and low intakes of some nutrients can impair fertility. The **gonadotrophic hormones** are affected by maternal protein intake, and diets very high or very low in protein can affect reproductive cycling and fertility. Rapid development in the early stages of pregnancy leads to high protein, energy and nutrient requirements. Nutrient levels may be sufficient for conception to occur but inadequate for the next stage, and this may affect embryonic growth. It is thought that excess intakes of some nutrients may affect the rate of genetic mutation.

> **Gonadotrophic hormones (gonadotrophins)** is the name for a family of protein hormones that includes follicle-stimulating hormone and luteinising hormone. They prepare the body for pregnancy.

Maternal nutritional stores at the time of conception are important in buffering fluctuations in nutrient intake. Some stores, such as iron, may be compromised if there is a short interval between pregnancies. For women who have closely spaced pregnancies, improving dietary quality or using vitamin and mineral supplements is particularly important.

Nutritional programming

Over the last couple of decades it has become clear that nutrition in very early life, a critical window of development, can permanently influence the long-term risk of disease in adult life, a phenomenon known as nutritional programming (Martin-Gronert & Ozanne 2012). Nutritional programming in the prenatal and early postnatal environment influences risk factors that are associated with insulin resistance and type 2 diabetes mellitus (T2DM), cardiovascular disease (CVD), hypertension, osteoporosis and other diseases.

Maternal undernutrition affects the genetic expression of the developing embryo by modifying the expression of DNA. These inheritable genetic changes allow the foetus to respond to environmental stressors including nutrient availability in a way which enhances the chances of survival in foetal and young life. A mismatch between the foetal environment and subsequent postnatal environment means that the foetal programming will be less appropriate, so the consequences are more severe. For instance, an individual exposed to nutrient deprivation in foetal life may be born small and metabolically prepared for food deprivation in later life; if these conditions are not met and the environment is nutrient rich, the individual is more likely to develop chronic diseases in adulthood. Maternal micronutrient deficiency, particularly in zinc, vitamin D and vitamin B12, has also been shown to detrimentally affect metabolic health in offspring.

The outcomes depend on the extent and the timing of the nutrient insult. The foetal period, early postnatal life and adolescence are particularly vulnerable periods. In addition, there is clear evidence

CASE STUDY 4.1 ***CULTURALLY AND LINGUISTICALLY DIVERSE COMMUNITIES AND GDM*** *Sharon Croxford*

Profile and problem

Mrs P., a 28-year-old woman of Indian descent, has just been diagnosed with GDM at 32 weeks' gestation. This is her second pregnancy; she did not have GDM in the first. She gave birth to healthy boy just over two years ago. Mrs P. is married to a supportive husband who works as a researcher at a university. She worked as a personal assistant for a manager at an accounting firm before their son was born. Mrs P. is a vegetarian and follows a traditional northern Indian diet. She is 160 centimetres tall and weighs 66 kilograms; she has gained 6 kilograms during her pregnancy.

Assessment

Mrs P. wants to delay the use of insulin as long as possible and is prepared to make necessary changes to her diet and activity to manage her blood glucose. Her current diet is high in carbohydrates, and she consumes full-fat dairy products. She uses ghee sparingly in cooking, but a number of dishes are moderately high in vegetable oil. Her mother and mother-in-law have been providing Mrs P. with sweet treats.

Treatment

Mrs P. has agreed to do the following:

- eat three small meals and three smaller snacks throughout the day
- reduce portion sizes of carbohydrate foods (rice, breads—e.g., chapatti) and include atta (chickpea flour) in making breads
- include foods such as tofu and legumes twice per day and nuts once per day
- reduce sweet treats to once per week
- reduce oil in cooking
- continue with full-fat dairy products to maintain energy and calcium levels
- maintain all other aspects of the traditional northern Indian diet.

Plan

Mrs P. will monitor her blood glucose, and her diet will be monitored by a dietitian for the remainder of her pregnancy. If her blood glucose is not controlled she will need to commence insulin. Following the birth, Mrs P. will be at risk of developing diabetes, and maintaining dietary changes will be important.

that **epigenetic changes** can be passed down through subsequent generations. Improved postnatal nutrition to promote accelerated or catch-up growth in LBW infants paradoxically appears to worsen the metabolic consequences of intra-uterine growth retardation. Although early studies identified a relationship between maternal undernutrition, LBW and risk of subsequent disease, it is also evident that maternal overnutrition and maternal obesity influence foetal development and that the paternal diet might also affect the health outcomes of offspring.

Energy

Energy requirements are increased in pregnancy to meet the energy costs of laying down maternal fat stores and developing the foetus, placenta and associated tissues. The additional energy requirements of an average pregnancy are about 340,000 kJ (or about 1200–1300 kJ) per day. Requirements for energy are highest between 10 and 30 weeks of pregnancy, when maternal fat stores are deposited.

Some of the increased energy requirements for pregnancy could be offset by physiological adaptations such as maternal metabolism being more efficient due to decreased basal metabolic rate (BMR), decreased energy expenditure or changes in diet-induced thermogenesis. Changes in BMR have been found to be variable; usually BMR increases, but it may decrease or not change. Women who are usually sedentary have little opportunity to reduce physical activity, but those who are normally more active, such as women involved in subsistence farming, can make significant savings in their energy expenditure and thus adapt to spare nutrients to support the energy cost of pregnancy. There may be other physiological adaptations, such as the changes in the circulatory system, that reduce energy expenditure.

Epigenetic changes are heritable changes to gene activity and expression that occur without altering DNA sequence.

Macronutrients

Carbohydrate

There are no recommendations for increased carbohydrate requirements in pregnancy because normal diets are usually adequate in carbohydrate. For some women, such as those who develop gestational diabetes mellitus (GDM; see below), it is important to control carbohydrate intake during pregnancy. Exposure to high levels of blood glucose in early development can cause neural tube defects in the developing embryo. This can be caused by pre-existing type 1 diabetes mellitus (T1DM) or T2DM in the mother. Good control of maternal glucose metabolism, including consumption of carbohydrate-rich foods with lower glycaemic index (GI), is thought to reduce the risk of congenital abnormalities.

As mentioned above, the high progesterone levels in pregnancy affect the smooth muscle tone of the gut, so gut motility is decreased and transit time for food through the gut is increased. This means the contents of the gut are in contact with sites of nutrient absorption for a longer time, making nutrient absorption more efficient. However, water absorption is also increased, which predisposes the woman to constipation. Pregnant women are recommended to increase their intake of both water and fibre. Increased intake of high-fibre carbohydrate foods that have lower GI tends to displace the intake of added sugar and fat. Dietary intake of lower GI carbohydrates is particularly important in women at risk of developing GDM and is associated with a decreased occurrence of large birthweight babies.

Protein

Additional protein requirements for the growth of the new maternal and foetal tissue are 500–700 grams in total. This means that the daily increase in protein synthesis averages at about 6 grams, which can be met from an additional 8–9 grams of dietary protein per day. There is some evidence that the efficiency of protein synthesis increases from early pregnancy (Duggleby & Jackson 2002).

Women who deliver babies of optimal birthweight tend to consume more protein in pregnancy than women who have smaller babies. The Australian and New Zealand nutrient reference values (NRVs) recommend an additional 14 grams of protein in women over nineteen years of age, bringing the recommended dietary intake (RDI) to 60 grams per day in the second and third trimesters (NHMRC & Ministry of Health 2006). The Institute of Medicine in the United States recommends a further 50 grams of protein for women with twin pregnancies (Rasmussen & Yaktine 2009).

Low protein intakes are associated with an increased risk of adverse outcomes of pregnancy, but protein intakes in developed countries are generally sufficient to meet the additional requirements of pregnancy. However, women following a vegan or vegetarian diet need to ensure they consume a range of legumes and whole grains in order to get adequate high-quality protein. Women must also consume adequate energy to ensure that the protein consumed is used for synthesis of new tissues rather than broken down to be used as a fuel for energy. Women with very low energy intakes are at risk of inadequate protein intake. It is recommended that the additional protein comes from food sources rather than from supplements, protein powders or high-protein formulated beverages, because these have been associated with adverse outcomes in intervention studies. High-protein diets, such as those promoted for weight loss, are not recommended, because the foetus has limited ability to filter and excrete the toxic by-products of excess protein intake, including ammonia and urea. Studies in rodent models suggest that high protein intakes resulting in high ammonia levels increase the rate of congenital abnormalities.

Blood protein levels fall from about 70 grams per litre to 60 grams per litre over the course of a pregnancy. Most of this change is due to dilution from the expansion in plasma volume. Protein in the blood has an important role in maintaining plasma osmotic pressure, which is responsible for the return of fluid into the blood vessels. During pregnancy, the lower level of protein in the blood results in increased movement of fluid out of the blood vessels, which increases the likelihood of swelling in the lower limbs.

Fat

In early pregnancy, levels of free fatty acids, triglyceride, cholesterol, lipoproteins and phospholipids transiently fall, reflecting the increase in maternal fat storage that occurs when foetal demands are relatively low. Maternal fat deposition is high in early pregnancy, effectively acting as an energy buffer for later pregnancy, when foetal requirements are maximal and maternal nutrient intake could be restricted by low availability of food or by restricted capacity for eating. Maternal fat deposition is on average about 3.5 kilograms of fat that could be mobilised to provide 132,000 kJ of energy.

Pregnant women have high requirements for long-chain polyunsaturated fatty acids, particularly in the third trimester, to meet the demands of the growth of the foetal brain and nervous tissue. Foetal demand for essential fatty acids is met from maternal dietary intake or through being released from maternal adipose tissue and transported across the placenta to the foetus. Oestrogen increases the expression of enzymes involved in maternal production of the long-chain polyunsaturated fatty acids. Other dietary factors influencing maternal synthesis of these fatty acids include trans fatty acids, deficiencies of B vitamins, iron, zinc, magnesium and calcium, low protein, high sucrose and alcohol. Low maternal intake of essential fatty acids is associated with reduced neonatal growth.

Consumption of oily fish rich in long-chain polyunsaturated fatty acids has been associated with a lower incidence of pre-eclampsia in the mother and LBW and preterm delivery of the infant. The best source of long-chain polyunsaturated fatty acids is oily fish, but some of the health warnings about fish consumption related to vitamin A levels in fish liver and mercury and other contaminants of older, long-lived fish often lead women to avoid fish during pregnancy.

High fat intakes are not recommended in pregnancy, because of concerns about fat intake and abnormal glucose metabolism and the effects on foetal development and intellectual performance. The recommendation is that a macronutrient intake that provides less than 30 per cent energy from fat and more than 50 per cent energy from carbohydrates is associated with a lower risk of glucose intolerance and GDM.

Micronutrients

Micronutrient deficiencies may have negative consequences for foetal development and pregnancy outcome; they appear to contribute to congenital abnormalities and compromised embryonic implantation and survival. Generally, micronutrient requirements increase during pregnancy. It is assumed that if the mother eats more of a good-quality diet to meet her energy requirements her micronutrient intake will increase adequately, with the possible exceptions of folate, iron, calcium and vitamin D. However, poor-quality and inadequate diets can adversely affect foetal development. Pregnancy is associated with increased susceptibility to oxidative stress, which may impair the development of the foetus, particularly in early brain development.

Generally, the maternal blood concentrations of fat-soluble vitamins rise and the levels of water-soluble vitamins decrease; however, vitamin A levels fall and carotenoids rise. Increased plasma volume can result in lowered circulating levels of nutrients, but it is suggested that the changes in pregnancy favour transfer to the foetus (Campbell-Brown & Hytten 1998). Lower maternal levels may limit uptake by the maternal cells. Fat-soluble vitamins readily cross the placenta, and their transfer to the foetus increases in parallel with time of gestation. The placenta can extract vitamins from the maternal circulation and transfer them to the foetus against a concentration gradient, maintaining foetal levels above maternal levels. Supplementation needs to be considered very carefully, because nutrient–nutrient interactions can be harmful, and some micronutrients are toxic in excess.

Oxidative stress occurs when free radicals (reactive oxygen) are not contained by the body's antioxidant capabilities causing oxidative damage.

B vitamins

There is significant interaction between the B vitamins such as folate, vitamins B6 and B12 and choline. They have a number of functions, including gene expression. In pregnancy, the need for B vitamins increases, because there is increased metabolism of nucleic and amino acids and increased synthesis of DNA, ribonucleic acid and protein. Inadequate levels of B vitamins are associated with adverse outcomes of pregnancy, including increased risk of neural tube defects and other congenital abnormalities such as **cleft lip**, pre-eclampsia, LBW, premature delivery, placental abruption and miscarriage.

Folate and folic acid

Folate requirements increase in pregnancy, and folate deficiency in pregnancy is associated with an increased risk of neural tube defects. Deficiency is

Cleft lip and palate is a congenital anomaly of the face caused when two sides of the face do not fuse properly during early prenatal development. A cleft lip results in a split in the lip, and cleft palate is a split in the roof of the mouth. The defect can be relatively minor or severe.

also associated with **megaloblastic anaemia** of pregnancy. Women planning pregnancy are advised to take folic acid supplements prior to conception and during the first trimester to reduce the risk of neural tube defects and other congenital abnormalities. The National Health and Medical Research Council of Australia recommends pregnant women take a folic acid supplement of 400 micrograms daily (NHMRC & Ministry of Health 2006), although the New Zealand Ministry of Health recommends 800 micrograms daily (Ministry of Health 2014).

Megaloblastic anaemia is low blood haemoglobin concentration that is characterised by larger than normal red blood cells. (Iron deficiency anaemia is characterised by hypochromic [pale] and microcytic [small] cells.) The larger cell size is caused by impaired production of DNA, usually due to inadequate folate and/or vitamin B12.

Neural tube defects result from failure of th neural tube to close effectively 22–27 days after fertilisation, which may be before many women realise they are pregnant. Women who are at higher risk for a pregnancy affected by neural tube defects (because they have a family history of neural tube defects, have had a previous affected pregnancy, have insulin-dependent diabetes or are taking certain types of anticonvulsant medications) are advised to have 5 milligrams of folic acid daily. In women who are obese, folic acid supplementation appears to be less effective.

The history of recommending folic acid supplementation in pregnancy has identified that, while many women are aware of the recommendations, a significant proportion do not follow them. See Chapter 3 for mandatory fortification of bread with folic acid.

CASE STUDY 4.2 ADOLESCENT PREGNANCY

Profile and problem

Miss T., an 18-year-old woman, has presented with her first pregnancy. She is unsure of the date of her last menstrual period; she thinks it was about five months ago. Miss T. is studying beauty therapy at college and has recently split up from her boyfriend. She gave up smoking and stopped drinking alcohol as soon as she realised she was pregnant. Miss T. is currently living in a shared house with other students and is living on a limited budget. She is 170 centimetres tall and today weighs 55 kilograms. She is unsure of her pre-pregnancy weight.

Assessment

Miss T. is worried, as she knows that alcohol is bad for the baby, but she thinks she drank in early pregnancy before she confirmed she was pregnant using a pregnancy testing kit. On college days, she skips breakfast and buys chips or wedges with sour cream in the student cafeteria at lunchtime. She knows sugary drinks are unhealthy and so has switched to diet cola. She enjoys fruit and vegetables, but she finds them expensive. She has seen a television advertisement recommending an expensive multivitamin supplement but is unsure if she really needs it.

Treatment

Miss T. has agreed to do the following:

- eat wholegrain breakfast cereal with dried fruit and low-fat milk daily
- take a packed lunch to college
- reduce chips and wedges to once per week
- drink water or low-fat milk instead of soft drinks
- buy tinned or dried fruit and frozen vegetables to reduce the cost
- increase her intake of iron-rich food (lean meat, fish, eggs, peas or beans)
- eat wholegrain bread and pasta
- take 150 micrograms of iodine supplement daily.

Plan

Miss T. will need to have her blood iron levels checked regularly for haemoglobin and serum ferritin (a measure of iron stores). She will need to be weighed regularly to check she is gaining sufficient weight. Her young age, low pre-pregnancy body weight and low pregnancy weight gain put her at risk of an LBW baby.

Vitamin B12

The requirements for vitamin B12 are increased in the first two trimesters, because it is involved in maternal and foetal **erythropoiesis**. Deficiency of vitamin B12 increases the risk of foetal neurological abnormalities and may exacerbate folate deficiency. The placenta concentrates vitamin B12 and transports it against a concentration gradient; foetal concentration of vitamin B12 is about double the maternal level. Newly absorbed dietary vitamin B12, rather than that mobilised from maternal stores in the liver, is preferentially transported across the placenta, so the mother may show no evidence of deficiency even if transfer to the foetus is limited. Strict vegetarians, vegans and women who avoid animal products are at higher risk of deficiency.

Choline

Placental transfer of choline to the foetus is significant, and foetal levels are about fourteen times maternal levels; choline is important for placental and foetal development, particularly **neurogenesis** and **angiogenesis** (Zeisel 2013). It may protect the foetus from **environmental insults** that contribute to birth defects. Women who do not eat rich sources of choline, such as meat, milk and eggs, may be advised to take choline supplements to reduce maternal choline depletion, because requirements in pregnancy are increased.

Vitamin C

Increased plasma volume during pregnancy means that the plasma vitamin C concentration falls, so pregnant women have increased requirements to ensure maternal needs are met and that there is adequate transfer to the placenta and foetus. Vitamin C protects the placenta from oxidative stress and cell death. Although foetal blood levels of vitamin C are higher than maternal levels even when the mother is insufficient, maternal vitamin C status is thought to be positively associated with pregnancy outcome, and the development of the foetus and placenta unit is sensitive to maternal vitamin C deficiency. Vitamin C deficiency is also associated with increased infection, premature rupture of the placental membranes and preterm delivery. Pregnant women who are exposed to increased oxidative stress—for instance, those who smoke, have a high intake of alcohol, use recreational drugs or regularly take aspirin—are advised to have a higher intake of vitamin C.

Erythropoiesis is the formation of red blood cells in the bone marrow.

Neurogenesis is the formation of new nerve cells in the brain.

Angiogenesis is the formation of new blood vessels from pre-existing blood vessels.

Environmental insults are external stressors that result in the foetus adapting in a way that negatively affects its health—for example, smoking or poorly controlled blood glucose of the mother resulting in hypoxia, or low oxygen, in the foetus.

Vitamin A

Vitamin A is important for embryonic development and foetal growth. Its role in gene expression means that both low and high intakes are associated with increased risk of congenital malformations. Maternal vitamin A status is positively correlated with infant size and length of gestation; low vitamin A is associated with increased maternal mortality. Vitamin A requirements are highest when foetal growth is highest, in the third trimester. Provitamin A from carotenoids is not efficiently converted into retinal, so pregnant women who do not consume sources of preformed vitamin A, because they avoid meat and dairy products, are advised to consume at least five portions a day of coloured fruit and vegetables, which are rich sources of carotenoids.

Teratogenicity of vitamin A in the first trimester has been demonstrated in animal models. The birth defects tend to derive from abnormalities affecting the central nervous system, heart, thymus and craniofacial regions, such as cleft lip and palate. Pregnant women are advised to avoid liver and liver products such as liver pate, because vitamin A accumulates in the livers of domesticated animals; levels can be very high in animals fed vitamin supplements to promote growth, a practice common in many countries. High doses of vitamin A (in the form of retinoic acid derivatives) are used in some dermatological agents for the treatment of acne, for instance. These are not recommended for pregnant women, and contraception may be advised for a washout period of up to twelve months following the cessation of treatment to ensure the early embryo is not exposed to high retinal levels. There is no concern

Teratogenicity is the capacity to cause birth defects. A teratogen is the agent that causes the damage to the embryo or foetus.

about dietary carotenoids in pregnancy, because retinal levels do not reach toxic levels. (Carotenoids slowly convert to retinal.)

Vitamin D

Vitamin D requirements increase significantly in pregnancy; deficiency and insufficiency are common in pregnancy (Hollis & Wagner 2013). Maternal vitamin D status may be important in determining the development and severity of GDM, because vitamin D affects insulin secretion and insulin sensitivity and reduces inflammation. This is important because of the high prevalence of vitamin D insufficiency in many countries; maternal hypovitaminosis D is potentially a modifiable risk factor for GDM. Vitamin D deficiency is also associated with decreased foetal growth, increased neonatal vulnerability to rickets and later increased risk of osteoporosis.

Women with darker skins and those who do not regularly expose their skin to ultraviolet light are more dependent on dietary sources of vitamin D or supplementation. Although many authorities recommend a daily supplement of 10 micrograms for pregnant women, emerging research suggests much higher dose supplements of 100 micrograms are associated with optimal infant health, including protection from autoimmune diseases and protection of the mother from comorbidities of pregnancy, such as pre-eclampsia, hypertension, GDM and need for surgical delivery (Hollis & Wagner 2013).

Vitamin K

The role of vitamin K in facilitating calcium delivery is important for the synthesis of blood-clotting factors and bone mineralisation. In pregnancy, vitamin K is not usually of concern. However, the use of broad spectrum antibiotics for prolonged periods can affect gut synthesis of vitamin K. The use of drugs that disrupt metabolism and recycling of vitamin K and can cross the placenta, such as warfarin, is associated with increased risk of foetal brain bleeding, intellectual disability or delay and altered bone mineralisation. Some treatments for epilepsy can inhibit placental transport of vitamin K and affect the foetal production of clotting factors and thus increase risk of bleeding.

Iodine

Maternal thyroid hormone production increases by about 50 per cent in early pregnancy, thus increasing the demand for iodine (Leung et al. 2011). The foetal thyroid starts to concentrate iodine towards the end of the first trimester. At about twenty weeks' gestation, foetal production of thyroid-stimulating hormone controls foetal thyroid activity, and iodine requirements are further increased in the later part of pregnancy when foetal thyroid hormone production increases, utilising maternal iodide, which can readily cross the placenta.

Elevated iodine excretion is common in early pregnancy, so plasma iodine levels fall. The higher thyroid hormone production, foetal requirements and increased excretion mean that iodine requirements are significantly increased in pregnancy.

Iodine is essential for thyroid hormone production, and thyroid hormones are required for foetal and neonatal neurodevelopment. Iodine requirements in pregnancy increase by 50 per cent to meet maternal and foetal demands. Severe maternal iodine deficiency can result in **endemic goitre**, **cretinism**, compromised intellectual development, growth retardation and neonatal hypothyroidism in the infant, increased pregnancy failure and higher infant mortality (Skeaff 2011). However, mild and moderate iodine deficiency may also have adverse effects. Iodine insufficiency in pregnancy and in the early postpartum period can affect the developing brain and result in impaired neurocognitive, psychomotor, behavioural and psychological development in children. Visual attention, visual processing and gross motor skills may also be affected by iodine deficiency. The intelligence quotients of children living in iodine-deficient regions are lower than those living in iodine-sufficient areas; iodine supplementation can positively affect intelligence quotients. There is also a higher prevalence of hyperactivity and attention deficient disorders in children of women living in iodine-deficient areas.

Iodine content of foods varies, and food iodine content is not a normal part of the nutritional information panels on packaging. As well as foods such as grain products, milk and cheese, which are

An **endemic goitre** is an enlargement of the thyroid gland (a goitre) that may develop in a person living in an area of low iodine concentrations in the soil, meaning foods grown in that area are low in iodine.

Cretinism is a condition characterised by profound mental function, severely retarded growth and/or deficits in speech, hearing, gait and stance in infants due to an underactive thyroid gland caused by iodine deficiency.

relatively rich in iodine, other sources of iodine include iodised salt, dairy foods (which may be contaminated with iodine from the iodophor cleaners used in the dairy industry) and bread dough (because some bread conditioners use iodate). However, population median intakes have tended to fall in recent years because iodophors in the dairy industry are being phased out, public health recommendations are driving a reduced salt intake, iodate bread conditioners are used less and the consumption of more processed food containing non-iodised salt is increasing.

Studies conducted since 2001 have shown low iodine intakes in New Zealand and Australia. In Tasmania, New South Wales and Victoria, pregnant women had median urinary iodine concentrations well below 100 micrograms per litre, and in New Zealand, median urinary iodine concentrations were less than 50 micrograms per litre (both well below the 150 micrograms per litre that defines sufficient intake). A New Zealand study in 2009 found mean intakes in a small sample of pregnant women to be 119 micrograms per day less than the estimated average requirement (EAR) during pregnancy (Brough et al. 2013).

In 2009 the mandatory fortification of all bread (except organic) with iodine was introduced in both Australia and New Zealand. This was predicted to meet the needs for the majority of the population, but pregnant and breastfeeding women would still require an additional iodine supplement. Both countries introduced recommendations for women to take 150 micrograms of iodine daily during pre-pregnancy, pregnancy and breastfeeding. Research in New South Wales found that, although median urinary iodine concentration was above the World Health Organization's (WHO) recommendation, only women using iodine supplements achieved sufficiency (Charlton et al. 2013). The upper limit (UL) of iodine is not universally agreed: the WHO (FAO 2001) recommends 500 micrograms per day, whereas the Institute of Medicine and the Australian and New Zealand NRVs recommend 1100 micrograms per day for adults (Institute of Medicine 2006; NHMRC & Ministry of Health 2006).

Calcium

Calcium requirements increase by about 600 milligrams in pregnancy to maintain maternal calcium homeostasis and to meet the demands of the growing foetus, which are highest in the third trimester when the foetal skeleton is calcified. The foetal skeleton incorporates about 30 grams of calcium, which is trivial compared to that in the maternal bones. Therefore, the maternal skeleton can act as a reservoir of calcium if the mother does not increase her dietary intake. The decrease in maternal bone density is restored after lactation is ceased. However, two groups of pregnant women may be particularly at risk: those who are close to menopause, who may have limited time for restoration of bone density, and adolescents, particularly if they conceive within two years of menarche, while peak bone mass is being accrued.

The greater requirement can be partially met by increasing intake, but both calcium absorption and calcium retention increase if vitamin D levels also rise appropriately. The placenta actively transports calcium. Although mobilisation of calcium from the maternal skeleton can occur, markedly inadequate maternal calcium intake can increase the risk of osteoporosis, osteopaenia, **tetany**, tremor, muscle cramps and **paraesthesia** and can also affect foetal bone mineralisation and foetal growth. In some studies, calcium supplementation in pregnancy has been shown to be beneficial in preventing pre-eclampsia and hypertensive disorders, particularly in those women with a low calcium intake, but it has not been shown to reduce preterm births or to affect the occurrence of LBW but may have a slight effect on mean infant birthweight.

Calcium is one of the significant nutrients that need to be considered in antenatal assessment; women of reproductive age often have calcium intakes that are below recommendations. Pregnant adolescents and women who do not consume dairy products are particularly at risk of inadequate intake; pregnant women are generally advised to consume at least three servings of calcium-rich foods a day. Although dietary sources of calcium are preferable, it may be necessary for women who avoid calcium-rich foods to be prescribed a calcium supplement. Women who are prescribed both iron and calcium supplements are advised to take them at different times of the day to maximise absorption of both as they compete for absorption.

Tetany is a condition involving involuntary contraction of muscles, resulting in cramps or spasms of the feet and hands or larynx.

Paraesthesia is a tingling, tickling, prickling sensation commonly known as 'pins and needles' or a limb 'falling asleep'.

Iron

Iron deficiency continues to be the most common nutritional deficiency in the world and is the most significant cause of anaemia in pregnancy. This is associated with an increased risk of adverse outcomes including LBW, prematurity, pre-eclampsia, perinatal mortality, maternal morbidity and mortality and reduced work capacity. Severe maternal anaemia may also be associated with intra-uterine growth restriction and high placental weight; both LBW and high placental weight are thought to confer an increased risk of hypertension in adulthood. Maternal iron status in pregnancy may also affect iron accumulation by the foetus and therefore neonatal iron status. LBW and premature infants have reduced iron stores and are at risk of iron deficiency. These problems are of particular concern in developing countries, where maternal anaemia can exceed 50 per cent. Those women who are poorer, more vulnerable and less educated are the most likely to be affected by iron deficiency.

A woman's iron stores at the time of conception and the amount of iron absorbed during pregnancy influence whether she will develop iron deficiency anaemia. In developing countries, pre-existing iron stores before pregnancy are often insufficient, so the physiological adaptations that increase iron absorption may be inadequate to meet the increased requirements, particularly if the diet has lower iron **bioavailability**. To meet the requirements for increased maternal and foetal red blood cell production, the RDI for women during pregnancy is increased to 27 milligrams per day from the normal requirement of 18 milligrams per day.

The efficiency of iron absorption increases in pregnancy, but a pregnant woman still requires significantly more dietary iron. In developing countries, where most diets have low iron bioavailability, these requirements are difficult to meet, and iron supplementation has been found to be beneficial.

Iron supplementation on its own or with folic acid has an impact on maternal and foetal wellbeing affecting birthweight, **Apgar scores**, infant iron stores in early life and neonatal mortality. The need for postpartum maternal blood transfusion is lower in women who take iron supplements in pregnancy. Oral iron supplementation, such as iron sulfate or iron fumarate, is usually bioavailable and effective, safe and of low cost.

Bioavailability is the extent to which a nutrient is absorbed and is available to the body, enabling physiological activity.

Apgar scores were introduced in the 1950s as a simple way of assessing an infant's health at birth, particularly in the transition to extra-uterine life. There are five criteria, each of which is scored from zero to two, giving a total score of zero to ten. The assessment is usually carried out at one and five minutes after delivery and may be repeated later (e.g., 10, 30, 60 minutes after birth) if the initial scores are low. The criteria are appearance, pulse rate, reflex response to stimuli, activity and respiratory effort. They are remembered using the mnemonic APGAR: appearance, pulse, grimace, activity, respiration.

The WHO promotes weekly iron and folic acid fortification in regions of the world where the prevalence of anaemia in women of reproductive age is greater than 20 per cent (FAO & WHO 2001). It is thought that weekly doses have more compliance and fewer side effects and are more effective. The International Nutritional Anaemia Consultative Group recommends 60 milligrams of iron per day for the last six months of gestation for pregnant women who are not anaemic and 120 milligrams per day for pregnant women who are anaemic or those who have a shorter possible period of supplementation before delivery. (More women in developing countries seek antenatal care later in pregnancy.) It is also recommended that these iron supplements are combined with 400 micrograms of folic acid (FAO & WHO 2001). High doses of supplemental iron consumed on an empty stomach are commonly associated with gastrointestinal symptoms including constipation, nausea, vomiting and diarrhoea.

Zinc

Zinc deficiency in pregnancy is associated with increased risk of congenital abnormalities, including neural tube defects, and other complications including pre-eclampsia, haemorrhage, hypertension, both short and prolonged gestation, protracted labour, foetal growth retardation and retarded neurogenesis and neurobehavioural and immunological development. Zinc affects embryonic and foetal development because over 200 enzymes contain zinc. These are involved in many reactions, including nucleic acid and protein synthesis, carbohydrate and protein metabolism, cell proliferation and cell death, protection from oxidative damage, hormone binding and gene transcription.

Zinc deficiency may affect the timing of onset of birth, because zinc is required for the synthesis of and responses to hormones involved in the pathways. Zinc is required for normal immune function, so deficiency can increase both intra-uterine and systemic infections that are associated with premature delivery.

Iron and zinc are transported across the gut by the same transporter, so iron supplementation can decrease zinc absorption, and zinc in excess may cause a secondary copper deficiency (Donangelo & King 2012). Zinc is transported across the placenta by active transport, so foetal levels are higher than maternal levels; but transfer of sufficient zinc to meet foetal demands depends on the maintenance of maternal serum zinc levels. Pregnant women retain about 100 milligrams of zinc during pregnancy; by the third trimester, zinc requirements are about double the non-pregnant requirements. Maternal plasma zinc levels decrease as pregnancy progresses. The primary mechanism to increase zinc retention is increased intestinal zinc absorption.

Although severe zinc deficiency is rare, the WHO estimates that about half the population of the world and over 80 per cent of pregnant women have suboptimal zinc status (FAO & WHO 2001). Zinc (and iron) absorption is inhibited by high phytate intake from cereal-based diets, which are often also low in animal-derived foods, which are zinc rich. Zinc absorption is further inhibited by high intakes of other divalent metals, such as iron supplements, and by gastrointestinal diseases. Physiological stress, such as infection or trauma, smoking and high alcohol intakes can affect maternal plasma zinc concentrations and thus transport of sufficient zinc to the foetus. Zinc levels in amniotic fluid are lower in pre-eclampsia. Some studies have suggested that zinc supplementation can reduce pre-eclampsia, preterm birth and LBW. Zinc supplementation is advised for pregnant women who have gastrointestinal diseases.

Copper

Copper is an essential cofactor of a number of enzymes involved in oxygen transport, metabolism, angiogenesis and antioxidant pathways, so it is essential for embryonic and foetal development. Plasma copper concentrations significantly increase in pregnancy, probably because increased oestrogen levels stimulate the copper binding protein.

Maternal copper deficiency is associated with increased risk of early embryonic death and gross structural and biochemical abnormalities; longer term, the risk of CVD is increased and there is a reduced fertility rate. Maternal deficiency may be due to a low dietary intake but may be secondary to other diseases or drugs which affect maternal or foetal copper metabolism. The effects on embryonic and foetal development appear to be due to impaired antioxidant defence mechanisms, altered energy metabolism and impacts on connective tissue. Maternal plasma copper levels are higher than foetal levels, suggesting that the placenta blocks transfer of copper to the foetus. In pre-eclampsia, maternal serum and placental copper levels are high.

Selenium

Selenium is required for the production of some proteins and hormones. Pregnant women therefore have increased selenium requirements to support increased protein synthesis and tissue accumulation. Maternal selenium levels fall in pregnancy. Selenium is transported across the placenta on its concentration gradient; foetal levels are lower than maternal levels.

Recurrent pregnancy loss is associated with lower maternal selenium levels that may result in reduced antioxidant protection of DNA and biological membranes. It is not clear whether low selenium status is associated with foetal growth restriction. There have been few supplementation studies, but it is suggested that selenium supplementation may be useful for women at risk of developing postpartum thyroid dysfunction and permanent hypothyroidism. Selenium requirements are increased with raised oxidative stress due, for instance, to smoking or intense exercise.

Magnesium

Magnesium deficiency in pregnancy has been associated with maternal leg cramps, pregnancy-induced hypertension, uterine irritability, preterm labour, foetal growth retardation, cerebral palsy and mental retardation. Pregnant women from poorer socio-economic backgrounds with low fruit and vegetable intakes tend to have lower magnesium intakes, and there is some evidence from intervention studies suggesting magnesium supplementation is associated with higher birthweight.

INFLUENCES ON PREGNANCY

Extremes of maternal age

Pregnancy during adolescence has increased risk for mother and infant, such as LBW, preterm birth and assisted delivery. Pregnant adolescents are at nutritional risk, as they are more likely to be still growing, be shorter, weigh less and have poorer diets. They are often from lower socio-economic groups and more likely than older women to engage in risky health behaviours such as cigarette smoking and using alcohol and drugs, which can be detrimental to the health of both mother and infant. Younger mothers are more likely to present for antenatal care later in pregnancy and thus miss out on early obstetric care. New Zealand has the second-highest rate of teenage pregnancy among member countries of the Organisation for Economic Co-operation and Development, with around 29 pregnancies per 1000 teenagers. Australia has a lower rate, with around 18 pregnancies per 1000 teenagers.

Women with high maternal age are also at increased risk of an adverse pregnancy outcome, such as preterm delivery and LBW. The reasons are not clear, but in recent years it has been exacerbated by the increased use of fertility treatment. Women over the age of 35 years are at increased risk of having a child with a chromosomal abnormality, such as Down syndrome.

Vegetarianism and veganism

Women who follow a vegetarian diet and consume dairy products and eggs should be able to attain all the nutrients required for a healthy pregnancy. They need to ensure they have an adequate intake (AI) of iron, as the only iron available is non-haem iron, which is less well absorbed that the haem iron available from animal sources. Women following a vegan diet are at risk of vitamin B12 deficiency and may require supplementation. There is currently controversy as to whether the omega-3 oils consumed as part of a vegetarian diet (for example, from walnuts and flaxseed) are equivalent to the omega-3 oils derived from fish in terms of their benefit regarding brain development.

Multiple pregnancies

Since the 1980s there has been a dramatic increase in the number of multiple pregnancies. This is primarily due to the widespread availability of fertility treatments—either stimulation of the ovaries, which can result in the release of more than one egg, or the implantation of more than one embryo through in-vitro fertilisation treatment. Another factor is the trend for women to have children later in life. Older women are at increased risk of multiple pregnancies, irrespective of fertility treatment. Although the number of multiple births is small (less than 2 per cent of births in Australia and New Zealand), the infants are more likely to require neonatal intensive care.

Nutrient requirements are higher in multiple pregnancies, and there is greater risk of preterm birth and LBW due to the increased requirement for maternal resources and the limitations of uterine size. The Royal Australian and New Zealand College of Obstetricians and Gynaecologists recommends supplementation with iron and 5 milligrams of folic acid throughout pregnancy for women with multiple pregnancies (RANZCOG 2013b).

Food safety

The food-borne bacteria *Listeria monocytogenes* can be especially dangerous if contracted during pregnancy, as it may cause preterm birth, miscarriage, stillbirth or serious health concerns for the infant. Foods that are more commonly found to be sources of *L. monocytogenes* include soft cheeses, unpasteurised dairy products, cold meats and pre-prepared cut fruit and vegetables (see Chapter 1). Mercury is also a concern because of its potential to damage the foetal nervous system. High levels of mercury can be found in some fish, especially longer lived and predatory species. Australia and New Zealand have specific recommendations regarding the safety of fish, and women are advised to check country-specific recommendations, as names of fish are not standard between the two countries.

Lifestyle

There is strong evidence to show that excessive alcohol intake during pregnancy can affect physical and mental development of the foetus, resulting in lifelong consequences for the offspring. The effects of low levels of alcohol during pregnancy are less well known, but in the interests of the safety of the foetus, both the Australian National Health and Medical Research Council and the New Zealand Ministry of Health recommend that women who are pregnant or trying to conceive should abstain from drinking alcohol.

Cigarette smoking during pregnancy can have serious health consequences for the foetus and has been associated with a two-fold increase in LBW. Paternal smoking pre-pregnancy has also been associated with sperm damage and developing cancer in the offspring. The Royal Australian and New Zealand College of Obstetricians and Gynaecologists recommends that both parents stop smoking prior to conception (RANZCOG 2013a).

Caffeine decreases the absorption of iron, and high intakes increase the risk of miscarriage and LBW.

Both the Australian Department of Health and the New Zealand Ministry of Health recommend caffeine consumption to be limited to 300 milligrams per day, which equates to no more than three cups of tea or instant coffee or no more than two single espresso drinks or one double espresso drink or 400 grams of plain chocolate. Pregnant women are advised to avoid energy drinks, as these can contain high amounts of caffeine as well as other stimulants. Pregnant women should also be cautious about drinking herbal teas.

While the nutrient intakes of pregnant women in Australia and New Zealand (see Table 4.1) are typically higher than those of non-pregnant women, both groups consume less than the RDI for many nutrients. Both pregnant and non-pregnant women consume less than the recommended daily intake of calcium, iron, magnesium and zinc. Pregnant women sufficiently increase their intake to meet vitamin A requirements, and both pregnant and non-pregnant women consume the recommended intake of vitamin B12 and vitamin C. These values apply to the overall population of women, and may not be reflective of the intakes of specific subgroups of women such as those following a vegetarian diet.

Health conditions

Overweight and obesity

High body fat before pregnancy and high weight gain during pregnancy are associated with an increased risk of developing GDM and hypertensive disorders.

Gestational diabetes mellitus

GDM is a form of diabetes that occurs during pregnancy and resolves after the baby is born. Mothers who develop GDM are more likely to develop it again in subsequent pregnancies and have an increased risk of developing T2DM later in life. Women with an increased risk of developing GDM include those over 30 years of age, those with a family history of diabetes and those who are overweight. Up to 8 per cent of Australian women will develop GDM. GDM occurs when placental hormones interfere with the action of insulin so the body cells are less able to respond to it; this is described as insulin resistance. Insulin is required to maintain normal blood glucose levels, and in uncontrolled diabetes blood glucose levels can rise.

Management of GDM to control blood glucose levels requires a specialised eating plan developed in consultation with a doctor, diabetes educator and dietitian. If poorly controlled, the high glucose levels will be passed on to the baby and can result in a large birthweight baby, which increases the risk of caesarean section, birth trauma and postoperative complications. Diabetes during pregnancy also increases the risk of neural tube defects and other birth defects, and offspring of mothers with GDM may be at risk of developing T2DM later in life. In Australia and New Zealand pregnant women are routinely screened for diabetes between 26 and 28 weeks' gestation. The current rate of GDM is 4–5 per cent in Australia and New Zealand.

Table 4.1 Selected micronutrient intakes for non-pregnant and pregnant adult females in Australia and New Zealand

Micronutrient	RDI (19–50 years)		Median intake		Mean intake, non-pregnant, 19–24 years (Aus)
	Non-pregnant	Pregnant	Pregnant, 3rd trimester (NZ)	Non-pregnant, 19–30 years (NZ)	
Calcium (mg)	1000	1000	790	704	750
Folate (µg)	400	600	190	–	233
Iron (mg)	18.0	27.0	12.1	10.2	11.9
Magnesium (mg)	310–20	350–60	270	–	273
Selenium (mg)	60	65	–	46	–
Vitamin A (µg)	700	800	1180	654	889
Vitamin B12 (µg)	2.4	2.6	3.1	3.0	–
Vitamin C (mg)	45	60	96	107	120
Zinc (mg)	14.0	11.0	10.1	8.9	10.2

Note: Aus: Australia; NZ: New Zealand; RDI: recommended dietary intake. – denotes lack of data.

Source: McKenzie-Parnell et al. (1993); McLennan & Podger (1998); NHMRC & Ministry of Health (2006); University of Otago & Ministry of Health (2011).

Foetal alcohol spectrum disorder

Foetal alcohol spectrum disorder refers to a group of conditions that may occur in individuals who were exposed to alcohol in utero. Alcohol consumed by pregnant women enters the foetal blood supply and leads to problems such as LBW, intellectual disability and behavioural problems, distinctive facial features and heart defects. There is no known safe level of alcohol consumption during pregnancy, so pregnant women are advised to abstain from consuming alcohol.

SUMMARY AND KEY MESSAGES

Pregnancy presents nutritional challenges which have to be met to ensure optimal maternal health and infant growth and development. Adequate energy intake is associated with optimal reproductive function and shorter time to conception. Nutrition in early embryonic life can permanently influence the offspring's chances of developing chronic diseases in later life. Optimal foetal growth is associated with lower infant morbidity and mortality.

Physiological adaptations to pregnancy result in enhanced fat deposition in early pregnancy and enhanced fat mobilisation in late pregnancy, when foetal growth is highest. Energy intake in pregnancy can be offset by changes in maternal energy metabolism; therefore, energy requirements tend to be lower than the theoretical energy costs. Although there are no explicit carbohydrate recommendations for pregnant women, there are concerns about high sucrose intake and high blood glucose and the risk of congenital abnormalities. Protein requirements are increased in pregnancy; recommendations take into account energy intake, protein sources and protein quality. Intake of long-chain polyunsaturated fatty acids is important for foetal brain development and optimal gestational length.

Micronutrient requirements tend to increase in pregnancy; maternal intakes of folate and other B vitamins, iodine, iron, calcium and vitamin D appear to be particularly important. B vitamins, such as folate, are important in reducing the risk of neural tube defects. Vitamin A affects gene expression, so both low and high intakes are associated with foetal abnormalities. Recent research suggests vitamin D may protect against maternal comorbidities. Iodine intake over recent years has been declining, which may have implications for foetal brain development and neurocognitive capability of the offspring. Iron deficiency is associated with an increased risk of adverse maternal and infant outcomes. Pregnant women who are at extremes of maternal age, have a restrictive diet such as being vegetarian or vegan, are obese or have multiple pregnancies have a greater risk of nutritional problems.

- Nutrient intake and body composition influence fertility in both males and females.
- Reduced ability to conceive when the nutritional environment is not optimal is protective to the survival of the species.
- Optimal birthweight is considered to be 3.01–3.6 kilograms.
- Women with a low pre-pregnancy weight and those with multiple foetuses are advised to gain more weight in pregnancy.
- Although maternal weight gain in the first trimester is small, it is associated with a positive outcome of pregnancy.
- Consumption of adequate high-quality protein during pregnancy is important for optimal birthweight.
- Micronutrient intakes of folate and other B vitamins, iodine, iron, calcium and vitamin D appear to be particularly important in foetal development.

REFERENCES

Brough, L., Jin, Y., Shukri, N.H., Wharemate, Z.R. et al. 2013. Iodine intake and status during pregnancy and lactation before and after government initiatives to improve iodine status, in Palmerston North, New Zealand. *Maternal and Child Nutrition.* <http://onlinelibrary.wiley.com/doi/10.1111/mcn.12055/abstract> (subscription required), accessed 2 October 2014.

Campbell-Brown, M. & Hytten, F. 1998. Nutrition. In Chamberlain, G., Broughton Pipkin, F., (eds). *Clinical Physiology in Obstetrics*, 165–91. Blackwell Science, Oxford.

Charlton, K.E., Yeatman, H., Brock, E., Lucas, C. et al. 2013. Improvement in iodine status of pregnant Australian women 3 years after introduction of a mandatory iodine fortification programme. *Preventive Medicine* 57: 26–30.

Donangelo, C.M. & King, J.C. 2012. Maternal zinc intakes and homeostatic adjustments during pregnancy and lactation. *Nutrients* 4: 782–98.

Duggleby, S.L. & Jackson, A.A. 2002. Protein, amino acid and nitrogen metabolism during pregnancy: How might the mother meet the needs of her fetus? *Current Opinion in Clinical Nutrition and Metabolic Care* 5: 503–9.

FAO (Food and Agriculture Organization) & WHO (World Health Organization) 2001. Human Vitamin and Mineral Requirements. FAO, Rome. <www.fao.org/docrep/004/Y2809E/y2809e00.htm#Contents>, accessed 13 November 2014.

Hollis, B.W. & Wagner, C.L. 2013. Vitamin D and pregnancy: Skeletal effects, nonskeletal effects, and birth outcomes. *Calcified Tissue International* 92: 128–39.

Institute of Medicine 2006. *Dietary Reference Intakes: The essential reference for dietary planning and assessment.* National Academy Press, Washington, DC. <www.iom.edu/Reports/2006/Dietary-Reference-Intakes-Essential-Guide-Nutrient-Requirements.aspx>, accessed 13 November 2014.

Leung, A.M., Pearce, E.N. & Braverman, L.E. 2011. Iodine nutrition in pregnancy and lactation. *Endocrinology and Metabolism Clinics of North America* 40: 765–77.

McKenzie-Parnell, J.M., Wilson, P.D., Parnell, W.R., Spears, G.F. et al. 1993. Nutrient intake of Dunedin women during pregnancy. *New Zealand Medical Journal* 106: 273–6.

McLennan, W. & Podger, A. 1998. *National Nutrition Survey: Nutrient intakes and physical measurements, Australia, 1995* new issue. Cat. no. 4805.0. ABS, Commonwealth of Australia. <www.ausstats.abs.gov.au/ausstats/subscriber.nsf/0/CA25687100069892CA25688900268A6D/$File/48050_1995.pdf>, accessed 28 September 2014.

Martin-Gronert, M.S. & Ozanne, S.E. 2012. Mechanisms underlying the developmental origins of disease. *Reviews in Endocrine and Metabolic Disorders* 13: 85–92.

Ministry of Health 2014. *Nutrition During Pregnancy.* New Zealand Government. <www.health.govt.nz/your-health/healthy-living/pregnancy/nutrition-during-pregnancy>, accessed 13 November 2014.

NHMRC (National Health and Medical Research Council) & Ministry of Health 2006. *Nutrient Reference Values for Australia and New Zealand: Including recommended dietary intakes.* Commonwealth of Australia & New Zealand Government. <www.nhmrc.gov.au/_files_nhmrc/publications/attachments/n35.pdf>, accessed 26 September 2014.

Rasmussen, K.M. & Yaktine, A.L. (eds) 2009. *Weight Gain During Pregnancy: Reexamining the guidelines.* National Academy Press, Washington, DC. <http://iom.edu/Reports/2009/Weight-Gain-During-Pregnancy-Reexamining-the-Guidelines.aspx>, accessed 14 November 2014.

RANZCOG (Royal Australian and New Zealand College of Obstetricians and Gynaecologists) 2013a. *Pre-pregnancy Counselling.* Statement C-OBS 3 (a). <www.ranzcog.edu.au/doc/pre-pregnancy-counselling.html>, accessed 13 November 2014.

——2013b. *Vitamin and Mineral Supplementation and Pregnancy.* Statement C-OBS 25. <www.ranzcog.edu.au/doc/vitamin-and-mineral-supplementation-in-pregnancy.html>, accessed 13 November 2014.

Skeaff, S.A. 2011. Iodine deficiency in pregnancy: The effect on neurodevelopment in the child. *Nutrients* 3: 265–73.
University of Otago & Ministry of Health 2011. *A Focus on Nutrition: Key findings of the 2008/09 New Zealand Adult Nutrition Survey*. New Zealand Government. <www.health.govt.nz/system/files/documents/publications/a-focus-on-nutrition-v2.pdf>, accessed 2 October 2014.
Zeisel, S.H. 2013. Nutrition in pregnancy: The argument for including a source of choline. *International Journal of Women's Health* 5: 193–9.

ADDITIONAL READING

Blackburn, S.T. 2007. *Maternal, Fetal, and Neonatal Physiology: A clinical perspective* 3rd edn. Elsevier, St Louis, MO.
Coad, J. 2011. *Anatomy and Physiology for Midwives* 3rd edn, with Dunstall, M. Elsevier, London.
Symonds, M.E. & Ramsay, M.M. 2010. *Maternal-Fetal Nutrition during Pregnancy and Lactation.* Cambridge University Press, Cambridge.
Williamson, C.S. 2006. Nutrition in pregnancy. *Nutrition Bulletin* 31: 28–59.

TEST YOUR UNDERSTANDING

Note: There is only one correct answer for each question.

1. Which pregnant women are particularly at risk of a low calcium intake?
   a. Older women who are approaching menopause
   b. Pregnant adolescents and women who do not consume dairy products
   c. Women who have a high intake of iron
   d. Women who have hypertensive disorders
   e. Women who have lightly pigmented skin colour
2. With what is deficiency of vitamin B12 in pregnancy associated?
   a. High intake of animal protein
   b. Lower risk of impaired erythropoiesis and maternal anaemia
   c. Increased risk of impaired angiogenesis
   d. Increased risk of foetal neurological abnormalities
   e. Exacerbation of maternal GDM
3. Which of the following statements regarding nausea and vomiting in pregnancy is true?
   a. It is usually associated with a poor outcome of pregnancy
   b. It occurs only first thing in the morning before breakfast
   c. It is not affected by foods
   d. It usually resolves after the first trimester and is positively associated with a healthy pregnancy
   e. It always requires hospitalisation and treatment with electrolytes
4. In pregnancy, what is the effect of raised progesterone levels on maternal physiology?
   a. Raised blood pressure
   b. Reduced chemoreceptor sensitivity to carbon dioxide, so pregnant women tend to breathe more slowly
   c. Relaxation of the smooth muscle tone, which affects blood vessels and the gut
   d. Increased gut motility and a faster transit time
   e. Reduced absorption of water in the colon, so diarrhoea is more likely
5. Why is a birthweight of 3.01–3.6 kilograms considered optimal?
   a. There is a lower risk of infant morbidity and mortality in this range of birthweights
   b. Larger infants tend to have an increased risk of impaired neurocognitive development
   c. Smaller infants are likely to be macrosomic
   d. There is a higher risk of instrumental delivery and birth trauma in babies of lower birthweight
   e. The risk of chronic disease in adulthood is highest in this range of birthweights

6. Maternal changes in carbohydrate handling in pregnancy mean which of the following?
   a. Early embryonic development always occurs in an environment which has high levels of glucose
   b. Embryonic pancreatic function controls glucose homeostasis in the developing embryo
   c. A maternal diet rich in foods with high GI will tend to be protective against maternal GDM
   d. The mother tends to be insulin sensitive in early pregnancy and insulin resistant in late pregnancy
   e. Glucose is never used as a foetal energy substrate
7. Why do pregnant women have high requirements for long-chain polyunsaturated fatty acids?
   a. Oestrogen inhibits the expression of the enzymes that produce long-chain polyunsaturated fatty acids
   b. Alcohol influences the maternal synthesis of long-chain polyunsaturated fatty acids
   c. They are advised to avoid consuming fish
   d. The placenta cannot transport long-chain polyunsaturated fatty acids to the foetus
   e. The foetus has a high requirement for long-chain polyunsaturated fatty acids for its developing brain and nervous system
8. With what is folate deficiency in pregnancy often associated?
   a. Increased risk of macrosomia
   b. Reduced levels of homocysteine in the blood
   c. Increased risk of neural tube defects such as spina bifida
   d. Extended gestational length
   e. Increased risk of insulin-dependent diabetes
9. Why are vegetarian and vegan women are advised to consume at least five portions of coloured fruit and vegetables a day?
   a. This will ensure their intake of vitamin C is adequate
   b. They have a low intake of haem iron
   c. The conversion of carotenoids into retinal is not efficient, and they might become deficient in vitamin A
   d. Antioxidant intake is important for sperm survival
   e. This will compensate for a low selenium intake
10. With what is iron deficiency anaemia in pregnancy associated?
   a. LBW
   b. Increased risk of maternal morbidity and mortality
   c. Increased risk of infant anaemia
   d. Pre-eclampsia
   e. All of the above

STUDY QUESTIONS

1. Discuss the implications of nutritional programming for public health nutrition.
2. Discuss the issues relating to the assessment of iron status in pregnant women.
3. In relation to the nutritional factors which affect conception, identify key nutritional and lifestyle advice for a couple planning pregnancy.
4. What are the implications of the increasing prevalence of obesity for maternal and infant health?
5. What are the reasons for low iodine intakes within Australia and New Zealand, and what are the implications for pregnant women?

5

Lactation

Jane Scott and Jacqueline Miller

LEARNING OUTCOMES

Upon completion of this chapter you will be able to:

- describe the anatomy of the breast and the physiological processes involved in lactation
- describe the short- and long-term health outcomes associated with lactation
- identify the maternal energy and nutrient requirements of lactation
- describe the composition of breast milk in terms of nutrients and bioactive components.

Breastfeeding is the best way to feed an infant, and the short- and long-term health benefits of breastfeeding to both mothers and their infants are well established. This chapter introduces current recommendations for breastfeeding and the evidence that supports these recommendations. It also describes the physiology of lactation and the unique nutrition and immunoprotective properties of breast milk, considers breastfeeding from the mother's perspective, identifies nutrient requirements for lactation and discusses the impact of maternal diet on breast milk composition. Some of the common problems experienced by lactating women are described, and strategies for supporting and encouraging women to breastfeed are presented.

BREASTFEEDING

Breastfeeding is an unequalled way to feed infants. Human breast milk is species specific and is tailored to meet the nutritional needs of an infant for the first six months or so of life. In addition, it is a 'living' medium that contains a wide variety of immunoprotective factors that augment the immature immune system of the infant. The composition and concentration of these factors are unique to breast milk and, in some instances, specific to the individual **mother–infant dyad**; as such, they cannot be mimicked by infant formula milk.

The World Health Organization (WHO) and the United Nations Children's Fund (also known as UNICEF) recommend that 'infants should be

Mother–infant dyad is the term used to describe a mother and her infant.

exclusively breastfed for the first six months of life to achieve optimal growth, development and health. Thereafter, to meet their evolving nutrient requirements, infants should receive nutritionally adequate and safe complementary foods while breastfeeding continues for up to two years of age or beyond' (WHO & UNICEF 2003: 7).

This recommendation is reflected in the Australian Infant Feeding Guidelines (NHMRC 2012) and the New Zealand Food and Nutrition Guidelines for Healthy Infants and Toddlers (Ministry of Health 2008), which recommend that infants be exclusively breastfed for around the first six months of life and that breastfeeding should continue until the baby is twelve months old, or for as long as both the mother and infant want to keep going. It should be noted that this is a population-based recommendation, and the timing of the introduction of complementary foods (and subsequently the duration of exclusive breastfeeding) for an individual infant may differ slightly from the recommendation (see Chapter 6).

Exclusive breastfeeding is not widely practised, even in traditional societies. It has been estimated that globally fewer than four in ten infants under the age of six months are exclusively breastfed and that in Australia and New Zealand fewer than one in four infants are exclusively breastfed until around six months of age. The majority of Australian mothers partially breastfeed, with almost 70 per cent of infants being given formula by six months of age (AIHW 2011; Ministry of Health 2008). The terms used to describe the levels of intensity of breastfeeding are defined in Table 5.1.

HEALTH OUTCOMES ASSOCIATED WITH BREASTFEEDING

Short-term outcomes

The immunoprotective factors in breast milk confer protection against a number of infections common in infancy, such as gastroenteritis, diarrhoea, respiratory tract infections and otitis media (inner ear infection), as well as **necrotising enterocolitis** in premature infants (Ip et al. 2007). These infections are major causes of infant morbidity and mortality, particularly in poor countries, and infants who are exclusively formula fed have a greater risk of contracting them than exclusively breastfed infants. The protection conferred by breast milk is dose related, which means that partially breastfed infants are at greater risk of these infections than exclusively breastfed infants but have a lower risk than exclusively formula-fed infants. This protection extends beyond the period that the infant is breastfed.

The key message is that while exclusive breastfeeding for six months followed by continued breastfeeding up to two years and beyond offers the greatest protection against ill-health, any breastfeeding, no matter what the intensity and duration, is beneficial to the infant. Furthermore, the protection provided by breastfeeding is enjoyed not only by infants born into **disadvantage** in developing countries but also

Necrotising enterocolitis is a medical condition primarily seen in premature infants, in which portions of the bowel undergo necrosis (tissue death).

Table 5.1 Infant feeding definitions

Feeding practice	Requires the infant to receive	Allows the infant to receive	Does not allow the infant to receive
Exclusive breastfeeding	Breast milk (including milk expressed or from a wet nurse)	Oral rehydration solution, drops or syrups (vitamins, minerals, medicines)	Anything else
Predominant breastfeeding	Breast milk (including milk expressed or from a wet nurse) as the predominant source of nourishment	Certain liquids (water and water-based drinks, fruit juice), ritual fluids and oral rehydration solution, drops or syrups (vitamins, minerals, medicines)	Anything else (in particular, non-human milk, food-based fluids)
Complementary feeding or 'partial' breastfeeding	Breast milk (including milk expressed or from a wet nurse)	Anything else: any food or liquid including non-human milk and formula	Not applicable
(Any) breastfeeding	Any of the above definitions	Any of the above definitions	Not applicable

Source: WHO (2008: 4).

Disadvantage is generally experienced by those who are less educated, have a low income and may be unemployed. A high proportion of people experiencing disadvantage may rely on some form of welfare, and those who live with disadvantage often experience poor health at an earlier age than people who do not experience disadvantage (see Chapter 18).

by infants born to mothers living in industrialised countries such as Australia and New Zealand.

Long-term outcomes

There is convincing evidence that breastfeeding is associated with improved cognitive performance and a reduced risk of sudden infant death syndrome and obesity (Ip et al. 2007). In addition, there is evidence linking breastfeeding with a reduced risk of long-term health conditions, including type 1 diabetes mellitus (T1DM), type 2 diabetes mellitus (T2DM), coeliac disease, inflammatory bowel disease, allergy and asthma.

Breastfeeding carries with it health benefits for the mother, and a history of lactation has been associated with a reduced risk of T2DM and breast and ovarian cancer. Breastfeeding may help women return to their pre-pregnancy weight, and on average breastfeeding mothers lose 2 kilograms more body weight over six months of lactation compared with mothers who do not breastfeed. Not breastfeeding and early cessation of breastfeeding have been associated with an increased risk of maternal postpartum depression.

Contraceptive

Breastfeeding has a contraceptive effect and is associated with delayed return of menstruation and suppressed fertility after menstruation has returned. The infant sucking on the nipple stimulates the release of the hormone prolactin and other pituitary hormones, which results in suppression of ovulation and menstruation. Therefore, frequent breastfeeding can help to delay a new pregnancy.

This method of birth control is known as the **lactational amenorrhoea method** and, while not promoted as a reliable form of contraception, it is an important method of natural contraception at the population level, especially in poor countries with limited or no access to more reliable forms of contraception. Breastfeeding in these countries contributes to child spacing through extending the interval between pregnancies. This in turn contributes to a reduction in infant mortality and malnutrition, as a short birth interval is associated with an increased risk of both.

Lactational amenorrhoea method is a natural birth control method based on the fact that lactation prevents menstruation in some women.

PHYSIOLOGY OF BREASTFEEDING

Breast development

Physiological preparation for lactation begins in **puberty**. Once ovulation and menstruation begin, the maturing of the breasts begins with the formation of secretory glands at the end of the ducts. The breasts and duct system continue to grow and mature with the development of many glands and lobules, and fibrous and fatty tissues increase around the ducts. These changes are driven by the production of oestrogen and progesterone. It usually takes from three to five years from the onset of **menarche** for the breast to fully develop; however, the system remains inactive until pregnancy (see Chapter 9).

The final preparations for lactation are completed during pregnancy, and breast changes (for example, tenderness and swelling) that are driven by hormonal changes are one of the earliest signs of pregnancy. In addition, the areolas (the dark areas of skin that surround the nipples of the breasts) begin to swell, followed by the rapid swelling of the breasts themselves. By the fifth or sixth month of pregnancy, the breasts are fully capable of producing milk.

Anatomy of the lactating breast

The breast structure (see Figure 5.1) includes the nipple and areola, mammary tissue, supporting connective tissue and fat, blood and lymphatic vessels, and nerves. The mammary tissue is a glandular tissue that makes and transports milk and includes the alveoli, which are small sacs made of milk-secreting cells, and the ducts

Puberty means sexual maturation, including increased production of sex hormones, development of secondary sex characteristics and reproductive system changes enabling sexual reproduction—for example, menarche in females.

Menarche is the first menstrual period, usually occurring during puberty.

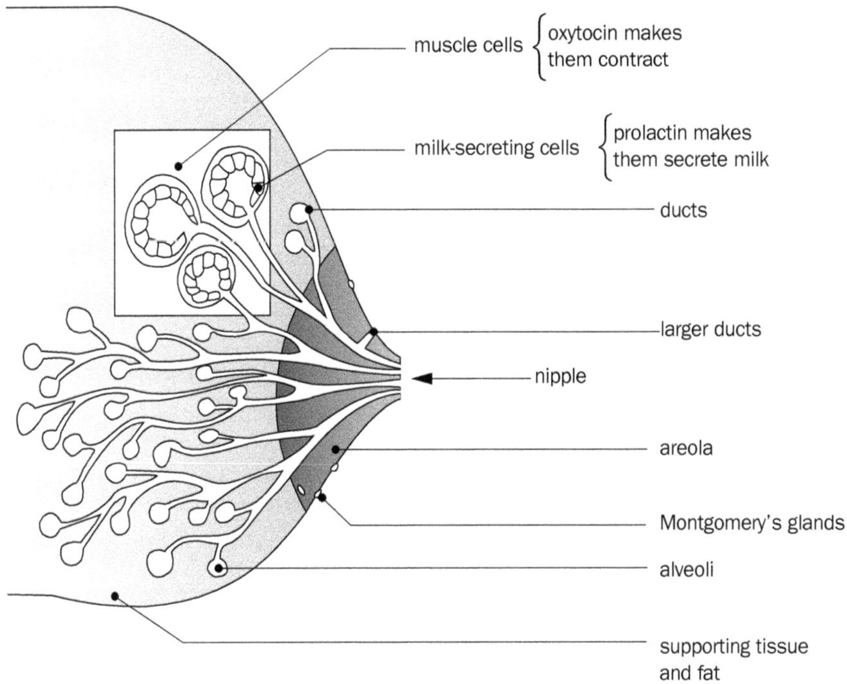

Figure 5.1 Anatomy of the lactating breast

Source: Wahlqvist (2011: 398) adapted from WHO (2009: 11).

that carry the milk to the outside. Between feeds, milk collects in the lumen of the alveoli and ducts. The alveoli are surrounded by a basket of myoepithelial, or muscle cells, which contract and make the milk flow along the ducts.

The nipple has an average of nine milk ducts passing to the outside, although this can range from four to eighteen ducts, and also muscle fibres and nerves. The ductal network is complex and not always arranged in a radial or symmetrical pattern as depicted. The nipple is surrounded by the circular pigmented areola. The ducts beneath the areola fill with milk and become wider during a feed, when the oxytocin reflex is active.

Blood nourishes the mammary tissue and provides the nutrients needed to make milk. Waste is removed via the lymphatic vessels. Nerves make the breast sensitive to touch and allow the baby's suck to stimulate the release of hormones that triggers the let-down and production of milk. Finally, the connective tissue supports the breast and the fatty tissue protects the mammary tissue.

Hormonal control of breast milk production

There are two hormones that directly affect breast-feeding: prolactin and oxytocin. When a baby suckles at the breast, sensory impulses pass from the nipple to the hypothalamus in the brain, which produces prolactin that is secreted via the anterior lobe of the pituitary gland and oxytocin that is secreted via the posterior lobe (see Figure 5.2). In addition to controlling breast milk production and milk flow, both hormones act on the reproductive organ with prolactin inhibiting ovulation and oxytocin promoting uterine contractions.

Prolactin stimulates the alveoli to produce milk. The prolactin level is highest about 30 minutes after the beginning of the feed, so its most important effect is to make milk for the next feed. The more a baby suckles and stimulates the nipple, the more prolactin is produced, and the more milk is produced. Thus, unrestricted suckling, or feeding on demand, is particularly important in the first six or so weeks after the birth of the infant, when lactation is becoming established. More prolactin is produced at night, so

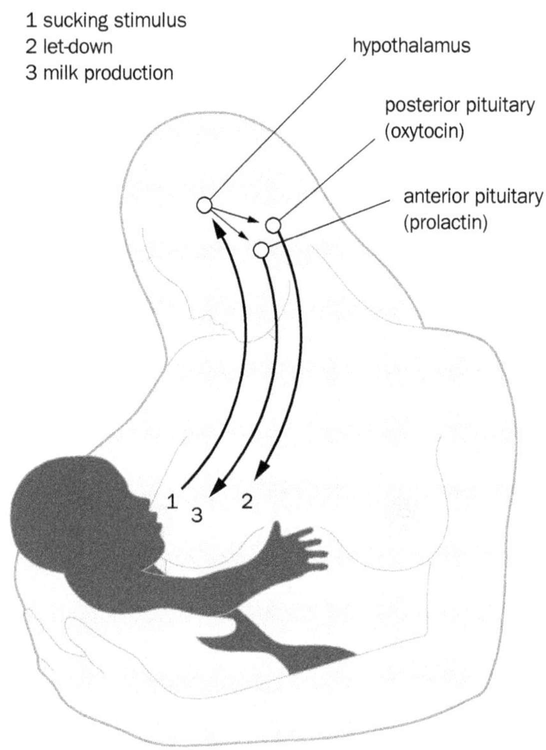

Figure 5.2 The milk production and let-down reflex

Source: Wahlqvist (2011: 399).

breastfeeding at night is especially helpful for keeping up the milk supply.

Oxytocin makes the muscle cells around th alveoli contract, causing the milk that has collected in the alveoli to flow along and fill the ducts. It makes the milk that is already in the breast flow for the current feed and helps the baby to get the milk easily. Oxytocin release can be conditioned and starts working when a mother anticipates a feed as well as when the baby is suckling, and the reflex can be triggered by the mother touching, smelling or seeing her baby or hearing her baby cry.

Position and attachment

Correct positioning and attachment are the secrets to establishing successful breastfeeding and if not achieved are the root causes of most problems experienced by breastfeeding women. First-time mothers need to be shown how to correctly position and attach their infant to the breast. Ideally, before leaving hospital mothers should be able to correctly position the baby at the breast and not experience any pain during feeding.

Examples of good and poor attachment are provided in Figure 5.3. The infant on the left has a good mouthful of breast, her chin is touching the breast, her mouth is wide open with the bottom lip curled back, and more of the areola is visible above her top lip than below her bottom lip. In contrast, the infant on the right has her chin away from the breast,

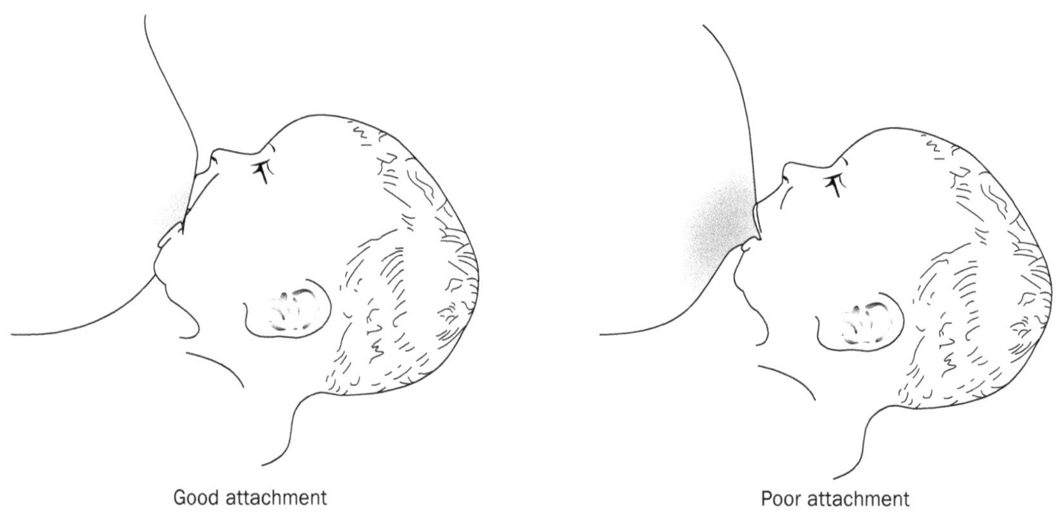

Figure 5.3 External signs of good and poor attachment

Source: Wahlqvist (2011: 414), adapted from WHO (2009: 14).

her mouth is not wide open, her bottom lip points forward or is turned inwards, and most of the areola is exposed, with more of it visible below her bottom lip than above her top lip.

BREAST MILK VOLUME AND NUTRITIONAL COMPOSITION

Infant demand is the main process for regulating milk production, with supply matching demand. Milk volume is not determined by breast size, as fatty tissue accounts for most of the difference in breast sizes between women. Women with smaller storage capacity (usually those with small breasts) will generally feed more frequently in order to produce a similar volume of milk as that produced by women with larger breast capacity. They may also need to feed from both breasts at a single breastfeed to satisfy their infant's appetite, whereas women with larger breast capacity may need to feed from only one breast at each breastfeed.

Traditionally, the average rate of milk production was considered to be 700–800 millilitres per 24 hours, and it was thought that the level remained relatively constant over the period of lactation. This rate, however, was based on studies using an indirect method of assessing infant milk intake. More recent studies, using **doubly labelled water**, a direct method, suggest that infant milk intake during exclusive breastfeeding is higher than previously thought, being in the range of 900–1000 millilitres per 24 hours, and that volume increases significantly over the period of lactation to match the needs of the growing infant.

The volume of milk produced by women is surprisingly consistent across population groups from a wide range of nutritional conditions. It appears that milk production is maintained at the expense of maternal stores, and even women in developing countries are able to produce about 700 millilitres of breast milk per day as long as they are not suffering severe protein energy malnutrition.

The composition of breast milk varies as milk matures. Colostrum is a thick yellow fluid that is secreted in the first two to three days after delivery. On the second or third day milk starts to be produced in larger amounts and at first appears thin and watery but becomes creamier over time. From day 7 to 14 the milk is called transitional and after two weeks it is called mature milk.

> **Doubly labelled water** has both oxygen and hydrogen partly or completely replaced with non-radioactive variants so that the elements or molecules can be traced.

Colostrum

Colostrum per 100 millilitres is lower in energy, lactose and fat but higher in protein, minerals and fat-soluble vitamins (A, E and K) than mature milk. It is produced in small amounts, about 40–50 millilitres on the first day, and infants may drink only 1–2 teaspoons per feeding; frequent feeding is therefore important. A healthy, term baby is well hydrated and nourished from being in the amniotic fluid and receiving all their sustenance from the placenta. The low volume of colostrum is specially designed not to overfill an infant's very small stomach, but it contains everything they need for the first few days of life.

Colostrum appears to play more of an immunoprotective than a nutritional role, as it is rich in cells and antibodies, in particular secretory immunoglobulin A (see Table 5.2). These and other bioactive components confer important immune protection when an infant is first exposed to microorganisms in the environment, and epidermal growth factor helps to prepare the lining of the gut to receive the nutrients in milk. It also eases the movement of meconium, a thick, sticky, greenish-black substance which constitutes the newborn infant's first bowel motion. For these reasons, it is important that infants receive colostrum and not other feeds at this time.

Macronutrients

Protein

The protein content of mature breast milk (0.8–1 per cent) is lower than that of milk from cows and other mammals and is tailored to support the relatively slow growth of human infants compared to the offspring of other species. Human milk also contains approximately 25 per cent of non-protein nitrogen (sources of nitrogen other than protein), which comprises free amino acids, peptides, nucleotides, N-acetyl sugars, glycosylated amines, urea, creatinine and traces of other organic compounds. The significance of all these compounds is unknown, although free amino acids supply a ready source of predigested protein for the infant. Nucleotides appear to play a role in gastrointestinal maturation and may be involved in the growth of *Bifidobacteria* (see Table 5.2) and also enhance lymphocyte proliferation and stimulate immunoglobulin production.

Table 5.2 Selected protective and bioactive factors present in human milk

Factor	Function
Alpha-lactalbumin	Expressed only in the lactating mammary gland and kills *Streptococcus pneumoniae* in vitro
Anti-giardia	Provides anti-giardia action
Antioxidants	Protect from free radical damage
Anti-staphylococcus factor	Provides anti-staphylococcal action
Beneficial bacteria	*Bifidobacteria* and *Lactobacillus* are beneficial bacteria that compete for and reduce colonisation of the gut with pathogenic bacteria
Bifidus factor	Stimulates lactic acid bacteria, such as *Bifidobacteria*, in the colon
Bile-salt-stimulated lipase	Improves fat digestion
Complement	Assists in bacterial lysis
DHA and arachidonic acids	Constituents of cell membranes in brain and neural tissue
Glycans	Human milk glycans act as 'decoys' for enterocyte glycans and bind to pathogens including *Escherichia coli*, *Shigella* and *Campylobacter jejuni*, rendering them harmless as they are passed out in the faeces
Immune cells	Cells such as macrophages engulf pathogenic bacteria and lymphocytes that secrete immunoglobulin (B cells) and lymphokines (T cells)
Interferon	Acts as an antiviral agent
Lysozyme	Antibacterial enzyme that damages the cell walls of pathogens
Oligosaccharides	Inhibits the binding of enteric and respiratory bacterial pathogens to epithelium, encourages the growth of *Bifidobacteria* and *Lactobacillus*
Secretory immunoglobulin A	Interferes with the attachment of microbial pathogens to epithelial surfaces and neutralises toxins or virulence factors, may actively prime the neonate's immune system
Trophic factors	Accelerate gut development
Vitamin B12 and folate binding proteins	Compete with bacteria for these vitamins

Note: DHA: docosahexaenoic acid.

Carbohydrate

Lactose is the primary carbohydrate in breast milk, and the lactose content of human milk is tightly controlled at 7 per cent. As well as being an important source of energy, lactose improves calcium absorption.

Oligosaccharides comprise the third largest component of breast milk, and concentration varies from 20–5 grams per litre in colostrum to 5–20 grams per litre in mature milk. Human milk oligosaccharides inhibit the binding of gut and respiratory bacterial pathogens to the lining of the intestine and encourage the growth of beneficial gut bacteria. It is worth noting that the structural characteristics of human milk oligosaccharides determine functionality, and oligosaccharides currently added to infant formula are structurally different from the oligosaccharides naturally occurring in human milk.

Fat

Fat provides about half the energy of human milk and is the most variable macronutrient in terms of both concentration (3–5 per cent in mature milk) and composition. Milk fat concentration is positively correlated with measures of body fatness; that is, the breast milk of overweight or obese women is higher in fat than that of healthy weight or underweight women. The fatty acid composition of the fat reflects the maternal diet and so varies between individuals. In addition, during the course of a breastfeed the fat content very gradually increases, such that the **hind milk** contains more fat than the **fore milk**.

Micronutrients

As previously discussed, unless a mother is deficient, breast milk will contain adequate levels of vitamins and minerals to meet the needs of a healthy, term infant. While iron and zinc are low in breast milk, both are highly bioavailable, and levels are sufficient to meet the

Hind milk is the breast milk secreted towards the end of a feed.

Fore milk is the breast milk secreted at the beginning of a feed.

CASE STUDY 5.1 GENERAL BREASTFEEDING

Profile and problem

K. is a 29-year-old woman who is 34 weeks pregnant with her first child. She is 165 centimetres tall, her pre-pregnancy weight was 63 kilograms, and her current weight is 74 kilograms. Up until the last few weeks K. was exercising regularly. She was formula-fed as an infant, and her sister had difficulties breastfeeding her baby and gave up a few days after leaving hospital.

Assessment

K.'s pre-pregnancy body mass index (BMI) was in the healthy weight range, and her gestational weight gain is acceptable. However, she is worried about the amount of weight she has gained and is keen to get back in shape as soon as possible. She is concerned about how much extra she will need to eat if she breastfeeds her baby. There is no history of breastfeeding among her immediate female relatives, and she is unsure if she will breastfeed given the difficulties that her sister had.

Treatment

K. has agreed to do the following:

- discuss her breastfeeding concerns with the hospital lactation consultant
- attempt to initiate breastfeeding
- breastfeed her baby on demand
- contact local support services when she has any difficulty with breastfeeding.

Plan

K. has been advised that she will receive practical breastfeeding assistance from hospital staff following the birth of her baby and has been given a list of breastfeeding support services to contact if necessary, including the Australian Breastfeeding Association and a local lactation consultant. She has also been encouraged to join a new mothers' breastfeeding support group once she has had her baby. K. has not gained an excessive amount of weight to date, and she will be able to exercise regularly without compromising her breast milk supply. Breastfeeding can help women return to their pre-pregnancy weight, and the amount of weight loss is determined by the duration and exclusivity of breastfeeding. Exercise and moderate weight loss of approximately 0.5 kilograms per week will not compromise breastfeeding success.

needs of healthy, term infants for the first six months or so of life.

Protective and bioactive components of breast milk

Previously, it was thought that breastfed infants suffered fewer infections simply because milk supplied directly from the breast is relatively free from bacteria. In comparison, formula, if made from contaminated water and/or under unhygienic household conditions, can be a vehicle for bacteria and thus infection. Breast milk, however, contains a variety of non-nutritive substances, including antimicrobial, anti-inflammatory and immunomodulatory agents, which are absent in formula, that not only offer passive protection but also stimulate the development of the infant's immature immune system (Labbok et al. 2004; see Table 5.2). The protection provided by these factors is uniquely valuable for an infant. For instance, secretory immunoglobulin A contains antibodies that are formed by the mother against the bacteria in her gut and against infections that she has encountered, so they protect against bacteria that are particularly likely to be in the baby's environment.

NUTRIENT REQUIREMENTS AND INTAKES

A mother's nutrient requirements are increased during lactation and are considerably greater than the requirements of pregnancy. Not only must a woman eat sufficiently to meet her own nutrient requirements, but she must meet all those of her growing infant. Some of the energy and many of the nutrients stored during pregnancy are available to support milk production. While a woman's diet does not affect appreciably the concentrations of lactose, protein and fat in breast milk, the concentration of a number of vitamins and minerals is determined by maternal intake.

The recommended intakes for energy and selected nutrients during lactation are listed in Table 5.3. Most

Table 5.3 RDIs of selected nutrients for lactating women and non-pregnant, non-lactating women in Australia and New Zealand

	Amount of nutrient, by lactation status	
Nutrient	**Lactating**	**Non-pregnant, non-lactating**
Protein (g/kg)	1.11 (14–18 years) 1.10 (19–30 years)	0.77 (14–18 years) 0.75 (19–30 years)
Vitamin A (retinol equivalents) (μg)	1100	700
Thiamin (mg)	1.4	1.1
Riboflavin (mg)	1.6	1.1
Niacin equivalents (mg)	17	14
Vitamin B6 (mg)	2.0	1.2
Vitamin B12 (μg)	2.8	2.4
Vitamin C (mg)	80 (14–18 years) 85 (19–30 years)	40 (14–18 years) 45 (19–30 years)
Iodine (μg)	220	150
Iron (mg)	10 (14–18 years) 9 (19–30 years)	15 (14–18 years) 18 (19–30 years)

Note: RDI: recommended dietary intake. Nutrient reference values are the same for females aged 14–18 years and females aged 19–30 years unless otherwise specified.

Source: NHMRC & Ministry of Health (2006).

of these requirements are based on the average amount of breast milk produced during the first six months of lactation to support an exclusively breastfed infant and what is known about the energy and nutrient content of breast milk. Nutritional demands are directly proportional to the intensity (level of exclusivity) of breastfeeding, and actual requirements will be less if a woman is only partially breastfeeding.

Energy

In determining the energy requirements for lactating women, Australia's National Health and Medical Research Council and the New Zealand Ministry of Health adopted the factorial method employed by the Food and Agriculture Organization of the United Nations, the WHO and the United Nations University. The theoretical energy cost of lactation is dependent on how much milk is produced and discharged, the energy content of the milk, and how efficiently the milk is produced (SACN 2012). While it is difficult to make a single recommendation for energy due to variations in milk production (individual variations between women, stage of lactation and intensity of breastfeeding), weight loss during lactation and changes in physical activity level (PAL), the recommended average additional requirement in lactation has been set at an extra 2000–2100 kilojoules per day, assuming exclusive breastfeeding in the first six months and partial breastfeeding thereafter (NHMRC & Ministry of Health 2006).

Macronutrients

Protein

Protein requirements during lactation have been calculated by estimating the amount of dietary protein needed to support production of a given amount of protein in milk. Achieving an adequate protein intake is unlikely to be a problem for women living in Australia and New Zealand, where average daily protein intakes already exceed requirements, but may be a problem for disadvantaged women living in the developing world. Despite this, there is not much difference in milk protein concentration even with varying levels of maternal protein intake, suggesting that the concentration is maintained at the expense of maternal stores.

Carbohydrate

Lactose is the primary carbohydrate in breast milk, and the concentration is affected little by maternal diet. There are more than 100 structurally different human milk oligosaccharides in breast milk, but not every woman synthesises the same set. The wide variations in the amount and composition of human milk oligosaccharides between women are explained in part by genetic variations, and there is speculation that other factors such as maternal diet and environment may also influence oligosaccharide amount and composition.

Fat

The fatty acid composition of breast milk is variable and influenced by the mother's diet. Fish-eating communities have higher docosahexaenoic acid (DHA) levels than the average Western diet; however, even vegetarians have some DHA in their milk, suggesting that it might be synthesised maternally from polyunsaturated fatty acid precursors. Breast milk levels of DHA can be elevated through supplementation. This is advantageous because evidence suggests that this translates to higher DHA levels in infants (Gibson et al. 1997), which has been linked to enhanced visual acuity maturation and cognitive function (Innis et al. 2001).

Micronutrients

During lactation the concentration of a number of vitamins—A, thiamin, riboflavin, niacin, B6, B12, C and D—and the mineral iodine are at least partially dependent on maternal diet and stores. The mammary gland can concentrate other nutrients (for example, calcium) from the mother's plasma into the milk, drawing from maternal reserves if necessary. Breast milk usually contains adequate amounts of vitamins and minerals to meet the needs of a healthy, term infant unless the mother herself is deficient.

Vitamin A

The content of vitamin A in human milk depends on maternal stores. In Australia and New Zealand low concentrations in breast milk are unlikely to be a problem, as it is widely available as retinol in animal-based foods and as its precursor beta-carotene in plant-based foods. Vitamin A deficiency, however, remains a problem for women and infants in many developing countries, and women lacking food sources of vitamin A demonstrate low concentrations in their breast milk. Postnatal supplementation of mothers may improve maternal nutritional status and infant stores of vitamin A.

B vitamins

Vitamins thiamin, riboflavin, niacin, B6 and B12 are readily transported across the mammary gland, and the concentration of these nutrients is unlikely to be low in the breast milk of women eating a varied diet that includes fortified grains (for example, breakfast cereals) and animal products. However, the concentration in breast milk of these vitamins is dependent on maternal diet, and deficiencies of these nutrients can result in a reduced secretion of them in breast milk, resulting in nutrient depletion in the infant. Women consuming primarily grain-based diets that are not fortified with B vitamins and diets low in animal products and thus in vitamin B12 intake may produce milk with low concentrations of these B vitamins. Similarly, women who have had an intestinal injury or gastric bypass or are consuming gluten-free diets that are not replacing the B vitamins usually obtained from wheat products could also be at risk.

Vitamin D

The vitamin D content of human milk is low even among women who are not deficient; nevertheless, the incidence of rickets among breastfed infants is rare, and most infants appear to receive adequate vitamin D through breast milk and casual exposure to sunlight. The vitamin D level in breast milk is closely related to maternal vitamin D status, and levels are particularly low in the milk of women with dark skin pigmentation, those who cover themselves for cultural or religious reasons and those living at high latitudes, due to limited conversion of vitamin D in the skin. It is prudent to recommend that women at high risk of deficiency take a supplement of vitamin D of 10 micrograms each day (Munns et al. 2006).

Iodine

Breast milk iodine concentrations are very sensitive to maternal status, and the re-emergence of mild to moderate iodine deficiency in pregnant women in some Australian states has led to renewed interest in this nutrient. In general, women who live where iodised salt programs exist appear to have reasonable iodine status; however, increased awareness of the association between high salt consumption and hypertension has resulted in the reduced use of iodised salt. The National Health and Medical Research Council recommends that pregnant and breastfeeding women take an iodine supplement of 150 micrograms each day (NHMRC 2010). Women with pre-existing thyroid disease should seek medical advice before taking a supplement.

Healthy eating plan for lactating women

The Australian Dietary Guidelines recommends the following daily servings from the five food groups:

- 7½ vegetables and legumes/beans
- 2 fruit
- 9 grain (cereals), mostly wholegrain and/or high-fibre varieties

- 2½ lean meats and poultry, fish, eggs, tofu, nuts and seeds, and legumes and beans
- 2½ milk, yoghurt and cheese, mostly low fat. (NHMRC 2013)

These servings will provide the nutrients and energy needed by an exclusively breastfeeding adult woman of average height with a sedentary to moderate activity level. (For more examples of serves see Chapter 1.) An example of a daily eating plan which provides these recommended servings across three meals and three snacks is provided in Table 5.4. It is worth noting that this eating plan, based on the Australian Dietary Guidelines, is just one example of how a lactating woman could meet her nutrient needs. However, some women can find it challenging to eat this many serves of vegetables and may choose, for example, to replace a serve of vegetables with a serve of fruit. Furthermore, this is the amount of food needed to provide the nutrient requirements of a woman who is exclusively breastfeeding, and fewer servings of grains and vegetables will be required if a woman is only partially breastfeeding.

FORMULA FEEDING

Economic and environmental costs

The higher rates of infection which accompany formula feeding, even in affluent countries, are associated with increased costs to the healthcare system in the form of increased frequency of primary health service utilisation and associated prescriptions of medications such as antibiotics, along with higher rates of hospitalisation for treatment of diarrhoea, respiratory infections and inner ear infections.

Human milk is a 'natural resource' and as such has an economic value. At the household level, in an affluent country the cost of exclusively formula feeding an infant is roughly double that of buying the extra food needed by a woman to meet the additional nutritional demands of breastfeeding and may account for one-third or more of the monthly wage of a salaried worker in poorer countries.

Breast milk is an environmentally friendly foodstuff and is produced and delivered to the infant without any pollution, unnecessary packaging or waste. On the other hand, there are environmental costs associated

Table 5.4 Sample daily menu for lactating adult women

Meal	Food	Vegetables	Fruit	Grain	Meat	Dairy
Breakfast	½ cup (45g) muesli			1½		
	1 banana		1			
	1 cup reduced fat milk					1
Mid-morning snack	¼ cup hummus (chickpea dip)	½				
	1 medium carrot	1				
	4 plain wholegrain crackers			1		
Lunch	1 small can tuna				1	
	2 cups leafy green and raw salad vegetables	2				
	2 slices wholegrain bread			2		
	¼ cup (30g) nuts				½	
Mid-afternoon snack	1 cup baked beans	2				
	2 slices wholegrain toast			2		
Dinner	1 cup cooked pasta			2		
	½ medium chicken breast				1	
	1 cup tomato, eggplant, capsicum and zucchini (ratatouille)	2				
	1 tbsp (20 g) grated cheese					½
Evening snack	1 piece of fruit		1			
	200 ml tub yoghurt (sprinkled with muesli)					1
	1 tbsp (15 g) muesli			½		
Total		**7½**	**2**	**9**	**2½**	**2½**

with the production and transport of infant formula milk and the production and destruction of formula packaging and the paraphernalia associated with formula feeding such as bottles, teats and sterilising equipment.

Acceptable medical reasons for use of infant formula

There are very few instances when a woman should be advised either to not breastfeed or to temporarily discontinue breastfeeding.

Maternal conditions

Medication

Typically, less than 1–2 per cent of a maternal dose of drugs is excreted into breast milk, which rarely poses a risk to the infant. Therefore, mothers can continue to breastfeed while taking most medications prescribed in Australia and New Zealand, notable exceptions being cytotoxic chemotherapeutic drugs and radioactive iodine-131. There are a number of relatively rare medical conditions for which the decision should be made on a case-by-case basis to either permanently or temporarily discontinue breastfeeding while a mother receives active drug therapy. These include the following:

- severe illness that prevents a mother from caring for her infant, e.g., sepsis
- active tuberculosis
- breast cancer treatment
- syphilis lesions of the breast or nipples
- herpes simplex virus type 1
- untreated brucellosis.

Human immunodeficiency virus/Acquired immune deficiency syndrome

While mother-to-child transmission is the primary route by which children acquire human immunodeficiency virus (HIV), the rate of this transmission is less than 20 per cent. With the use of antiretroviral therapy given either to an HIV-infected mother or an HIV-exposed infant the risk of postnatal transmission of HIV through breastfeeding is reduced significantly. There is evidence that exclusive breastfeeding carries less of a risk of transmission than mixed feeding and it is thought that components in infant formula milk increase intestinal permeability to the HIV virus.

In countries where antiretroviral therapy is unavailable and replacement feeding with infant formula milk is not acceptable, feasible, affordable, sustainable and safe, then consideration has to be given to protecting the infant from the risk of death associated with not breastfeeding (see above). All breastfeeding should stop once a nutritionally adequate and safe diet without breast milk can be provided. Key recommendations from the WHO's Guidelines on HIV and Infant Feeding are summarised in the box above.

Breastfeeding and transmission of human immunodeficiency virus

Mothers known to be HIV infected should only give commercial infant formula milk as a replacement feed to their HIV-uninfected infants or infants who are of unknown HIV status, when these specific conditions are all met:

- Safe water and sanitation are assured at the household level and in the community.
- The mother or other caregiver can reliably provide sufficient infant formula milk to support normal growth and development of the infant.
- The mother or caregiver can prepare it cleanly and frequently enough so that it is safe and carries a low risk of diarrhoea and malnutrition.
- The mother or caregiver can, in the first six months, exclusively give infant formula milk.
- The family is supportive of this practice.
- The mother or caregiver can access health care that offers comprehensive child health services.

In countries such as Australia and New Zealand, where replacement feeding (that is, infant formula milk) and antiretroviral therapy are available, it is recommended that mothers either breastfeed and receive antiretroviral interventions, or avoid all breastfeeding, as the strategy will most likely give infants the greatest chance of HIV-free survival. (WHO 2010: 7)

Infant conditions

With guidance and careful monitoring by a team of specialist health professionals—typically a metabolic consultant and specialist dietitian—infants with some inborn errors of metabolism such as phenylketonuria or maple syrup urine disease can receive some breast milk supplemented with a special formula specific to their condition. The decision to breastfeed should be made on a case-by-case basis.

ALCOHOL, SMOKING AND OTHER DRUGS

Maternal use of nicotine, alcohol, ecstasy, amphetamines, cocaine and related stimulants has been demonstrated to have harmful effects on breastfed babies. While mothers should be encouraged not to use these substances, smoking and the moderate use of alcohol, in particular, should not prevent a mother from breastfeeding. Mothers taking illicit drugs and planning to breastfeed should be advised and managed on a case-by-case basis, depending on their plans and ability to abstain from illicit drug use and their access to consistent antenatal and postnatal care.

Alcohol enters the breast milk and may persist in the milk for several hours after consumption. Drinking two standard drinks or more a day while breastfeeding has been shown to negatively affect lactation performance, leading to earlier cessation of breastfeeding, and has been associated with disrupted infant sleep–wake behavioural patterns and deficits in infant psychomotor development.

The Australian Infant Feeding Guidelines (NHMRC 2012) provide practical advice for breastfeeding women who may encounter a social situation where they wish to drink (see box).

Breastfeeding and alcohol consumption

Breastfeeding mothers should be advised that not drinking is the safest option and, specifically, to consider not drinking alcohol in the first month after birth, until breastfeeding is well established.

For women who choose to drink after this time, advice should be provided on a recommended maximum level of consumption, the length of time that alcohol is excreted in breast milk and the optimal timing of breastfeeding in relation to intake. For instance:

- Women should breastfeed before drinking.
- Women should limit their consumption to two standard drinks or less in any one day.
- As a general rule it takes two hours for an average woman (60 kilograms) to clear the alcohol from one standard drink and therefore four hours for two drinks, six hours for three drinks and so on.
- The option of expressing prior to consuming alcohol should also be discussed. (NHMRC 2012: 67)

While maternal smoking has been consistently associated with a shorter duration of breastfeeding it has also been linked with a reduced likelihood to intend to breastfeed and to initiate breastfeeding. It is unclear, therefore, if there is a biological mechanism for the association with breastfeeding duration—for example, if some ingredient in cigarettes reduces breast milk production or if the mechanism is psychological, behavioural and/or cultural. Nevertheless, mothers should be advised of the general health risks to both themselves and their infant, and the potential impact on breastfeeding duration associated with smoking, and encouraged to quit.

However, mothers who choose to continue to smoke should still be encouraged to breastfeed, as studies have shown that breastfeeding modifies the adverse effects of smoking. For instance, infants exposed to environmental smoke by their mother's smoking have higher rates of respiratory tract infections than infants of women who do not smoke, but breastfeeding reduces the risk of respiratory tract infections in infants whose mothers smoke. Furthermore, an infant's exposure to environmental smoke should be minimised by asking smokers to smoke outside.

DETERMINANTS OF BREASTFEEDING INITIATION AND DURATION

Most women in Australia and New Zealand initiate breastfeeding, but by about six months roughly four in ten infants are no longer receiving any breast milk. In general, younger, less-educated women are the most likely to discontinue breastfeeding early. While in Australia there is no difference in breastfeeding rates between rural and urban women, in developing countries poorer, less-educated and rural women are more likely to initiate breastfeeding and to breastfeed for longer than urban and advantaged women in these countries. Breastfeeding rates among Māori and Pacific Islander mothers are lower than overall New Zealand rates, and similarly in Australia the initiation of breastfeeding by Aboriginal and Torres Strait Islander mothers living in non-remote areas is lower than that of non-Indigenous mothers.

PROMOTING AND SUPPORTING BREASTFEEDING

Australia was the first country to include a breastfeeding guideline in its dietary guidelines for adults. This recognises the public health importance of breastfeeding

CASE STUDY 5.2 MASTITIS IN BREASTFEEDING *Adrienne Forsyth*

Profile and problem

Mrs T., a 29-year-old woman, recently gave birth to her first child. Her daughter, Little T., was born at 40 weeks' gestation, 3.8 kilograms in weight and 50 centimetres in length. The hospital midwives taught Mrs T. to breastfeed Little T. while they were in hospital, and Little T. regained the weight she had lost since birth before going home on day 5. By day 7, Little T. was refusing to feed, and Mrs T. was experiencing flu-like symptoms and red swollen breasts with cracked nipples. Mrs T. contacted the lactation consultant, who identified mastitis, an infection in the breast tissue.

Assessment

Mrs T. wants to continue to breastfeed but is finding it too painful due to cracked nipples. Little T. is refusing to feed, and this is common for babies with mothers experiencing mastitis, possibly due to changes in the taste of the milk. Little T. needs a readily available source of nutrition and fluids, and Mrs T. needs rest and an opportunity for her nipples to heal.

Treatment

Mrs T. has agreed to do the following:

- continue to breastfeed if possible
- hand express or use a breast pump to express milk if Little T. refuses to feed
- provide expressed milk to Little T. in a dropper, cup or bottle
- apply warm compresses such as a warm damp cloth to the breasts prior to feeding or expressing, and massage hard red patches to release milk from blocked ducts while feeding
- recommence feeding, paying special attention to ensure that Little T. is attached well
- feed Little T. on demand
- use gel nipple pads between feeds to help soothe and heal cracked nipples.

Plan

Mrs T. will be in contact with the lactation consultant daily and return for follow-up within one week. If she has been unable to resume normal breastfeeding with Little T. they will consider other options, including the use of nipple shields, continuing to express or complementary feeding. Mrs T. will be provided with information about local breastfeeding groups where she can receive ongoing support.

and the role that we all have to play in encouraging and supporting women to breastfeed.

Social support for breastfeeding extends beyond that provided by a woman's partner and immediate social network and includes community support for breastfeeding. This might, for example, take the form of the following support measures:

- positive media promotion of breastfeeding both in entertainment and news programs
- workplace support in the form of maternity leave provision, workplace childcare facilities and breastfeeding breaks and comfortable, hygienic and private facilities where women can express and store their breast milk
- policies and legislation which recognise an infant's right to be breastfed in public.

Typically, we refer to the mother–infant dyad when we talk about breastfeeding, but this should be expanded to a triad that recognises the father's role in encouraging and supporting his partner to breastfeed. Studies have repeatedly shown paternal support for breastfeeding to be a strong predictor of breastfeeding success. Fathers have an important role to play in providing both the practical and the emotional support needed by their partners to successfully establish and continue breastfeeding.

The first few days after delivery are critical, and there are a number of hospital practices that can either support or undermine the establishment of breastfeeding. The Ten Steps to Successful Breastfeeding, which underpin the WHO and United Nations Children's Fund's Baby Friendly Hospital/Health Initiative are outlined in the box below. Supportive practices of particular importance are early skin-to-skin contact, avoidance of **prelacteal feeding** and supplementary feeds, 24-hour rooming-in and unrestricted feeding on demand.

Prelacteal feeding is the giving of any liquid or food item (except breast milk) to a newborn within the first three days after birth—that is, prior to the establishment of milk flow.

Ten Steps to Successful Breastfeeding

Step One: Develop a breastfeeding policy
Have a written breastfeeding policy that is routinely communicated to all healthcare staff.

Step Two: Ensure staff receive training
Train all healthcare staff in skills necessary to implement this policy.

Step Three: Provide antenatal information
Inform all pregnant women about the benefits and management of breastfeeding.

Step Four: Help initiate breastfeeding
Place babies in skin-to-skin contact with their mothers immediately following birth for at least an hour and encourage mothers to recognise when their babies are ready to breastfeed, offering help if needed.

Step Five: Teach breastfeeding
Show mothers how to breastfeed and how to maintain lactation even if they should be separated from their infants.

Step 6: Avoid supplementation
Give newborn infants no food or drink other than breast milk, unless medically indicated.

Step 7: Practice rooming-in
Practise rooming-in, allow mothers and infants to remain together 24 hours a day.

Step 8: Encourage breastfeeding on demand
Encourage breastfeeding on demand.

Step 9: No artificial teats or dummies
Give no artificial teats or dummies to breast-feeding infants.

Step 10: Provide access to support groups
Foster the establishment of breastfeeding support and refer mothers on discharge from the facility. (Baby Friendly Health Initiative, n.d.)

COMMON CONCERNS AND HEALTH CONDITIONS

While breastfeeding is the 'natural' way to feed an infant, it does not necessarily 'come naturally' to all women. Most women will experience some degree of difficulty with breastfeeding in the early stages, and it may take up to six weeks for some women to successfully establish breastfeeding. Unfortunately, many women who expect breastfeeding to be easy give up prematurely if they experience problems and label themselves as being unable to breastfeed or to produce sufficient breast milk. With support and encouragement from hospital staff, their partner and other family members, and a lactation consultant if necessary, most women can successfully breastfeed their infant.

Sore nipples

Many women experience sore or cracked nipples, which can lead to painful breastfeeding and cessation of breastfeeding if not correctly managed. The most common causes of sore nipples are poor positioning, poor attachment and an incorrect sucking action. Positioning and attachment should be checked and corrected and the mother encouraged to continue breastfeeding unless the pain is intolerable, in which case the mother should temporarily stop feeding from the affected breast and feed her baby with expressed breast milk until the pain and trauma subside.

Breast fullness and engorgement

Between days 2 and 4 the breast milk 'comes in' and the breasts feel heavy, hot and hard. This is normal, and frequent feeding will help reduce the fullness until milk production adjusts to infant demand. If, however, the breasts are not frequently and adequately emptied, particularly in the first few days, they can become engorged, in which case they appear swollen and the skin looks shiny and red. The nipples may become stretched tight and flat, which makes it difficult for the baby to attach and remove the milk. The milk does not flow well, compounding the problem. The common reasons why milk is not removed adequately are delayed initiation of breastfeeding, infrequent feeds, poor attachment and ineffective suckling.

If an infant is having trouble attaching to the breast, the mother can express some milk so that the breast is soft enough for the infant to attach. If a mother is separated from her infant for any length of time, say overnight, complete expression of the milk, which can be stored for later use, is necessary.

Mastitis

If not managed effectively, engorgement can lead to mastitis, and roughly one in five breastfeeding women will experience at least one episode of mastitis. It is usually unilateral but may affect both breasts and is characterised by a painful, red, swollen breast. Women may also suffer flu-like fever. Mastitis is more common among first-time mothers, with the majority of cases occurring in the first month. Women who suffer mastitis may stop breastfeeding because of the pain associated with the condition or because they have been inappropriately advised to do so. However, the most effective treatment of mastitis is frequent and effective milk removal. Most women can effectively self-manage their mastitis but may need to take a course of antibiotics if symptoms persist. It is safe to continue to breastfeed when taking antibiotics.

Self-perceived breast milk insufficiency

Women everywhere doubt the adequacy of their breast milk, both in terms of quality and quantity, and commonly give inadequate milk supply or milk insufficiency as their reason for stopping breastfeeding. True breast milk insufficiency is relatively uncommon, with very few women being physiologically incapable of producing sufficient milk due to severe illness or inadequate mammary tissue. However, stress and a lack of confidence in the ability to breastfeed may inhibit the release of oxytocin and affect the flow of a mother's milk.

Some women mistakenly try to 'train their baby' by imposing a feeding schedule (for example, four hourly) or forcing their baby to sleep through the night, both of which result in less stimulation to the breast and therefore reduced supply. Newborn infants have a small stomach capacity and are physiologically programmed to ingest small amounts at frequent intervals. As they mature and their stomach capacity grows, it is often the mother's breast capacity that limits the amount of milk that can be delivered at any one feed.

The best way for a woman to increase her milk production is to remove milk from the breast by feeding frequently or expressing her milk. However, mothers usually self-diagnose breast milk insufficiency on the basis of changes in infant behaviour (for example, the infant wanting to feed more often, increased fussiness or crying between feeds), which they interpret as an indication that their infant is not getting enough breast milk. These are not necessarily signs of infant hunger; the infant may be tired or going through a growth spurt.

Educating parents about realistic expectations with feeding frequency is important. The two reliable signs that a baby is not getting enough milk are poor weight gain and low urine output (fewer than six to eight very wet nappies), and the cause of these should be identified and managed appropriately.

SUMMARY AND KEY MESSAGES

From the available evidence and current recommendations it is clear that, wherever possible, breastfeeding is the best way to feed an infant and the short- and long-term health benefits of breastfeeding to both mothers and their infants are well established.

- The unique nutrition and immune-protective properties of breast milk make it the most suitable food for the infant from birth and for the first six months of life, and where possible continuing to breastfeed to at least twelve months is recommended. There are a number of common problems that can be experienced by lactating women and it is critical that strategies for supporting and encouraging women to continue to breastfeed are well understood and provided to women when needed. Breast milk contains all of the nutrients needed to support the growth and development of a healthy, term infant for the first six months of life.
- Breast milk contains bioactive factors that boost an infant's immature immune system. These factors are lacking in infant formula milk, and formula-fed infants have a higher risk of gastrointestinal, respiratory and ear infections than breastfed infants.
- A woman's diet does not affect appreciably the concentrations of lactose, protein, fat and major minerals in breast milk, but the fatty acid profile and concentration of a number of vitamins and minerals are determined by maternal intake.
- There are few instances when a woman should be advised either to not breastfeed or to temporarily discontinue breastfeeding.
- With support and encouragement from hospital staff, their partner and other family members, and a lactation consultant if necessary, most women can successfully breastfeed their infant.

REFERENCES

AIHW (Australian Institute of Health and Welfare) 2011. *2010 Australian National Infant Feeding Survey: Indicator results*. Cat. no. PHE 156. Commonwealth of Australia. <www.aihw.gov.au/WorkArea/DownloadAsset.aspx?id=10737420925>, accessed 26 September 2014.

Baby Friendly Health Initiative n.d. <www.babyfriendly.org.au/about-bfhi/ten-steps-to-successful-breastfeeding>, accessed 9 November 2014.

Gibson, R.A., Neumann, M.A. & Makrides, M. 1997. Effect of increasing breast milk docosahexaenoic acid on plasma and erythrocyte phospholipid fatty acids and neural indices of exclusively breast fed infants. *European Journal of Clinical Nutrition* 51: 578–84.

Innis, S.M., Gilley, J. & Werker, J. 2001. Are human milk long-chain polyunsaturated fatty acids related to visual and neural development in breast-fed term infants? *Journal of Pediatrics* 139: 532–8.

Ip, S., Chung, M., Raman, G., Chew, P. et al. 2007. *Breastfeeding and Maternal and Infant Health Outcomes in Developed Countries*. Evidence Report/Technology Assessment 153, AHRQ Publication 07-E007. Tufts-New England Medical Center Evidence-based Practice Center, Boston, MA.

Labbok, M.H., Clark, D. & Goldman, A.S. 2004. Breastfeeding: Maintaining an irreplaceable immunological resource. *Nature Reviews. Immunology* 4: 565–72.

Ministry of Health 2008. *Food and Nutrition Guidelines for Healthy Infants and Toddlers (Aged 0–2): A background paper* 4th edn, partially revised December 2012. New Zealand Government. <www.health.govt.nz/system/files/documents/publications/food-and-nutrition-guidelines-healthy-infants-and-toddlers-revised-dec12.pdf>, accessed 25 September 2014.

Munns, C., Zacharin, M.R., Rodda, C.P., Batch, J.A. et al. 2006. Prevention and treatment of infant and childhood vitamin D deficiency in Australia and New Zealand: A consensus statement. *Medical Journal of Australia* 185: 268–72.

NHMRC (National Health and Medical Research Council) 2010. *Iodine Supplementation for Pregnant and Breastfeeding Women*. NHMRC Public Statement. Commonwealth of Australia. <www.nhmrc.gov.au/_files_nhmrc/publications/attachments/new45_statement.pdf>, accessed 26 September 2013.

——2012. *Eat for Health: Infant Feeding Guidelines; Information for health workers*. Commonwealth of Australia. <www.nhmrc.gov.au/_files_nhmrc/publications/attachments/n56_infant_feeding_guidelines.pdf>, accessed 28 September 2014.

——2013. *Eat for Health: Australian Dietary Guidelines; Providing the scientific evidence for healthier Australian diets*. Commonwealth of Australia. <www.eatforhealth.gov.au/sites/default/files/files/the_guidelines/n55_australian_dietary_guidelines.pdf>, accessed 28 September 2014.

NHMRC (National Health and Medical Research Council) & Ministry of Health 2006. *Nutrient Reference Values for Australia and New Zealand: Including recommended dietary intakes*. Commonwealth of Australia & New Zealand Government. <www.nhmrc.gov.au/_files_nhmrc/publications/attachments/n35.pdf>, accessed 26 September 2014.

SACN (Scientific Advisory Committee on Nutrition) 2012. *Dietary Reference Values for Energy*. <www.gov.uk/government/uploads/system/uploads/attachment_data/file/339317/SACN_Dietary_Reference_Values_for_Energy.pdf>, accessed on 5 November 2014.

Wahlqvist, M. (ed.) 2011. *Food and Nutrition: Food and health systems in Australia and New Zealand* 3rd edn. Allen & Unwin, Sydney.

WHO (World Health Organization) 2008. *Indicators for Assessing Infant and Young Child Feeding Practices: Conclusions of a consensus meeting held 6–8 November 2007 in Washington, DC, USA*. WHO. <http://whqlibdoc.who.int/publications/2008/9789241596664_eng.pdf>, accessed 25 September 2014.

——2009. *Infant and Young Child Feeding: Model chapter for textbooks for medical students and allied health professionals*. WHO. <http://whqlibdoc.who.int/publications/2009/9789241597494_eng.pdf?ua=1>, accessed 25 September 2014.

——2010. *Guidelines on HIV and Infant Feeding 2010: Principles and recommendations for infant feeding in the context of HIV and a summary of evidence.* WHO. <http://whqlibdoc.who.int/publications/2010/9789241599535_eng.pdf?ua=1>, accessed 25 September 2014.

WHO (World Health Organisation) & UNICEF 2003. *Infant and Young Child Nutrition.* Global strategy for infant and young child feeding. Fifty-fifth World Health Assembly, <http://apps.who.int/gb/archive/pdf_files/WHA55/ea5515.pdf?ua=1?>, accessed 13 November 2014.

ADDITIONAL READING

Australian Breastfeeding Association (website) 2014. <www.breastfeeding.asn.au/>, accessed 29 September 2014.

La Leche League New Zealand (website) 2014. <www.lalecheleague.org.nz/>, accessed 29 September 2014.

TEST YOUR UNDERSTANDING

Note: There is only one correct answer for each question.

1. For how long does the WHO recommend that an infant be exclusively breastfed?
   a. Four months
   b. Four to six months
   c. Six months
   d. Twelve months
   e. None of the above
2. What does exclusive breastfeeding allow the baby to receive?
   a. Water
   b. Oral rehydration solution
   c. Small amounts of formula
   d. Sugar solution
   e. None of the above
3. Which of the following medical conditions are breastfed children less likely to have?
   a. Diarrhoea
   b. Ear infections
   c. Pneumonia
   d. Allergies
   e. All of the above
4. When does a woman's body produce hormones that stimulate the growth of the milk duct system in the breasts?
   a. Before pregnancy
   b. From the first month of pregnancy
   c. From the fourth month of pregnancy
   d. From the sixth month of pregnancy
   e. When pregnancy ends
5. Which hormone is responsible for the milk let-down reflex?
   a. Oxytocin
   b. Prolactin
   c. Oestrogen
   d. Testosterone
   e. Progesterone
6. What is the recommended average additional energy requirement in lactation?
   a. 1000–1200 kJ
   b. 1500–1600 kJ
   c. 2000–2100 kJ
   d. 2500–2600 kJ
   e. 2800–2900 kJ
7. What can be affected appreciably by what a woman eats?
   a. Concentration of lactose in her breast milk
   b. Fatty acid composition of her breast milk
   c. Concentration of calcium in her breast milk
   d. All of the above
   e. None of the above
8. What is the usual sequence of breast milk production?
   a. Foremilk—colostrum—mature milk
   b. Transitional milk—colostrum—mature milk
   c. Colostrum—transitional milk—mature milk
   d. Colostrum—mature milk—hind milk
   e. Foremilk—transitional milk—mature milk

9. Which of the following statements correctly describe human milk oligosaccharides?
   a. They inhibit the binding of bacterial pathogens to the gut epithelium
   b. They encourage the growth of beneficial gut bacteria
   c. They are structurally different from oligosaccharides in infant formula
   d. They are a food source for gut bacteria
   e. All of the above
10. Breastfeeding is not recommended for mothers who have which of the following medical conditions?
   a. HIV
   b. Hepatitis
   c. Tuberculosis that has been treated
   d. Influenza
   e. All of the above

STUDY QUESTIONS

1. Why do key authorities (such as the WHO, the National Health and Medical Research Council and the New Zealand Ministry of Health) emphasise exclusive breastfeeding for the first six months of life?
2. How do colostrum and mature breast milk differ in composition?
3. Which nutrients in breast milk are affected by a mother's diet, and what dietary advice would you give a healthy breastfeeding woman?
4. What are four key immunoprotective functions of breast milk and the constituents involved?
5. What are the primary causes of nipple pain and mastitis among breastfeeding women?
6. Insufficient milk supply is a reason given by many women for stopping breastfeeding. What advice would you give to a new mother in the early postpartum period to help her develop confidence in her milk supply?

6

Newborn and infant (0–12 months)

Jacqueline Miller and Jane Scott

> **LEARNING OUTCOMES**
>
> Upon completion of this chapter you will be able to:
>
> - appreciate the impact of infant feeding practices on health outcomes
> - describe infants' nutrient requirements
> - understand how to assess and monitor infants' growth
> - describe the differences in composition between infant formulas and breast milk
> - advise how to safely introduce complementary feeding
> - describe infant nutrition issues regarding preterm infants, allergies and dental caries.

Good nutrition in infancy is critical for optimal growth and development. This chapter describes the links between infant feeding practices and short- and long-term health outcomes, outlines infant nutrient requirements and introduces recommendations for infant feeding and the evidence to support these. Breast milk and infant formula are compared and contrasted, and the introduction of complementary foods is also discussed. Finally, the chapter describes some common special conditions related to infant feeding. A focus of the chapter is practical information on aspects of infant feeding. As growth in infancy is the main indicator of nutritional status, this chapter aims to give a thorough overview of how to assess and monitor infant growth appropriately.

INFANT FEEDING PRACTICES

Good nutrition in infancy is essential to ensure that children meet their full potential for growth and development. The first six months of life is a unique lifecycle stage: it is the time of most rapid growth throughout life, including growth of organs such as the brain. A typical infant gains around 150–200 grams per week and doubles their birthweight by six months. And yet it is the only time throughout life when we are dependent on a single food: breast milk. Little wonder that it is described as the perfect food.

Infant nutrition affects not only immediate growth but the quality of organ development and so impacts on growth and health outcomes far into the future. The 'first 1000 days', which includes the nine months of pregnancy and the first two years of life, is seen as a critical window of opportunity to impact on a child's growth and learning. Undernutrition is associated with infant death, particularly in poorer countries, but is also a major restrictor of children reaching their full potential.

While overall infant mortality is low in Australia and New Zealand, the rate for Aboriginals and Torres Strait Islanders is three to four times higher than for non-Indigenous Australians in some regions, and the rate for Pacific and Māori infants is approximately double that of non-Indigenous New Zealanders. The proportion of these deaths related to poor nutrition is not known, but globally malnutrition is responsible for approximately one-third of all deaths in children under five years. Malnourished children grow into adults who generally are shorter and have lower intellectual capability and possibly reduced work capacity. Therefore, they have less chance of well-paid employment, which in poor countries reinforces poverty. Women who were malnourished as children tend to have infants with lower birthweight, and so the consequences are passed on to the next generation. In response to this, in 2002 the World Health Organization (WHO) and the United Nations Children's Fund (also known as UNICEF) adopted the Global Strategy for Infant and Young Child Feeding. Its recommendations for infant feeding are exclusive breastfeeding for six months and nutritionally adequate and safe complementary feeding starting from the age of six months, with continued breastfeeding up to two years of age or beyond.

Breastfeeding initiation rates in Australia and New Zealand are high but decline rapidly in the first few weeks of life. Over 90 per cent of Australian mothers and 95 per cent of New Zealand mothers initiate breastfeeding, but this drops to 50–60 per cent at six months of age in both countries. Worldwide, only about four in ten infants are exclusively breastfed for six months. Both the Australian National Health and Medical Research Council (NHMRC 2012) and New Zealand Ministry of Health (2008) recommend exclusive breastfeeding for around six months. This is a population guideline so allows for some individual variation around it, depending on circumstances.

Complementary foods are introduced at a median age of 4.7 months in Australia (AIHW 2011) and 5 months in New Zealand (Ministry of Health 2008), with almost all infants receiving solids within a few weeks of six months of age. Some culturally and linguistically diverse populations, including Māori and Pacific peoples, as well as young mothers and first-time mothers, tend to introduce solids earlier than four months.

PHYSIOLOGICAL CHANGES

Most babies born after 36 weeks' gestation display the feeding reflexes necessary to successfully nourish themselves. The rooting reflex causes an infant to turn their head towards the direction of being touched on the cheek, opening their mouth quite wide and putting their tongue down and forward. The sucking reflex can be stimulated when a finger, breast, bottle or dummy is placed into the baby's mouth and pressure is applied to their palate. When the baby's mouth fills with milk, they swallow. This is the swallowing reflex. The gag reflex is triggered when a baby swallows too much milk. In this case the baby's throat closes off and their tongue pushes the excess milk out of their mouth. Coordination of suckling, swallowing and breathing appears between 32 and 35 weeks of gestation.

As the infant's central nervous system matures, they start to gain more control over movement of their body. Head control is achieved first, followed by the ability to sit without support. Both of these are crucial to safe introduction of complementary foods. As shown in Figure 6.1, there is wide individual variation in when these milestones are achieved. Once the gross motor skills are mastered, fine motor skills follow, so at some stage during the second six months of life, infants are able to pick up finger foods, or a spoon, albeit messily!

Growth is an important indicator of an infant's health and nutritional status so should be measured and monitored accurately and consistently. Very often, decisions about infant feeding, such as introducing formula or complementary foods, are based on growth measurements, meaning that it is essential that these decisions, with their far-reaching implications, are based on sound measurement and interpretation. An infant's growth can occur in spurts, and variations can occur due to factors such as illness or teething. Therefore, an assessment should be made on a number of anthropometric indices and measurements taken on multiple occasions to ensure a big-picture look at growth.

Measuring growth

All growth measurements should be made using standardised techniques and appropriate equipment. Weight should be assessed on accurate scales (zeroed before starting) that can be calibrated at regular intervals. The infant needs to be weighed under similar conditions each time—that is, either naked or with minimum clothing. Factors such as proximity to a feed and bowel action can account for as much as one-third

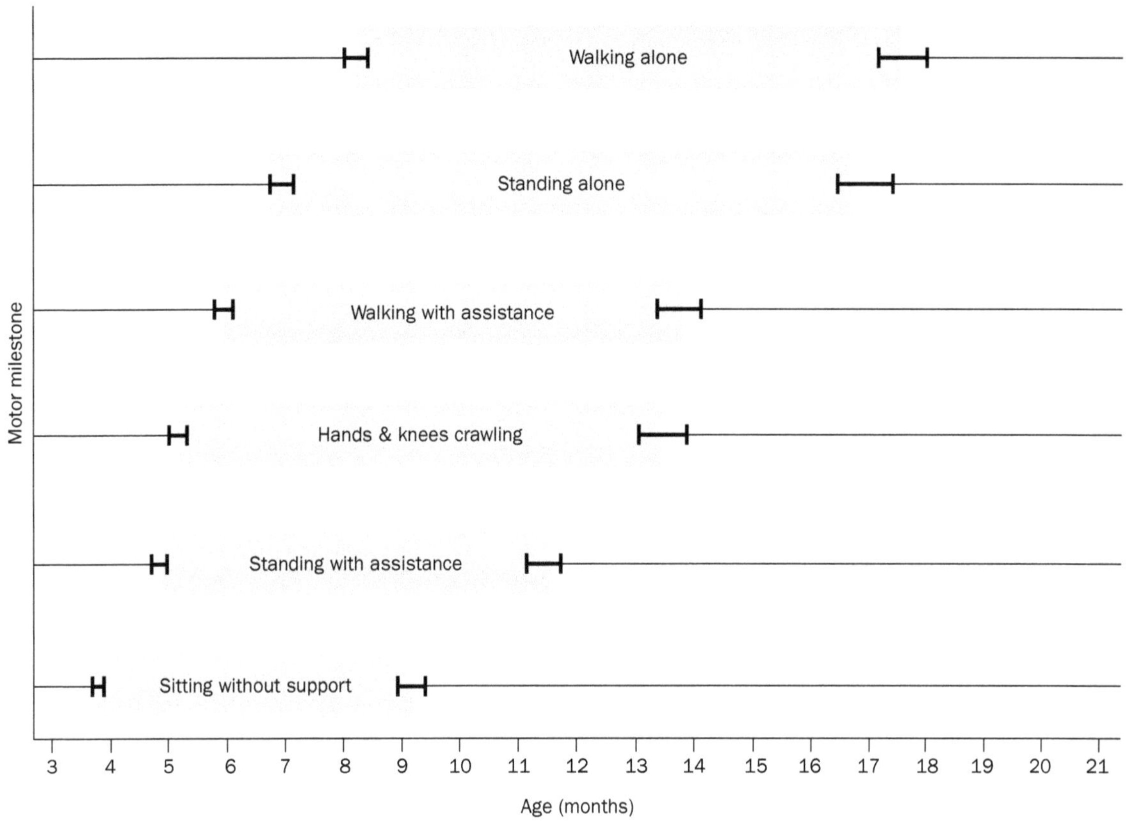

Figure 6.1 Windows of achievement for six gross motor milestones

Source: Adapted from WHO MGRSG (2006).

of the expected weekly weight gain and should be taken into consideration.

Length is measured in infants as opposed to standing height, which is measured after two years of age. A **recumbent length** board should be used: this is a board with a headpiece at one end at a right angle to the board, against which the infant's head is placed, and a moveable foot piece, also at a right angle to the board, which can slide up against the soles of the infant's flexed feet (see Figure 6.2). Head and other circumference measurements, such as mid arm, should be measured using a non-stretchable tape.

Recumbent length refers to stature when lying down. A recumbent length board is specialised equipment used to measure infant length.

Monitoring growth

Growth is monitored against growth charts which compare a child's length or height and weight with those of other children of the same age. Head circumference is also often monitored as an indicator of brain growth. A head circumference at either extreme (very small or very large) could be an indicator of brain development problems, but these are rare. If infants are not growing due to insufficient nutrient intake then a drop in weight and length percentiles will usually be noticed rather than a drop in the head circumference percentile.

Growth charts are pooled measurements from a large number of children, which are statistically smoothed and presented as percentiles or **Z-scores**. If a child's measurement is plotted on the tenth percentile it means that 90 per cent of children in the reference group are above that measurement (taller or heavier)

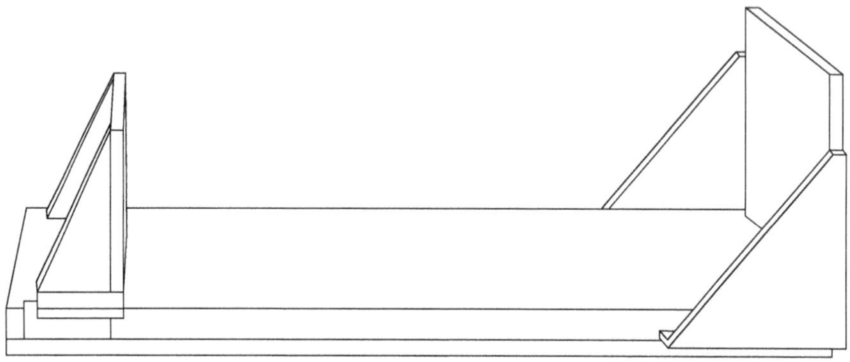

Figure 6.2 A recumbent length board

Source: Adapted from WHO (2008a: 19).

Z-scores show the number of standard deviations above or below the mean; i.e., a negative Z-score is below the mean, while a positive Z-score is above the mean. They are also referred to as normal or standard scores.

and 10 per cent are below. There are two main growth charts used in the Australasian region: the US Centers for Disease Control and Prevention's Individual Growth Charts, developed in 2000 and widely used in the early 2000s (CDC 2009), and the WHO's (n.d.) Child Growth Standards, introduced in 2006.

The Centers for Disease Control and Prevention's charts are based on data from over one million children and adults aged from birth to twenty years from five cross-sectional surveys nationally representative of the population in the United States. The population sampled was ethnically diverse but included predominantly formula-fed infants (Kuczmarski et al. 2002). The charts relevant for infants cover the age range 0–36 months and consist of eight percentile charts, one for each gender, on weight-for-age, length-for-age, weight-for-length and head-circumference-for-age. The charts are limited in that they are representative of growth only in the United States, where formula feeding was the norm at the time, and there were limited data for the first two years of life. Therefore, the charts do not represent the normally growing breastfed infant and may lead to misdiagnosis of growth failure in other populations.

Between 1997 and 2003 the WHO (n.d.) undertook the large Multicentre Growth Reference Study to assess growth of infants and young children around the world. The resulting Child Growth Standards were published in 2006.

Racial differences in growth seen early in the 21st century could be largely attributed to suboptimal conditions in different countries, since if children less than five years of age are provided with optimum conditions for growth—abundant food and nutrients and a clean, safe environment—then growth is quite similar across the various races. Therefore, an international standard—that is, how children *should* grow rather than a reference describing *how* children *do* grow—was considered desirable. The Multicentre Growth Reference Study collected data from approximately 8500 children from healthy privileged families in Brazil, Ghana, India, Norway, Oman and the United States who were breastfed and whose mothers did not smoke—children who were therefore likely to achieve their full growth potential.

The WHO (n.d.) Child Growth Standards relevant to infancy cover the age range 0–24 months and are presented as both percentiles and Z-scores. The charts available are—for age for each gender—weight, length, body mass index (BMI), head circumference, arm circumference, subscapular skin fold and triceps skin fold, as well as weight-for-length and motor developmental milestones. It is recommended that the WHO charts be used in preference to the Centers for Disease Control and Prevention charts to assess infant growth, as the former were based on breastfed infants, used a larger sample size and included more frequent measurements in the first year of life, which better captured the rapidly changing rate of growth in an infant's first few months.

Z-scores

Z-scores are a method of assessing growth and express growth as standard deviations, either positive or negative, from the reference mean. They are commonly used for population nutritional status surveys, as they are linear and can therefore be summarised as a mean and standard deviation for a population, thereby allowing comparisons. However, they are also useful for describing individual growth and classifying malnutrition. The reference mean is given the Z-score of zero, and normal growth is considered to be plus or minus one standard deviation around this. A Z-score lower than minus one signals possible malnutrition, while a Z-score higher than plus one indicates possible overnutrition (see Table 6.1). A low length-for-age Z-score is called stunting and reflects chronic malnutrition. A low weight-for-age Z-score is termed underweight and can reflect both acute and chronic malnutrition, as well as illness. A low weight-for-length Z-score is defined as wasting and is the most sensitive measure of acute malnutrition, widely used to identify individuals at risk.

Figures 6.3–6.5 show photos of infants with different anthropometric Z-scores. Stunting and wasting are terms commonly used to describe nutritional status in poor countries, and prevalence statistics of these statuses are frequently reported. In Australia and New Zealand, poor growth in infancy is more commonly termed **failure to thrive** and describes weight loss or poor weight gain, which can be coupled with reduced length growth velocity if prolonged.

Failure to thrive describes not achieving expected growth as assessed by measurements of height and weight.

Failure to thrive is caused by inadequate dietary intake and can be either organic (due to a medical illness) or non-organic (not due to a medical illness). Organic failure to thrive occurs when nutrients are lost through vomiting or diarrhoea, malabsorption or reduced intake due to **gastro-oesophageal reflux**

Figure 6.3 Underweight boy

Note: One-year, one-month-old, 70.3 centimetres, 7.5 kilograms. Z-scores: weight-for-age minus 2.5, weight-for-length minus 1.5.

Source: WHO (2008b: 4).

Gastro-oesophageal reflux (also known as gastro-oesophageal reflux disease, and shortened to GORD or GERD) is a condition in which the acid contents of the stomach flow back into the oesophagus.

Table 6.1 Anthropometric indicators and their interpretation, using percentiles and Z-scores

Anthropometric indicator	Percentile	Z-score	Interpretation
Length-for-age	< 15th	< −2	Moderate stunting
	≤ 3rd	≤ −3	Severe stunting
Weight-for-age	< 15th	< −2	Moderately underweight
	≤ 3rd	≤ −3	Severely underweight
Weight-for-length	< 15th	< −2	Moderate wasting
	≤ 3rd	≤ −3	Severe wasting
	85th –to < 97th	+2 to < +3	Overweight
	≥ 97th	≥ +3	Obese

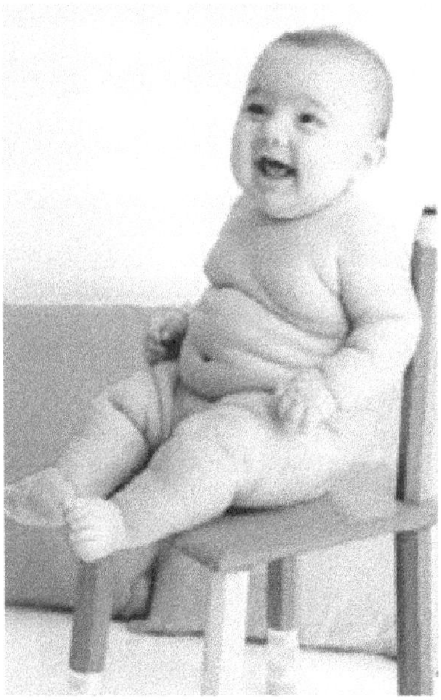

Figure 6.4 Obese boy

Note: Three-and-a-half-month-old, 63 centimetres, 10 kilograms. Weight-for-length Z-score: +4.6.

Source: WHO (2008b: 4).

Figure 6.5 Stunted girl

Note: 1-year-old, 67.8 centimetres, 7.6 kilograms. Z-scores: length-for-age −2.4, weight-for-length −0.2.

Source: WHO (2008b: 5).

or disabilities resulting in poor oral motor skills. Identifying the cause of non-organic failure to thrive may be more complex, as it is often related to a poor feeding relationship between carer and infant, maternal depression or inadequate feeding practices.

Plotting growth measurements

Ensure that the charts you have selected are appropriate for the infant being measured (correct gender and age range). WHO (n.d.) Child Growth Standards charts come in a number of different versions—0–6 months, 6–24 months and 0–24 months—with different scales between charts, so vigilance is required with plotting. Calculate the child's age accurately in complete weeks or months. The more measurements you have for an infant, the better overview of growth you will see and the easier it will be to identify mistakes in measuring or variations in growth over short periods of time. What is important is the overall growth trend.

It is not uncommon for infants to move from one weight percentile to another shortly after birth, as birthweight is more dependent on conditions in utero than final weight destiny. Infants normally have a weight loss of 5–7 per cent in the first three to four days after birth, which is mostly due to fluid. Once the mother's milk has come in, weight starts to be regained, and most infants will be back to their birthweight by ten to fourteen days of age. After this time, the infant should track along a percentile line or in a channel between percentile lines, as shown in the example depicted in Figure 6.6. There is a broad range of normal growth. Crossing two or more percentile lines or plateauing on a chart can be a warning sign of abnormal growth and warrants close attention.

Nutrient requirements and intakes

Breast milk is the ideal infant food and provides all the nutrition an infant needs in the first six months of life in an easily digested form. Some nutrients, such as iron and zinc, are at low levels in breast milk but in a form that is highly bioavailable. Some of the non-nutritive components of breast milk are enzymes, which start the digestion process, and factors that enhance the absorption of nutrients (see Chapter 5). This metabolic efficiency protects the mother as well as the infant: if iron is a limiting nutrient in the environment, then providing it to the infant in just the right amount, with factors to help its absorption, will protect the mother from deficiency. The infant also has to expend less energy to absorb the iron.

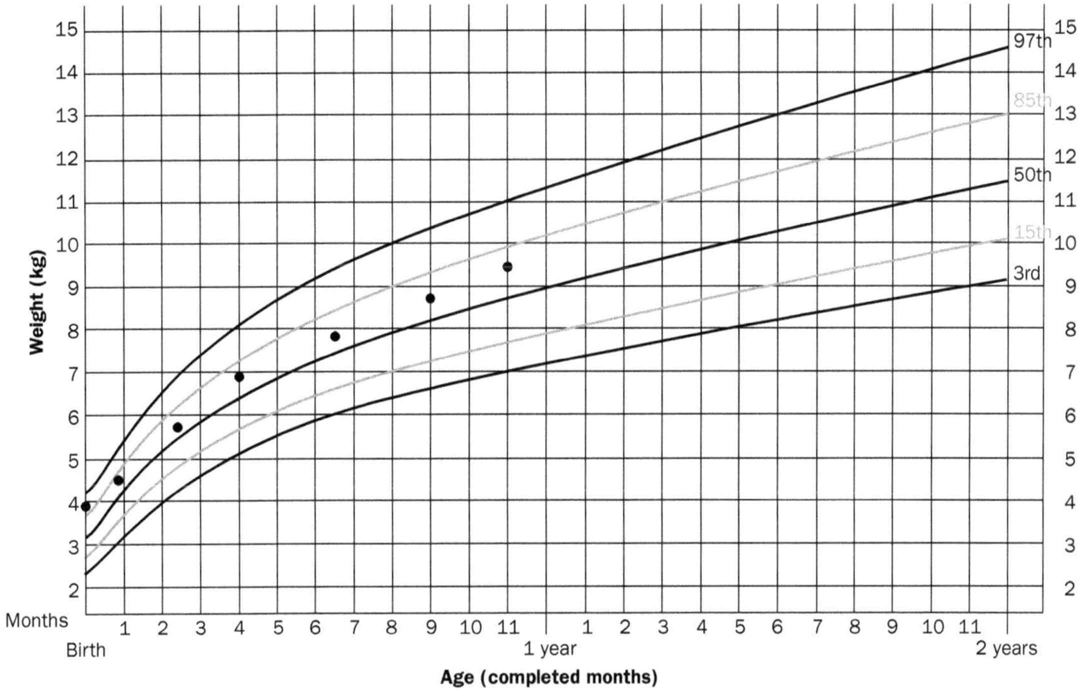

Figure 6.6 Healthy growth in an infant

The Australian and New Zealand nutrient reference values (NRVs) for infants are based on the composition of human milk (NHMRC & Ministry of Health 2006). The premises underlying the recommended adequate intakes (AIs) for infants aged from birth to six months are based on the content of breast milk in healthy mothers, with 780 millilitres breast milk per day assumed and rounding where appropriate, and a reference body weight of 7 kilograms. For infants aged seven to twelve months, the recommended AIs are set for most nutrients based on estimates of the content of breast milk in healthy mothers at an assumed 600 millilitres per day and based on the content of complementary foods in the usual daily amounts consumed, or where intake data are not available or are unreliable, by extrapolation from younger infants or adults.

Water

Water requirements for infants are 0.7 litres for birth to six months and 0.8 litres for seven to twelve months (NHMRC & Ministry of Health 2006). Breast milk has a high water content (87 per cent) and meets all the requirements for infants from birth to six months of age. There is no need for supplemental water, even in hot weather, providing the infant is allowed unrestricted access to the breast. The requirement for seven- to twelve-month-old infants assumes an intake of 600 millilitres from breast milk, and the rest can be met by breast milk, formula, plain water or other beverages and complementary foods. A way of calculating fluid requirements often used in the clinical setting is to measure an amount of fluid per kilogram of body weight. Infants require 140–60 millilitres of water per kilogram per day, depending on environmental and clinical conditions. Infants may need extra fluid if the weather is very hot or they are ill with a fever.

Energy

Energy requirements are the highest per kilogram of body weight in infancy than at any other stage of life. Growth accounts for approximately one-third of the energy requirement for the first three months. Requirements vary quite widely in infants and, apart from growth, depend on the same factors that influence differences in energy expenditure in adults—that is, body size, temperature, health status and physical activity. A small placid infant who sleeps a lot will have

CASE STUDY 6.1 FAILURE TO THRIVE

Profile and problem

T. is a 32-year-old woman with one child, a nine-month-old baby boy, S. She was diagnosed with postnatal depression after the birth of S. and remains exhausted, tired and emotionally drained. S. has recently been diagnosed with inorganic failure to thrive. He is described as a placid baby who is not overly interested in either bottle feeds or solid foods.

Assessment

S. currently has six 150-millilitre feeds of a standard infant formula daily. He eats two meals a day of commercially pureed vegetables or fruit. His growth history is as follows:

At birth	3.8 kilograms, 52 centimetres
At 4 months	6.5 kilograms, 64 centimetres
At 9 months (current)	7.4 kilograms, 69.5 centimetres

Appropriate growth standards for plotting these measures can be found at the WHO (n.d.).

Treatment

T. has agreed to do the following:

- offer S. more formula, either by increasing the volume in each bottle to 200 millilitres or by adding one or two extra bottles a day
- offer S. more solids—i.e., three meals a day
- introduce snacks in between meals
- incorporate higher energy/protein foods such as meats, egg, custards, yoghurts, cheese
- add more variety into S.'s solids by giving more than one food at each meal and gradually moving towards a family meal of meat and two to three types of vegetables
- gradually increase the consistency of the food to mashed, soft cooked lumps and then finger foods
- provide more stimulation for S. by having some set play periods throughout the day for mother and son to interact.

Plan

S.'s growth and length will be monitored over the next few months with regular visits to the healthcare facility. T. has agreed to keep a diary of what S. eats and drinks the day before she comes to each appointment. T. has been referred to a parenting group at the local child health facility.

lower energy needs than a fretful infant who is often awake. Healthy infants are adept at regulating their own energy intake provided they are allowed to demand feed—that is, to dictate when and how much they feed. This is easier to achieve with breastfeeding which is infant led. Formula feeding is usually caregiver led, and there is a tendency for caregivers to encourage the infant to finish the bottle; this may result in overriding natural satiety cues, a mechanism that protects breastfed infants from later obesity. Average estimated energy requirements for infants are given in Table 1.5.

Macronutrients

Protein

Like energy, protein requirements per kilogram of body weight are highest in infancy: 1.43 grams per kilogram in infants from birth to six months and 1.6 grams per kilogram in infants from seven to twelve months (see Table 1.9a). Protein is essential for lean tissue accretion—that is, growth. The protein in breast milk has higher biological value compared to protein in infant formulas; hence, formulas need to have a higher protein concentration to achieve a better amino acid profile. The higher protein content in infant formulas has been implicated as contributing to higher risk of obesity in formula-fed infants, and modern formulas have a reduced protein concentration which is closer to breast milk.

Fat

Fat is an important nutrient for infants, and approximately 55 per cent of the energy in breast milk is from fat. It is a good source of essential fatty acids, linoleic and alpha-linolenic acids. Breast milk is also a source of omega-3 and omega-6 long-chain polyunsaturated fatty acids, which are incorporated into brain and eye tissue. Some infant formulas have added long-chain polyunsaturated fatty acids (see below). The AI for total fat is based on that provided by breast milk and is set at 30–1 grams per day. Essential fatty acid requirements are given in Tables 1.10a and 6.2. Fat restriction is not recommended for infants because of their high need for fat as a substrate for brain and other organ development.

Micronutrients

NRVs for selected nutrients are given in Table 6.2. As discussed earlier, breast milk provides an adequate amount of vitamins and minerals for the first six months of life. Iron is low in breast milk but highly bioavailable. Iron absorption from formula is much lower, so higher concentrations are required. Infants are born with body stores of iron laid down in the last 12–13 weeks of gestation, and these are usually sufficient for the first six months of life. After that, iron sourced from complementary foods is required. Infants born prematurely, who have not had the full benefit of in utero accretion of iron, may need iron supplementation at earlier than six months.

Vitamin D status of the infant at birth depends on maternal status during pregnancy. If the mother has good vitamin D status the infant will have some stores which will last a few months. It would seem prudent that infants deemed at high risk of vitamin D deficiency should receive a daily supplement. To reduce the risk of rickets, sun exposure for a total of around two hours a week is required if just the face is exposed or 30 minutes a week with just a nappy on. Regular small doses of sunshine are recommended and will minimise the risk of sunburn.

The vitamin A content of breast milk is dependent on maternal diet. Colostrum is high in beta-carotene, which gives it the characteristic yellowish colour. After that, levels in mature milk decrease. While this is not a problem in Australia and New Zealand, areas at high risk of vitamin A deficiency may benefit from maternal supplementation.

MILK FEEDING

Breast milk is the optimal feed for infants, and breastfeeding should be encouraged, supported and promoted. Provided mothers are given the encouragement and support they need, there are very few occasions when they will be unable to breastfeed for physical reasons or when breastfeeding will be contraindicated (see Chapter 5). Any breastfeeding is beneficial, and partial feeding should be encouraged if the mother is unable to meet all her infant's requirements with breast milk alone. Parents should make an informed choice about feeding method, which includes knowing the hazards of not breastfeeding. But once that choice is made, it should be respected, and support and education without prejudice given about bottle feeding.

Human milk, cow's milk and infant formula

While breastfeeding is the optimal choice for infants, there will always be parents who cannot or choose not to breastfeed. Infant formula is the only recommended alternative to breast milk for infants until twelve months of age. In Australia and New Zealand, Food

Table 6.2 NRVs for selected nutrients for infants (0–12 months) in Australia and New Zealand

Nutrient	Daily AI (UL) of nutrients, by age (months)	
	0–6	7–12
Protein (g)	10 (BM)	10 (BM)
Fat		
Linoleic (omega 6 (n-6)) (g)	4.4 (BM)	4.6 (BM)
Alpha-linolenic (omega 3 (n-3)) (g)	0.5 (BM)	0.5 (BM)
Carbohydrate (g)	60.0 (BM)	60.0 (BM)
Vitamin A (retinol equivalents) (μg)	250[a] (600)	430 (600)
Vitamin D (μg)	5 (25)	5 (25)
Iron (mg)	0.2 (20.0)	–[b] (20.0)

a As retinol.

b No AI for iron for seven to twelve months. EAR: 7 milligrams per day; RDI: 11 milligrams per day.

Note: AI: adequate intake; BM: amount normally received from breast milk; EAR: estimated average requirement; NRV: nutrient reference values; RDI: recommended dietary intake; UL: upper limit.

Source: NHMRC & Ministry of Health (2006).

Standards Australia New Zealand is the statuary agency that regulates infant formula through the Food Standards Code for infant formula products (FSANZ 2013). Most formulas available are based on cow's milk, and the composition is modelled on human milk, as this is the gold standard, although it is not possible to replicate many of the bioactive components of human milk (see Table 5.2).

Cow's milk undergoes considerable modification in the manufacture of infant formula: the protein content is reduced, the fraction of used cow's milk protein—either casein or whey—is manipulated to achieve a more favourable amino acid content, and vitamins and minerals are added. Many formulas also have added extras, including long-chain polyunsaturated fatty acids, oligosaccharides and nucleotides, which aim to mimic human milk composition. These formulas are usually more expensive, and there is a lack of evidence that they replicate breast milk better than the standard formulas. Table 6.3 compares the composition of human milk with that of cow's milk and infant formulas, as specified by Food Standards Australia New Zealand (FSANZ 2013). Formulas for full-term infants usually contain approximately 290 kilojoules per 100 millilitres, which is similar to both human milk and cow's milk.

Macronutrients

The protein content of human milk is much lower than that of other milks, and the amino acid content is tailored to the infant's requirements. This is ideally suited to the infant's immature kidneys, as it means minimal stress is caused through eliminating nitrogenous products. Overall, human milk proteins form a curd that is softer and easier to digest than cow's milk proteins.

Milk proteins are often classified into casein and whey fractions. Casein is the insoluble fraction of the protein, or the curd, which precipitates out in the cheese-making process or when acid is added to milk. It is tougher and more difficult to digest than the whey fraction, which is soluble in water. Human milk contains less casein than cow's milk, and consequently a number of infant formulas are manufactured from the whey fraction of cow's milk to more closely resemble human milk.

However, the actual proteins that make up the casein and whey fractions differ in their molecular structure between species. Beta-casein is the predominant casein in human milk, compared with alpha-casein, which is the main bovine casein. Human milk whey contains alpha-lactalbumin and is free of beta-lactoglobulin, while the latter is the predominant whey protein in cow's milk and is one of the main protein allergens in infants.

Human milk, cow's milk and standard formulas all contain lactose, although there are special lactose-free formulas available for infants who are lactose intolerant. Some infant formulas use a mix of lactose and maltodextrin to bring the carbohydrate level to about 7 per cent, equivalent to human milk. Human milk also contains oligosaccharides which promote the growth of *Lactobacillus bifida* and thereby protect against pathogens (see Chapter 5).

Infant formulas usually either partially or fully replace dairy fat with blends of vegetable oils to meet Food Standards Australia New Zealand's Food Standards Code for infant formula products (FSANZ

Table 6.3 Nutrients in human milk, cow's milk and infant formula

Milk source	Carbohydrate (g)	Energy (kJ)	Fat (g)	Iron (mg)	Protein (g)	Vitamin A (μg)	Vitamin D (μg)
Human	6.9[a]	~290	4.4[c]	0.3	1.0[f]	610	1.00
Cow	6.3[b]	~290	3.5[d]	Negligible	3.5[g]	530	5.20
Cow's milk–based infant formula	7.2–7.5[b]	250–355 (usually ~290)	2.6–5.3[e]	5.0–17.7	1.1–2.5[h]	350–1526	6.25–22.30

a Lactose, oligosaccharides.
b Lactose.
c Reflects maternal diet, essential fatty acids, long-chain polyunsaturated fatty acids.
d Butter fat.
e Blended vegetable oils, essential fatty acids, some have added long-chain polyunsaturated fatty acids.
f Casein:whey approximately 40:60, beta-casein, alpha-lactalbumin.
g Casein:whey approximately 80:20, alpha-casein, beta-lactoglobulin.
h Casein:whey varies, beta-lactoglobulin.

Source: Adapted from NHMRC (2012: Table 2.1).

2013). Long-chain polyunsaturated fatty acids such as docosahexaenoic acid (DHA) and arachidonic acid are present in human milk, and in the absence of these in formula the infant must manufacture them from their precursors linolenic acid and alpha-linolenic acid. Some infant formulas have added DHA and arachidonic acid to replicate human milk; however, the structure of these fatty acids, which are derived from fish, algae or fungal sources, is different from that of human milk. Clinical trials investigating the efficacy of formulas supplemented with long-chain polyunsaturated fatty acids have not been conclusive, although some have shown a benefit on visual acuity, particularly for preterm infants, possibly due to these infants missing out on the placental transfer of long-chain polyunsaturated fatty acids that occurs in the last 12–13 weeks of gestation.

Micronutrients

As discussed in Chapter 5, in general, breast milk contains adequate vitamins and minerals to meet the requirements of the infant unless the mother's nutritional status is deficient (Butte et al. 2002). Some vitamins and minerals are low in breast milk but highly bioavailable, as discussed above. Infant formulas generally have higher concentrations of vitamins and minerals to account for the lower bioavailability. Concentrations of vitamins and minerals in formula are dictated by Food Standards Australia New Zealand's Food Standards Code for infant formula products (FSANZ 2013).

Choice of formula

There are many infant formulas available, and the choice can be overwhelming for parents, especially since many 'specialised' formulas, such as soy and anti-reflux, are available in supermarkets without trained staff to offer advice. These specialised formulas are usually more expensive. Food Standards Australia New Zealand allows optional ingredients to be added on the basis of safety (that is, they are not harmful) rather than on the basis of efficacy (that is, they have a proven benefit) (FSANZ 2013). This can be misleading for some parents who may feel they should pay the higher price for formulas with added extras. 'Designer formulas' are increasingly available and marketed for common infant problems such as colic, constipation and diarrhoea. There is very little evidence that one formula is better than another for healthy infants.

Formulas suitable for use from birth are often called starter, newborn or step 1 formulas. Manufacturers also market follow-on formulas for infants aged from six to twelve months, promoted as step 2 or progress formulas, but these are not necessary to use. The main difference between step 1 and step 2 formulas is that the latter contain more protein and iron, therefore making them unsuitable from birth. However, from six months of age infants obtain extra protein and iron from solids.

All infant formulas contain similar amounts of macronutrients and micronutrients, as regulated by Food Standards Australia New Zealand, and the main differences occur between various additives designed to mimic human milk. Table 6.4 describes some infant formula additives, their use and the evidence for their efficacy.

Specialised infant formulas

A range of specialised infant formulas for disease requirements are available. These should be used in consultation with a paediatrician and a paediatric dietitian. In general, one or more of the macronutrients is modified; many specialised formulas manipulate more than one macronutrient at a time. The main modifications are described below:

- *Protein*—In hydrolysed formulas the protein is treated with an enzyme to break it down to peptides of various lengths and some free amino acids. The base protein is usually cow's milk or a particular fraction of cow's milk, such as whey. The extent of hydrolysis dictates the use of the formula: extensively hydrolysed formulas with peptide length of two to three amino acids are the treatment of choice for cow's milk allergy. These differ from partially hydrolysed formulas (often labelled as HA, meaning hypo-allergenic), which have larger peptide fragments and are not recommended for cow's milk allergy.
- *Amino acid*—These formulas are manufactured from synthetic amino acids and do not contain peptide fragments. They are used for allergy and malabsorption conditions.
- *Carbohydrate*—Several lactose-free formulas are available without prescription, such as soy formula and lactose-free cow's milk–based formula. These usually replace the lactose with corn syrup solids (glucose). Low-lactose formula contains about 30 per cent lactose.
- *Fat*—This can be modified to include a higher quantity of medium-chain triglycerides, usually from fractionated coconut oil. These fats are more readily absorbed than long-chain fats in the absence of bile acids and pancreatic lipase. They are useful for infants with malabsorption conditions.

Table 6.4 Additives to infant formula and their interpretation

Label	Description and use	Evidence
AR	Anti-reflux or anti-regurgitation. Thickened with corn starch or carob bean gum.	Helps some infants with relief of symptoms of reflux and vomiting, but there may be other medical ways to treat this. Consult with doctor.
HA	Hypo-allergenic. Partially hydrolysed whey protein (peptides). Marketed for infants at risk of allergy (i.e., those with at least one first-degree relative with allergy).	Not better than breast milk, and of no benefit for low-risk infants. Meta-analyses show some prophylactic benefit for high-risk infants in reducing atopy when compared with cow's milk (Osborn & Sinn 2006). Not suitable for cow's milk protein allergy.
Lactose free, low lactose Colic	Lactose replaced with glucose or corn syrup solids. Colic formula has reduced lactose but is not lactose free.	True lactose intolerance quite rare in infancy. Need to discriminate between lactose overload and lactose intolerance. Not suitable for cow's milk protein allergy.
Added DHA and AA Gold	Added DHA and arachidonic acid. Potential benefits for visual acuity and brain development.	Cochrane Review showed inconclusive evidence in full-term infants (Simmer et al. 2011). ESPGHAN supported their optional inclusion where amount and ratio are controlled (Koletzko et al. 2005).
Pre- or probiotics Bifidus	Probiotics and bifidus formula contains beneficial bacteria.	ESPGHAN concluded that there was insufficient evidence to recommend their routine use (Braegger et al. 2011).
Oligosaccharides	Plant-based oligosaccharides encourage the growth of beneficial bacteria.	Insufficient evidence that plant-based oligosaccharides function in the same way as human milk oligosaccharides.
Diarrhoea	Designed for short-term use. Casein based. Low lactose (replaced with fructose and glucose). Higher levels of electrolytes. Fibre increased with apple and banana powder and cream of rice.	Limited.
Nucleotides	Benefits gastrointestinal tract and immune system.	ESPGHAN concluded that there was some evidence for benefit but inconclusive (Koletzko et al. 2005). Evidence of harm at very high levels.
Constipation	Whey based. Lactose slightly increased to increase osmolality. Ca:PO4 ratio increased to soften stool.	Limited.
Sweet dreams	Promoted to help infants sleep through the night by delaying gastric emptying. Casein based. Some lactose replaced with maltodextrin and starch.	Limited.

Note: DHA: docosahexaenoic acid; ESPGHAN: European Society for Paediatric Gastroenterology, Hepatology and Nutrition.

Soy and goat's milk formulas

Formulas based on soy provide an alternative for infants who are vegan. Soy formula is often advocated for use with cow's milk intolerance; however, between 30 and 60 per cent of infants with cow's milk protein allergy also react to soy protein, so an extensively protein-hydrolysed formula is more appropriate. Soy formulas are lactose free and therefore suitable for infants with

Safe preparation, storage and transport of infant formula

- Clean the preparation area.
- Wash hands with soap and water.
- Wash feeding and preparation equipment in hot soapy water. Rinse with clean water. (Sterilising equipment by boiling for five minutes is the preferred method. Other suitable methods include sterilising with a commercial antibacterial solution or using a steam or microwave steriliser. For these methods, follow the manufacturers' instructions.)
- Boil fresh water. Tap water is recommended in Australia and New Zealand as it contains fluoride. (If using an electric kettle, wait until the kettle switches itself off. Otherwise, bring the water to a rolling boil. A hot water urn can be used, provided the water supply is safe and the water dispensed is very hot.)
- Cool the water. (The WHO recommends cooling to no lower than 70 degrees Celsius before adding powdered formula (i.e., cooling for less than 30 minutes), as this decreases the risk of *Cronobacter sakazakii* (WHO & FAO 2007). However, this temperature also destroys some vitamins and is hot enough to cause scalds. This is an appropriate recommendation for infants at high risk of infection. The Australian government's Infant Feeding Guidelines recommend cooling the water for 30 to 60 minutes (NHMRC 2012), and the New Zealand Ministry of Health (2008) recommends boiling a 24-hour supply of water and keeping it in the fridge to use for each bottle as it is prepared.)
- Add the right amount of water to the bottle. It is preferable to make up one bottle at a time just before feeding.
- Add the exact amount of formula to the bottle according the instructions on the can. Use only the scoop provided in the formula can.
- Place the cap and seal on the bottle and shake gently until dissolved. (If using a cup instead of a bottle, apply the same step using a sterile spoon to mix the powder into the water.)
- Allow the milk to cool to an appropriate feeding temperature. This is particularly important if you have used very hot water. Test the temperature before feeding the infant.
- Discard any leftover prepared formula once feeding is completed. In addition:
  - Prepared formula held at room temperature for over an hour should be discarded.
  - Formula prepared in advance should be refrigerated immediately after preparation.
  - Prepared formula to be transported should be cold and held in a cool bag with ice packs. Either use the feed within two hours or transfer it to a refrigerator on arrival, where it can be stored for 24 hours from preparation.

lactose intolerance or **galactosaemia**. There is no evidence of allergy prevention with the use of soy-based formula.

Goat's milk is similar in composition to cow's milk. The main protein allergen in cow's milk, beta-lactoglobulin, is also present in goat's milk, so it is not a suitable alternative for cow's milk protein allergy.

Lactose intolerance is an absence or deficiency of the enzyme lactase, which results in the inability to digest lactose, the sugar in milk.

Galactosaemia is a rare genetic disorder that affects the body's ability to metabolise a sugar called galactose, which is a component of lactose.

Practical aspects of formula feeding

In Australia and New Zealand, most infant formula is available as powder, although a limited amount of liquid or ready-to-feed infant formula is also marketed. Care must be taken to correctly mix and store powdered formula to avoid contamination (see box above). Access to a clean water supply, facilities for boiling water and refrigeration, and soap for cleaning utensils and bottles as well as hand-washing are essential for the safe preparation of infant formula. While these facilities are taken for granted in most of Australia and New Zealand, there are some communities, both within Australia and New Zealand and in developing countries, where these facilities are not available.

In addition to potential contamination during preparation, powdered formula is unable to be produced sterile and is a potential source of pathogens.

Cronobacter sakazakii and *Salmonella* are of most concern in powdered formula; both can cause serious illness and death, particularly in the most vulnerable infants—that is, those who are premature, have low birthweight (LBW) or are immune compromised. Contamination with these bacteria is rare, and the risks of pathogen contamination can be reduced by proper handling and preparation.

A more common concern, identified in a number of studies, is problems with incorrect preparation, such as under- or over-concentrating formula and the addition of additives such as baby cereal or crushed biscuits. Under-concentration may happen when there are insufficient funds to buy more formula and the caregiver attempts to make it last longer. It may also occur simply from error: confusion can arise over scoop sizes and rates of mixing, as different formula companies mix at different rates (for example, one scoop per 30, 50 or 60 millilitres of water). Problems arise if parents switch formula but retain and use the scoop from the previous formula.

How much to feed

As with breastfeeding, healthy infants should be allowed to dictate the volume of milk they take at a feed. Encouraging an infant to finish the bottle forces them to override their natural satiety cues. As a general rule of thumb infants will take approximately 150 millilitres per kilogram of body weight per day of milk from a few days after birth. Before this, 30–60 millilitres per kilogram is sufficient, and this can be gradually graded up. It may be divided into smaller, more frequent bottles (for example, eight per day) for very young infants or larger, less frequent feeds as the infant matures.

COMPLEMENTARY FOODS

Complementary feeding is the name given to the introduction of soft and semi-solid foods as a complement to breastfeeding, and this generally happens at about six months of age. Other terms used for this are introducing solids or weaning, as it coincides with the time when breast milk intake starts to gradually decrease; however, the term complementary feeding is the most appropriate definition. Infants need complementary foods for two reasons: to meet their nutritional needs, which become increasingly difficult to achieve from breast milk or infant formula alone after six months of age, and to prepare them developmentally for family foods.

Ideal complementary foods provide nutrients that are becoming depleted in the body's stores (for example, iron and zinc) and can no longer be supplied in adequate amounts in breast milk. The term complementary indicates that solid foods are given in addition to milk rather than as a replacement, and that breast milk or infant formula should still form the basis of the diet until about twelve months of age.

Feeding an infant is a time of nurturing and an opportunity for positive interaction between the caregiver and the child. The infant is learning new skills and being exposed to different and stimulating tastes and experiences, so it is important to feed patiently and provide encouragement during this process. The caregiver should respond to the infant's cues regarding hunger. Ideally, the infant should have their own bowl in order to monitor the amount of food taken.

When to introduce complementary foods

Both the Australian government's Infant Feeding Guidelines (NHMRC 2012) and the New Zealand Ministry of Health's (2008) Food and Nutrition Guidelines for Healthy Infants and Toddlers recommend exclusive breastfeeding until, and introduction of complementary foods at, around six months of age, with continued breastfeeding after that. As mentioned earlier, these are population recommendations; consistent with other development milestones, infants will vary by a few weeks on either side of six months in their readiness for solids. In places where sanitation is an issue, complementary foods are best delayed in breast-fed infants until six months to avoid exposure to pathogens. After this time, exposure to environmental microbes is inevitable as the infant explores their environment.

Six months of age is around the time that an infant will show signs of readiness for food such as increased appetite, chewing and biting on fists and objects, less tongue thrust, better manoeuvrability of the tongue and excitement or reaching for foods. The infant can sit and hold their head stable, and their digestive system has matured enough to digest starches.

Early introduction of complementary foods can lead to displacement of breast milk, an increased risk of food allergies and exposure to pathogens. Failure to introduce at around six months of age is also associated with some problems. Body stores of iron and zinc that the infant is born with are often depleted by this stage, and there is a risk of micronutrient deficiencies as well as poor weight gain due to inadequate intake. Oral motor development may also be delayed.

What foods to introduce

Between six and twelve months of age, the complementary foods offered should be progressively graded up in both the texture and the amount offered. As a guiding principle, complementary foods should include foods from the core food groups and be based on what the family usually eats, meaning they are locally available, affordable and culturally acceptable. Excessive sugar and salt consumption should be avoided.

As the infant approaches twelve months of age, a wide variety of nutritious foods, according to national dietary guidelines, should be introduced, as this is the time that food preferences are formed. It is quite usual for infants to reject new foods because of **neophobia**; this is thought to be an evolutionary protective response, though parents often interpret food refusal as dislike of the food. However, repeated exposure will overcome neophobia. Flavours that the infant has been previously exposed to through the breast milk after the mother has eaten the food tend to be more readily accepted; this is why healthy family foods are a good choice.

Commercial baby foods are not necessary but may be convenient on occasion, especially when away from home or travelling. Food hygiene and careful food-handling practices are essential to avoid contamination, which can cause diarrhoea. Utensils and infants' and parents' hands should be washed thoroughly before food preparation and eating. Food should be refrigerated or eaten within two hours if refrigeration is not possible.

There is no particular order in which foods should be introduced, but it is important that the first foods are high in iron and have a greater energy density than breast milk or formula, as they tend to displace the amount of milk taken. Initially, they should also be smooth or well mashed in texture. Suitable choices include iron-fortified cereals or thick porridge, and well-mashed or finely minced moist family foods such as meat, chicken, fish, vegetables, fruits, custards and yoghurts. Vegetarian alternatives such as legumes, lentils and tofu are also suitable, provided the consistency is appropriate and some foods rich in vitamin C are given concurrently to enhance iron bioavailability. However, plant foods do not usually provide enough iron or zinc and they should be supplemented with acceptable animal foods, such as egg yolk if available, or consideration should be given to using fortified complementary foods, such as iron-fortified cereals. Parents should start with small amounts of food and increase them as the infant grows.

Neophobia is a fear of new things (including unfamiliar foods).

Introducing new foods often initiates a change in the appearance, texture and frequency of bowel actions in the infant. This is quite normal and should not necessarily be interpreted as intolerance, although severe gastrointestinal symptoms should be investigated. At around six to eight months, an additional 840 kilojoules are needed, and this requirement can be met by one-quarter to one-half of a cup of mashed foods given two to three times a day.

As well as meeting the infant's nutrient requirements, appropriately textured complementary foods help develop oral motor skills. Infants should progress from smooth foods to soft lumpy, minced and chopped foods. There is some evidence that if infants are not introduced to lumps before ten months of age they may develop feeding difficulties later on.

By about eight months of age, most infants will enjoy trying to self-feed and most will have the manual dexterity to manage finger foods such as pieces of cooked vegetable, cubes of soft cheese and pasta. At around nine to eleven months, an additional daily 1250 kilojoules are required from complementary foods, which can be met by three meals a day of approximately half a cup of chopped, mashed or finger foods, with one to two snacks depending on appetite.

By twelve months of age, the infant should progress to family foods, mashed or chopped if necessary, with plenty of variety, according to the Australian Dietary Guidelines (NHMRC 2013a) or the Food and Nutrition Guidelines of New Zealand (Ministry of Health 2008). At this age an infant will require an additional 2300 kilojoules per day and will usually eat three meals of three-quarters to one cup a day with one or two snacks depending on appetite.

Table 6.5 summarises the process of complementary feeding and shows the gradual increases in quantity, variety and texture that should occur over this period. Note that age ranges, rather than specific ages, are given, as there can be wide individual variation.

Table 6.6 shows the Australian Dietary Guidelines' recommendations for daily serves from the five good groups for infants from seven to twelve months (NHMRC 2013b). An allowance for unsaturated spreads or oils or a half-serve of nut or seed paste (4–15 grams) per day is included. A sample menu is provided in Table 6.7.

Table 6.5 Suitable complementary foods for infants (6–12 months)

Age (months)	Developmental stage	Suitable complementary foods	Amount
~6	Has head control Sits supported Shows interest in food Follows food with eyes Does not push food out of mouth with tongue Can move food around mouth	Pureed or well mashed in texture Spoon feeding Fruits, vegetables, iron-fortified cereal Pureed meat, poultry, fish Pureed or well-mashed tofu, legumes	2–3 tablespoons twice a day
7–10	Sits alone easily Chews Bites soft foods Uses finger and thumb to pick up food Begins to drink from a cup	Soft cooked mashed foods Start to introduce some lumps Introduce finger foods Soft cooked meat, poultry, fish, tofu, legumes Cooked vegetables mashed or cut into pieces Well-cooked rice, pasta Toast, crackers Yoghurt, cheese, custards Mashed soft ripe fruit, pieces of peeled soft fruit (e.g. banana, peach, avocado)	Grade up to 3 meals a day of approximately ½ cup plus 1 or 2 snacks
9–12	Better motor control Picks up food May attempt to use spoon for self-feeding	Finger foods, soft cooked family foods Breakfast cereal with milk Finely cut-up or flaked meat, poultry, fish Tofu, legumes, eggs Soft cooked vegetables Yoghurt, cheese, custards Pasta, rice, other grains Toast, crackers, bread Soft raw fruits, stewed fruit	Three meals a day of ¾–1 cup plus 1 or 2 snacks depending on appetite

Table 6.6 Recommended daily servings of foods for infants (7–12 months) in Australia

Food group	Amount per serve	Number of serves
Vegetables and legumes/beans	1 heaped tablespoon (20 g)	1½–2
Fruit	1 heaped tablespoon (20 g)	½
Grain (cereal) foods	1 slice (40 g) bread or equivalent	1½
Dried infant cereal	2 heaped tablespoons (20 g)	1
Lean meats, poultry, fish, eggs, tofu	30 g	1
Breast milk or formula	Just under 2½ cups (600 mL)	1
Yoghurt/cheese or alternative	1 tablespoon (20 mL) yoghurt or 10 g cheese	½

Source: NHMRC (2013b).

Foods to avoid

Infants are at higher risk of choking than adults because they have not yet developed back teeth for grinding up food and are still learning the skill of manoeuvring food within their mouths. Foods need to be soft enough for infants to squash with their gums or suck until softened (such as cracker biscuits). Hard foods such as pieces of carrot, celery, whole nuts and seeds and large pieces of crunchy fruit should be avoided, as they can cause choking. Parents can modify the consistency of these foods before giving them to infants (for example, by grating carrots). Small round foods such as whole grapes and cherry tomatoes also pose a risk of choking, as they can slip easily to the back of the throat

Table 6.7 Sample daily menu for infants (7–12 months)

Meal	Food	Vegetables	Fruit	Grain	Meat	Dairy
Breakfast	2 heaped tablespoons (20 g) infant cereal mixed with breast milk, formula or cow's milk			1		
	½ slice wholemeal bread or toast with fat spread			½		
Mid-morning snack	½ banana		½			
	20 mL yoghurt					½
Lunch	30 g minced meat				1	
	1 heaped tablespoon (20 g) soft mashed vegetables	1				
Mid-afternoon snack	½ slice wholemeal bread with fat spread			½		
Dinner	¼ cup baked beans	½				
	½ slice wholemeal bread or toast			½		
Throughout day	600 mL breast milk or formula					1
Total		**1½**	**½**	**2½**	**1**	**1½**

without chewing. These foods are best cut in half first. Infants should not be left unattended to feed. Other unnecessary or possible harmful foods to avoid include juice and sweetened beverages, because of the risk of dental caries, and honey, as it may contain *Clostridium botulinum*, a toxin-producing bacterium.

COMMON CONCERNS AND HEALTH CONDITIONS

Preterm Infants

Preterm infants are defined as those born before 37 weeks' gestation and will include the smallest, sickest infants, born at around 24 weeks, as well as late preterm infants born at 36 weeks. These infants have special requirements and are best managed by a paediatric dietitian. Very preterm infants are unable to suck and require their nutrition to be delivered via a feeding tube. The ability to suck effectively develops at approximately 34 weeks. A comprehensive discussion of the nutrient requirements of preterm infants and how best to meet these is outside of the scope of this chapter; however, breast milk is best for any infant, and this is even more the case for small vulnerable preterm infants, who are at risk of infection and whose gastrointestinal tract and other organs involved in digestion and metabolism have not yet fully matured.

It is difficult to make recommendations for such a diverse group of infants as a whole; hence, they are often further classified into LBW of less than 2000 grams, very low birthweight of less than 1500 grams and extremely low birthweight of less than 1000 grams. The same general feeding guidelines apply to all preterm infants, with some modifications depending on birthweight. Requirements for protein are much higher if preterm infants are to meet the in utero growth rates. The milk from mothers who deliver preterm infants is higher in protein content than that from mothers who deliver a full-term infant. However, even though initially higher, the protein content of the mother's milk falls rapidly, and for very low birthweight infants it is generally not enough to meet requirements. It is therefore standard practice to use a **human milk fortifier** for these infants, which is added to the expressed breast milk and delivered via a feeding tube.

In the absence of enough mothers' own milk, many neonatal health professionals believe that milk from another human mother is the best choice for preterm infants. In Australia there is a resurgence of interest in breast milk banks and the use of donor human milk for preterm infants. In the absence of human milk, there are special preterm infant formulas available. These are higher in protein, calcium, phosphorous and iron than regular formulas. They are normally used only during the neonatal admission and are available as ready-to-feed

Human milk fortifier is a bovine-based supplement containing protein, some carbohydrate or fat and selected vitamins and minerals that is added to breast milk for very preterm infants to meet their increased nutrient requirements.

CASE STUDY 6.2 INDIGENOUS INFANT

Amanda Lee

Profile and problem

Miss K. is a seven-month-old infant who has had diarrhoea for two days and is only 37 per cent standard weight-for-length. Her mother, Mrs D., is a nineteen-year-old Aboriginal woman and Miss K. is her first child. Mrs D. usually resides in a remote community in the north of South Australia but has been living in a town camp in Alice Springs with her cousins for the past three months, following the death of her mother from renal disease. Mrs D. and her cousins are unemployed and receive welfare payments. Since the oven broke they have been living on takeaway foods, mainly hot chips and pizzas.

Mrs D. has breastfed Miss K. from birth but started her on formula two weeks ago as she was worried that she was not producing enough breast milk. Miss K. still breastfeeds at night and sometimes during the day. When they can afford to buy baby food, Mrs D. and her cousins have been feeding Miss K. directly from the jar since she was about five months of age. Mrs D. is worried that Miss K. has diarrhoea because she is allergic to the baby food.

Assessment

In the house where they are living, there is a kitchen sink and hot water, a functioning electric kettle, two saucepans, a fry pan and some other cooking utensils, but the fridge and the electric stove are not working. Mrs D. has been making up Miss K.'s formula with water directly from the tap. She knows it is important to correctly measure the right amount of dried powder but has lost the measuring cup that came in the tin she bought at the supermarket; she thinks a camp dog may have taken it.

Now she is living in the Northern Territory, Mrs D. realises that she may need to reorganise her welfare payments due to income management under the Northern Territory intervention. She says she is too shy to go to the Department of Community Services to sort things out and has been living on what little her cousins can share with her.

Treatment

Miss K. has been referred to the clinic doctor and is being treated for her diarrhoea. The health service social worker is assisting Mrs D. with her financial arrangements and has organised interim emergency food relief. The housing council has been alerted and will fix the fridge and oven this week. Mrs D. has agreed to do the following:

- continue to breastfeed Miss K. and feed her as frequently as required
- discontinue the infant formula
- boil all water to be given to Miss K., put it in a clean covered container and allow it to cool, preferably in the fridge when it is mended
- start feeding Miss K. iron-fortified infant cereal and other suitable foods when her diarrhoea has resolved
- try to follow a more nourishing diet herself
- accept provision of practical support until Miss K. is thriving again.

Mrs D. has agreed to home visits from health professionals, including a health worker/elder to whom she is distantly related, at least every day in the first instance.

Plan

Miss K.'s growth and progress will be monitored at least weekly, and early intervention will be provided if required.

formula, although some manufacturers now market a post-discharge powdered preterm formula.

Preterm infants, especially very low birthweight infants, often have a complicated clinical course initially, and these effects frequently last into childhood and even beyond. They are more prone to developmental delays, and their growth after discharge appears not to catch up to that of infants born at term. It is therefore important to monitor growth carefully.

Growth of preterm infants is plotted on special preterm growth curves until two weeks after the estimated date of delivery—that is, at 42 weeks gestation. After the estimated date of delivery, preterm infants can be plotted on the WHO Child Growth Standards charts, but allowances need to be made for the prematurity, and this is done using corrected age: the infant's current age minus how many weeks prematurely they were born. For example, if an infant was born at 34 weeks' gestation, they are six weeks early (40 – 34 = 6), and each growth measurement needs to be adjusted for this. So if a premature infant born at 34 weeks' gestation is now twelve weeks postnatal, their corrected age is 12 – (40 – 34) = 6 weeks.

To avoid any misunderstanding about what the plotted point on a growth chart is representing, especially for other people looking at the chart, it is usual to first plot the infant's measurement at their actual age and then draw an arrow back to their corrected age (see Figure 6.7). Corrected age is used for the first twelve months for infants born 32–36 weeks' gestational age and for two years for infants born under 32 weeks' gestation.

Infants born very prematurely miss out on the accretion of body stores which usually occurs in utero in the last 12–13 weeks of gestation. In particular, they have not accumulated the long-chain polyunsaturated fatty acids in brain and eye tissue so benefit from breast milk or a formula fortified with long-chain polyunsaturated fatty acid. Most preterm formulas contain long-chain polyunsaturated fatty acids.

Iron is also an issue, as most iron is accreted in the third trimester; hence, body stores in the premature infant may deplete before six months, when solids are usually introduced. These infants may benefit from a slightly earlier introduction of solids (but not before seventeen weeks of age) with an emphasis on iron-fortified and iron-rich foods. Before introducing solids it is important to ascertain whether the infant is developmentally ready to safely receive solid foods. Premature infants can have significant developmental delays, and good head control and a degree of oral manoeuvrability is required for safe oral feeding. Safety must be balanced with nutritional needs.

Dental caries

Dental caries occurs when sweetened liquids or foods come into contact with teeth for a prolonged period of time. Mouth bacteria convert the sugar into acids which attack the tooth enamel. The sugars can be natural (that is, lactose found in milk and fructose in fruit juice) or added as sucrose.

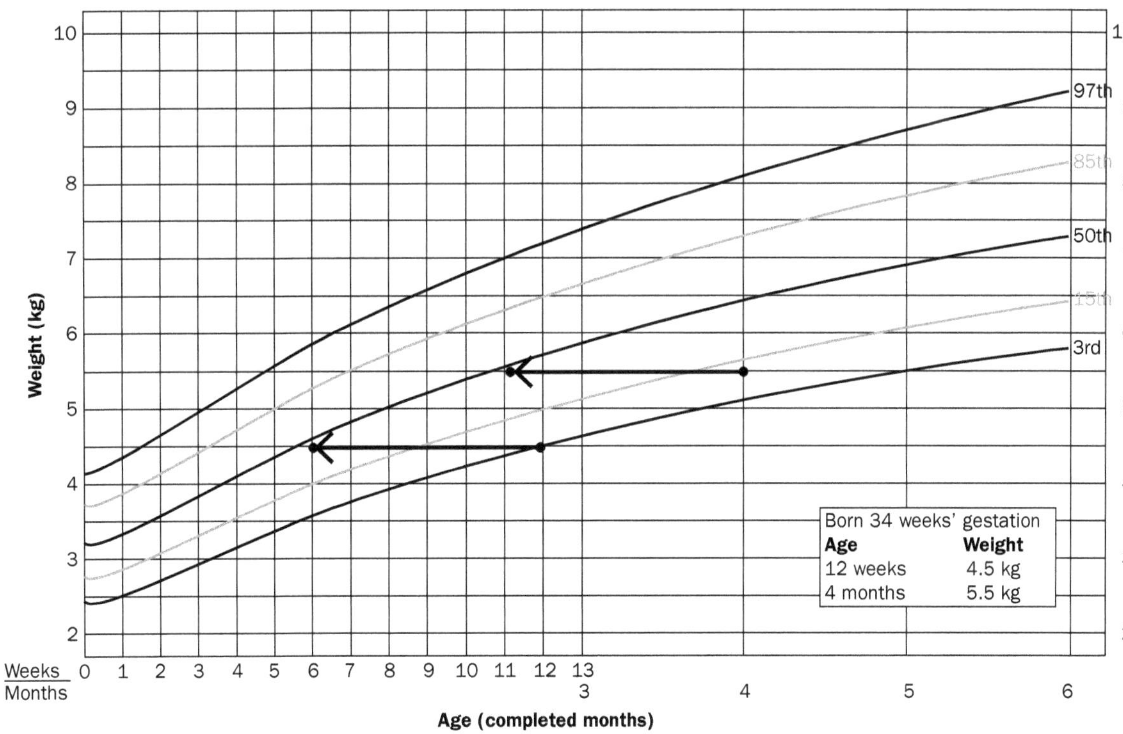

Figure 6.7 Plotting corrected age for preterm infants

Source: Adapted from WHO (n.d.).

Bottle caries occurs when infants are put to bed with a bottle, which allows the liquid to drip and pool around the infants' teeth. Saliva flow decreases when infants sleep, so the normal cleansing of teeth that occurs is reduced. Infants who are breastfed to sleep seem to be at lower risk of, but not immune to, developing caries than infants who are put to bed with a bottle. This is partly because a bottle will continue to slowly leak after the infant has stopped sucking, whereas a breast will release milk only when being actively sucked. There may also be some protection in the breast milk from antibodies, which may impede bacterial growth.

Another common cause of caries is the use of dummies or pacifiers that have been dipped in something sweet—for example, jam or honey. Cavity-causing bacteria can also be transferred from the caregiver's mouth to the infant by the caregiver cleaning the dummy in their own mouth before giving it to the infant or by sharing feeding spoons.

To reduce dental caries in infants, they should not be put to bed with a bottle. During the day, only milk (either breast milk or infant formula) or water should be given via a bottle. If infant formula is used, it should be made with tap water, which, for most of Australia, is fluoridated. If fluoridated water is not available, supplementation is not required for the first six months. After that, parents should seek advice from a dentist. Fruit juice is not necessary, and sweetened drinks such as cordials and soft drinks should be avoided. Dummies or pacifiers, if used at all, should be cleaned under running water and never coated with anything. Caregivers need to avoid saliva transfer to the infant. Nutrition educators should encourage healthy eating habits.

Allergies

In Australia, New Zealand and other developed countries, the prevalence of allergy has increased dramatically over the past three or four decades. Early diet appears to play a role in this, as infants' gastrointestinal tracts are immature and may allow the passage of peptide fragments large enough to cause an immune response. The most common food allergies are to eggs, cow's milk and peanuts. Not surprisingly, breastfeeding exclusively for around six months is associated with the lowest risk of developing allergies. If infant formula is used, infants classified as high risk (that is, at least one first-degree relative with allergy) may benefit from the prophylactic use of a partially hydrolysed formula. However, if a cow's milk allergy has been diagnosed, an extensively hydrolysed formula is indicated. Continuing breastfeeding while potential allergens are introduced also seems to confer some benefit for tolerance.

There is still some debate about the optimum time to introduce potential food allergens to high-risk infants. Very early introduction of solids is associated with a higher risk of allergy. It was previously thought that delaying the introduction of high-allergen foods beyond six months would decrease the risk of allergy, but in fact this may be associated with an increased risk. It seems that around six months is the perfect time to promote tolerance of foods, particularly if this is accompanied by breastfeeding. Some mothers who are anxious to spare their children from allergies restrict their own diet during pregnancy or lactation. There is no evidence that this prevents allergies; indeed, it may put the mothers' own nutritional intake at risk, as many of the commonly eliminated foods, such as dairy and wheat, contribute significant nutrients to our diet. For infants with diagnosed atopic disease the evidence for maternal diet restriction of the allergen is less conclusive. Advice should be sought from a paediatrician and paediatric dietitian to ensure that the mother's high nutrient requirements during lactation are met.

SUMMARY AND KEY MESSAGES

Nutrition in infancy impacts on health throughout the entire life cycle, and optimising feeding practices is the most effective intervention to improve child health.

- Growth is an indicator of health and should be monitored appropriately using the WHO Child Growth Standards.
- Breast milk is the ideal food for infants. Infant formula should be used for the first twelve months if breast milk is not available.
- Complementary foods should be introduced at around six months of age and should include iron-rich foods. It is important to gradually change the consistency from smooth purees to mashed foods and to incorporate some lumps or foods that require chewing by ten months of age.
- By twelve months of age, infants should be eating a range of family foods from the core food groups, according to either the Australian Dietary Guidelines (NHMRC 2013a) or the New Zealand Food and Nutrition Guidelines (Ministry of Health 2008).

REFERENCES

AIHW (Australian Institute of Health and Welfare) 2011. *2010 Australian National Infant Feeding Survey: Indicator results*. Cat. no. PHE 156. Commonwealth of Australia. <www.aihw.gov.au/WorkArea/DownloadAsset.aspx?id=10737420925>, accessed 26 September 2014.

Braegger, C., Chmielewska, A., Decsi, T., Kolacek, S. et al. 2011. Supplementation of infant formula with probiotics and/or prebiotics: A systematic review and comment by the ESPGHAN committee on nutrition. *Journal of Pediatric Gastroenterology and Nutrition* 52: 238–50.

Butte, N.F., Lopez-Alarcon, M.G. & Garza, C. 2002. *Nutrient Adequacy of Exclusive Breastfeeding for the Term Infant during the First Six Months of Life*. WHO. <http://whqlibdoc.who.int/publications/9241562110.pdf>, accessed 26 September 2014.

CDC (Centers for Disease Control and Prevention) 2009. *Individual Growth Charts*. CDC. <www.cdc.gov/growthcharts/charts.htm >, accessed 29 September 2014.

FSANZ (Food Standards Australia New Zealand) 2013. *Standard 2.9.1: Infant formula products*. Commonwealth of Australia. <www.health.gov.au/internet/main/publishing.nsf/Content/4DCF744789D1AF64CA257BF0001C9622/$File/standard2.9.1-infant-formula-products.pdf>, accessed 28 September 2014.

Koletzko, B., Baker, S., Cleghorn, G., Neto, U.F. et al. 2005. Global standard for the composition of infant formula: Recommendations of an ESPGHAN coordinated international expert group. *Journal of Pediatric Gastroenterology and Nutrition* 41: 584–99.

Kuczmarski, R.J., Ogden, C.L., Guo, S.S., Grummer-Strawn, L.M. et al. 2002. 2000 CDC Growth Charts for the United States: Methods and development. *Vital Health and Statistics* 11(246). CDC. <www.cdc.gov/nchs/data/series/sr_11/sr11_246.pdf>, accessed 29 September 2014.

Ministry of Health 2008. *Food and Nutrition Guidelines for Healthy Infants and Toddlers (Aged 0–2): A background paper* 4th edn, partially revised December 2012. New Zealand Government. <www.health.govt.nz/system/files/documents/publications/food-and-nutrition-guidelines-healthy-infants-and-toddlers-revised-dec12.pdf>, accessed 16 March 2014.

NHMRC (National Health and Medical Research Council) 2012. *Eat for Health: Infant Feeding Guidelines; Information for health workers*. Commonwealth of Australia. <www.nhmrc.gov.au/_files_nhmrc/publications/attachments/n56_infant_feeding_guidelines.pdf>, accessed 28 September 2014.

——2013a. *Eat for Health: Australian Dietary Guidelines; Providing the scientific evidence for healthier Australian diets.* Commonwealth of Australia. <www.eatforhealth.gov.au/sites/default/files/files/the_guidelines/n55_australian_dietary_guidelines.pdf>, accessed 28 September 2014.

——2013b. *Eat for Health: Educator guide; Information for nutrition educators.* Commonwealth of Australia. <www.eatforhealth.gov.au/sites/default/files/files/the_guidelines/n55b_educator_guide_140321.pdf>, accessed 30 September 2014.

NHMRC (National Health and Medical Research Council) & Ministry of Health 2006. *Nutrient Reference Values for Australia and New Zealand: Including recommended dietary intakes.* Commonwealth of Australia & New Zealand Government. <www.nhmrc.gov.au/_files_nhmrc/publications/attachments/n35.pdf>, accessed 26 September 2014.

Osborn, D.A. & Sinn, J. 2006. Formulas containing hydrolysed protein for prevention of allergy and food intolerance in infants. *Cochrane Database Systematic Review.* 2006:CD003664.

Simmer, K., Patole, S.K. & Rao, S.C. 2011. Longchain polyunsaturated fatty acid supplementation in infants born at term. *Cochrane Database Systematic Review.* 7;(12):CD000376. doi: 10.1002/14651858.CD000376.pub3.

WHO (World Health Organization) 2008a. *Training Course on Child Growth Assessment: WHO Child Growth Standards; B, Measuring a child's growth.* WHO. <www.who.int/childgrowth/training/module_b_measuring_growth.pdf>, accessed 29 September 2014.

——2008b. *Training Course on Child Growth Assessment: WHO Child Growth Standards; E, Photo booklet.* WHO. <www.who.int/childgrowth/training/module_e_photo_booklet.pdf?ua=1>, accessed 29 September 2014.

——n.d. *The WHO Child Growth Standards.* WHO. <www.who.int/childgrowth/standards/en/>, accessed 28 September 2014.

WHO (World Health Organization) & FAO (Food and Agriculture Organization) 2007. *Safe preparation, storage and handling of powdered infant formula: Guidelines.* WHO. <www.who.int/foodsafety/publications/micro/pif_guidelines.pdf>, accessed 14 November 2014.

WHO MGRSG (World Health Organization Multicentre Growth Reference Study Group) 2006. WHO Motor Development Study: Windows of achievement for six gross motor development milestones. *Acta Paediatrica* Suppl. 450: 86–95.

ADDITIONAL READING

1,000 Days (website) 2014. <www.thousanddays.org/>, accessed 28 February 2014.

World Health Organization 2009. *Infant and Young Child Feeding: Model chapter for textbooks for medical students and allied health professionals.* WHO. <http://whqlibdoc.who.int/publications/2009/9789241597494_eng.pdf?ua=1>, accessed 25 September 2014.

——n.d. Multicentre Growth Reference Study. *Training Course and Other Tools.* WHO. <www.who.int/childgrowth/training/en/>, accessed 4 September 2013.

World Health Organization & United Nations Children's Fund 2003. *Global Strategy for Infant and Young Child Feeding.* WHO. <http://whqlibdoc.who.int/publications/2003/9241562218.pdf?ua=1>, accessed 28 September 2014.

TEST YOUR UNDERSTANDING

Note: There is only one correct answer for each question.

1. On what are the nutrient requirements for infants from birth to six months of age based?
   a. Experimental studies that establish the dietary intake level needed to avoid deficiency in this age group
   b. Experimental studies that establish the dietary intake level needed to avoid deficiency in older children whose data are extrapolated for this age group
   c. The concentration of the nutrient found in a breast milk volume of 780 millilitres per day
   d. The concentration of the nutrient found in a breast milk volume of 1000 millilitres per day
   e. None of the above
2. Which of the following is not true of Z-scores?
   a. Z-scores are defined as the number of standard deviations above or below the mean of the population
   b. Z-scores are defined as the rank position of an individual compared to a reference population
   c. Z-scores can be used to calculate the prevalence of malnutrition in a population group
   d. Z-scores can be both positive or negative values
   e. A Z-score of 0 means the score is equal to the mean.
3. What is the recommended dietary intake (RDI) for protein for infants from birth to six months?
   a. 0.8–1 gram per kilogram of body weight per day
   b. 1.4 grams per kilogram of body weight per day
   c. 1.6 grams per kilogram of body weight per day
   d. 2.4 grams per kilogram of body weight per day
   e. None of the above
4. According to the relevant Australian and New Zealand authorities, complementary foods should be introduced at what stage?
   a. Three months
   b. Around six months
   c. Delayed as long as possible to optimise breast milk intake
   d. No specific age but should be based on infant's readiness to feed, such as losing gag reflex, being able to sit up, able to grasp for food
   e. None of the above
5. Which of the following factors does not impact on a child's growth?
   a. Ethnicity
   b. Socio-economic status
   c. Nutrition
   d. Genetics
   e. All of the above
6. What is the most important factor when monitoring infant growth?
   a. Plotting both weight and length measurements
   b. A series of measurements over time
   c. Appropriate equipment
   d. A sleeping child
   e. None of the above
7. Which of the following can cause failure to thrive?
   a. An illness that causes nutrients to be lost through vomiting and diarrhoea
   b. Maternal depression
   c. Delayed oral motor skills
   d. Inadequate intake of food and drink
   e. All of the above
8. What should the main drink for infants from birth to twelve months be?
   a. Water
   b. Cow's milk
   c. Goat's milk
   d. Breast milk or infant formula
   e. Fruit juice
9. Why is cow's milk unsuitable as the main drink for infants less than twelve months old?
   a. It is low in iron
   b. It is high in protein compared with breast milk or infant formula
   c. It is difficult to digest
   d. It is low in essential fatty acids compared with breast milk or infant formula
   e. All of the above
10. To what should infants be introduced at around six months of age?
    a. Baby rice
    b. Mashed fruit
    c. Mashed vegetables
    d. Pureed meat
    e. Food that is high in iron and smooth in texture (there is no particular required order of foods)

STUDY QUESTIONS

1. What 'nutritional' additives are currently added to commercial infant formulas? Which of these do you think have a legitimate function? What role do these formulas play in our society?
2. Discuss the appropriate age to introduce solids to an infant and the rationale for this.
3. A mother and her four-month-old baby daughter have come to see you for advice about the daughter's growth. The mother is concerned because her baby was on the 50th percentile for weight at birth but then immediately lost weight and dropped to the 25th percentile at three weeks of age and has tracked along this percentile since then. The mother expected that her daughter would catch back up to the 50th percentile. The mother enjoys breastfeeding and it is going well, but she feels her breast milk may not be good enough and is wondering if she should switch to formula. What is your assessment of this infant's growth and what would you say to this mother to reassure her about breastfeeding?
4. What advice would you give regarding formula selection to mothers who are not able to breastfeed their infants and have a family history of allergy to cow's milk?
5. A young breastfeeding vegan mother wants to introduce complementary foods to her six-month-old infant. She supplements the baby with almond milk that she makes herself from a recipe a naturopath provided. She does not want to use commercial baby foods and avoids fortified foods such as infant cereal. What are the key nutrients of concern and what advice can you give her about complementary foods and suitable breast milk substitutes?

7

Toddler and preschooler (1–5 years)

Evelyn Volders

LEARNING OUTCOMES

Upon completion of this chapter you will be able to:

- state the nutrient requirements of this age group
- describe the developmental tasks of toddlers and preschoolers
- identify some of the behavioural issues associated with eating in this age group
- recommend some techniques to assist in managing behavioural issues
- identify the common health issues in this age group
- describe the nutritional issues and standards associated with formal child care.

This chapter explores the eating practices and nutritional needs of children between the ages of one and five years. This is a time of development when children learn to communicate and are increasingly physically able and active. The chapter examines the growth and development of toddlers and preschoolers and the outside influences that might impact on what a child eats and outlines some of the common nutritional problems observed in this age group, including fussy eating, constipation, iron deficiency and dental caries. Finally, the division of responsibilities and nutrition in child care is explored.

A toddler is generally considered to be a young child between the ages of one and two and a half years. During this period young children develop independence and autonomy as they master various developmental skills including walking and talking. Toddlers strive to establish themselves as independent from parents and are very interested in exploring their environment. These developmental tasks can influence behaviour and cause anxiety as they explore limits and practise exerting control over emotions; this might be exhibited as temper tantrums or neophobia and can be reflected in food intake. Toddlers have high nutritional needs relative to their size and can have variable appetites related to their growth and physical activity.

The preschool years follow on from toddlerhood and include children until the time they commence school, usually between five and six years of age. Many preschoolers attend formal school preparation programs and develop relationships with friends and teachers that further influence their attitudes towards and practices with food and eating.

PHYSIOLOGICAL AND PSYCHOLOGICAL CHANGES

Growth rates slow considerably after the first year, with the trunk and limbs becoming longer and the head becoming proportionally less large when compared to length. The World Health Organization's (WHO) Child Growth Standards charts give the standard growth for children tracking the 50th percentile, as shown in Figure 7.1. Both sexes continue to experience a slow growth rate after passing two years of age; for example, boys gain only 2.1 kilograms in the year from age two to three (WHO n.d.). Similarly, growth in length also progresses at a slower rate. It is recommended to measure recumbent length until two years of age and then standing height from two years on. However, it can be difficult to obtain an accurate measurement of wriggling toddlers regardless of whether length or height is measured.

The skills developed in the first year of life are consolidated and built upon as children grow through the toddler and preschool years. During these stages children begin to run, jump, hop and throw with increasing coordination. Their fine motor skills also improve as they learn to dress themselves, draw, use scissors and tie shoe laces. Language skills rapidly increase in early childhood, allowing children to express their hunger, thirst and food preferences. They also develop cognitively and socially.

The toddler years are a time of transition regarding food and eating. Toddlers change to a diet comprising primarily solid foods rather than liquids. Feeding skills acquired at this age include competent use of a cup, efficient finger feeding and mostly efficient spoon feeding with the improvement in fine motor skills and visual coordination. Generally, by two years of age children can eat with toddler utensils that have short but enlarged-width handles to make them easier to grasp and manipulate, but it can take until four years of age to eat skilfully with a knife and fork. Many toddlers prefer to feed themselves independently rather than being spoon fed as they develop cognitively and emotionally. Toddlers and preschoolers may also enjoy helping to prepare food, which in turn may increase their acceptance of the food provided. Toddlers can help with simple tasks such as layering a sandwich, while older preschoolers can assist with most aspects of meal preparation.

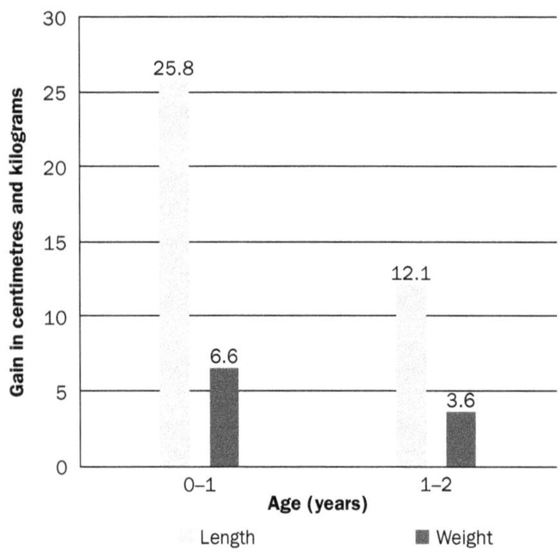

Figure 7.1a Standard weight and height gain for boys on 50th percentile (0–2 years)

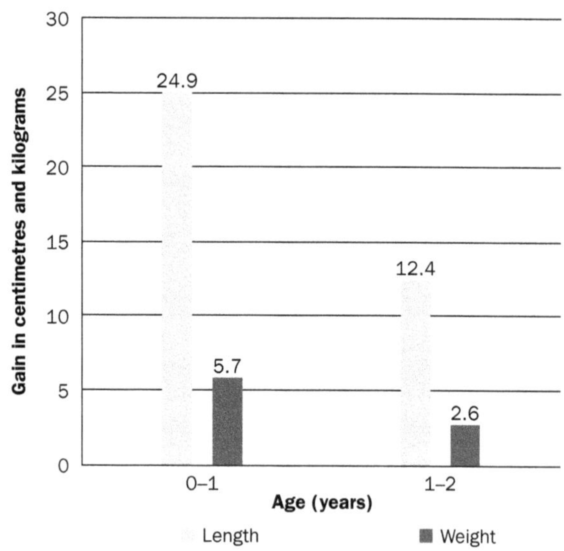

Figure 7.1b Standard weight and height gain for girls on 50th percentile (0–2 years)

Note: The WHO (n.d.) Child Growth Standards charts for weight and length are recommended for use in Australia to two years of age. After this age common practice is to change to the US Centers for Disease Control and Prevention's Clinical Growth Charts of weight, height and body mass index for monitoring growth in children and adolescents aged two to eighteen years (CDC 2009). Slight differences in percentiles might be seen when changing from the WHO to the Centres for Disease Control and Prevention charts.

Source: Adapted from WHO (n.d.).

The first set of molar teeth erupt between thirteen and nineteen months of age, followed by the incisors for biting and then the second set of molars between two and three years, making chewing hard textures more effective. Toddlers are able to cope with most textures of foods as rotary chewing skills develop but still require supervision to avoid choking (see below) as they may not be capable of dealing with large pieces of hard foods. Most of these foods can be offered in a form that is easier to manage, such as chopped or grated.

By one year of age, children have the strength and coordination required to drink from a cup. Parents and caregivers may choose to provide non-spill lids for cups as this skill is developed. Bottles should not be used for drinks after one year of age, as their continued use is linked with both dental caries and obesity. Bottle-fed children may have become particularly attached to their bottle as part of a comforting routine, and it may be helpful to engage children in choosing their own 'big' cup to create a positive transition.

NUTRIENT REQUIREMENTS AND INTAKES

Although growth rates slow after the first year of life, toddlers and preschoolers still demand nutrient-dense foods to support ongoing growth and development and increasing levels of activity. Gradual increases in nutrient needs at this age are proportional to body size and growth rates.

Energy requirements change over time; the requirements for toddlers and preschoolers are given in Tables 1.5 and 1.6. The estimated energy requirements (EERs) to two years of age consider age, gender, weight, length and a growth factor. After two years of age it is recommended to calculate the basal metabolic rate (BMR) using body weight and then to apply the appropriate physical activity level (PAL; NHMRC & Ministry of Health 2006; see Chapter 1).

Due to continued growth and development of all of the body's organs, including the brain, optimal nutrient intake continues to be important during the toddler and preschool years. Young children require sufficient amounts of a variety of nutritious foods to meet their nutrient requirements.

Nutrients of particular importance for this age group include protein, B vitamins, iron, calcium and vitamin D. Protein is needed for muscle growth as well as the production of collagen, enzymes, hormones, antibodies and a number of other essential functions. The B vitamins (including thiamin and folate) and iron are required for energy metabolism, and calcium and vitamin D for bone development. The current Australian nutrient reference value (NRV) for vitamin D (adequate intake [AI] of 5 micrograms) is deemed to be out of date by some researchers, who instead suggest an estimated average requirement (EAR) of 10 micrograms per day for this age group (Paxton et al. 2013).

The NRVs are based for one- to three-year-olds on a body weight of 13 kilograms, and for four- to eight-year-olds on a body weight of 22 kilograms. They are mostly calculated using adult EARs and then extrapolated proportionate to body weight with the addition of a small factor for growth.

The nutrients included in Table 7.1 are examples showing differences between infants, toddlers and preschoolers due to growth rates, energy needs and activity levels. The NRVs displayed in the table are expressed as recommended dietary intakes (RDIs) and are expected to meet the needs of 98 per cent of healthy individuals within this age bracket. AIs are provided for infants where there is insufficient evidence to determine an RDI. They are based on intake from breast milk and complementary foods. For more general information on NRVs for Australians and New Zealanders see Chapter 1.

How to avoid choking

- Always supervise young children when eating.
- Never force children to eat.
- Ensure children are seated while eating.
- Avoid known choking risks such as whole nuts, whole grapes, seeds and pips, foods that can break off into hard lumps (e.g., raw carrot and apple), popcorn, corn chips, hard lollies, sausages and frankfurters.

How much should toddlers and preschoolers eat?

In 2011 the Dietitians Association of Australia used a modelling system to translate the NRVs into food consumption patterns (DAA 2011). This led to the Australian Dietary Guidelines including a recommended dietary pattern for toddlers and preschoolers in which the EARs for most nutrients and the EERs are met (see Table 7.2; NHMRC 2013). It should be remembered that this is a guide only, with the varied

Table 7.1 Daily RDIs (AIs) for selected nutrients for children (1–8 years) in Australia and New Zealand

Nutrient	Amount of nutrient, by age (years)		
	1[a]	1–3	4–8
Protein (g)	(14.0)	14.0	20.0
Thiamin (mg)	(0.3)	0.5	0.6
Folate (μg)	(80.0)	150.0	200.0
Vitamin C (mg)	(30.0)	35.0	35.0
Calcium (mg)	(270.0)	500.0	700.0
Iron (mg)	11.0	9.0	10.0
Vitamin D (μg)	(5.0)	(5.0)	(5.0)

[a] Many requirements for one-year-old infants are designated adequate intakes (AI) and are based on intake from breast milk and complementary food.

Note: AI: adequate intake; RDI: recommended dietary intake.

Source: NHMRC & Ministry of Health (2006).

Table 7.2 Recommended serves per day for children (1–8 years)

Food	Number of serves, by age (years)		
	1	2–3	4–8
Vegetables	2–3	2½	4½
Fruit	½	1	1½
Grains	4	4	4
Meat & alternatives	1	1	1½
Dairy	1–1½	1½	1½ (girls) 2 (boys)
Extras[a]	–	0–1	0–1 (girls) 0–2½ (boys)

[a] Includes an allowance for unsaturated spreads or oils, nuts or seeds (½ serve or 4.5 grams per day for children two to three years of age, one serve or 7–10 grams per day for four to eight years of age). An allowance for unsaturated spreads or oils or nut or seed paste of one serve or 7–10 grams per day is included for one- to two-year-olds.

Source: NHMRC (2013).

needs of individuals and their activity levels influencing the adequacy of their diet. Monitoring growth and development will indicate if intake is adequate.

New Zealand recommendations are very similar to Australian recommendations, with the amount suggested by the former for two- to five-year-olds being two serves of vegetables, two fruit, four grains, two to three dairy and one meat and alternatives each day (Ministry of Health 2012).

The amount of food that makes a standard serve is the same for adults and children (for example, one slice of bread, one piece of fruit and half a cup of vegetables); however, it is recommended that the amount of each serve presented at a single eating occasion is much less for young children than for older children and adults. It is better to serve smaller amounts at meals initially: a suggested amount is roughly 1 tablespoon of each food per year of age. The child can then ask for more if they are still hungry.

Most young children are able to self-select an appropriate amount of food to meet their appetite and energy demands. The amount consumed from meal to meal or day to day can vary considerably, so assessing the adequacy of a diet from a period of a few days to a week is more realistic than assessing from a single day's intake. Most children's needs for growth will be met by the Australian Guide to Healthy Eating if foods are nutrient and energy dense (NHMRC n.d.).

Low-fat milks and dairy products can be used from two years of age. Prior to this, for growth and brain development children need the energy and fat provided by full-fat varieties. Breastfeeding can continue through toddlerhood if the mother–infant dyad so chooses. Small frequent meals are the usual pattern, with three meals and two mid-meals offered each day. While no **discretionary foods** are preferable for this age group, it is important to avoid strict restriction of these foods, as it may increase their desirability and lead to overconsumption of them in later years.

How much do toddlers and preschoolers eat?

Food is an opportunity for learning for young children—colours and shapes, cooking, shopping and social aspects of meals—but can also be a source of great distress, frustration, worry and anxiety for parents. The Australian National Children's Nutrition and Physical Activity Survey, conducted in 2007, showed that most toddlers consumed adequate diets for energy and macronutrients but that diet quality was variable (CSIRO et al. 2008). About two-thirds of children ate more than one serve of fruit per day, and only 5 per cent had more than two serves of vegetables each day. Similarly, only 5 per cent had more than three serves of cereal-based foods each day. High intakes of saturated fats and sugar were reported, with about one-third of three-year-olds consuming more than the EER. In the toddler age group, milk and milk products and cereal-based products provided almost half of children's energy needs, and intake of salt was higher than the recommended upper limit (UL). Fruit juice was also a significant contributor to energy intake. Almost 10 per cent of toddlers were given dietary supplements, mostly as multivitamins.

Daily intakes of energy were on average 6000 kilojoules, of protein 60 grams, of calcium 800 milligrams and of iron 8 milligrams. These intakes exceeded the requirements for a three-year-old child in energy (5600 kilojoules for boys, 5300 kilojoules for girls), protein (14 grams) and calcium (500 milligrams). Iron intakes were lower than the RDI of 9 milligrams, but above the EAR of 4 milligrams for three-year-old children. These nutrient intakes are consistent with the food groups consumed, showing a high consumption of milk and cereal-based foods to provide energy, protein and calcium, but insufficient serves and variety in vegetables and meat or alternatives to meet iron requirements.

Discretionary foods are foods and drinks that are not an essential part of a healthy diet. They are high in energy, saturated fat, added sugar, salt or alcohol. They should be consumed infrequently and in small amounts.

A similar examination of the eating habits of infants and toddlers in the United States showed that a few foods provided the majority of the energy intake and that there was little variety in the range of fruits and vegetables eaten. Milk provided about 25 per cent of energy needs, with almost 20 per cent of energy derived from sweetened biscuits, cereals and drinks. Almost 30 per cent of children consumed dietary supplements regularly (Briefel et al. 2006; Butte et al. 2010).

Parents anxious about their children's poor eating habits may provide vitamin and mineral supplements in an attempt to compensate for a poor diet. While there are some special circumstances in which children may benefit from supplementation, this is best discussed with a dietitian or a maternal and child health nurse. Supplement use can lead to excessive intakes of certain nutrients, with potential toxic effects. In most cases a nutritious diet can supply all nutrient requirements.

Nutrition in child care

It is estimated that around 50 per cent of three-year-old children attend child care of some type (ABS 2011). Childcare settings are important influences on young children's eating habits. Children attending **long day care** may consume 50 per cent or more of their daily intake while in care. Some childcare centres provide food for children, and others require parents to send their children with food from home. The Australian government introduced the National Quality Framework for Early Childhood Education and Care (often referred to as the National Quality Framework) in 2012 with a legislative framework and a 'quality area' covering seven key topics, one of which was health and safety (ACECQA 2013). The framework and standards were designed to enhance the standards of care in childcare centres. Standard 2.2 relates to nutrition and physical activity (see box below).

The Ministry of Education in New Zealand has developed licensing criteria for early childhood education in which nutrition fits within the Health and Safety Practices Standard (see box below).

Childcare centres are expected to have policies in place to promote and support healthy eating habits in order to meet these standards. Issues for childcare centres include accessing nutrition training, providing

Long day care are day-care services that look after children under school age for up to ten hours a day on weekdays.

> **National Quality Framework for Early Childhood Education and Care, Australia**
>
> **Standard 2.2 Healthy eating and physical activity are embedded in the program for children**
>
> **Element 2.2.1** Healthy eating is promoted and food and drinks provided by the service are nutritious and appropriate for each child.
>
> **Element 2.2.2** Physical activity is promoted through planned and spontaneous experiences and is appropriate for each child. (ACECQA 2012)
>
> **Centre-based Early Childhood Education Services, New Zealand**
>
> **Health and Safety Practices Criterion 19**
>
> Food is served at appropriate times to meet the nutritional needs of each child while they are attending. Where food is provided by the service, it is of sufficient variety, quantity, and quality to meet these needs. Where food is provided by parents, the service encourages and promotes healthy eating guidelines. (ECE Lead 2014)

a nutritious menu within budget constraints, setting policies in terms of food safety with regard to the provision of food from home, setting the mealtime environment and communication with parents about food intake. A number of nutrition resources have been designed for the childcare setting and include books for centre directors, childcare workers and parents, as well as posters and brochures. These are available through the Department of Health (2014) as the Get Up and Grow resources and include translations into several languages common in Australia and a specific resource set for Aboriginals and Torres Strait Islanders.

INFLUENCES ON EATING

Children are exposed to a range of influences relating to eating from a very young age. The most important influence is family, with the lifestyle, culture, attitudes and knowledge of caregivers impacting on the foods available and how they are offered. Childcare centres and other settings can also influence food intake, with children being increasingly aware of the food choices of other people. Advertising and marketing of foods can also manipulate young children to desire unsuitable high-fat, high-sugar or high-salt foods that are significant factors in the rise of obesity. Television advertising in Australia is regulated in order to minimise exposure to marketing during children's peak viewing times; however, marketing is often designed to appe to very young children and can build brand recognition and drive the 'pester power' that parents find difficult to handle.

There are several sources available about the various psychosocial theories of child development, which of course influences behaviour. The key theorists on development in young children include Jean Piaget, Erik Erikson and Lev Vygotsky, and many psychology texts discuss their theories in detail. In essence, however, they all provide slightly varied explanations of the influences of developmental change through childhood.

Vegetarianism

A vegetarian diet can be appropriate for a toddler if managed carefully. Many staple foods used in vegetarian diets, such as legumes and whole grains, can be quite bulky, so it may be difficult for a child with a small appetite to consume enough for good health. It can be difficult to achieve AIs of iron when following a vegetarian diet. Absorption of iron is quite low on a vegetarian diet, and it is suggested that the intake of iron needs to be about 80 per cent higher than the NRV for age. Absorption of iron can be boosted by consuming a food containing vitamin C, such as tomato, capsicum or citrus fruit, at the same meals as non-meat sources of iron, such as wholegrain cereals, fortified breakfast cereals and green leafy vegetables. If no dairy products are consumed, a calcium-fortified alternative such as soy should be consumed in similar amounts to those recommended for milk and milk products.

Zinc and vitamin B12 also need careful attention, as they are most commonly found in animal foods. Zinc can be sourced from some plant foods.

Physical activity

The Australian Department of Health (2014) has made recommendations regarding physical activity for all Australians. The recommendation for toddlers and preschoolers is that they should be physically active for three hours per day—not all at once, but spread throughout the day. This includes a suggestion that they should not be inactive, apart from sleeping, for more than one hour at a time. Inactivity commonly occurs when young children are restrained in high chairs or strollers as well as when watching television. It is recommended that children under two years

CASE STUDY 7.1 FUSSY EATING IN TODDLERS

Profile and problem

R. is two and a half years old. His parents are Indian and have been in Australia for the last ten years. R. is their first child. His mother, J., is worried about his nutrition and growth. R. attends child care on two days a week while J. works at a bank.

Assessment

A diet history reveals the following intake:

9 a.m.	Cow's milk with Aptamil Toddler (2 scoops, added to 150 millilitres)
10 a.m.	½ Weet-Bix with milk
11 a.m.	Cheese slice
1 p.m.	⅓ cup dahl and rice (if fed by mother), banana, milk (as at 9 a.m.)
At child care	Usually pasta with meat-based sauce, chicken and vegetables or meatballs with rice and salad (eats most of what is served)
Afternoon sleep	
3 p.m.	Milk (as 9 a.m.)
5 p.m.	Curry (meat and vegetables most nights) and rice (2–3 teaspoons), yoghurt (3–4 tablespoons)
7 p.m.	Pureed fruit (pouch)
9 p.m.	Milk (as at 9 a.m.)

Questions about meal times reveal that J. usually feeds R. and he is reluctant to feed himself. He has to watch the television or his favourite DVD. He will mostly sit on the couch or on his mother's knee to eat, but often J. follows him around to feed him. J. and her husband usually eat their own meals at the table at a separate time to R.

Growth assessment shows R.'s height is progressing along the 25th percentile, but his weight is falling to just below the 10th percentile. His bowels open every two or three days and he passes a hard pebbly stool.

Treatment

J. agrees to do the following:

- have the family eat together
- sit at the table to eat meals, with no chasing and no distractions
- take a relaxed, neutral approach to feeding
- encourage self-feeding
- increase texture in food to encourage chewing
- give only two bottles per day and to change from the bottle to a cup
- consider removing the Aptamil from the milk (it may be a source of iron)
- consider the timing of milk feeds to avoid filling up Ravi before breakfast
- start Ravi on some bread (preferably wholemeal or wholegrain) for fibre
- include meat in meals
- encourage R. to drink water after and between meals.

Plan

J. and R. will be monitored and receive advice from the dietitian on a fortnightly basis until R. is eating an improved diet independently. Consultation with the maternal and child health nurse will determine whether referral for speech pathology or other health professionals are warranted to assist R. in adopting new feeding practices.

should not have any screen time (television, DVDs or computer games) and that those over two years should have only one hour per day.

A simple assessment of activity can include questions about how long young children spend in high chairs or if they walk or are in a stroller when out of home and if there is any regular screen time. Health professionals can help parents increase their children's activity by suggesting practical ideas to encourage active play such as a walk to a playground, playing with balloons, outdoor play like in a sandpit and so on, as well as ways to limit screen time. Sport New Zealand (2014) has made similar suggestions for encouraging active movement in young children.

COMMON CONCERNS

Fussy eating

Early food experiences are important, as children develop food preferences and eating patterns that promote good nutrition for the future. A large proportion of parents of toddlers are concerned about their food intake, with studies suggesting 50 per cent of mothers of two-year-olds consider their children 'fussy eaters' (Mitchell et al. 2013). As food is an inherent part of nurturing, parents can extend the rejection of food to rejection of themselves and become very stressed and upset. Worries include limited intakes of food groups, food fads, rejection of foods, reluctance to try new foods and extreme likes and dislikes; these worries occur at a reported frequency of 20–30 per cent (Wright et al. 2007). This behaviour is often referred to as fussy eating and happens for a number of reasons as toddlers grow. The slowing growth rate, increased independence, neophobia and parental influences can all affect eating behaviours.

Food fads are often seen in the toddler years when a child prefers to eat one or two key foods over and over, but this behaviour is also seen in other areas of their life, with a favourite DVD, book or toy being highly preferred for a period of time before it changes. If the food fad continues for an extended period of time it can impact on nutrition.

Parental influences can be strong, and children are more likely to eat increased amounts of dairy, fruit, vegetables and even breakfast if at least one parent is present at meals. However, extreme pressure to eat healthy foods, over-restriction of snacks or even the use of food as rewards can have a negative impact on food intake. Parenting style that is overly restrictive (referred to as authoritarian—high control) is less effective than authoritative parenting that is responsive to the child's needs but still sets limits and expectations. Permissive parenting, on the other hand, allows the child to control what, where and how much food is eaten (Hubbs-Tait et al. 2008). The authoritarian—high control style of parenting is associated with more food pickiness in young children, while healthier eating behaviours are linked to the authoritative style (Moroshko & Brennan 2013).

Children have their own food preferences that relate to the flavours of food. Young children seem to prefer foods that are energy dense, which is probably an instinctive choice relating to the historical scarcity of foods. Hence, sweet foods are highly preferred and bitter tastes are usually disliked. Maternal food preferences also seem to correlate well with children's food preferences, as this will influence which foods are offered to children. Those who have been breast fed have been exposed to flavours in the breast milk that relate to the foods consumed by the mother, so they tend to be less picky and more willing to try new foods. Additionally, young children may reject food on the basis of smell or colour.

Toddlers reject vegetables for many reasons. It may be that they simply do not like the taste. Some people are 'super tasters' and find the taste of some vegetables very strong and unpleasant. Others will grow into enjoying the taste of these foods as they gain experience. Some children may not have good role models who show them that eating vegetables can be pleasant, or they relish the extra attention they receive when they say no to vegetables or new foods. For others this behaviour is a way of establishing independence.

Neophobia is a normal part of childhood development and protects a child from eating a new food that might be dangerous, so repeated exposure to new foods induces a learned safety, as does watching an adult eat the food first (Shutts et al. 2013). Avoiding unfamiliar foods protects children from consuming dangerous things but can also restrict variety.

When is fussy eating a concern?

Fussy eating is very common and for most children will not impact on health. Health professionals may need to be consulted when there is a significant change in growth patterns, such as a fall of two percentiles or if one of the major food groups has been rejected for long period of time. For example, refusing all meats over a period of a couple of months may impact on iron status. Many children will refuse cooked vegetables but are happy to eat some raw vegetables at snack times or to eat a range of different coloured fruit, so this food group is less likely to cause a health concern. Issues with oral motor skills such as difficulty chewing or swallowing, and gagging or vomiting when eating, also need further investigation.

Management strategies

The first step when seeing a 'fussy eater' is to undertake a thorough assessment. This needs to include an assessment of growth patterns and diet to determine if physical health or nutritional status is affected. Exploring not only what food is eaten but when, how and where it is offered can be valuable. The child's **diet history** needs to look at the forms in which foods are offered, the timings, and the amounts of fluids consumed. Questioning about the use of distractions like toys or television at meal

A **diet history** is a narration or documentation of any and all food and drink consumed in a set time period.

times and what happens when the child refuses food can help determine strategies for the individual family. The eating environment for young children should make it easy for them to eat safely, so being well supported in a high chair or booster seat will make it possible for them to sit at the table. In addition, eating with other family members offers valuable role modelling of eating skills. Simple strategies that are often recommended for managing fussy eating include those listed below:

- *Repeated exposure*—Parents often stop presenting new foods after they have been rejected once or twice, but it is thought that it is necessary to offer a food eight to ten times before it is happily accepted by a child. This is probably due to a learned safety and the need for familiarisation through the sight, smell, touch and taste of the food before significant quantities are eaten. In order to allow children to familiarise themselves with food, allowing them to handle foods is an important step despite the associated mess.
- *Limited choices*—This strategy allows the child some control over food choices, but those choices are made from within the range of foods offered by the parent. Asking the child if they want carrots or corn allows some autonomy for the child but means the child is still having vegetables served.
- *Positive language and role modelling*—Keeping the focus on the preferred behaviour for the child and not on what is not being eaten allows for some success and encouragement. Too much praise, however, can also negatively affect food intake. Role modelling is an important way of encouraging appropriate mealtime behaviour and consumption of family meals for this age group.
- *Routines*—The child should understand that food and drink will be made available at regular times through the day. Families who set up expectations that children will be seated at the table at meal times will probably be more successful in achieving the goal of consumption of family meals. Meal times should be of limited duration, as most children lose concentration after about twenty minutes. Extended meals are unlikely to increase food intake and will just raise stress levels. Parents need to recognise signs that the child has had enough and remove uneaten foods.
- *Fluids*—Young children often find it quicker to fill up with drinks than to chew through an entire meal. Many toddlers may drink large quantities of milk or juice throughout a day that limit the appetite and impact on foods eaten. Milk is an important food, but no more than 500 millilitres per day is needed for this age group. Milk may be better presented at mid-meals. Other drinks such as juice, cordial and soft drinks are not necessary in the diet and can contribute to tooth decay and overweight. A simple suggestion is for parents to offer water as the main drink and to limit drinks for an hour before meal times.
- *Screen time*—Many children use screens (computer tablets and television) during meal times. These provide a distraction and reduce the importance of meals and nutrition as well as detracting from social interactions and family relationships. A focus on the screen rather than the plate can be a distraction from hunger and fullness cues and can mean less or more food is consumed than is needed.

Educating caregivers is a vital role the health professional plays in ensuring children obtain a healthy balanced diet. A few ways to encourage parents when dealing with fussy eaters are listed below:

- Be a healthy-eating role model. If you eat a food, a child is more likely to try the food.
- Children have small stomachs, so smaller frequent meals are better than large amounts of food on the plate at one time. The latter can be very off-putting to children.
- Children thrive on routine. Try to keep meal times regular.
- Do not bribe children with dessert or other food treats if they eat their vegetables. Children may eat the vegetables but still not necessarily like them. In fact, such bribes can cause children to intensely dislike the food they are being bribed to eat and to increase their preference for the prize food.
- Do not fill children up on juice, milk, cordial or even water just before a meal.
- Children tend to eat if they are allowed to serve themselves. Having children help with meal preparation or setting the table and other roles can encourage them to try foods.
- Do not give up on foods a child does not like; remember it sometimes takes ten presentations for a child to try a new food or retry an old food. Just put it on the side of the plate and do not force them to eat it. At various stages, encourage them to try the food and praise their efforts to do so.
- Do not ignore problems that interfere with eating, such as teething, a sore throat, a blocked nose or an upset tummy.

Division of responsibility

An important point to raise with families is the concept of division of responsibility. This was devised by a US dietitian, Ellyn Satter, who has authored a number of books about nutrition for families. She describes the division of responsibility as parents taking leadership and children becoming self-directed with feeding. Her theories include increased anxiety in children who do not have their feeding cues recognised and poor self-regulation of intake in children who do not learn to interpret and trust their own internal cues. The theories allow parents to step back and avoid confrontation and encourages children to discover and respond to their own needs and internal cues for hunger, satiety and food preferences. It places a degree of trust in the child and can be difficult for caregivers to implement but is ultimately more successful in the longer term in encouraging the development of healthy eating habits and an enjoyment of food and eating.

Division of responsibility for toddlers through adolescents

The parent is responsible for *what, when, where*. The child is responsible for *how much* and *whether*. Fundamental to parents' jobs is trusting children to decide *how much* and *whether* to eat. If parents do their jobs with *feeding*, children do their jobs with *eating*.

Parents' feeding jobs

- Choose and prepare the food.
- Provide regular meals and snacks.
- Make eating times pleasant.
- Show children what they have to learn about food and mealtime behaviour.
- Be considerate of children's food inexperience without catering to likes and dislikes.
- Not let children have food or beverages (except for water) between meal and snack times.
- Let children grow up to get bodies that are right for them.

Children's eating jobs

- Children will eat.
- They will eat the amount they need.
- They will learn to eat the food their parents eat.
- They will grow predictably.
- They will learn to behave well at meal times. (Ellyn Satter Institute 2014; see her website for more about stage-related feeding and for Ellyn Satter's books, videos, and other resources.)

- Fussy eaters are often slow eaters who dawdle over their plates. Trying to hurry children to eat can cause them to become stressed and put them off their food. Be patient and let children eat in their own time.
- If children refuse a food, remove the food without a fuss (it can be offered again later), but remember that they cannot have a snack an hour afterwards!

Extreme fussy eating or problem feeders

Some children may exhibit fussy eating behaviour for a long period of time but show no progress with increasing their range of foods and may also refuse entire categories of food textures or show extreme reactions when presented with new foods. This is termed as extreme fussy eating, or the children as problem feeders. They generally eat only twenty foods or fewer. This may be due to an inborn disposition and high level of anxiety around food or to a lack of experience. The behaviour might be seen in children who are disorganised feeders due to some physiological issues or who were very premature or suffering from an autism spectrum disorder.

This form of eating has nutritional consequences and needs to be managed by an interdisciplinary feeding team working with the child and family to make the diet acceptable to the child as well as being nutritionally adequate. Parents are encouraged to be sensitive to the child's likes and dislikes and not to force feed the child. There are different programs available; one of the most common is the Sequential Oral Sensory approach to feeding. A number of practitioners offer group or individual programs based on this method, which uses six steps to eating food through which the child moves: tolerates, interacts with, smells, touches, tastes and eats. This systematic desensitisation is combined with a focus on social modelling, structured meal and snack times, reinforcement and accessing cognitive aspects particular to each individual in order to develop skills and increase experience with a wide range of foods.

Toddler formulas

Formulas for toddlers (aged one to two years) and juniors (aged over two years) are similar in composition to infant formula. As the advertising of these is not restricted, they are heavily marketed. The main differences in composition are that these formulas are higher in protein than infant formula and that toddler formulas are not nutritionally complete but are similar to a fortified cow's milk drink (with iron, calcium and other components like long-chain polyunsaturated fatty acids, prebiotics and probiotics). Many parents are won over by both the advertising and their perception of poor food intake, which makes them worry that their child is not consuming an adequate range of nutrients.

Use of toddler formula can in fact limit food intake, as it can reduce appetite. This then spirals to reinforce the poor intake but reassures parents that essential nutrients are being consumed. The use of these formulas is rarely necessary, and they do not solve the problem of fussy eating. The use of normal foods, milk and water as the preferred drink is a more sustainable solution. Multivitamins are also often given to children by parents as a safety net, in a similar way to the fortified drinks.

HEALTH CONDITIONS

Iron deficiency anaemia

Iron is very important for toddlers, and they rely on solid foods for most of their iron intake. Iron deficiency is commonly seen in young children who fill up on large quantities of cow's milk and then eat small or minimal amounts of other foods. Studies of Australian populations have suggested that up to 19 per cent of toddlers have depleted iron stores and 4 per cent suffer from iron deficiency anaemia, with a higher proportion in Aboriginal and Torres Strait Islander populations and Asian populations in Australia. A study in New Zealand indicated a prevalence of iron deficiency of 14 per cent of the population (Grant et al. 2007).

Iron deficiency anaemia in young children can cause behavioural changes and long-term cognitive issues (Pasricha et al. 2010). Meat and meat alternatives are the main sources of iron in children's diets. When these foods are excluded by fussy eaters, iron intake may be limited. While gently encouraging the reintroduction of these foods, parents should be provided with tips recommended for vegetarians (described above). Parents may also try providing a variety of different iron-rich foods or preparing meat in different ways

CASE STUDY 7.2 ***CONSTIPATION IN TODDLERS*** *Adrienne Forsyth*

Profile and problem

Mrs J. presents with her three-year-old daughter A.J., whom she says is constipated. A.J. has not passed a bowel movement for four days but had previously had regular bowel movements at least once daily. A.J. has no other known health problems but can be a 'difficult' child who likes to have her own way.

Assessment

A.J. has recently started three-year-old kinder. She has been toilet trained for two months. Mrs J. says that while she tries to provide a variety of nutritious foods, A.J. avoids most fruit and vegetables. A.J. prefers crackers, cheese, sweet biscuits and 'treats'. She drinks full-cream milk with each meal and usually drinks her milk before eating her food.

Treatment

Mrs J. has agreed to do the following:

- talk to A.J. about when and how to use the toilet at kinder and encourage her to avoid 'holding on'
- ensure A.J. has an adequate fluid intake throughout the day by providing a drink bottle for kinder and offering frequent drinks of water at home
- provide reduced-fat milk, limiting milk intake to 500 millilitres per day, and offer milk after or between meals
- continue to provide fruit and vegetables at meals and use different preparation methods to appeal to A.J. (e.g., raw carrot sticks instead of cooked carrots)
- provide fruit and vegetables as snack options
- encourage A.J. to be physically active.

Plan

If A.J. has not had a bowel movement within a week, Mrs J. will consult her general practitioner for further advice and investigation. If constipation continues to be a problem for A.J., Mrs J. should attend an additional session with the dietitian to discuss appropriate amounts and choices of soluble and insoluble fibre and fluids.

to suit the needs of a fussy eater. Some children may prefer moist meat in a casserole or stew, while others may insist on dry roast meat with no sauce.

Constipation

This is often seen in young children, particularly around the time of toilet training. It may be linked with diet but more often is associated with 'holding on' or resisting the urge to pass a bowel motion. Diet can be explored to ensure adequate fibre and fluid are being consumed, but generally, increasing fibre above normal intakes is not necessary. In fact, large amounts of fibre supplements like bran can cause further problems. Excessive intake of milk may also be linked to constipation, usually because children who drink a lot of milk eat very little fibre-containing food.

Diarrhoea

Healthy toddlers sometimes pass frequent loose stools containing identifiable food matter such as peas or corn. This is generally a harmless occurrence thought to be related to gut immaturity in absorbing water, particularly of the colon, and often resolves by about three or four years of age. Large amounts of fruit juice seem to make this condition worse.

Dental caries

Although toddlers still have their deciduous baby teeth, it is important to care for these teeth, as they help with development of speech and of the permanent teeth. Dental caries in toddlers are often caused by having sweet drinks or foods that are sticky and stay in contact with the teeth for a period of time. These might be foods like muesli bars and dried fruit as well as sugar-containing drinks like juice, cordial and soft drinks. Night-time bottles can also contribute to dental caries, as can frequent eating (more than seven times per day). Parents should brush young children's teeth twice a day with a soft toothbrush until they are seven or eight years old. Fluoride strengthens teeth enamel and is added to the water supply in most of Australia and about 50 per cent of council regions in New Zealand, but people who rely on tank or bore water may need to discuss fluoride with their dentist.

SUMMARY AND KEY MESSAGES

The toddler and preschool years are a time of growth and development. It is in these years that children form their earliest memories of food and eating. The eating habits that are developed during the toddler and preschool years will influence an individual's eating habits into adulthood. Key considerations during the toddler and childhood years include the following:

- Toddlers grow rapidly in the first twelve months, slowing considerably in the following years.
- Toddlers and preschoolers are able to feed themselves and enjoy doing so but require adult supervision.
- Toddlers and preschoolers require a sufficient amount and variety of nutritious foods to meet their energy and nutrient requirements.
- Young children may require several exposures to a food before it is accepted.
- Fussy eating is common at this age and is not a serious concern unless whole food groups are omitted for long periods of time.
- Fussy eaters are at risk of deficiency in iron, zinc and vitamin B12 from avoidance of meat and in fibre from avoidance of whole grains.
- Parents and caregivers can help children develop healthy eating habits by acting as role models, creating a positive eating environment and providing a variety of healthy foods from each of the core food groups.

REFERENCES

ABS (Australian Bureau of Statistics) 2011. *Child Care: Australian social trends, June 2010.* Cat. no. 4102.0. Commonwealth of Australia. <www.abs.gov.au/AUSSTATS/abs@.nsf/Lookup/4102.0Main+Features50Jun+2010>, accessed 6 November 2014.

ACECQA (Australian Children's Education and Care Quality Authority) 2012. *Quality Area 2: Children's health and safety.* Commonwealth of Australia. <www.acecqa.gov.au/Childrens-health-and-safety>, accessed 30 September 2014.

——2013. *Guide to the National Quality Standard.* Commonwealth of Australia. <http://files.acecqa.gov.au/files/National-Quality-Framework-Resources-Kit/NQF03-Guide-to-NQS-130902.pdf>, accessed 6 November 2014.

Briefel, R., Hanson, C., Fox, M.K., Novak, T. et al. 2006. Feeding Infants and Toddlers Study: Do vitamin and mineral supplements contribute to nutrient adequacy or excess among US infants and toddlers? *Journal of the American Dietetic Association* 106(1): S52–S65.

Butte, N.F., Fox, M.K., Briefel, R.R., Siega-Riz, A.M. et al. 2010. Nutrient intakes of US infants, toddlers and preschoolers meet or exceed dietary reference intakes. *Journal of the American Dietetic Association* 110(12 Suppl.): S27–S37.

CDC (Centers for Disease Control and Prevention) 2009. *Clinical Growth Charts.* CDC. <www.cdc.gov/growthcharts/clinical_charts.htm>, accessed August 2013.

CSIRO (Commonwealth Scientific and Industrial Research Organisation), Preventative Health National Research Flagship & University of South Australia 2008. *2007 Australian National Children's Nutrition and Physical Activity Survey: Main findings.* Commonwealth of Australia, Canberra.

DAA (Dietitians Association of Australia) 2011. *A Modelling System to Inform the Revision of the Australian Guide to Healthy Eating.* Commonwealth of Australia. <www.eatforhealth.gov.au/sites/default/files/files/the_guidelines/n55c_dietary_guidelines_food_modelling.pdf>, accessed 28 September 2014.

Department of Health 2014. *Get Up and Grow: Healthy eating and physical activity for early childhood resources.* Commonwealth of Australia. <www.health.gov.au/internet/main/publishing.nsf/Content/phd-early-childhood-nutrition-resources>, accessed 28 September 2014.

ECE Lead (Early Childhood Education Leadership, Management and Administration) 2014. *HS19 Food and Nutrition.* New Zealand Government. <www.lead.ece.govt.nz/ServiceTypes/CentreBasedECEServices/HealthAndSafety/FoodAndDrink/HS19FoodAndNutrition.aspx>, accessed 30 September 2014.

Ellyn Satter Institute 2014. *Ellyn Satter's Division of Responsibility in Feeding.* Ellyn Satter Institute. <http://ellynsatterinstitute.org/dor/divisionofresponsibilityinfeeding.php>, accessed 28 September 2014.

Grant, C.C., Wall, C.R., Brunt, D., Crengle, S. et al. 2007. Population prevalence and risk for iron deficiency in Auckland. *New Zealand. Journal of Paediatric and Child Health* 43: 532–8.

Hubbs-Tait, L. Kennedy, T., Page, M., Topham, G. et al. 2008. Parental feeding practices predict authoritative, authoritarian and permissive parenting styles. *Journal of the American Dietetic Association* 108: 1154–61.

Ministry of Health 2012. *Food and Nutrition Guidelines for Healthy Children and Young People (Aged 2–18 years): A background paper.* New Zealand Government. <www.health.govt.nz/system/files/documents/publications/food-and-nutrition-guidelines-for-healthy-children-and-young-people-aged-2-18-years.pdf>, accessed 16 March 2014.

Mitchell, G., Farrow, C., Haycraft, E. & Meyer, C. 2013. Parental influences on children's eating behaviour and characteristics of successful parent-focussed interventions. *Appetite* 60: 85–94.

Moroshko, I. & Brennan, L. 2013. Maternal controlling feeding behaviours and child eating in preschool-aged children. *Nutrition & Dietetics* 70: 49–53.

NHMRC (National Health and Medical Research Council) 2013. *Eat for Health: Australian Dietary Guidelines; Providing the scientific evidence for healthier Australian diets*. Commonwealth of Australia. <www.eatforhealth.gov.au/sites/default/files/files/the_guidelines/n55_australian_dietary_guidelines.pdf>, accessed 28 September 2014.

——n.d. *Australian Guide to Healthy Eating*. Commonwealth of Australia. <www.eatforhealth.gov.au/guidelines/australian-guide-healthy-eating>, accessed 28 September 2014.

NHMRC (National Health and Medical Research Council) & Ministry of Health 2006. *Nutrient Reference Values for Australia and New Zealand: Including recommended dietary intakes*. Commonwealth of Australia & New Zealand Government. <www.nhmrc.gov.au/_files_nhmrc/publications/attachments/n35.pdf>, accessed August 2013.

The Parents' Jury (website) 2014. <www.parentsjury.org.au>, accessed 30 September 2014.

Pasricha, S.S., Flecknoe-Brown, S.C., Allen, K.J., Gibson, P.R. et al. 2010. Diagnosis and management of iron deficiency anaemia: A clinical update. *Medical Journal of Australia* 193(9): 525–32.

Paxton, G.A., Teale, G.R., Nowson, C.A., Mason, R.S. et al. 2013. Vitamin D and health in pregnancy, infants, children and adolescents in Australian and New Zealand: A position statement. *Medical Journal of Australia* 198(3): 142–3.

Shutts, K., Kinzler, K.D. & DeJesus, J.M. 2013. Understanding infants' and children's social learning about foods: Previous research and new prospects. *Developmental Psychology* 49(3): 419–25.

Sport New Zealand 2014. *Active Movement Activity Guides for Children 0–5 Years*. New Zealand Government. <www.sportnz.org.nz/managing-sport/guides/active-movement-activity-guides-for-children-0-5-years->, accessed 28 September 2014.

WHO (World Health Organization) n.d. *The WHO Child Growth Standards*. WHO. <www.who.int/childgrowth/standards/en/>, accessed August 2013.

Wright, C., Parkinson, K., Shipton, D. & Drewett, R. 2007. How do toddler eating problems relate to their eating behaviour, food preferences and growth? *Pediatrics* 120:e1069-75.

ADDITIONAL READING

Australian Children's Education and Care Quality Authority (website) 2014. <www.acecqa.gov.au/>, accessed 28 September 2014.

Centre for Learning Innovation 2006. A basic introduction to child development theories. Department of Education and Training, New South Wales Government. <http://lrrpublic.cli.det.nsw.edu.au/lrrSecure/Sites/LRRView/7401/documents/theories_outline.pdf>, accessed 14 November 2014.

Department of Health 2014. *Australia's Physical Activity and Sedentary Behaviour Guidelines*. Commonwealth of Australia. <www.health.gov.au/internet/main/publishing.nsf/content/health-pubhlth-strateg-phys-act-guidelines>, accessed 28 September 2014.

Innovative learning (website) 2011. <http://innovativelearning.com/educational_psychology/development/index.htm>, accessed 14 November 2014.

Satter, E.M. 1986. The feeding relationship. *Journal of the American Dietetic Association* 86: 352–6.

TEST YOUR UNDERSTANDING

Note: There is only one correct answer for each question.

1. What are the suggested serving sizes of foods for a toddler?
   a. ½ an adult serve
   b. ½ cup per year of age
   c. 1 tablespoon per year of age
   d. 1 small bowl
   e. None of the above
2. What is the recommended number of serves of dairy foods each day for a three-year-old?
   a. 1 serve
   b. 1½ serves
   c. 2 serves
   d. 2½ serves
   e. 3 serves
3. Which of the following is a nutritious mid-meal for a toddler?
   a. Baked beans on toast and a glass of water
   b. Cheese and crackers and a glass of juice
   c. Muesli bar and a glass of milk
   d. Avocado and corn chips and a glass of cordial
   e. All of the above
4. The iron requirement for a vegetarian toddler is what percentage higher than the NRV?
   a. 20 per cent
   b. 40 per cent
   c. 60 per cent
   d. 80 per cent
   e. 100 per cent
5. To what is constipation in toddlers is usually due?
   a. Inadequate fibre
   b. Inadequate fluid
   c. Behavioural issues
   d. Emotional issues
   e. All of the above
6. What is neophobia?
   a. Fear of new things
   b. Fear of the unknown
   c. Fear of gagging
   d. Fear of vomiting
   e. Fear of food
7. When is fussy eating a major concern?
   a. No fruit is eaten
   b. A major food group is not eaten for a week
   c. A child gags on all lumpy food
   d. A child drinks 500 millilitres of milk per day
   e. None of the above
8. How many hours of physical activity are recommended for a three-year-old child?
   a. 1 hour
   b. 2 hours
   c. 3 hours
   d. 4 hours
   e. 5 hours
9. At what age can children usually manage a knife and fork competently?
   a. 2 years
   b. 3 years
   c. 3½ years
   d. 4 years
   e. 5 years
10. Until what age is it recommended that a parent should brush a child's teeth?
    a. 4 years
    b. 5 years
    c. 5½ years
    d. 6 years
    e. 7 years

STUDY QUESTIONS

1. What are the advantages of using fruit and vegetables instead of vitamin supplements in the diet of toddlers?
2. Suggest some good food sources of iron that would be easy for toddlers to eat.
3. List four ways in which young children can be physically active when indoors at home.
4. Devise some strategies for parents to manage children who drink large amounts of milk.
5. Compare the energy and protein content of 100 millilitres of breast milk, cow's milk, infant formula and toddler milk.
6. List three simple suggestions for parents to help them manage fussy eating.
7. What are the current restrictions relating to advertising during children's peak viewing times? Use the internet to explore this. The Parents' Jury (2014) is a good place to start.

8

Child (6–12 years)

Clare Collins and Annette Murphy

> **LEARNING OUTCOMES**
>
> Upon completion of this chapter you will be able to:
>
> - define normal growth and development for a child in this age group
> - summarise the current dietary intake recommendations and requirements of children in this age group to optimise growth, health and development
> - identify common behavioural and environmental factors that impact on dietary intake in children in this age group
> - describe common nutritional issues and health conditions in children in this age group, including overweight and obesity, diabetes and heart disease.

The period from six to twelve years of age is important for the initiation, development and reinforcement of healthy eating habits to last throughout life. Significant change occurs during this time in both physical and psychological factors that benefit from habits such as healthy eating. This chapter outlines the growth and development stages of children in relation to nutrient requirements specific to this age group. The dietary intake of children aged six to twelve years in Australia and New Zealand is examined and compared to current recommendations. Behavioural and environmental factors that impact on child dietary intake are also explored, including fussy eating, school lunches and discretionary foods. Eating patterns and growth over a period of time give a more accurate view of how a child is developing and whether they are meeting their recommended intakes at both the whole food level and the macronutrient and micronutrient levels.

It is important to acknowledge that long-term dietary habits and health status can be determined in childhood. The Cardiovascular Risk in Young Finns Study, commenced in 1980, has been investigating dietary intake and disease outcomes in 3596 subjects initially aged three to eighteen years. The results showed that overweight children are more likely to become overweight or obese adults. In addition, the research found that elevated childhood blood pressure and serum **lipid** levels correlated strongly with elevated values measured in middle age. The study discovered that dietary habits in a child's formative years can develop into lifelong eating habits. According to the

Lipids include fats, such as cholesterol and triglycerides (fatty acids and glycerol).

findings, childhood nutrition plays a significant role in the progression of disease states into adulthood and that the major adult cardiovascular disease (CVD), coronary artery disease and hypertension, begin in childhood (Kaikkonen et al. 2013).

The European Youth Heart Study, a long-running epidemiological project which began in 1997, initially tracked the dietary intake of children in Denmark over six years. This study's results also showed that general dietary habits established in childhood often track into adulthood (Patterson et al. 2009).

CVD risk factors have been shown to track over a fifteen-year period and are predictive of adult levels (Berenson et al. 1997). The ongoing Bogalusa Heart Study has shown that the major **aetiologies** of adult heart disease, atherosclerosis, coronary heart disease (CHD) and essential hypertension, begin in childhood. Documented anatomic changes occur as early as five to eight years of age (Bao et al. 1995; Berenson et al. 1998). This evidence suggests that interventions to improve the health outcomes of children need to occur at an early age so as to minimise risk of later **poor health** status.

PHYSIOLOGICAL AND PSYCHOLOGICAL CHANGES

Many factors influence a child's growth, including genetics, environment and health aspects. For a child to grow at their expected rate they must be healthy and have access to adequate amounts of nutritious food. Children's rate of growth should align with population norms for age, sex and stage of physiological maturity.

A child's nutrient and energy needs (relative to their body size) are greater than those of an adult. From the age of six, a child's growth rate and nutritional needs slow relative to their body weight compared to those in their infancy and toddler years (NHMRC 2013c). A child between the age of six and the onset of puberty (eleven in girls and twelve in boys, on average) will gain an average of 2 kilograms per year (NHS 2013). Physical growth in this age group usually involves 'bursts' that appear as small fluctuations on their growth curves. Nutrient needs vary depending on their changing growth rate. Changes in height and weight can occur in a short period of time and can bring with them changes in a child's appetite and food consumption (Better Health Channel 2013). There can be vast differences in height, weight and body composition among children aged from six to twelve years, but a healthy child will follow their own pattern of growth, roughly tracking along a single growth percentile on the growth charts, despite some variation in dietary intake. As a child approaches adolescence, nutrient requirements start to increase again relative to their body size, and this differs between boys and girls.

Aetiologies (aetiology) are the causes (or cause) of a disease or condition.

Poor health is experienced by people with higher rates of illness and disability and higher death rates than might be expected.

Since 2000 an increasing number of studies have identified significant sex differences in body fat and lean body mass (muscle) between the ages of three and ten (Kirchengast 2010). At the age of three, girls and boys do differ in relative fat mass and lean body mass, adjusted for height. During puberty, sex differences in relative fat mass and relative lean body mass increase markedly, with boys having greater lean body mass and lower fat mass compared with girls.

Although this is completely normal, there can be a vast difference in the age at which children begin to mature. In general, girls tend to mature both physically and emotionally earlier than boys. Children start to gain a sense of body image in this age group due to their changing physiology and exposure to outside influences including peers and the media.

Six to twelve years is the ideal time for children to start learning more about food and how their bodies use this fuel for physical activity and growth. It is a time when children start to develop some independence, engage more with others socially away from the family environment and begin making their own decisions on aspects of their lifestyle.

Appropriate growth is an indicator of health and nutritional status. Children's growth in the six- to twelve-year-old age group is assessed using growth percentile charts and body mass index (BMI). As discussed in previous chapters, in Australia, the Centers for Disease Control and Prevention's Clinical Growth Charts (CDC 2009) or the World Health Organization's (WHO n.d.) Child Growth Standards are used to assess growth in children. Specialised growth charts can be used for children with particular growth assessment needs, such as those with cerebral palsy. Australia's National Health and Medical Research Council's Clinical Practice Guidelines for the Management of Overweight and Obesity in Adults, Adolescents and Children in Australia encourage the use of the Centers for Disease Control and Prevention charts for children aged six to twelve years (NHMRC 2013a). New Zealand's Ministry of Health (2009) also encourages the use of these growth charts, in the Clinical Guidelines for Weight Management in New Zealand Children and Young People.

NUTRIENT REQUIREMENTS AND INTAKES

Regular consumption of a variety of healthy food choices from the five food groups and less frequent consumption of discretionary food choices that are energy dense and nutrient poor are required for children to perform at their best. Children in this age group need to be consuming foods including vegetables, fruit, cereal and grain foods, lean meat and alternatives and dairy. (See Table 1.2 for serving quantities and sizes for Australian children and Table 1.3 for New Zealand children.) It is important to note that there is not a significant difference in the servings recommended for both countries, but each has its own specific guidelines endorsed by its national directives. Dietary restrictions are not recommended for growing children. Suspected food intolerance and allergy need to be confirmed by a medical practitioner. Children with health conditions that require dietary restriction need to be reviewed by a dietitian.

Knowledge of current dietary intake of Australian children comes from the 1995 National Nutrition Survey (ABS 1995), the 2010 NSW Schools Physical Activity and Nutrition Survey (Hardy et al. 2010) and the 2007 Australian National Children's Nutrition and Physical Activity Survey (CSIRO et al. 2008). However, new data is available with results starting to emerge from the Australian Health Survey 2011–13 (ABS 2012). Most of the dietary information for New Zealand children is derived from the 2002 National Children's Nutrition Survey (Ministry of Health 2003), the 2006–07 New Zealand Health Survey (Ministry of Health 2008) and the 2010 National Survey of Children and Young People's Physical Activity and Dietary Behaviours in New Zealand (CTRU & Synovate 2010).

In Australia, child intake of vegetables, fruit, grain (cereal) foods and milk, yoghurt and cheese products and alternatives is below recommended levels, while intakes of saturated fat, sugar and salt exceed recommendations by the Australian Dietary Guidelines. In the 1990s, children between the ages of two and eighteen consumed 41 per cent of their total energy intake as discretionary foods (NHMRC 2013c: 8, citing ABS 1995). While there is no directly comparable data from New Zealand, the New Zealand Health Survey found that 70 per cent of children between the ages of two and fourteen years ate fast food at least once a week, 14 per cent twice a week and 7 per cent three or more times a week (Ministry of Health 2008). New Zealand children are also consuming higher than recommended intakes of fat, sugar and salt on a regular basis.

Studies show that diet quality throughout life is linked to dietary habits that are established in childhood. Poor diet quality increases the risk of morbidity and adverse medical conditions including heart disease and diabetes (Kaikkonen et al. 2013; Wirt & Collins 2009). Research suggests that in children the higher intakes of energy, total fat, saturated fat and sodium are related to increased consumption of takeaway food as they get older. In addition, older children eat breakfast less often and this is associated with lower calcium and dietary fibre intakes (NHMRC 2013c: 8).

Energy

Children aged six to twelve years require energy for heat generation, muscle function, metabolism, physiological functions and growth and development. During this age, 1–2 per cent of energy is required to supply the fuel for growth, compared to 35 per cent of energy used for the same function by a child up to the age of three months (NHMRC & Ministry of Health 2006).

For all children, there are vast differences in rates of growth and development; therefore, there are large variations in the 'normal' range of energy requirements. This is why they are presented as estimates based on representative heights and weights. The estimated daily energy requirements of children from Australia's National Health and Medical Research Council and New Zealand's Ministry of Health nutrient reference values (NRVs) are shown in Table 8.1 (NHMRC & Ministry of Health 2006).

In Australia, children aged six to twelve years are meeting the requirements for energy, although many children are consuming more than their estimated energy requirements (EERs) for their specific age (CSIRO et al. 2008; see Table 8.1). This is one of the major factors contributing to an increase in overweight and obesity in children (see below).

Macronutrients

Protein

In Australia and New Zealand, children aged six to twelve years are meeting the requirements for protein

Table 8.1a Reference body weights and heights, daily EERs, by PAL, and proportion of population consuming above the EERs for boys (6–12 years) in Australia and New Zealand

Age (years)	Height (m)	Weight (kg)	EER (kJ), by PAL			% consuming above the EER
			1.6 (light activity)	1.8 (moderate activity)	2.0 (heavy activity)	
6	1.15	20.7	6600	7400	8200	25
7	1.22	23.1	7000	7800	8700	23
8	1.28	25.6	7300	8200	9200	26
9	1.34	28.6	7800	8800	9700	29
10	1.39	31.9	8300	9300	10,400	16
11	1.44	35.9	8800	9900	11,000	19
12	1.49	40.5	9300	10,500	11,600	27

Table 8.1b Reference body weights and heights and daily EERs, by PAL, and proportion of population consuming above the EERs for girls (6–12 years) in Australia and New Zealand

Age (years)	Height (m)	Weight (kg)	EER (kJ), by PAL			% consuming above the EER
			1.6 (light activity)	1.8 (moderate activity)	2.0 (heavy activity)	
6	1.15	20.2	6100	6900	7600	21
7	1.21	22.8	6500	7300	8100	32
8	1.28	25.6	6900	7700	8600	12
9	1.33	29.0	7300	8200	9100	16
10	1.38	32.9	7600	8500	9500	19
11	1.44	37.2	8000	9000	10,000	23
12	1.51	41.6	8500	9500	10,600	7

Note: EER: estimated energy requirement; PAL: physical activity levels. More accurate estimations of the energy requirements for children can be calculated using the Schofield (1985) equation (see Chapter 1).

Source: NHMRC & Ministry of Health (2006).

consumption. In Australia, children aged four to eight years consumed 74.3 grams of protein per day, and children aged nine to thirteen years consumed 95.2 grams per day (CSIRO et al. 2008). In New Zealand, boys and girls aged five to six years consumed respectively 62 and 52 grams of protein per day. Boys and girls aged seven to ten years consumed respectively 71 and 61 grams per day (Ministry of Health 2003). The protein requirements for children aged one to eighteen are shown in Table 1.9b.

Fat

In the six- to twelve-year-old age group in Australia and New Zealand, the average total fat intake is approximately 30 per cent of total energy intake, which meets the recommendations for total fat intake. Of concern is that saturated fat intake (which contributes to CVD and other chronic diseases) contributed 13–14 per cent of children's energy intake in Australia and 14–15 per cent in New Zealand (CSIRO et al. 2008; Ministry of Health 2003). The Australian and New Zealand guidelines recommend that saturated fat needs to contribute less than 10 per cent of total energy intake.

Dietary recommendations for fat intake are that children should aim for approximately 30 per cent of total energy from fat and less than 10 per cent from saturated fat. Tips for decreasing saturated fat intake are listed below:

- Eat low-fat or reduced-fat dairy foods for all family members over the age of two years.
- Swap butter for a margarine spread made from canola, sunflower, olive or dairy blends.

- Trim all visible fat from meat, remove skin from chicken and avoid processed meat (e.g., sausages and salami, ham, Devon, Fritz).
- Choose healthier foods for snacks. Cakes, pastries and biscuits are among the main sources of saturated fat in our food supply.
- Limit takeaway and fast foods such as pastries, pizza, fried fish, hamburgers, hot chips and creamy pasta, as these can be high in saturated fat.

Sugar

Sugars such as lactose, fructose and glucose contributed to nearly 25 per cent of Australian six- to twelve-year-old children's total energy intake (CSIRO et al. 2008), much higher than the recommendation in the Australian Dietary Guidelines, which suggest a moderate sugar intake of no more than 20 per cent of overall energy intake from total sugars (NHMRC 2013c). The Better Health Channel (2011) recommends that refined or

CASE STUDY 8.1 OBESITY IN CHILDHOOD

Profile and problem

S. is a nine-year-old girl who has been referred to a dietitian by her general practitioner regarding weight management. S.'s current weight is 55 kilograms, and she is 130 centimetres tall.

Assessment

S. has been diagnosed with impaired glucose tolerance (prediabetes). Her maternal grandfather has type 2 diabetes mellitus.

S.'s father works full time and her mother undertakes home duties and part-time work as a checkout operator at a local store. S. has a younger sister who is at the upper end of the healthy weight range. According to S.'s mother, she has always had 'puppy fat' and she herself was 'chubby' at the same age, but she reports that she grew out of it.

S. participates in no organised sporting activities. Her mother noted she was quite proud of her daughter, as she was a 'very good girl' and could sit in front of the television for up to three hours without making a sound. This occurred on most days of the week, both before and after school. For her birthday S. was given gifts of three DVDs, and the family bought a PlayStation last Christmas. They are planning to get her a television for her room when she turns ten. They feel that push bikes are too dangerous, because children might ride on the road or they might not know where the children have gone.

S.'s mother noted she often gave her daughter food as a reward for good behaviour (such as an ice cream or potato crisps in front of the television). S. also only receives dessert if she eats all of her dinner. The family have takeaway one or two times per week but have 'easy' dinners in front of the television three to four times per week to 'keep the peace'. The family favourites are oven chips, pizza, spaghetti bolognaise with garlic bread, packet pastas with extra cheese and Chinese takeaway. No one is particularly keen on vegetables, and they rarely buy fruit, because no one eats it.

Treatment

S.'s mother has agreed to do the following:

- eat dinner as a family at the dinner table
- include at least one vegetable at every dinner
- participate with S. in a children's cooking class at the local community centre
- do one physical activity with the family every week
- create an opportunity for S. to be active for at least one hour per day
- limit S.'s screen time to one hour per day excluding homework.

Plan

S. and her mother will see the dietitian monthly to monitor her eating habits and progress on the goals agreed to for healthy eating for the family. The dietitian will encourage S.'s mother to attend a local community course for healthy eating for families. If improvements in S.'s physical activity level (PAL) are not made at home, she will be referred to a children's activity program designed to improve motor skills, fitness and confidence. Over time, changes made to diet and physical activities should enable S. to grow into her current weight by limiting further weight gain as she grows.

processed sugar, such as white sugar, brown sugar and caster sugar, needs to be limited to no more than 10 per cent of total dietary intake per day. The 2007 Children's Survey has set the 'ideal' for all forms of sugar intake (which includes those sugars added, not sugars found in fruit, milk or yoghurt) as being less than 20 per cent of total energy.

The Australian Dietary Guidelines have included a specific recommendation on limiting sugar intake, with the guideline 'Limit intake of foods containing saturated fat, added salt, added sugars and alcohol' (NHMRC 2013c). The New Zealand Food and Nutrition Guidelines also state, 'Prepare foods or choose pre-prepared foods, snacks and drinks that are low in sugar, especially added sugar (Ministry of Health 2012a). Evidence has noted a strong association between the consumption of sugar-sweetened drinks and the risk of excessive weight gain and dental caries in children.

Fibre

Fibre intake averaged 18.6 grams per day for four- to eight-year-old and 22.3 grams per day for nine- to thirteen-year-old Australian children (CSIRO et al. 2008). In New Zealand, fibre intakes were between 15 and 17 grams per day for children in the six- to twelve-year-old age group (Ministry of Health 2003). Comparison with the NRVs for Australia and New Zealand indicate children are only just meeting the recommendations in Australia, while consumption of dietary fibre in New Zealand is slightly under that recommended. The fibre requirements for children are shown in Table 1.12a. Tips for increasing fibre intake are listed below:

- Eat high-fibre breakfast cereals that contain barley, wheat, rye or oats, such as porridge (rolled oats) or breakfast biscuits.
- Switch to wholemeal or multigrain breads and brown rice. If children eat only white bread, ensure it is a high-fibre white bread.
- Snack on fruit, dried fruit, nuts or wholemeal crackers.

Fruit and vegetables

Preliminary results from the Australian Health Survey indicate that younger children (5–11 years) are closer to reaching the recommended one serve of fruit per day (95.3 per cent) compared with 20.3 per cent of 12–17-year-old children meeting their recommended three serves per day. Vegetable intakes are poorer, with just over half (56.1 per cent) of children aged 5–7 years eating two serves, 30.8 percent of 8–11-year-olds eating three serves, and 15.2 per cent of 12–17-year-olds eating four serves per day, as recommended (ABS 2012).

In New Zealand, 55 per cent of girls and 46 per cent of boys aged five to six years ate two or more serves of fruit per day, meeting the national recommendations. In the seven- to ten-year-old age group, 47 per cent of girls and 42 per cent of boys were eating two serves of fruit per day. This figure decreased in the eleven- to fourteen-year-old age group to 36 per cent of girls and 39 per cent of boys consuming two or more serves of fruit per day. Fifty-two per cent of boys aged five to six years ate three or more serves of vegetables per day, and this increased to 53 per cent of boys aged seven to ten and 59 per cent aged eleven to fourteen. For females, the intakes of vegetables were higher, being 60 per cent for five- to six-year-olds, 56 per cent for seven- to ten-year-olds and 60 per cent for eleven- to fourteen-year-olds (Ministry of Health 2003).

Data from Australian and New Zealand nutrition surveys indicate that children need to be eating more fruit and vegetables and less energy-dense, nutrient-poor food. Compared to current intakes, vegetable consumption in children aged six to twelve years needs to approximately double, and the intake of a variety of vegetables needs to increase.

For children aged six to twelve years, dietary recommendations are to aim for five serves of vegetables per day and two serves of fruit. For a full list of serving sizes and age-specific recommendations on fruit and vegetable intake see Chapter 1. Tips for increasing fruit and vegetable intake are listed below:

- Pack one to two pieces of fruit in children's lunchboxes daily.
- Camouflage vegetables by adding them to casseroles and stews or grating them into rissoles and patties.
- Add vegetables to a stir-fry with onion, garlic and ginger or add a dressing of olive oil, balsamic vinegar and black pepper.
- Cut down the 'stewed' smell of some vegetables by cooking for as short a time as possible. Use a microwave or a large saucepan with the water boiling well before you drop them into it to cook.
- Increase the variety of vegetables presented to children. Try beans, beetroot, carrots, corn, egg-plant, lettuce, onion, peas, pumpkin, sweet potato (kumara), spinach and tomato.
- Try giving children raw vegetables as an afternoon snack or school snack; e.g., cut-up carrots with mild salsa or low-fat cream cheese.
- Never give up! Offer little tastes on a regular basis. Praise all efforts to try them.

Micronutrients

Folate

Nearly all Australian children aged four to eight years met the requirements for folate intake (CSIRO et al. 2008). Just over 10 per cent of girls aged nine to thirteen years were at risk of folate deficiency, with 89 per cent meeting the estimated average requirement (EAR) for folate intake. The prevalence of inadequate folate intake in New Zealand children and young people aged five to fourteen years was estimated to be 37 per cent overall but was considerably higher, at 64 per cent, for females aged nine to fourteen years (Ministry of Health 2003). Daily requirements for folate for children are shown in Table 8.2. Tips for increasing folate intake are listed below:

- Eat a wide variety of green vegetables (e.g. asparagus, spinach, lettuce and broccoli) and fruits
- Add chickpeas, dried beans and lentils to soups, casseroles and stews.
- Eat wholemeal cereals, nuts and yeast extracts such as Vegemite.
- Choose foods that have been fortified with folate, including some breakfast cereals and bread.

Calcium

Adequate calcium intake is vital for children aged six to twelve years. During these years, the body uses calcium to build strong bones, a process that is nearly complete by the end of the teen years. Bone calcium begins to decrease in young adulthood, and progressive loss of bone occurs during ageing, especially in females. If children are not provided with adequate calcium, they are at greater risk of developing osteoporosis in later life, which increases the risk of fractures from weakened bones.

Australian children are not meeting their requirements for calcium. For both boys and girls aged four to eight years, 89 per cent were meeting their calcium requirements (CSIRO et al. 2008). As children get older, the proportion meeting their requirements for calcium intake decreases. In nine- to eleven-year-olds, 65 per cent of boys and 45 per cent of girls met requirements for calcium; in twelve- to thirteen-year-olds, 50 per cent of boys and 11 per cent of girls met the requirements. In New Zealand, in children aged five to fourteen years, only 35 per cent were estimated to be consuming adequate calcium (Ministry of Health 2003). Daily requirements for calcium for children are shown in Table 8.3. Tips for increasing calcium intake are listed below:

- Choose reduced-fat, low-fat or skim milk and milk products, such as low- and reduced-fat cheese, yoghurt and custard, or calcium-fortified milk alternatives, such as soy or rice milk.
- Eat green leafy vegetables, including bok choy, broccoli, cabbage and spinach.
- Eat calcium-fortified foods, including breakfast cereals and bread.
- Consume salmon and tuna with the small bones included.
- Eat nuts and seeds, including Brazil nuts, almonds and sesame seed paste (tahini).

Iron

In Australia, most children met the dietary requirements for iron intake. Those most at risk were girls over the age of twelve (CSIRO et al. 2008). In New Zealand, the recommended dietary intake (RDI) for iron for boys and girls aged six to twelve years was primarily met, although evidence indicates that 1.3 per cent of children aged five to fourteen years did not meet their iron requirements (Ministry of Health 2003). Daily requirements for iron for children are shown in Table 8.4. Tips for increasing iron intake are listed below:

- Consume moderate amounts of lean beef, pork, chicken, salmon and eggs.

Table 8.2 Daily EARs and RDIs for folate for children (4–13 years) in Australia and New Zealand

Age (years)	Amount of folate (as folate equivalents), by NRV (µg)	
	EAR	RDI
4–8	160	200
9–13	250	300

Note: EAR: estimated average requirement; RDI: recommended dietary intake

Source: NHMRC & Ministry of Health (2006).

Table 8.3 Daily EARs and RDIs for calcium for children (6–12 years) in Australia and New Zealand

Age (years)	Amount of calcium, by NRV (mg)	
	EAR	RDI
6–8	360	500
9–11	800	1000
12	1050	1300

Note: EAR: estimated average requirement; RDI: recommended dietary intake

Source: NHMRC & Ministry of Health (2006).

Table 8.4 Daily EARs and RDIs for iron for children (4–13 years) in Australia and New Zealand

Age (years)	Amount of iron, by NRV (mg)	
	EAR	RDI
4–8	4	10
9–13	6	8

Note: EAR: estimated average requirement; RDI: recommended dietary intake

Source: NHMRC & Ministry of Health (2006).

- Have wholemeal bread and iron-fortified cereals for breakfast.
- Add lentils and kidney beans to casseroles and salads.

Sodium

Australian children's sodium intakes ranged from 1658 to 3672 milligrams per day, the latter clearly exceeding the upper limit (UL) for all children aged six to twelve years. Sodium intake was found to increase with age (CSIRO et al. 2008). A comprehensive measure of sodium intake has not been undertaken in recent national nutrition surveys in New Zealand due to the difficulty in quantifying discretionary salt use (CTRU & Synovate 2010). Sodium has a UL imposed, as the adverse effects of higher levels of sodium intake on blood pressure have been scientifically identified. Daily requirements and limits for sodium for children are shown in Table 8.5. Tips for decreasing sodium intake are listed below:

- Avoid adding salt to cooking or at the table.
- Avoid giving high-salt foods (e.g. potato crisps) to children.
- Choose reduced-salt bread, canned foods, margarine and breakfast cereals.
- Eat less processed and preserved foods.
- Reduce fast food and takeaways.
- Use herbs and spices such as garlic, oregano and lemon juice to add flavour to meals.

Table 8.5 Daily AIs and ULs for sodium for children (4–13 years) in Australia and New Zealand

Age (years)	Amount of sodium, by NRV (mg)	
	AI	UL
4–8	300–600	1400
9–13	400–800	2000

Note: AI: adequate intake; UL: upper limit

Source: NHMRC & Ministry of Health (2006).

For a complete list of NRVs for all macronutrients and micronutrients in Australia and New Zealand, see Chapter 1.

INFLUENCES ON EATING

Lifestyle habits that increase the risk of developing poor eating habits that were prevalent among surveyed New South Wales school students were skipping breakfast, eating dinner in front of the television, eating food from fast food outlets and offering sweets to younger children for good behaviour (Hardy et al. 2010).

The role of family

Family meals are an important time during which children learn and consolidate their healthy eating practices, especially children aged six to twelve years. It is a valuable time for families to share news of the day and to model healthy eating behaviour. Evidence has shown that interventions aimed at the family have improved overweight and obesity outcomes and therefore risk factors for other chronic diseases (Collins et al. 2011). A meta-analysis of seventeen studies with 182,836 children found the frequency of shared family meals is significantly related to positive nutritional health in children. Children who share meals with the family at least three or more times per week are more likely to be in a normal weight range and have healthier dietary and eating patterns than those who share fewer than three family meals together. In addition, they are less likely to engage in disordered eating (Hammons & Fiese 2011). Parents have important roles to play in their children's eating habits; some of these are listed below:

- Parents are the gatekeepers of the family's food supply. Children can only eat the food that is made available to them. If it is not there, they cannot eat it. If it is available and healthy, like fruit, they will eat it. It is important for parents to set boundaries and to monitor children's dietary intake and eating patterns.
- Parents should shop the way they want their children to live in the future. To grow healthy children parents have to plan, purchase, cook and offer them healthy food.
- Parents should provide a healthy eating role model. If children see parents, caregivers and all the household members enjoying vegetables, fruit and healthy foods on a regular basis, they are more likely to eat the same foods.
- Have food rules. Plan regular meal and snack times and close the kitchen in between times. Tips include

starting the day with breakfast. Children who eat breakfast perform better at school. Parents should also pack a healthy lunchbox for school to fuel children through their day.

- Learn to read food labels and look for snacks that have 400 kilojoules or less and 3 grams of fat or less per 100 grams. Cut up fruit for a platter and offer it around for snacks or dessert.
- Eat at the table *every* night. This is an important time for families to communicate and model healthy eating behaviours. Plan to have as many meals together as possible without television, telephone or social media interruptions. Children need to be encouraged to listen to their bodies and determine how much food they wish to eat, while the caregiver is responsible for providing the child with healthy food options.
- Persevere with vegetables and other healthy foods. It can take up to ten repeated exposures to the same vegetable for a child to accept it.
- Adopt positive attitudes towards food by focusing on efforts to try healthy food. For example, say, 'I like the way you tried that zucchini'. Choose to ignore the behaviour of family members who refuse to try specific foods.
- Involve children in cooking and meal preparation. Children are more likely to eat foods that they have prepared and cooked.
- Make the main drink water or reduced- or low-fat milk. Avoid soft drinks and sugar-sweetened drinks. Limit 100 per cent fruit juice to no more than half a glass per day (125 millilitres).
- Parents should practise positive weight management strategies. Never comment on your own or a family member's body, unless you are saying something positive.

Breakfast

More than 15 per cent of Australian primary school children were not eating breakfast. This figure increased for boys in lower socio-economic groups (Hardy et al. 2010). A series of nationally representative surveys in New Zealand presented in 2008 showed that approximately 10 per cent of New Zealand children did not eat breakfast regularly (CTRU & Synovate 2010; Ministry of Health 2003). The prevalence of skipping breakfast tended to increase with age, especially in girls.

Breakfast is the most important meal of the day. Early studies noted that consuming breakfast has a positive effect on performance indicators including memory recall, attention span and creativity (Benton & Parker 1998). Other studies have noted breakfast consumption may improve cognitive function related to memory, test grades and school attendance (Rampersaud et al. 2005). A 2009 systematic review examined 45 laboratory and school breakfast studies to determine if breakfast really does have an impact on children's performance at school. Evidence showed that eating breakfast is more beneficial than skipping breakfast among schoolchildren. The type of breakfast consumed is less important than whether the breakfast is consumed at all. Children who do not eat breakfast are more likely to eat more energy-dense, nutrient-poor foods throughout the remainder of the day due to increased hunger (Hoyland et al. 2009). A European systematic review conducted in 2010 also noted that eating breakfast is associated with a reduced risk of becoming overweight or obese and with a reduction in BMI in children (Szajewska & Ruszczynski 2010).

Breakfast ideas

- Porridge or wholegrain cereal with cut-up banana, stewed pears or fresh or frozen berries
- Raisin bread, wholegrain English muffins or fruit loaf, either plain or with ricotta cheese or low-fat cream cheese
- Baked beans with wholegrain toast
- Scrambled or poached eggs with toast and a glass of low-fat milk
- Fruit smoothie—low-fat milk, yoghurt, cut-up soft fruit
- For an on-the-go breakfast, a piece of fruit, a tub of low-fat yoghurt or a piece of raisin toast

School lunches and school canteens

As obesity rates are of major concern in Australia (see below), targeting school canteens is one population health strategy geared at improving the eating choices of children aged six to twelve years. The school environment is well recognised as an essential site for nutrition intervention. One Australian study estimated that more than one-third of total energy intake is consumed at school (Hands et al. 2004). In New Zealand, almost all students aged five to fourteen years eat or drink something while at school, and nearly 60 per cent of students purchase some of the food consumed at school from the school canteen or tuckshop (Ministry of Health 2003). Many of these foods can be quite high in energy and cost.

Sales data from a study in New Zealand primary schools showed less healthy food choices (such as chips and pies) dominated sales by more than two to one (Carter & Swinburn 2004). A study by Bell and Swinburn (2004) investigating the proportion of daily energy intake children obtained at school suggested that Australian children using the canteen consumed approximately 200 kilojoules more energy at school than those who obtained energy only from food in their lunchbox. Additional results suggested canteen users obtained significantly more energy from fast food, packaged snacks, desserts, chocolate and confectionery and milk than non-canteen users.

Research undertaken in Australia and New Zealand has shown that many foods consumed at lunch time either from school lunchboxes or from the school canteen contain food that is high in fat, sodium and sugar. There also tends to be a lower consumption of fruit and vegetables (Sanigorski et al. 2005). The Hunter Region study of 2423 Australian primary school children found that the commonly purchased canteen items at recess and lunch are less healthy choices, with food and drink high in fat and sugar dominating sales and conflicting with the Australian Dietary Guidelines for children and adolescents (Finch et al. 2006). Findings are consistent with those of Cleland et al. (2004), who suggest that children most commonly use the canteen to purchase less healthy foods like hot chips, pies and pastries.

A New Zealand study of five- to twelve-year-old children found that about 10 per cent of children used the school canteen and consumed more total energy, and more energy from cakes, fast foods and soft drinks, than non-canteen users (Dresler-Hawke et al. 2009).

CASE STUDY 8.2 FUSSY EATING IN CHILDHOOD

Profile and problem

J. is an eight-year-old boy and the youngest of four children. He has frequent tantrums regarding food choices and will often refuse outright to eat any food on his plate. J.'s parents noted that for years it was a struggle to get him to eat anything that wasn't frozen processed 'chicken' pressed into the shape of a dinosaur. They recalled one particularly intense public meltdown when 'something green' (a small piece of broccoli) was placed on his plate. J. has been a fussy eater since he was a toddler, but the family assumed he would 'grow out of it', like some of their other children. As his parents both work full time, they feel too exhausted to challenge J. with his eating. If he does not like a particular dish, his mother indicated she would reheat chicken nuggets from the freezer and he would at least eat those. Even though he is in the healthy weight range, his parents are concerned that he is not consuming enough nutrients from the limited food he eats, as he refuses all forms of vegetables (except hot chips), and the only fruit he eats is apple a few times a week. He also dislikes all dairy foods, and his main drink is cordial.

Assessment

J.'s eating behaviours are limiting his dietary intake. He does not consume sufficient amounts of fruit, vegetables, whole grains or dairy foods. Consequently, his diet is low in fibre and several vitamins and minerals.

Treatment

J.'s parents have agreed to do the following:

- eat meals together as a family and model eating a variety of foods
- allow J. to select from a variety of foods at each meal, including some foods that he is already willing to eat
- present new foods on at least ten occasions
- offer water and milk instead of cordial
- allow J. to 'play' with his food to explore its texture and other sensory qualities.

Plan

J. and his parents will follow the new strategies to try to increase J.'s dietary variety. It is likely to take several weeks or months to make a substantial improvement, as new foods are slowly introduced and accepted. If the relaxed approach to offering new foods does not result in any improvement, J. and his parents will be referred to a community program for fussy eaters run by a psychologist and a dietitian.

Lunchbox ideas for children aged 6–12 years

Packing lunches can be a difficult task when children get tired of eating the same food every day. Being organised in the process of deciding what to pack can help make it easier and faster. Parents should try to make each part of a packed lunch varied; for example, bread can be a great source of energy, B vitamins and fibre. Varieties to include are wholemeal, high fibre, multigrain wraps, rolls, focaccia or pita pockets. Some examples of different sandwich fillings and snacks that children might like are given below.

Sandwich fillings

Chicken and avocado
Drained pineapple and ham
Chicken, hard-boiled egg and mayonnaise
Baked beans and grated cheese
Grated carrot, sultanas and low-fat mayonnaise
Low-fat cream cheese and sultanas

Snacks

Fruit (fresh, canned, frozen or dried)
Vegetable sticks (carrot, cucumber, capsicum) with salsa or low-fat dip (e.g., hummus)
Raisin bread or fruit loaf
Wholegrain breakfast cereal
Wholegrain rice cakes topped with vegetables and cheese or sliced banana
Small tin of baked beans, spaghetti or creamed corn
Reduced-fat yoghurt or custard
Corn on the cob
A frozen bottle of water to keep everything cool

The New Zealand Food and Nutrition Guidelines also note that foods brought from home are likely to be more nutritious and cost-effective than foods bought either at school or on the way to school (Ministry of Health 2012a).

In 2010, the Australian Department of Health and Ageing developed the National Healthy School Canteens Project. This provided guidelines for the implementation and running of healthy school canteens. Policies are encouraged but not mandatory for each state. The New Zealand Ministry of Health (2007) also has guidelines on providing healthy choices in school canteens. Future dietary surveys will determine the effectiveness of these strategies to improve childhood nutrition in the school environment.

Discretionary foods

Foods that are not necessary for a healthy diet and are too high in saturated fat, added sugars and/or added salt and low in fibre are classified by the Australian Dietary Guidelines as discretionary foods (NHMRC 2013c). The New Zealand Food and Nutrition Guidelines name these foods 'high fat, sugar and salt foods' (Ministry of Health 2012a). These foods and drinks are often also high in kilojoules but low in essential nutrients, making them energy-dense but nutrient-poor foods. They need to be consumed in moderation, as they are often eaten in place of other more nutritious foods. Discretionary foods are often associated with increased risk of obesity and chronic disease such as heart disease, stroke, type 2 diabetes mellitus (T2DM) and some forms of cancer (NHMRC 2013b).

The NSW Schools Physical Activity and Nutrition Survey noted that discretionary foods were commonly consumed by children aged six to twelve years. Hot chips, one of the most popular discretionary foods, were consumed at least once per week by two-thirds of students, with 15 per cent consuming hot chips at least three times per week. Discretionary snack foods, such as crisps and salty snacks, biscuits, cakes, donuts and muesli bars, confectionery and ice cream or ice blocks, were consumed on an even more frequent basis. In brief, all of these discretionary snack foods were reportedly consumed by approximately one-third of students as often as three to six times per week and by approximately 10 per cent of students on a daily basis (Hardy et al. 2010).

Nearly 60 per cent of primary school children are either sometimes or usually rewarded with sweets for good behaviour. Nearly one-quarter of primary school children eat takeaway meals or snacks from a fast food outlet one or more times per week (Hardy et al. 2010). In New Zealand, foods and drinks that are high in fat, sugar and salt contributed to 20 per cent of total energy intake for five- to fourteen-year-olds, indicating that healthier foods are being displaced in the diet (Ministry of Health 2012a).

At this age, children are exposed to many factors that can impact on their food choices, including peer pressure, television and other forms of media, parties and other environmental influences. It is normal that a child will eat crisps, lollies and other discretionary foods at parties and special events, but these foods need to be eaten only occasionally. Overconsumption of these foods can lead to overweight and obesity and underconsumption of valuable nutrients. These foods also tend to be more costly.

Soft drinks and sugar-sweetened beverages

The 2007 Australian National Children's Nutrition and Physical Activity Survey found that almost half (47 per cent) of children from two to sixteen years of age consumed sugar-sweetened beverages (including energy drinks) daily, with one-quarter (25 per cent) consuming sugary soft drinks daily (CSIRO et al. 2008). In the 2007 New Zealand Children's Food and Drink Survey of children aged five to nineteen years, just over half (51 per cent) consumed sugar-sweetened beverages, including cordial, fizzy or soft drinks, at least once per week (Health Sponsorship Council 2007). High levels of soft drink consumption have been linked to a range of illnesses, including dental caries, overweight and obesity, which can lead to T2DM, metabolic syndrome (MetS), osteoporosis. The displacement of healthier food and beverage options (particularly milk) from the diet is also a risk factor (Gill et al. 2006; Rangan et al. 2009).

The consumption of sugar-sweetened beverages is also a risk factor for development of heart disease. The Telethon Kids Institute (previously called the Telethon Institute for Child Health Research) in Perth has followed the health of more than 1400 teenagers as part of its ongoing long-term Western Australian Pregnancy Cohort Study into children's health. Those children who drank more than one can of soft drink a day were more likely to develop T2DM, heart disease or a stroke in adulthood (Ambrosini et al. 2013).

A study of nearly 2000 twelve-year-old children by the Centre for Vision Research at the University of Sydney showed that there was a narrowing of the retinal arteries (and therefore increase in narrowing of blood vessels throughout the body) in those children with an intake of over 278 grams of carbohydrate per day. A major source of that carbohydrate came from soft drinks or cordial, with high-risk children consuming one or more glasses a day (Gopinath et al. 2012).

COMMON CONCERNS AND HEALTH CONDITIONS

Fussy eating

Fussy eating is a common issue during childhood and can cause considerable parental anxiety. It can be defined as aversive behaviours related to eating. Often, fussy eating issues resolve themselves with time or with caregiver education, but for some children fussy eating can lead to development of short-term and/or long-term mild nutritional imbalances including constipation, iron deficiency anaemia, failure to thrive and obesity. Various studies have estimated fussy eating prevalence as between 13 and 22 per cent of children (Mascola et al. 2010). If aversive behaviours surrounding failure to eat such as refusing to sit at the table and throwing food are included, this percentage increases to between 20 and 40 per cent (Queensland Health 2009).

Fussy eating habits are more likely to develop before the age of five, but new cases can emerge later in childhood. Those children who do have fussy eating habits persisting beyond two years are more likely to have strong likes and dislikes of food and are not likely to accept new foods in later years. Over half of all fussy eaters recover over a two-year period irrespective of their age at onset, but a smaller number of fussy eaters continue to be a problem for some parents for many years (Mascola et al. 2010).

Factors in the family environment that influence fussy eating may include parental food preferences, parental–child feeding practices, parental knowledge and beliefs and children's exposure to food. Fussy eating may be a result of medical complications (for example, gastro-oesophageal reflux) or non-organic reasons (such as aversive feeding experiences including force feeding or choking). Developmental delay, speech delay and behavioural problems are also associated with feeding difficulties in children.

Overweight and obesity

Prevalence

The prevalence of overweight and obesity in children has significantly increased over the past two decades. Latest survey results in Australia have shown that nearly one in four Australian children are overweight or obese (CSIRO et al. 2008; see Table 8.6). Figures from The Health of New Zealand Children 2011–12 report shows that the rates of overweight and obesity are even higher in New Zealand (see Table 8.7; Ministry of Health 2012b).

Examination of Australian children's weight over the previous century has shown that overweight prevalence among children was relatively low and constant throughout most of the century, until it accelerated from the early 1970s (Norton et al. 2006). This trend has continued over the last 40 years. In 1995, around 21 per cent of Australian children aged seven

Table 8.6 Weight status for children (4–13 years) in Australia, 2007 (%)

Age (years)	Underweight	Normal weight	Overweight	Obese
4–8	4	78	13	4
9–13	6	69	18	7

Source: CSIRO et al. (2008).

Table 8.7 Weight status for children (5–14 years) in New Zealand, 2010 (%)

Age (years)	Underweight	Normal weight	Overweight	Obese
5–9	3.9	66.6	19.0	10.6
10–14	4.5	62.2	22.5	10.8

Source: Ministry of Health (2012b).

to fifteen years were considered to be overweight or obese. This is nearly double the prevalence recorded in 1986 (Magarey et al. 2001). Between 1995 and 2007 the rate increased another 4 percentage points, to nearly 25 per cent, in nine- to thirteen-year-olds (ABS 2009).

Making a direct comparison between Australian and New Zealand children in terms of overweight and obesity is difficult, due to different data collection methodologies and reporting age ranges. But in summary, approximately one-quarter of children are overweight or obese in both countries.

Causes

Overweight and obesity are the result of an imbalance between energy intake (diet) and energy output (physical activity). Evidence has shown that excess energy intake is occurring at a population level, contributing to excess weight gain (NHMRC 2013c). Other contributing factors include living in an obesogenic environment. Factors that have contributed to this environment include the following:

- changes in the food supply, making more energy-dense, nutrient-poor options more marketable and cost-effective
- changes in urban design, resulting in less incidental activity
- reduced perceptions of safety, leading to fewer opportunities to expend energy
- changes in the social and economic environment, including an increase in dual-income families and a higher disposable income, which may make parents time poor in relation to food preparation
- decreased food literacy and cooking skills and greater reliance on convenience and takeaway foods
- increased use of small screen recreation by children.

In Australia, one study determined that 58.9 per cent of boys and 48 per cent of girls were not meeting the daily guideline for less than two hours of screen time per day (Hardy et al. 2010).

Complications

A relatively large and consistent body of evidence has shown that overweight and obesity in childhood are associated with increased risk of premature morbidity and mortality in adulthood. In a systematic review, four out of five eligible studies found a strong association between overweight and/or obesity and premature mortality. All eleven studies with cardiometabolic morbidity (diabetes, hypertension, ischaemic heart disease and stroke) as outcomes reported that overweight and obesity were associated with significantly increased risk of later cardiometabolic morbidity in adult life. Analysis shows that child overweight and obesity were also associated with significantly increased risk of later disability pension, asthma and polycystic ovary syndrome symptoms (Reilly & Kelly 2011).

Another consequence of overweight and obesity in childhood may include psychological distress. By the age of seven years, children may already be experiencing teasing and social isolation as a result of their obesity. Orthopaedic problems such as slipped capital epiphysis, foot pronation and abnormal plantar pressures can result from childhood obesity. Children who are obese are also more prone to respiratory conditions such as **obstructive sleep apnoea** and gastrointestinal

Obstructive sleep apnoea is the cessation of air flow at the mouth and nose for at least ten seconds during sleep.

problems including gastro-oesophageal reflux, fatty liver and **cholelithiasis** (NHMRC 2013c).

Prevention and treatment

Overweight and obese children increase their risk of being overweight in later life and therefore their risk factors for other lifestyle-related diseases. It is important that the management of overweight and obesity in childhood is addressed as a priority. In children it important to maintain appropriate growth and development while at the same time encouraging weight management. A number of strategies have been recommended to address rising overweight and obesity and their associated health risks:

- Behavioural interventions addressing diet and exercise in children are more effective when they are family based. Families are recommended as the agents of change for weight loss in overweight and obese children; family approaches are more effective than those that focus on the child alone (Jull & Chen 2013).
- Decreased consumption of sugar-sweetened beverages is associated with decreased risk of weight gain in children.
- School-delivered healthy eating and physical activity interventions have been shown to improve weight outcomes in children.

In Australia, recommendations for action to address overweight and obesity at the individual level include the following:

- not making soft drinks available in the home and offering water as a beverage
- finding alternatives to using sweets as a reward or treat, such as offering a choice from a rewards box containing stickers, non-permanent tattoos, pencils, erasers and bookmarks
- packing foods appropriate for healthy lunchboxes
- consuming a healthy breakfast
- not eating in front of the television and having evening meals at the table as a family
- limiting children's screen time. (Hardy et al. 2010)

Future trends

Based on past trends and the assumption that current interventions will not be effective, BMI is predicted to continue to increase for both male and female children. This would result in around one-third of five- to nineteen-year-old Australians being overweight or obese by 2025 (37 per cent in males and 33 per cent in females), compared with 21 per cent in both males and females in 1995 (Department of Human Services 2008).

Research conducted in New Zealand has indicated that the rates of obesity in children may be reaching a plateau. Reports from the 2006–07 New Zealand Health Survey found that there has been no change in the average (mean) BMI or the prevalence of obesity for children aged five to fourteen years since 2002. It is important to note that this data is for obesity only and was measured over just a four-year period, as compared to long-term prevalence studies (Ministry of Health 2008). Regardless, the issue of the growing rate of childhood overweight and obesity needs to be addressed.

> **Cholelithiasis** is the presence or formation of gallstones.

Metabolic syndrome

MetS can be defined as a condition in which a group of risk factors for diabetes and CVD (heart disease and stroke) occur together. The main characteristics include abdominal obesity, high blood pressure, decreased high-density lipoprotein levels ('good' cholesterol), high low-density lipoprotein levels ('bad' cholesterol), increased triglycerides and insulin resistance. MetS is affecting an increasing number of children and adolescents as the epidemic of obesity spreads. Estimates of MetS occurrence in primary school children are hard to determine due to its varying definition. The Bogalusa Heart Study estimated that 3–4 per cent of children had MetS (Chen et al. 2000), and the Cardiovascular Risk in Young Finns study also determined that approximately 4 per cent of children have the syndrome (Raitakari et al. 1995).

Diabetes

Type 1 diabetes mellitus (T1DM) is the most common form of diabetes in children aged six to twelve years. It occurs when the pancreas is unable to make enough insulin (a hormone which allows the body to use glucose in the blood as energy). Glucose in the blood is mainly derived from the food we eat and is then stored in the liver and muscle. When a child has diabetes, their blood glucose levels increase because there is a lack of insulin. As a result, the child can experience diabetic symptoms such as an increase in thirst, excessive urination, weight loss and dehydration. T1DM requires insulin injections for management.

T2DM is less common in children than T1DM. The pancreas still produces insulin, but the body is resistant to the insulin working. T2DM is often associated with being overweight. Children with T2DM do not need insulin treatment in most cases (Ambler & Cameron 2010).

Prevalence

In 2008, it was estimated that over 5700 children aged from birth to fourteen years had T1DM in Australia. Among the 34 countries that are current members of the Organisation for Economic Co-operation and Development, Australia had the seventh highest prevalence and sixth highest incidence of T1DM in children aged from birth to fourteen years (AIHW 2010). If current trends continued, it was estimated that the prevalence of T1DM in children would increase by 10 per cent between 2008 and 2013 (AIHW 2011). Accurate figures for the numbers of youth with T2DM are not yet available for Australia, but many of the reported cases have been among Aboriginal and Torres Strait Islander children (Baker IDI 2012).

Causes

The region on chromosome 6 containing the human leucocyte antigen is responsible for approximately 50 per cent of the genetic risk of T1DM. The exact cause of T1DM initiation is unknown, but it is proposed that an environmental factor such as a viral infection may cause the immune system to attack the insulin-producing cells of the pancreas in these genetically prone children. Although other family members may carry the same 'at risk' genes for developing T1DM, the overall risk is low. The risks for family members are estimated as follows:

- Mother who has T1DM: 1–2 per cent risk for child
- Father who has T1DM: 4–6 per cent risk for child
- Identical twin who has T1DM: at least 35 per cent
- Brother or sister who has T1DM: 3–6 per cent. (Australian Diabetes Council 2007)

Management of diabetes requires commitment for life as to date there is no cure. Several insulin injections throughout the day or use of an insulin pump that delivers insulin continuously is required in T1DM to maintain ideal blood glucose levels. People with T1DM often need to test their blood sugars from six to eight times a day, using a blood sugar monitor. The prevalence of T1DM in Australia is approximately 10 per cent, which is among the highest rates in the world (AIHW 2009). The onset of T1DM typically occurs at an early age, although it can occur at any age.

Cardiovascular disease

University of Oxford researchers analysed 63 studies (2000–11) involving more than 49,220 healthy children aged five to fifteen years and determined that obese children have increased risk factors for heart disease. These increased risks include higher cholesterol, blood pressure and blood sugar levels and a thickening of the heart muscle (increase in left ventricular mass) compared with healthy weight children. If a child's obesity tracks into adulthood, they will be at a 30–40 per cent higher risk of stroke and heart disease than their normal weight counterparts. Additionally, fasting insulin levels and insulin resistance (known markers for diabetes) were significantly higher in obese children but not in overweight children (Friedemann et al. 2012). The Bogalusa Heart Study has found that anatomic changes occur as early as five to eight years of age (Berenson et al. 1998).

The Western Australian Pregnancy Cohort Study identified 25 per cent of children were at increased risk of future obesity, CVD and diabetes at the age of eight years (Huang et al. 2012). A 2009 study conducted by the Children's Hospital at Westmead, Sydney, determined that there were many risk factors for CVD already common in all young Australian children. In all children, overweight and obesity have the strongest association with hypertension (Haysom et al. 2009).

Based on 21 years of follow-up data from children aged from three to eighteen years, the Cardiovascular Risk in Young Finns Study demonstrated that if healthy dietary patterns were developed in childhood, the cardiovascular health benefits accrued from such patterns would track into adulthood (Kaikkonen et al. 2013).

Risk factors for CVD in children include being overweight or obese, physical inactivity, genetic predisposition (for example, familial hypercholesterolaemia), MetS, hypertension, dyslipidaemia and an atherogenic diet (AIHW 2013). Lifestyle and eating habits are vitally important for protection against the development and progression of CVD risk. Ensuring children follow the Australian or New Zealand dietary guidelines would help reduce risk factors for heart disease, including limiting saturated fat intake, reducing salt intake, consuming a diet mainly plant based (fruit, vegetables, pulses and a wide range of wholegrain foods) and consuming moderate amounts of lean unprocessed meats, poultry and fish (NHFA & CSANZ 2012).

SUMMARY AND KEY MESSAGES

A large number of physical and psychological changes occur over this life stage that impact on the initiation, development and reinforcement of healthy eating habits. There are also a number of behavioural and environmental factors, such as the family and school environment, that impact on a child's eating behaviours and dietary intake and need to be considered.

- There can be vast differences in height, weight and body composition among children aged six to twelve years, but a healthy child will follow their own pattern of growth, roughly tracking along a growth percentile on the growth charts, despite some variation in dietary intake.
- In Australia and New Zealand, most children's intake of vegetables, fruit, grain (cereal) foods and milk, yoghurt and cheese products and alternatives is below recommended levels, while their intakes of saturated fat, sugar and salt exceed those recommended by the Australian and New Zealand dietary guidelines.
- Children aged six to twelve years in Australia and New Zealand are consuming more energy than their requirements, which may lead to weight gain.
- Habits that increase the risk of developing poor eating habits are skipping breakfast, eating dinner in front of the television, eating food from fast food outlets and offering sweets to younger children for good behaviour.
- Poor eating habits in childhood are often tracked into adulthood.
- Children with poor nutritional habits are at increased risk of overweight and obesity, T1DM and T2DM and cardiovascular risk factors.
- The family plays an important role in establishing healthy eating habits in children.

REFERENCES

ABS (Australian Bureau of Statistics) 1995. *National Nutrition Survey: Selected highlights, Australia, 1995*. Cat. no. 4802.0. Commonwealth of Australia. <www.abs.gov.au/ausstats/abs@.nsf/mf/4802.0>, accessed August 2013.

——2009. Children who are overweight or obese. *Australian Social Trends, Sep 2009*. Cat. no. 4102. Commonwealth of Australia. <www.abs.gov.au/AUSSTATS/abs@.nsf/Lookup/4102.0Main+Features20Sep+2009>, accessed August 2013.

——2012. *Children's Risk Factors. Australian Health Survey: First Results, 2011–12*. Cat. no. 4364.0.55.001. Commonwealth of Australia. <www.abs.gov.au/ausstats/abs@.nsf/Lookup/27D7FFFD3AEE46CCCA257AA30014BFFA?opendocument>, accessed 6 November 2014.

AIHW (Australian Institute of Health and Welfare) 2009. *Insulin-treated Diabetes in Australia, 2000–2007*. Diabetes Series 11, cat. no. CVD 45. Commonwealth of Australia. <www.aihw.gov.au/WorkArea/DownloadAsset.aspx?id=6442455100>, accessed 28 September 2014.

——2010. *Australia's Health 2010: The twelfth biennial health report of the Australia Institute of Health and Welfare*. Australia's Health Series 12, cat. no. AUS 122. Commonwealth of Australia. <www.aihw.gov.au/WorkArea/DownloadAsset.aspx?id=6442452962>, accessed 3 October 2014.

——2011. *Australia's Welfare 2011: The tenth biennial welfare report of the Australia Institute of Health and Welfare*. Australia's Welfare Series 10, cat. no. AUS 142. Commonwealth of Australia. <www.aihw.gov.au/WorkArea/DownloadAsset.aspx?id=10737420589>, accessed 3 October 2014.

——2013. *Risk Factors for Cardiovascular Disease, Type 2 Diabetes and Chronic Kidney Disease*. Commonwealth of Australia. <www.aihw.gov.au/cardiovascular-health/risk-factors/>, accessed August 2013.

Ambler, G. & Cameron, F. 2010. *Caring for Diabetes in Children and Adolescents* 3rd edn. Royal Children's Hospital, Melbourne. <http://video.rch.org.au/diabetes/Diabetes_Book_Third_Edition.pdf>, accessed August 2013.

Ambrosini, G.L., Oddy, W.H., Huang, R.C., Mori, T.A. et al. 2013. Prospective associations between sugar-sweetened beverage intakes and cardiometabolic risk factors in adolescents. *American Journal of Clinical Nutrition* 98: 327–34.

Australian Diabetes Council 2007. *About Type 1*. Australian Diabetes Council. <www.diabeteskidsandteens.com.au/about_type_1.html>, accessed 1 October 2014.

Baker IDI (Baker IDI Heart & Diabetes Institute) 2012. *Diabetes: The silent pandemic and its impact on Australia*. Baker IDI, Diabetes Australia & Juvenile Diabetes Research Foundation. <www.diabetesaustralia.com.au/Documents/DA/What's%20New/12.03.14%20Diabetes%20management%20booklet%20FINAL.pdf>, accessed August 2013.

Bao, W., Threefoot, S.A., Srinivasan, S.R. & Berenson, G.S. 1995. Essential hypertension predicted by tracking of elevated blood pressure from childhood to adulthood: The Bogalusa Heart Study. *American Journal of Hypertension* 8: 657–65.

Bell, A.C. & Swinburn, B.A. 2004. What are the key food groups to target for preventing obesity and improving nutrition in schools? *European Journal of Clinical Nutrition* 58: 258–63.

Benton, D. & Parker, P.Y. 1998. Breakfast, blood glucose, and cognition. *American Journal of Clinical Nutrition* 67: 772S–778S.

Berenson, G.S., Srinivasan, S.R. & Bao, W. 1997. Precursors of cardiovascular risk in young adults from a biracial (black-white) population: The Bogalusa Heart Study. *Annals of the New York Academy of Sciences* 817(1): 189–98.

Berenson, G.S., Srinivasan, S.R., Bao, W., Newman, W.P. et al. 1998. Association between multiple cardiovascular risk factors and atherosclerosis in children and young adults: The Bogalusa Heart Study. *New England Journal of Medicine* 338: 1650–2.

Better Health Channel 2011. *Sugar*. In consultation with Royal Children's Hospital Melbourne. Better Health Channel. <www.betterhealth.vic.gov.au/bhcv2/bhcarticles.nsf/pages/Sugar>, accessed August 2013

——2013. *Growth and Development: Primary school children*. Better Health Channel. <www.betterhealth.vic.gov.au/Bhcv2/bhcpdf.nsf/ByPDF/Growth_-_assessing_primary_school_children/$File/Growth_-_assessing_primary_school_children.pdf>, accessed July 2013.

Carter, M. & Swinburn, B. 2004. Measuring the 'obesogenic' food environment in New Zealand primary schools. *Health Promotion International* 19: 15–20.

CDC (Centers for Disease Control and Prevention) 2009. *Clinical Growth Charts*. CDC. <www.cdc.gov/growthcharts/clinical_charts.htm>, accessed 29 September 2014.

Chen, W., Bao, W., Begum, S., Elkasabany, A. et al. 2000. Age-related patterns of the clustering of cardiovascular risk variables of syndrome X from childhood to young adulthood in a population made up of black and white subjects: The Bogalusa Heart Study. *Diabetes* 49: 1042–8.

Cleland, V., Worsley, A. & Crawford, D. 2004. What are grade 5 and 6 children buying from school canteens and what do parents and teachers think about it? *Australian Journal of Nutrition & Dietetics* 61: 145–50.

Collins, C.E., Okely, A.D., Morgan, P.J., Jones, R.A. et al. 2011. Parent diet modification, child activity, or both in obese children: An RCT. *Pediatrics* 127: 619–27.

CSIRO (Commonwealth Scientific and Industrial Research Organisation), Preventative Health National Research Flagship & University of South Australia 2008. *2007 Australian National Children's Nutrition and Physical Activity Survey: Main findings*. Commonwealth of Australia, Canberra.

CTRU (Clinical Trials Research Unit) & Synovate 2010. *A National Survey of Children and Young People's Physical Activity and Dietary Behaviours in New Zealand: 2008/09; Key findings*. New Zealand Government. <www.health.govt.nz/system/files/documents/publications/cyp-physical-activity-dietary-behaviours-08-09-keyfindgs.pdf>, accessed 28 September 2014.

Department of Health and Ageing 2010. *National Healthy School Canteens: Guidelines for healthy foods and drinks supplied in school canteens*. Commonwealth of Australia. <www.health.gov.au/internet/main/publishing.nsf/Content/5FFB6A30ECEE9321CA257BF0001DAB17/$File/Canteen%20guidelines.pdf>, accessed 28 September 2014.

Department of Human Services 2008. *Future Prevalence of Overweight and Obesity in Australian Children and Adolescents, 2005–2025*. State Government of Victoria. <http://docs.health.vic.gov.au/docs/doc/768FD9A0683F9259CA2578EC0081AD6A/$FILE/future_overweight_prevalence_report.pdf>, accessed 28 September 2014.

Dresler-Hawke, E., Whitehead, C. & Coad, J. 2009. What are New Zealand children eating at school? A content analysis of 'consumed versus unconsumed' food groups in a lunch-box survey 2003–2004. *Health Education Journal* 68: 3–13.

Finch, M., Sutherland, R., Harrison, M., Collins, C. 2006. Canteen purchasing practices of year 1–6 primary school children and association with SES and weight status. *Australian and New Zealand Journal of Public Health* 30: 247–51.

Friedemann, F., Heneghan, C., Mahtani, K., Thompson, M. et al. 2012. Cardiovascular disease risk in healthy children and its association with body mass index: Systematic review and meta-analysis. *British Medical Journal* 345: e4759.

Gill, T., Rangan, A.M. & Webb, K.L. 2006. The weight of evidence suggests that soft drinks are a major issue in childhood and adolescent obesity. *Medical Journal of Australia* 184: 263–4.

Gopinath, B., Flood, V.M., Wang, J.J., Smith, W.et al. 2012. Carbohydrate nutrition is associated with changes in the retinal vascular structure and branching pattern in children. *American Journal of Clinical Nutrition* 95: 1215–22.

Hammons, A.J. & Fiese, B.H. 2011. Is frequency of shared family meals related to the nutritional health of children and adolescents? *Pediatrics*, 127: e1565–e1574.

Hands, B., Parker, H., Glasson, C., Brinkman, S. et al. 2004. *Child and Adolescent Physical Activity and Nutrition Survey 2003: Report*. Western Australia Government, Perth.

Hardy, L.L., King, L., Espinel, P., Cosgrove, C. et al. 2010. *NSW Schools Physical Activity and Nutrition Survey (SPANS) 2010: Full report*. NSW Government. <www.health.nsw.gov.au/pubs/2011/pdf/spans_full.pdf>, accessed August 2013.

Haysom, L., Williams, R.E., Hodson, E.M., Lopez-Vargas, P. et al. 2009. Cardiovascular risk factors in Australian Indigenous and non-Indigenous children: A population-based study. *Journal of Paediatrics and Child Health* 45: 20–7.

Health Sponsorship Council 2007. *2007 New Zealand Children's Food and Drink Survey Tables*. New Zealand Government. <http://archive.hsc.org.nz/publications/2012/2007-CFDS-Data-Tables.html?destination=/researchpublications.html>, accessed August 2013.

Hoyland, A., Dye, L. & Lawton, C.L. 2009. A systematic review of the effect of breakfast on the cognitive performance of children and adolescents. *Nutrition Research Reviews* 22: 220–43.

Huang, R.C., Mori, T.A. & Beilin, L.J. 2012. Early life programming of cardiometabolic disease in the Western Australian pregnancy cohort (Raine) study. *Clinical and Experimental Pharmacology and Physiology* 39: 973–8.

Jull, A. & Chen, R. 2013. Parent-only vs. parent-child (family-focused) approaches for weight loss in obese and overweight children: A systematic review and meta-analysis. *Obesity Reviews* 14: 761–8.

Kaikkonen, J.E., Mikkilä, V., Magnussen, C.G., Juonala, M. et al. 2013. Does childhood nutrition influence adult cardiovascular disease risk? Insights from the Young Finns Study. *Annals of Medicine* 45: 120–8.

Kirchengast, S. 2010. Gender differences in body composition from childhood to old age: An evolutionary point of view. *Journal of Life Sciences* 2: 1–10.

Magarey, A.M., Daniels, L.A. & Boulton, J.C. 2001. Prevalence of overweight and obesity in Australian children and adolescents: Reassessment of 1985 and 1995 data against new standard international definitions. *Medical Journal of Australia* 174: 561–4.

Mascola, A.J., Bryson, S.W. & Agras, W.S. 2010. Picky eating during childhood: A longitudinal study to age 11 years. *Eating Behaviors* 11: 253–7.

Ministry of Health 2003. *NZ Food NZ Children: Key results of the 2002 National Children's Nutrition Survey*. New Zealand Government. <www.moh.govt.nz/NoteBook/nbbooks.nsf/0/658D849A2BAC7421CC256DD9006CC7EC/$file/nzfoodnzchildren.pdf>, accessed 28 September 2014.

——2007. *Catering Guide: Food and beverage classification system for Years 1–13*. New Zealand Government. <www.health.govt.nz/system/files/documents/pages/heha-schools-catering-guide-contents.pdf>, accessed 6 November 2014.

——2008. *A Portrait of Health: Key results of the 2006/07 New Zealand Health Survey*. New Zealand Government. <www.health.govt.nz/system/files/documents/publications/portrait-of-health-june08.pdf>, accessed 28 September 2014.

——2009. *Clinical Guidelines for Weight Management in New Zealand Children and Young People*. New Zealand Government. <www.health.govt.nz/system/files/documents/publications/weight-management-children-guidelines_0.pdf>, accessed 28 September 2014.

——2012a. *Food and Nutrition Guidelines for Healthy Children and Young People (Aged 2–18 years): A background paper* 1st edn. New Zealand Government, Wellington.

——2012b. *The Health of New Zealand Children: Key findings of the New Zealand Health Survey 2011/12*. New Zealand Government. <www.health.govt.nz/publication/health-new-zealand-children-2011-12>, accessed 30 November 2014.

NHFA (National Heart Foundation of Australia) & CSANZ (Cardiac Society of Australia and New Zealand) 2012. *Reducing Risk in Heart Disease: An expert guide to clinical practice for secondary prevention of coronary heart disease*. NHFA. <www.heartfoundation.org.au/SiteCollection Documents/Reducing-risk-in-heart-disease.pdf>, accessed August 2013.

NHMRC (National Health and Medical Research Council) 2013a. *Clinical Practice Guidelines for the Management of Overweight and Obesity in Adults, Adolescents and Children in Australia 2013*, Commonwealth of Australia. <www.nhmrc.gov.au/_files_nhmrc/publications/attachments/n57_obesity_guidelines_131204_0.pdf>, accessed 28 September 2014.

——2013b. *Discretionary Food and Drink Choices*. Commonwealth of Australia. <www.eatforhealth.gov.au/food-essentials/discretionary-food-and-drink-choices>, accessed August 2013.

——2013c. *Eat for Health: Australian Dietary Guidelines; Providing the scientific evidence for healthier Australian diets*, Commonwealth of Australia. <www.eatforhealth.gov.au/sites/default/files/files/the_guidelines/n55_australian_dietary_guidelines.pdf>, accessed 28 September 2014.

——n.d. *Australian Guide to Healthy Eating*. Commonwealth of Australia. <www.eatforhealth.gov.au/guidelines/australian-guide-healthy-eating>, accessed 28 September 2014.

NHMRC (National Health and Medical Research Council) & Ministry of Health 2006. *Nutrient Reference Values for Australia and New Zealand: Including recommended dietary intakes*. Commonwealth of Australia & New Zealand Government. <www.nhmrc.gov.au/_filesnhmrc/publications/attachments/n35.pdf>, accessed 28 September 2014.

NHS (National Health Service) 2013. *Puberty*. NHS. <www.nhs.uk/Conditions/Puberty/Pages/Introduction.aspx>, accessed August 2013.

Norton, K., Dollman, J., Martin, M. & Harten, N. 2006. Descriptive epidemiology of childhood overweight and obesity in Australia: 1901–2003. *International Journal of Pediatric Obesity* 1(4): 232–8.

Patterson, E., Wärnberg, J., Kearney, J. & Sjöström, M. 2009. The tracking of dietary intakes of children and adolescents in Sweden over six years: The European Youth Heart Study. *International Journal of Behavioral Nutrition and Physical Activity* 6:91 <www.ijbnpa.org/content/6/1/91>, accessed 28 September 2014.

Queensland Health 2009. *Fun Not Fuss with Food*. Queensland Government, <www.health.qld.gov.au/health_professionals/food/funnotfusswithfood.asp>, accessed August 2013.

Raitakari, O.T., Porkka, K.V., Ronnemaa, T., Knip, M. et al. 1995. The role of insulin in clustering of serum lipids and blood pressure in children and adolescents. The Cardiovascular Risk in Young Finns Study. *Diabetologia* 38: 1042–50.

Rampersaud, G.C., Pereira, M.A., Girard, B.L., Adams, J. et al. 2005. Breakfast habits, nutritional status, body weight, and academic performance in children and adolescents. *Journal of the American Dietetic Association* 105: 743–60; quiz 761–2.

Rangan, A., Hector, D., Louie, J., Flood, V.M. et al. 2009. *Soft Drinks, Weight Status and Health: Health professional update*, NSW Centre for Public Health Nutrition (now known as the Cluster of Public Health Nutrition). <http://sydney.edu.au/medicine/public-health/panorg/pdfs/HealthProf_softdrinks_update.pdf>, accessed 28 September 2014.

Reilly, J.J. & Kelly, J. 2011. Long-term impact of overweight and obesity in childhood and adolescence on morbidity and premature mortality in adulthood: Systematic review 1. *International Journal of Obesity* 35: 891–8.

Sanigorski, A.M., Bell, A.C., Kremer, P.J. & Swinburn, B.A. 2005. Lunchbox contents of Australian school children: Room for improvement. *European Journal of Clinical Nutrition* 59: 1310–16.

Schofield, W.N. 1985. Predicting basal metabolic rate, new standards and review of previous work. *Human Nutrition. Clinical Nutrition* 39: 5–41.

Szajewska, H. & Ruszczynski, M. 2010. Systematic review demonstrating that breakfast consumption influences body weight outcomes in children and adolescents in Europe. *Critical Reviews in Food Science and Nutrition* 50: 113–19.

WHO (World Health Organization) n.d. *The WHO Child Growth Standards*. WHO. <www.who.int/childgrowth/standards/en/>, accessed 28 September 2014.

Wirt, A. & Collins, C.E. 2009. Diet quality: What it is and does it matter? *Public Health Nutrition* 12: 2473–92.

ADDITIONAL READING

Akerblom, H.K., Viikari, J., Raitakari, O.T. & Uhari, M. 1999. Cardiovascular risk in Young Finns Study: General outline and recent developments. *Annals of Medicine*. 31 Suppl 1:45–54.

TEST YOUR UNDERSTANDING

Note: There is only one correct answer for each question.

1. A child between the ages of six and the onset of puberty will gain an average of how many kilograms per year?
   a. 5 kilograms
   b. 4 kilograms
   c. 3 kilograms
   d. 2 kilograms
   e. 1 kilogram
2. In Australia and New Zealand, what are the most frequently used growth charts?
   a. Centers for Disease Control and Prevention's Clinical Growth Charts
   b. National Health and Medical Research Council's Clinical Guidelines growth charts
   c. Australian Paediatric Endocrine Group's growth charts
   d. Global population growth chart
   e. None of the above
3. The recommendation for saturated fat intake in Australia and New Zealand is what percentage of total energy?
   a. 25 per cent
   b. 20 per cent
   c. 15 per cent
   d. 10 per cent
   e. 5 per cent
4. Eating breakfast has been shown to improve which of the following?
   a. Memory recall
   b. Creativity
   c. School attendance
   d. Concentration
   e. All of the above

5. According to national health surveys approximately what percentage of children in Australia and New Zealand do not eat breakfast?
   a. 25–30 per cent
   b. 20–5 per cent
   c. 15–20 per cent
   d. 10–15 per cent
   e. 5–10 per cent
6. Which of the following is a study that has examined the long-term tracking of diet and health?
   a. Bogalusa Heart Study
   b. Demetrius Study
   c. Caronia Study
   d. Scania Study
   e. All of the above
7. Which of the following are a parent's role in developing their children's eating habits?
   a. Providing a healthy eating role model
   b. Eating at the table every night
   c. Involving children in cooking
   d. Providing water as a main drink
   e. All of the above
8. Based on future trends in overweight and obesity, what is the predicted percentage of children who will be overweight or obese by 2025?
   a. 20 per cent
   b. 25 per cent
   c. 33 per cent
   d. 40 per cent
   e. 50 per cent
9. According to research, children who are obese have a 30–40 per cent higher risk of which longer term disease?
   a. Stroke and heart disease
   b. Eating disorders
   c. Diabetes
   d. Chronic fatigue
   e. All of the above
10. Which is appropriate for managing overweight and obesity in children?
    a. Involving the family
    b. Limiting screen time to less than two hours per day
    c. Using school-delivered healthy eating interventions
    d. Limiting consumption of discretionary foods
    e. All of the above

STUDY QUESTIONS

1. What are the sex differences in nutrient requirements for children aged six to twelve years and why are they required?
2. In children aged six to twelve years, what nutrients are consumed at levels below the recommended levels in the NRVs?
3. Discuss the major nutritional, environmental and behavioural factors affecting the diet and eating habits of children aged six to twelve years.
4. What are the risk factors for T2DM in children and how can these be addressed at the family level to reduce risk?

9

Pre-adolescent and adolescent (13–18 years)

Kelly Bobridge and Therese O'Sullivan

LEARNING OUTCOMES

Upon completion of this chapter you will be able to:

- describe the physical and cognitive development of adolescence and how this affects nutrient requirements and intakes
- identify the key macronutrients and micronutrients requiring consideration in adolescent diets
- discuss why obesity and metabolic health are important in adolescence and summarise potential ways to address these issues
- describe how dietary intake can affect adolescent issues such as academic performance and acne
- explain why adolescents are particularly vulnerable to eating disorders and what factors can encourage and discourage the development of such disorders.

The adolescent period is a time of significant physical, cognitive, emotional and social development. Adolescents become more independent and take on increased responsibilities both inside and outside their homes, such as preparing meals for themselves and their family and taking on a part-time job. **Adolescence** is also an important time for establishment of a unique self-identity. Peers tend to have a big influence at this age, and there are opportunities for new behaviours and experiences. This chapter examines the growth and development occurring from thirteen to eighteen years, the accompanying nutritional needs at this time and food behaviours commonly established. Metabolic health and obesity are also discussed, as these factors set the scene for future adult health. Adolescent-specific topics covered in this chapter include academic performance, energy drinks, acne and disordered eating.

PHYSIOLOGICAL AND PSYCHOLOGICAL CHANGES

Weight and height

Unlike the uniform growth rate observed through childhood, growth in adolescence is much faster. Most growth occurs over a span of 1.5–3 years and is referred to as a growth spurt. In adolescence, on average boys grow 20 centimetres in height and gain 20 kilograms, and girls grow approximately 16 centimetres and gain 16 kilograms: 1 kilogram weight gain for each additional centimetre of height (Kuczmarski et al. 2000).

Adolescence is the stage of life spanning from the initial development of secondary sexual characteristics to the attainment of full adult stature.

The average height of adolescents is increasing. Factors contributing to this gain may include better nutrition and health in childhood, safer environments and better health of the mother in pregnancy (including less toxic substances like alcohol and nicotine). From 1995 to 2012 the average height of Australians increased by 0.8 centimetres for men and 0.4 centimetres for women, which was followed by a disproportionate increase in weight of 3.9 kilograms and 4.1 kilograms respectively (ABS 2012). As growth and development is highly individual, growth charts are commonly used to monitor development in adolescents.

Sexual maturation

The commencement of sexual maturation precedes this period of rapid growth. Sexual maturation can be stressful for an adolescent, as there are many new physiological changes occurring. For females, menstruation begins, breasts develop and body composition changes with the widening of the hips. The average age of menarche in Australia and New Zealand is twelve to thirteen years. For males, genitals increase in size and the testes descend, the voice deepens, and testosterone production causes a widening of the shoulders and back. Both sexes develop pubic hair and areas of thickened, darkened hair.

Five stages of sexual development have been defined and are referred to as the Tanner Stages, or **sexual maturity rating** (Tanner 1962; see Table 9.1). As gains in weight and height are closely linked with stage of sexual development, it can be useful to determine adolescents' sexual maturity rating to allow for a more accurate estimation of nutrition requirements. Peak-velocity growth occurs during Stages 2 and 3 for females and in Stage 4 for males.

Changing body composition

Levels of circulating sex hormones increase with sexual maturation, which influences body composition. Increased androgen production, particularly testosterone, by the testes in males supports greater muscle development, which results in proportionately less fat than muscle mass. Females produce less testosterone and more oestrogen via the ovaries; although muscle development continues there is a marked increase in fat deposition. This results in females having a higher fat to muscle ratio. It has been estimated that the proportic of body fat increases from 19 per cent to 22–26 per cent in adolescent females and from 15 per cent to 16–18 per cent in males (Mahan & Escott-Stump 2008).

There have been links made between the higher fat mass of overweight and obese girls with earlier menstruation. One theory behind this relationship is that adipose cells produce a portion of the body's oestrogen. Girls with a greater proportion of body fat may have greater oestrogen levels, which may sequentially initiate earlier menstruation. This may help explain why menstruation age is decreasing with increasing overweight and obesity rates in childhood and adolescence.

Skeletal growth and density

Another significant physiological change occurring during adolescence is rapid bone growth. Puberty causes an increase in sex hormones, thyroid hormones and growth hormones which stimulate the bones to grow. Once full adult stature is attained, bones are at their longest, and the **epiphyseal cartilage** at the end of the long bones is replaced with bone. This is referred to as **epiphyseal closure** and occurs at approximately 17 years for girls and 21 years for males, ranging approximately 5 years (Roche & Davilia 1972). During this period the bones also widen and become denser, with over half of total adult bone mass laid down in adolescence. Although bone density increases until approximately 25–30 years of age (a stage referred to as **peak bone mass**), once the length of the bones are set, it will not change.

Cognitive and psychosocial development

In addition to the physiological changes, there are significant cognitive, social and emotional changes in

The **sexual maturity rating** or **Tanner Stages** classify adolescents into one of five stages of sexual development, allowing for a better understanding of the adolescent's physical and psychological developmental stage.

Epiphyseal cartilage or **epiphyseal plate** is the layer of cartilage at the end of bones which is produced by chondrocytes. As long as this continues to grow the bones will lengthen.

Epiphyseal closure occurs when osteoblasts completely replace the epiphyseal cartilage with bone.

Peak bone mass is the greatest bone mineral density achieved in an individual's lifetime, usually occurring around 25–30 years of age.

Table 9.1a Tanner's sexual maturity rating for girls

Stage	Pubic hair	Breast development
1	Pre-pubertal—none	Nipple elevation only
2	Light, downy hair growth along labia	Small breast buds
3	Thicker, curlier and darker hair, increasing amount	Increased size of breast and areola
4	Thicker and coarser hair	Increased size of breast; nipple and areola form secondary mound
5	Full covering of pubic hair to medial thighs	Full-size breast with breast and areola on the same contour, nipple protruding

Table 9.1b Tanner's sexual maturity rating for boys

Stage	Pubic hair	Genital development
1	Pre-pubertal—none	Pre-pubertal—no change
2	Light growth at penis base	No or small increases in penis size, reddening of the scrotum and change in skin texture, increased size of scrotum and testes (5 ml volume)
3	Increased hair covering pubis; darker, coarser and curlier	Increased penis size, continued growth of scrotum and testes (8–10 ml)
4	Fully developed pubic hair but not yet spread to medial thighs	Increased length and width of penis, continued growth of testes (12 ml), darker scrotum
5	Increased quantity, spread to medial thighs	Complete penile growth, testes increase to 15 ml

Source: Adapted from Tanner (1962).

adolescence. Although the vast majority of Australian adolescents aged twelve to seventeen years live with their parents, adolescents commonly become more involved with their friends and less with their family. Three stages of cognitive and psychosocial development have been defined, being early, middle and late adolescence (Ingersoll 1992).

Early adolescence involves adjustment to a developing body and body image. Thinking processes are concrete and moral concepts are just beginning to develop. In early adolescence, peers have a marked influence and family influence begins to decline. At this stage the adolescent will be more receptive to nutritional changes that would benefit them in the present (for example, increasing iron intake for more energy and better sports performance), than to encouragement to change behaviour for perceived benefits gained in the future (for example, decreasing saturated fat intake to reduce the risk of cardiovascular disease [CVD] in adulthood).

Middle adolescence is signalled by emotional independence from family. Cognition becomes more abstract and morals continue to develop. Socially, peer influence is more specific, usually involving a group of close friends. Sexual interest increases alongside health risk behaviours such as experimentation with drugs and alcohol. Adolescents become more comfortable with their physical development and sexual maturation, reducing body image concerns. Financial independence increases, further reducing reliance on family.

By late adolescence, physical development is almost complete and cognition is verging on adult level with abstract, complex thinking processes. The adolescent has a better understanding of who they are as an individual, with their own beliefs, morals and values. They will often commence full-time employment or continue education at a tertiary level. At this stage the adolescent can better understand and respect the link between present health behaviours and future health outcomes. The adolescent becomes more independent from parents, even if they continue to live at home.

The increasing level of independence in all three stages of development is likely to impact nutrient intake, as foods are more commonly eaten away from home, with increased consumption of convenience foods. Decreased food quality may result in reduced nutrient intake, which could adversely affect the adolescent's capacity for growth and development. As the stage of cognitive, emotional and social development influences adolescents' perceived importance of nutrition, it is beneficial for health professionals to have an understanding of the stages of development and how to use these to promote positive health behaviour changes.

INFLUENCES ON EATING

Energy drinks and sugar-sweetened beverages

Energy drinks differ from other soft drinks in that they have a high caffeine content, ranging from 80 milligrams (similar to a cup of coffee) to over 200 milligrams. Caffeine is generally recognised as a 'safe' drug and is widely used in Australia. It acts via stimulation of the nervous system and has been shown to enhance endurance performance and concentration in adults at moderate intakes. Clever marketing around increased performance and links with extreme sports and music concerts makes these drinks especially attractive to adolescents. Australian high school students report consuming the drinks to help with sports performance and give them an energy boost or 'buzz' (O'Dea 2003). Although caffeine has some beneficial effects in terms of being an ergonomic aid, most children and adolescents gain little benefit from habitual caffeine intake.

Reports of caffeine toxicity from energy drinks increased five-fold in Australia between 2004 and 2010, particularly through recreational use in the adolescent age group (Gunja & Brown 2012). Palpitations, agitation, tremor and gastrointestinal upset were most commonly reported as symptoms, and cases of serious cardiac and neurological effects such as cardiac ischaemia and seizures were also reported. There has been debate over whether energy drinks should be banned for children under eighteen, with some experts suggesting that problems occur only with overconsumption. But even at low to moderate intakes, there are some issues of particular concern for teenagers.

Disruptive and hyperactive behaviour has been noted with low doses of energy drinks. Caffeine can result in disruption to sleep, with adolescent users of energy drinks reporting daytime sleepiness and reliance on caffeine to get through the day. Nutritionally, the high sugar content of energy drinks, generally similar to soft drink, may replace more nutrient-dense forms of energy intake or contribute to risk of obesity. Caffeine has no nutritive value, although vitamins may be added in some energy drinks.

Children who consume energy drinks on a regular basis may become dependent on them. Caffeine acts on parts of the brain that mediate reward and addiction, and this may result in preferences for foods or drinks that were paired with caffeine at the time. In the case of energy drinks this means sugary drink, which may have future implications for obesity and type 2 diabetes mellitus (T2DM).

The practice of mixing alcohol with energy drinks may also be an issue for some teens, as the stimulation effect from energy drinks can mask feelings of being drunk. This can cause individuals to underestimate their level of intoxication and may lead to dangerous practices that would otherwise be avoided, such as driving while over the alcohol limit.

Given the potential health issues with caffeine intake for this age group, adolescents who are seeking specific benefits from energy drinks such as improved sports performance or more energy over the day may be better off making healthier nutritional and lifestyle changes (see Chapter 17). When reducing caffeine intake, it is useful to note that withdrawal symptoms, including decrease in reaction time in a task requiring sustained attention, have been reported in school-age children who consumed drinks with as little as 120–45 milligrams per day of caffeine (Bernstein et al. 1998). Overall, it is important for health professionals and parents to use opportunities to promote water as the preferred drink for everyone, including adolescents.

Vegetarianism

Vegetarian or vegan diets may be adopted during adolescence for a number of reasons. As adolescents mature they become more aware of how their behaviours affect the environment, including animal welfare. An adolescent may also choose to become vegetarian for health purposes, to assist with weight loss or to further exert their independence from their family. Vegetarianism is most common in females.

A well-planned vegetarian diet can provide adequate nutrition for growth and development and has been associated with reduced risk of obesity, hypertension, CVD and cancer. A study of Australian adolescents aged fourteen to fifteen years showed significantly better markers of cardiovascular health in those consuming a vegetarian diet, including body mass index (BMI), waist circumference and cholesterol (Grant et al. 2008). However, it can be more difficult to meet dietary requirements secondary to the bioavailability, quality and volume of some nutrients in plant foods, particularly for protein, iron, zinc, calcium, vitamin D, omega-3 fatty acids and vitamin B12. Parents of vegetarian or vegan adolescents are often concerned about nutritional adequacy and should be educated alongside their adolescent on how to achieve a well-balanced diet. There are many meat alternatives available, including tofu, tempeh, soy sausages and burgers, imitation meats and Quorn, which is based on a mycoprotein extracted from a fungus.

Some adolescents adopt a vegetarian diet in order to lose weight, putting them at increased risk of developing an eating disorder. The vegetarian diet, generally being higher in plant foods, is usually lower in fat and therefore energy, which may promote weight loss. Adolescents may use a vegetarian diet as a mask for reducing intake and limiting certain foods that they believe cause weight gain. Alternatively, for adolescents who are overweight or obese, it has been suggested that a vegetarian diet could be used as a method to encourage maintenance of a healthy body weight.

NUTRIENT REQUIREMENTS AND INTAKES

As adolescence is the final period of growth and a significant time for development, adequate nutrient intake is important for the adolescent both in the short-term and for long-term health into adulthood. Ninety-five per cent of people aged 12–24 years did not meet the dietary recommendations outlined in the Australian Dietary Guidelines in 2007–08 (AIHW 2011; NHMRC 2013). These guidelines advocate the consumption of a variety of foods, maximising the chance of meeting required intakes of the macronutrients and micronutrients. Some key nutrients to consider in adolescence are protein, carbohydrate, fibre, fat, iron, zinc, calcium, vitamin D and the B vitamins, including B12 and folate.

Energy

Development during adolescence requires energy not only for physical activity, metabolism and cell maintenance but also for growth. Approximately 1–2 per cent of energy is used to synthesise tissue specifically for growth. This often makes the energy requirements in adolescence as high as those in adulthood.

The estimated daily energy requirements for Australian and New Zealand children are shown in Table 1.6 (NHMRC & Ministry of Health 2006). Energy requirements for light activity are shown under a physical activity level (PAL) of 1.6, for moderate activity under 1.8 and for heavy activity under 2.0. This range accounts for different heights and PALs. The 2007 Australian National Children's Nutrition and Physical Activity Survey found that boys aged fourteen to sixteen years on average consumed approximately 11,800 kilojoules per day, which was substantially higher than the approximate 8600 kilojoules per day consumed by girls (CSIRO et al. 2008).

Peak energy intake has been found to align with peak growth rate in adolescents. The macronutrients carbohydrate, protein and fat are the main sources of energy in the diet. If energy deficits occur in adolescence, maximum growth and height may not be attained, often referred to as stunting. Common risk factors for stunting include dieting to reduce body weight, eating disorders, certain health conditions such as cystic fibrosis, living conditions (for example, homelessness) and financial restraints reducing access to nourishing foods in sufficient quantities.

Macronutrients

Protein

Adequate protein intake is required in adolescence to support the synthesis of body tissue allowing optimal growth and development. Adequate intake also improves iron and zinc absorption from the diet. Since growth in adolescence is highly variable, it may be more appropriate to follow the recommended grams of protein per kilogram of body weight when working with adolescents than use an amount based on age. The Australian and New Zealand recommendations are 0.99 grams per kilogram for boys and 0.77 grams per kilogram for girls aged fourteen to eighteen years. However, if looking at a population group, the acceptable macronutrient distribution range (AMDR) for protein can be useful, with a recommended 15–25 per cent of total energy derivation from dietary protein (NHMRC & Ministry of Health 2006).

The Australian National Children's Nutrition and Physical Activity Survey determined that fourteen- to sixteen-year-old adolescents consumed 16–18 per cent of their energy in the form of protein, which was slightly higher than that found in fifteen- to eighteen-year-old New Zealanders, at 15–16 per cent (CSIRO et al. 2008; University of Otago & Ministry of Health 2011). These intakes fall within the AMDR.

Protein can be derived from both animal and plant foods. Animal protein foods include meat, chicken, fish, seafood, eggs and dairy. They provide 'high-quality' protein, meaning that they contain all nine essential amino acids which cannot be synthesised by the human body, and the protein is effectively digested and absorbed for use in the body. Individual plant-based protein foods are not digested and absorbed to the same extent as animal foods and often lack one or more essential amino acid, which is why a variety of plant-based protein foods should be consumed, including grains, legumes (including beans and lentils), vegetables, nuts and seeds. This is particularly important for vegan adolescents to consider in their diets.

Protein drinks and bars are currently popular, but are these good for growing adolescents? Growing male adolescents, particularly those who are very active, have high protein requirements and may be tempted to use protein drinks or bars with the aim to increase muscle mass. However, as mentioned earlier, Australian and New Zealand adolescents usually receive adequate protein from their diets. In addition, although protein is essential for building muscle, this does not necessarily mean that bigger muscles will result from eating more.

Excessive protein intake contributing to energy intakes above the individual's requirements will increase fat mass, not lean mass. Excess protein is also associated with increased urinary excretion of calcium (Linkswiler et al. 1981), potentially reducing bone mass if calcium intake does not replace losses. Protein metabolism also increases circulating nitrogenous wastes, which need to be filtered out and excreted by the kidneys. Some adolescents may reduce their protein intakes either purposefully or incidentally as increased independence leads to new food behaviours.

CASE STUDY 9.1 ACADEMIC PERFORMANCE AND VEGETARIAN DIETS

Profile and problem

Miss L. is a sixteen-year-old girl who is studying for her year 11 exams. A few months ago Miss L. decided to adopt a vegetarian diet because she disagrees with the treatment of farmed animals. Her usually supportive family feel that a vegetarian diet is not healthy and are pushing for Miss L. to eat meat. Recently, Miss L. has been feeling tired and is having trouble concentrating at school. She knows that vegetarian diets can be healthy but wonders if she is eating the right kinds of vegetarian foods. Miss L. is 168 centimetres tall and weighs 65 kilograms. She has lost 3 kilograms over the past few months.

Assessment

Miss L. doesn't eat breakfast because she feels that she doesn't have enough time. She buys an energy drink and a bag of mixed lollies at the train station on her way to school. This makes her feel alert for a couple of hours but lethargic by mid-morning. She usually has a white-bread salad sandwich for lunch, some fruit for afternoon tea and vegetables and gravy from the dinner that her mother makes at home. After dinner Miss L. usually has a small bowl of ice cream.

Treatment

Miss L. saw her general practitioner, who gave her a referral to meet with a dietitian after seeing her iron and vitamin B12 levels were below recommended levels. After her dietitian appointment, Miss L. has realised that she has been choosing high-GI carbohydrate foods, was becoming reliant on her energy drink for caffeine in the morning and is not getting enough protein, iron and calcium, and potentially many other nutrients. With assistance from the dietitian, Miss L. has agreed to do the following:

- grab a yoghurt and banana to eat on her way to school instead of buying an energy drink and lollies
- add cheese or egg and avocado to her salad sandwich at lunchtime and swap from white to multigrain bread
- talk with her mother to see if she will buy her some tofu and legumes to have with her vegetables at dinner
- have a glass of milk or calcium-fortified soy milk with her ice cream at supper time or make a smoothie with fruit, milk and ice cream.

Outcomes

A few weeks after making the dietary changes, Miss L. feels that her concentration levels have increased markedly at school. Her family have noticed the change in Miss L.'s behaviour and have become more supportive, with her mother providing a vegetarian-protein food for dinner each night, though her father still tries to sneak a piece of steak on her plate every now and again.

Plan

Miss L. is going to see the dietitian again in two months' time for monitoring.

For example, adolescents may eliminate certain foods such as dairy, or they may adopt vegetarian or vegan diets. Dietary changes often require an adjustment of the overall diet in order to meet nutrient requirements (for example, replacing meat with a vegetarian protein source such as tofu or replacing cow's milk with calcium-fortified soy milk); otherwise, deficiencies may occur. Inadequate protein intake can lead to reduced tissue and muscle development and muscle wasting, affecting not only the skeletal muscles but also muscles and tissues forming vital organs such as the heart, brain and gastrointestinal system. Inadequate dietary protein also reduces the capacity at which the body can synthesise haemoglobin, enzymes, hormones and immune factors, which may predispose adolescents to illness or infection or delay development.

Carbohydrate

Carbohydrate is essential as the primary energy source for adolescent bodies, providing glucose for optimal brain function and preventing the metabolism of protein that is required for other more important bodily functions, including growth. Requirements for adolescents vary in accordance with activity levels and rate of growth. The nutrient reference values (NRVs) for Australia and New Zealand recommend carbohydrate intake at levels adequate to provide 45–65 per cent of total energy requirements (AMDR), with an emphasis on low glycaemic index (GI) and wholegrain carbohydrate (NHMRC & Ministry of Health 2006).

Wholegrain products, undergoing less processing, retain a greater amount of fibre and micronutrients, because the outer fibrous layers of the grain are left intact. In adolescents, an increasing amount of carbohydrate is being derived from refined carbohydrate food and drinks, such as sugar-sweetened beverages (soft drinks and energy drinks) and food with added sugar (takeaway foods, confectionery, cakes and biscuits), which are often low in micronutrients. The Australian National Children's Nutrition and Physical Activity Survey estimated that approximately 39 per cent of energy was derived from total sugars, although the Australian Dietary Guidelines recommend less than 20 per cent (CSIRO et al. 2008; NHMRC 2013). Excess sugar consumption may lead to dental caries and higher risk of obesity and CVD.

Fat

Body-conscious adolescents may have a negative view of dietary fats due to their high energy content (37 kilojoules per gram). What they might not know is that fat is an essential nutrient for the body, with many important functions (see Chapter 1). The recommended intake for fat is 20–35 per cent of total energy (NHMRC & Ministry of Health 2006).

Thirty per cent of energy was derived from total fat in Australian adolescents aged fourteen to sixteen years according to the Australian National Children's Nutrition and Physical Activity Survey (CSIRO et al. 2008). This was slightly higher in fifteen- to eighteen-year-old New Zealanders, at 34–35 per cent (University of Otago & Ministry of Health 2011). Saturated fat contributed 13–14 per cent of total daily energy, which is above the recommendation of less than 10 per cent.

Replacement of saturated fats with unsaturated fats is considered to be beneficial to health, while replacement of high-saturated-fat foods in the diet with refined carbohydrate foods may be detrimental to health (Jakobsen et al. 2009). Saturated fat is thought to increase low-density lipoproteins ('bad' cholesterol) in the blood; however, individual saturated fatty acids and different food sources may have different effects on health outcomes (O'Sullivan et al. 2013). For example, the type of saturated fat found in dairy products may be protective for chronic disease.

Although there is not yet enough evidence to give dietary recommendations for individual saturated fatty acids, adolescents should aim to derive the majority of their fat intake from polyunsaturated and monounsaturated fatty acid food sources, which increase high-density lipoproteins ('good' cholesterol). Omega-3 fatty acids, a polyunsaturated fat, have also been associated with improved cholesterol profiles and numerous other health benefits, including cognitive development, growth, eye health and anti-inflammatory actions supporting heart health. Adolescents should regularly consume omega-3 food sources such as oily fish, seafood, soy and soy products, seeds (linseeds), nuts (walnuts), dark green vegetables and canola oil or spread. Diets omitting fish and eggs are likely to be deficient in the long-chain omega-3 fatty acids docosahexaenoic acid (DHA) and eicosapentaenoic acid (EPA). Microalgae, krill or fish oil supplements are an option for increasing omega-3 intake, as well as consuming enough plant sources of linolenic acid to allow conversion in the body to long-chain omega-3 fatty acids (although conversion rates are small).

Fibre

Adequate fibre intakes have been associated with a healthier bowel, reduced cholesterol, improved regulation of blood glucose levels and blood pressure and lower body weights, which are beneficial leading into adulthood for the prevention of chronic disease. Adolescent boys aged thirteen years have a recommended fibre intake of 24 grams per day, which increases to 28 grams per day for fourteen to eighteen years. Girls require 20 grams and 22 grams per day respectively (NHMRC & Ministry of Health 2006).

In Australia and New Zealand the majority of dietary fibre is derived from breads, cereals, fruit and vegetables. The average intake of fibre for boys according to the 1995 National Nutrition Survey was approximately 24 grams for twelve- to fifteen-year-olds and 27 grams for sixteen- to eighteen-year-olds; for girls it was approximately 19 grams for both age groups (McLennan & Podger 1998). The New Zealand Adult Nutrition Survey reported an average intake of 22 grams for boys and 16 grams in girls aged fifteen to eighteen years (University of Otago & Ministry of Health 2011).

Fibre intakes below estimated requirements are likely to result from increasing consumption of refined carbohydrate foods and inadequate fruit and vegetable intake in adolescents. The Australian Dietary Guidelines recommend five serves of vegetables per day for girls and five and a half serves for boys aged twelve to eighteen years. Recommendations for fruit are two per day for both girls and boys (NHMRC 2013). Just over half of Australian adolescents consume two serves of fruit per day, compared to approximately 40 per cent reported in New Zealand adolescents. For vegetables, just over half of New Zealand adolescents ate three serves a day, and only 6 per cent of Australian adolescents consumed five or more serves (McLennan & Podger 1998; University of Otago & Ministry of Health 2011).

Micronutrients

Iron

The nutrient requirements for iron are high in adolescence, to meet the physiological changes of growth and sexual development. As the adolescent grows, the increase in body size requires a greater volume of blood and therefore red blood cells. Red blood cells contain haemoglobin, which relies on iron at its centre for carrying oxygen. Requirements also increase in adolescent females with menarche secondary to the cyclic shedding of the endometrial layer of the uterus and associated blood loss.

Thirteen-year-old adolescents require 8 milligrams of dietary iron per day, which increases in the fourteen to eighteen years age group to 11 milligrams per day for boys and 15 milligrams per day for girls. The NRVs estimate requirements in vegetarians to be approximately 80 per cent higher than for non-vegetarians (NHMRC & Ministry of Health 2006).

Inadequate iron intakes are common in Australian and New Zealand adolescents, particularly in the Māori adolescent population, who appear to be at greater risk of both iron and zinc deficiency secondary to an earlier peak growth and the associated onset of menarche (Gibson et al. 2002). The 1997–98 Adult Nutrition Survey in New Zealand found that 45 per cent of adolescent girls aged fifteen to nineteen years were at risk of inadequate iron intake (Ferguson et al. 2001).

The use of oral contraceptives has been associated with decreased blood loss and length of menstruation, resulting in reduced iron losses (ESHRE Capri Workshop Group 2005). Iron and zinc may be more difficult to obtain from a plant-based diet, so vegetarian and vegan adolescents should ensure they consider their iron intakes carefully. The bioavailability of iron from plant foods is lower, so a greater volume of iron-rich vegetarian foods is required. Vegetarians and vegans should consider having a source of vitamin C with their iron-rich foods, to aid the absorption. Plant foods also contain **phytates**, dietary fibre and polyphenols (for example, **oxalates**), which bind some minerals, such as iron, zinc and calcium, causing their excretion. Avoiding foods rich in phytates and oxalates may assist in increasing mineral absorption.

Zinc

Zinc is required for protein synthesis and acts as a cofactor in many enzymatic reactions and is therefore essential for normal growth and sexual development in adolescence. Adolescents aged thirteen years require 6 milligrams per day, fourteen- to eighteen-year-old girls need 7 milligrams per day, and fourteen- to eighteen-year-old boys need 13 milligrams per day (NHMRC & Ministry of Health 2006). Estimated zinc requirements increase by 50 per cent in vegetarians and vegans secondary to reduced absorption. Similarly to iron, zinc from plant foods is less bioavailable; however, some adaption can occur in the human body

Phytates are components of plant seeds found in the husks of legumes, seeds and cereals which bind minerals including iron, zinc, calcium, copper and magnesium, forcing them to be excreted rather than absorbed.

Oxalates are naturally occurring substances found in some plants, such as spinach and rhubarb, which form an insoluble salt with some minerals, including calcium and iron, causing them to be excreted rather than absorbed.

to allow for increased absorption of iron and zinc in vegetarians.

The body has tight control over its level of zinc. When zinc is deficient, absorption increases, and vice versa. Iron and zinc compete for absorption in the body, which may explain associations found between high-dose iron supplementation and reduced zinc status (Solomons 1986). Inadequate zinc intake has been estimated to occur in approximately 17 per cent of boys and 7 per cent of girls aged fifteen to eighteen years (University of Otago & Ministry of Health 2011), while 2 per cent of girls and 7 per cent of boys aged fourteen to sixteen years are not meeting estimated average requirements (EARs; CSIRO et al. 2008).

Folate

The rapid growth in adolescence requires sufficient energy and therefore the B group vitamins, which assist in releasing the energy for use in the body. One of the B group vitamins, folate, has been found to be low or deficient in many adolescents. Folate is essential for the production of DNA, required in turn for the production of proteins for growth and development. Adequate folate intake is particularly important for pregnant adolescent females, with deficiencies in folate strongly associated with neural tube defects in the offspring.

Adolescents aged thirteen years require 300 micrograms of folate per day, which increases to 400 micrograms per day for fourteen- to eighteen-year-olds (NHMRC & Ministry of Health 2006). Estimated folate equivalent intakes in Australian fourteen- to sixteen-year-olds in 2007 was 440 micrograms for girls and 578 micrograms for boys (CSIRO et al. 2008). The marked difference in intake is likely related to the different techniques used to measure dietary intake and analysis, and whether fortification was considered.

Due to the prevalence of folate deficiency and the adverse health effects of inadequate folate intake, folic acid fortification of flours used in non-organic bread-making is mandatory in Australia. It remains voluntary in New Zealand. Flour is also fortified with thiamin.

Vitamin B12

Adolescent vegans who exclude all animal-derived foods need to consider how they will attain vitamin B12, which is not naturally found in any plant foods. Options include fortified foods such as meat alternatives (for example, tofu products), cereals and milks, supplementation either orally or by injection or nutritional yeasts.

Calcium

Humans synthesise over half their bone mass during adolescence, more than in childhood or adulthood. Good calcium intake during this period is essential in reaching peak bone mass and to reduce the risk of developing osteoporosis in later life. The Australian Dietary Guidelines recommend three and a half serves of dairy or dairy alternatives per day (NHMRC 2013). This should allow adolescents to meet their recommended dietary requirements of 1000–1300 milligrams per day for thirteen-year-olds and 1300 milligrams for fourteen- to eighteen-year-olds (NHMRC & Ministry of Health 2006).

Adolescent girls are suggested to be at the highest risk of inadequate calcium intake, with over 80 per cent of girls and just under half of boys aged fourteen to sixteen years in the Australian National Children's Nutrition and Physical Activity Survey not meeting EARs (CSIRO et al. 2008). In New Zealand, almost 90 per cent of fifteen- to eighteen-year-old girls are considered to have inadequate calcium intakes, compared with just over half of boys (Ministry of Health 2012a).

Milk and milk products provide a large proportion of calcium in Australian and New Zealand diets, but milk intake typically decreases from childhood to adolescence. The recommended dietary intake (RDI) of calcium can be difficult to reach for those who do not consume dairy. However, in Australia and New Zealand there are dairy alternatives available, not only for vegans but also for those who are lactose intolerant or allergic to cow's milk protein. These include oat milk, rice milk, nut milks, soy milk and their products (for example, soy cheese), which are commonly fortified with calcium. Not all of these types of products are fortified, however, so it is important to read the nutrition information panel to check: calcium content should be approximately 100 milligrams per 100 millilitres.

Vitamin D

In Australia and New Zealand, sunshine year-round is sufficient to synthesise adequate vitamin D, as long as enough time is spent outdoors. However, this needs to be balanced against the risk of skin cancer from excessive sun exposure. The amount of sunlight exposure needed varies according to location, season and time of day. In summer, most people need only a few minutes a day of sun exposure to an area of skin equivalent to the face, arms and hands. In winter, it is the same for northern areas like Darwin, but in more southern parts (for example, Sydney and Perth), most people need about two to three hours spread over each week. People with naturally very dark skin need three to six times this amount.

Adolescents who spend minimal time outdoors should consider food sources of vitamin D or a supplement if serum levels are low. Most dietary vitamin D in a vegetarian diet is derived from fortified milk, margarine and eggs. If the adolescent is vegan, a supplement may be required.

COMMON CONCERNS AND HEALTH CONDITIONS

Academic performance

What makes some students bright eyed, attentive and successful in exams, while others are distracted, tired and don't seem to reach their potential? Academic performance at high school can be influenced by a variety of factors, including study skills, access to resources, adequate sleep, motivation and nutrition. Regular eating, a varied diet and consuming breakfast are important nutritional factors in this regard.

Regular supply of glucose

Glucose is the brain's preferred fuel source, and a constant supply is important for optimal brain functioning. Blood glucose concentrations are closely regulated by the body, but small variations in glucose supply can have effects on the brain. When blood glucose concentrations fall below normal, hormones such as adrenalin and cortisol, which are associated with feelings of agitation and irritability, are released. Low blood glucose can also result in difficulty concentrating and becoming more frustrated—these are all factors that can hinder learning and the ability to produce good schoolwork. Falls in blood glucose may be due to a lack of food or to foods that result in a high glucose peak: insulin is released in response to the high peak, and the body can then overcompensate and cause a return to low blood glucose.

All core food groups except meat and meat alternatives supply carbohydrate that is converted to glucose. Regular meals throughout the day and use of low-GI foods can help ensure a constant supply of glucose to the brain (see Chapter 1).

Breakfast

Known as the most important meal of the day, a good breakfast is recommended as part of a healthy diet. Breakfast consumers tend to have higher intakes of micronutrients overall. This can be attributed to the foods chosen at breakfast, but it is also because people who skip breakfast are more likely to choose less nutritious foods over the course of the day. Eating breakfast is also associated with healthier body weight and other lifestyle factors.

In comparison with skipping breakfast, starting the day with breakfast has positive effects on cognitive performance. This has been shown to be particularly important for children and adolescents whose nutritional status is compromised. Having nutrients delivered at the start of the day is important after the overnight fast during the sleep period. The liver is responsible for gradual release of nutrients overnight. Eating a good breakfast may be particularly important for children and adolescents who are not fully grown, as they have a larger brain to liver ratio than adults. This means that children do not have the same ability to store required nutrients for the overnight fast. In addition, children and adolescents have a greater metabolic rate per unit of brain weight, which means there are greater demands on nutrient stores in periods of fasting (Hoyland et al. 2009).

A good-quality breakfast with a variety of foods from different food groups may further impact positively upon cognitive function and academic performance (Rampersaud et al. 2005). Increasing the number of food groups at breakfast has been shown to be associated with better mental health in Australian adolescents (O'Sullivan et al. 2009), which may contribute to a more positive school experience. An example of increasing the number of food groups at breakfast could be moving from toast with jam (one food group) to baked beans on toast plus a glass of milk (three food groups). Milk, fortified breakfast cereals and bread are common breakfast foods reported by Australian adolescents. These foods are sources of nutrients that affect brain function, including carbohydrate, calcium, B vitamins, iron and folate (see Table 9.2).

Lunch

Encouraging children and adolescents to organise and pack their own lunch for school helps instil good habits

Table 9.2 Selected nutrients affecting cognition

Nutrient	Effect	Examples of food
B vitamins	B6, B12 and folate have positive effects on memory. Thiamin improves clarity of thought.	Wholegrain cereals, seeds, nuts, fortified breakfast cereals, green leafy vegetables, eggs, meat (B12 from animal sources only). Easily destroyed by cooking, alcohol, food processing.
Vitamin D	Deficiency linked to brain dysfunction.	Liver, mushrooms, generated from sunlight.
Vitamin E	Improves cognitive impairment after brain trauma in rat models.	Avocado, nuts, seeds, olives, oils.
Choline	Protection against adverse effects of some neurotoxic agents (including alcohol) and improved performance in tests in rat models.	Egg yolk, meat.
Curcumin	Stops cognitive decline in animal models.	Turmeric.
Flavonoids	Enhance cognitive function.	Green tea, dark chocolate, red wine.
Iron	Iron treatment improves cognitive performance in young women.	Red meat, legumes.
Omega-3 fatty acids, particularly DHA	Required for maintenance of normal brain function. Deficiency associated with poorer learning.	Fatty fish (e.g. mackerel, salmon), algae sources, walnuts, canola oil.

Note: DHA: docosahexaenoic acid

Source: De Jesus Moreno Moreno (2003); Gómez-Pinilla (2008); McCann et al. (2006); Wainwright (2000).

for adulthood—and means their lunch is more likely to be eaten rather than swapped or binned! Some easy and nutritious lunchbox ideas for high schoolers are suggested in the box below.

Exam day

The following tips may be helpful for students to put in place on exam days. They should:

- Eat a breakfast consisting of at least three different food groups.
- Pack a decent lunch, including some nuts and fruit for a light snack between exams or during breaks.
- Look for low-GI foods to assist in gradual release of energy over the day.
- Avoid trying any new foods, drinks or supplements on exam day in case they have an unexpected adverse effect.
- Avoid caffeine if you are not a regular consumer, as intake can increase nervousness.
- Stay hydrated and take a bottle of water into the exams with you (if permitted). Dehydration can lead to poor concentration and even at mild levels can impact upon school performance.
- Get to bed early the night before; sleep-deprived people score lower in memory tests.
- Chew on some gum; studies suggest chewing improves memory, concentration on long tasks and alertness (Hirano et al. 2008; Morgan et al. 2013; Smith 2009).

Overweight and obesity

Overweight and obesity are growing concerns for Australian and New Zealand adolescents, with excess body weight predisposing adolescents to adverse health effects and psychosocial issues such as bullying and depression. In Australia and New Zealand the prevalence of overweight and obesity in adolescence is following an upward trend. The Australian Health Survey 2011–13 reported that of the twelve- to fifteen-year-olds surveyed approximately 19 per cent were overweight and 7 per cent were obese. For sixteen- to seventeen-year-olds the proportions were 18 per cent and 8 per cent respectively. Therefore, more than a quarter of Australian adolescents surveyed were overweight or obese (ABS 2012). The New Zealand Health Survey 2011–12 reported that 33–4 per cent of ten- to seventeen-year-old New Zealanders were overweight or obese (Ministry of Health 2012a). The prevalence of overweight and obesity is higher again for the Aboriginal, Torres Strait Islander, Māori and

Lunchbox ideas

- Small can of baked beans
- Handful of cherry tomatoes
- Nut bars (check school policy on nuts first—peanut-free bars may be allowed)
- Snack packs of canned fruit
- Slices of cheddar cheese
- Individual tubs of yoghurt (better to eat at morning tea time if refrigeration or ice pack is not available)
- Hard-boiled eggs
- Sushi
- Fruit cake
- Sandwiches
  - For bread, try wholegrain, sourdough, rye. Wraps, wholegrain crackers or mini bread-sticks can also be used.
  - For fillings, try curried egg; tuna, grated cheese and mayonnaise; bean sprouts, avocado and grated carrot with roast meats, chickpea patties.

Make the following on the weekend and freeze. Take them to school frozen, and they should defrost by lunch time:

- Savoury muffins (e.g. spinach and feta, corn and zucchini)
- Pizza (use a vegetarian frozen pizza and add extra toppings using vegetables in the fridge, roast meats, canned or smoked salmon, or pineapple and banana for the more adventurous!)
- Wholemeal pikelets with grated apples or sultanas in the mixture
- Mini quiches or zucchini slice.

Pacific populations of Australia and New Zealand. This is concerning for health, as weight-related chronic diseases that were once limited to adult populations are now being seen in adolescence.

Being overweight or obese with excess abdominal fat is commonly linked with insulin resistance and high blood glucose levels, high blood pressure and high cholesterol or dyslipidaemia, a group of factors commonly referred to as metabolic syndrome (MetS), which has been linked with significant increased risk of CVD and T2DM. Obesity in adolescence is also linked to obesity in adulthood and the associated comorbidities. However, it is important to note that adolescents who hold excess fat mass are not necessarily unfit or unhealthy, just as lean adolescents are not necessarily fit and healthy. In addition to the adverse physical health effects of the conditions, adolescents who are overweight or obese may be exposed to bullying and have low self-esteem, which may contribute to psychological health conditions such as depression or disordered eating.

How can we help reduce overweight and obesity in adolescence? Excess weight gain in adolescence is usually multifactorial and is likely to be a combination of lifestyle factors and genetics rather than a simple matter of eating too much and not exercising enough. Research from the fields of biochemistry and physiology indicates that energy input and expenditure are interdependent and regulated at several levels. For example, things like an increase in physical activity might be offset by later fatigue or raised appetite, or a reduction in food intake might be offset by a change in metabolic rate and reduced physical activity. Factors which may contribute to overweight and obesity risk in adolescence include decreasing physical activity, increasing screen time, skipping breakfast, not getting enough sleep and high consumption of discretionary foods.

It is recommended that thirteen- to seventeen-year-old Australians and New Zealanders should do 60 minutes of moderate to vigorous activity daily and limit screen time to two hours per day to stay healthy (Department of Health 2014; Ministry of Health 2012b). Research suggests that around 80 per cent of ten- to fourteen-year-olds meet these recommendations, with the number halving to 40 per cent for fifteen- to nineteen-year-olds. Girls are less likely than boys to meet the recommendations (CTRU & Synovate 2010).

Eating breakfast has been shown to reduce dietary fat and minimise impulsive snacking and therefore may be an important part of a weight control (Schlundt et al. 1992). An Australian study of adolescents reported that 15 per cent reported skipping breakfast on one or more days of a three-day study period (O'Sullivan et al. 2009).

Total sleep time in adolescence is reported to drastically decrease from eleven to fifteen years of age (Leger et al. 2012), creating a significant sleep debt. Adequate sleep is important for physical health, and a short sleep duration may increase obesity risk in adolescence (Weiss et al. 2010), particularly in females (Garaulet et al. 2011).

Foods that are nutritionally poor but nutrient dense—discretionary foods—may contribute to over-

CASE STUDY 9.2 MENTAL HEALTH

Profile and problem

Ms N., an eighteen-year-old woman, was diagnosed with bipolar disorder three years ago. She has tried several different medications to control her symptoms and also eats when she is feeling 'low'. She has gained 30 kilograms since her diagnosis and now weighs 85 kilograms, with a height of 161 centimetres. Ms N. currently lives with her parents and younger sister and would like to study to become a beauty therapist.

Assessment

Ms N. would like to lose enough weight to return to a healthy BMI. She has tried restrictive diets in the past and found them difficult to stick with. She has food readily available at home but has little say in which foods are purchased. Ms N. usually has a poor appetite in the morning and doesn't eat before noon. She snacks on dry biscuits and ice cream through the afternoon. She enjoys these foods but mostly chooses them because they are easy to prepare. Ms N. eats dinner with her parents and usually snacks on potato chips in the evening while watching television.

Treatment

Ms N. has agreed to do the following:

- eat breakfast within one hour of waking up each morning
- attend a cooking class provided by the local community centre to develop cooking skills
- aim to consume meals consistent with the Australian Dietary Guidelines, starting by adding one piece of fruit per day
- incorporate physical activity into her daily routine by walking for twenty minutes each day before dinner.

Plan

Ms N. will see a dietitian on a regular basis to assist with motivation and planning for dietary changes. They will work together to set goals, evaluate progress and address barriers.

consumption and obesity in the long term. They have been estimated to contribute 41 per cent of daily energy intake in Australian children and adolescents, with the greatest food contributors being fried potatoes, sugar-sweetened soft drinks, ice cream or ice confection and cordials (Rangan et al. 2008).

BMI, growth charts or waist to height ratio can be used to assess obesity in adolescents. BMI provides an estimate of whole body mass in relation to height and can be biased in adolescents with greater muscle mass. This is why waist to height ratio is becoming more popular, with abdominal obesity being a better predictor for cardiovascular risk factors. Waist circumferences greater than 50 per cent of height may indicate excess abdominal fat (Garnett et al. 2008). Waist to height cut-offs for obesity have been established in the Australian population, based on the 85th percentile for percentage body fat; for eleven- to sixteen-year-olds, the cut-offs for overweight are 46 per cent for boys and 45 per cent for girls (Nambiar et al. 2010). The Australian National Children's Nutrition and Physical Activity Survey found that 13 per cent of fourteen- to sixteen-year-olds had waist circumferences greater than 50 per cent of their height, indicating excess abdominal fat (CSIRO et al. 2008).

Eating disorders and disordered eating

The physical and psychosocial changes during adolescence can be accompanied by concerns about body image and peer acceptance, which may influence how the adolescent behaves. Girls in particular may not understand that gains in fat mass during puberty are normal and healthy and are greater than gains experienced by boys. In a small but increasing proportion of Australian and New Zealand adolescents, food behaviour changes such as dieting or overeating develop into an eating disorder.

An eating disorder is a diagnosable psychological illness in which food intake, exercise and body shape become preoccupations and place the individual at risk of adverse health effects. Eating disorders as defined by the *Diagnostic and Statistical Manual of Mental Disorders* (fifth edition, known as *DSM-5*) include anorexia nervosa, bulimia nervosa, binge-eating disorder, avoidant/restrictive food intake disorder and eating disorders not otherwise specified (APA 2013).

Approximately 9 per cent of Australians will develop an eating disorder during their lifetime, with onset most common between the ages of 12 and 25 years. The estimated prevalence for anorexia nervosa is 0.3–1.5 per cent, bulimia nervosa 0.9–2.1 per cent and binge-eating disorder 2.5–4.5 per cent. Boys can also experience issues around body image, with the estimated prevalence of eating disorders in males reported to be between 0.1 and 3.0 per cent (NEDC 2010).

Causes

Adolescence is a common time for the development of psychological issues such as depression and anxiety, as well as disordered eating. It is estimated that up to three-quarters of psychological conditions commence in adolescence. Why is this the case? Some factors may include hormonal changes, emotion and mood changes, stress, greater deposition of body fat, increased self-awareness, reduced self-esteem, poor body image or body dysmorphia, peer influence, increased risk-taking behaviour, media influence, increased independence from family, changing dietary behaviours and dieting to reduce body weight. It has been noted that girls post-menarche are at the greatest risk. Within this puberty stage, those most at risk include girls who begin seriously dieting, those with higher anxiety levels and perfectionist character traits and often higher achieving students.

The 1995 National Nutrition Survey in Australia found that by sixteen to eighteen years of age just under 20 per cent of girls reported being on a particular diet, and 6 per cent were aiming for weight loss (ABS 1995). Dieting will often be related to poor body image or body satisfaction, which is a combination of the adolescent's thoughts, feelings and perception of their physical self.

Body dysmorphia is different from poor body image and is a warped perception of the body. For example, a person with body dysmorphia may see a person in the mirror heavier than the true reflection shows. It has been found that satisfaction with body weight in adolescent girls drastically declines post-menarche; however, in adolescent boys body satisfaction commonly increases post-puberty, with greater gains in height and musculature. This may help to explain why eating disorders are more prevalent in adolescent females.

Effects

Physically, adolescents with eating disorders are often malnourished and are therefore at increased risk of stunting, reduced and/or impaired sexual and cognitive development and lower bone density, with some of these deficits being chronic and resulting in permanent disability. Note, though, that adolescents with bulimia nervosa may have a relatively normal body weight (see Chapter 19). Bulimia nervosa and anorexia nervosa are in the top ten leading causes of **burden of disease** or injury to females aged 15–24 years in Australia. Risk of premature death is higher in those with eating disorders, resulting from medical problems or suicide. The suicide rate for people with eating disorders is higher than that for those with any other psychological illness. Emotionally, those with eating disorders may withdraw from others, display mood swings and have feelings of failure and guilt. Signs and symptoms of eating disorders are listed below:

- Underweight/fluctuating weight
- Wasted appearance/muscle loss
- Downy hair growth
- Feeling cold/wearing several layers of clothing
- Purple/blue extremities
- Dry and thinning skin
- Dry chapped lips
- Bruising easily
- Headaches
- Fatigue (anaemia)
- Thinning hair
- Irregularity or loss of menstruation
- Food hoarding
- Tooth decay
- Sore throat
- Abdominal pain
- Constipation or diarrhoea
- Sores on the back of hands.

Contrary to those with anorexia nervosa, people with bulimia nervosa are often worried about their condition. The physical, psychological and behavioural signs and symptoms of anorexia nervosa and bulimia nervosa are listed below:

- Depression
- Low self-esteem
- Frustration with being unable to lose weight
- Perfectionism
- Inflexible thinking
- Mood swings
- Restrained emotional expression
- Secretive behaviour
- Compulsivity

Burden of disease is a measure used to compare the relative impact of illnesses on populations by calculating health lost due to illness.

- Anxiety
- Obsessive-compulsive behaviour
- Distorted body image
- A preoccupation with food
- Limited social spontaneity
- Insomnia
- Guilt
- Self-disgust
- Self-harm
- Suicidal ideations/attempts.

Prevention and treatment

Effective treatments are available for eating disorders, but unfortunately only a minority of adolescents who suffer from eating disorders ask for help. Adolescents with an eating disorder may feel embarrassed or afraid of treatment or may feel like they have their problem under control.

Anorexia nervosa is preferably treated in a specialist centre with access to inpatient facilities and multidisciplinary care from a range of health professionals. Binge-eating disorder may respond to a number of psychotherapeutic approaches, particularly cognitive behaviour therapy, which can involve monitoring of thoughts and discovering how they influence behaviours, the use of problem-solving skills and relapse prevention strategies. A selective serotonin reuptake inhibitor may be prescribed, which may help through an effect on satiety mechanisms. Although pharmacotherapy can be effective in the short term, the ability of drugs to assist with long-term change is currently unknown.

Early detection of an eating disorder can greatly improve the treatment outcome. There are many organisations in Australia and New Zealand designed to support those with disordered eating, with many informative websites, online support groups and contacts. One example is the National Eating Disorders Collaboration. Kids Helpline is also a useful service for adolescents experiencing or supporting a friend or family member with disordered eating. General practitioners are a good first point of call for health advice, as they can make a referral depending on the individual situation.

Disordered eating has a significant social and environmental component. Family, friends and the media are some key factors influencing adolescents and their eating behaviours. A New Zealand study of adolescents reported a positive link between happy and supportive peer and family environments and body satisfaction and wellbeing (Wood et al. 2011). Greater frequency of family meals has been associated with reduced risk of disordered eating practices, better nutrition intake and reduced consumption of takeaway foods.

It has been suggested that the prevalence of eating disorders is higher in adolescents from Westernised backgrounds who are exposed to media messages supporting a thin figure for females or lean, muscular figure for males. Adolescents are inundated with media messages via radio, television, magazines and the internet. This is why it is so important for the media to advocate for healthy body image; for example, the Dove Campaign for Real Beauty promotes a range of different body shapes, and some magazine companies have chosen to reduce the use of airbrushing to show 'real' women.

Acne

Almost all teenagers experience acne at some stage. This skin condition results in pimples, cysts, blackheads or whiteheads, and it accompanies the hormonal changes during puberty. Hormones can stimulate the **sebaceous glands** in the skin's hair follicles to produce too much oil, or sebum. Some sebum is useful to keep the skin from drying out, but too much in conjunction with a build-up of dead skin cells can result in the hair follicles (also known as pores) becoming clogged. Bacteria penetrating these hair follicles results in inflammation and acne. The most common form of acne in adolescents is acne vulgaris, which usually affects the face, neck, upper back, chest and shoulders.

Can the way we eat help this common adolescent ailment? It appears that diet does not cause acne, but it could improve or worsen the severity. The number of good-quality studies in the area of diet and acne is relatively small, and more research is warranted before comprehensive evidence-based guidelines can be established. However, there is a growing body of evidence to suggest there is a relationship between the diet and acne. A high-carbohydrate diet containing high-GI foods may contribute to acne by elevating serum insulin concentrations, which can stimulate sebum production. Acne is more prevalent in people who have insulin-resistant conditions such as polycystic ovarian syndrome (PCOS). Omega-3 fatty acid intake

Sebaceous glands are situated in the skin, at the bottom of the hair follicle, and secrete a thick oil, or sebum, composed of fat and cellular wastes to lubricate the skin and hair.

can help reduce inflammation in the body and has been suggested to have a role in reducing the severity of acne. Drinking water to maintain adequate hydration can help skin health, as can fruits, vegetables and good fats.

Developed countries have much higher rates of acne, and it has been hypothesised that a Western-type diet may contribute to acne. The Western diet is typically high in processed foods, meat and sugars and low in whole grains, vegetables, legumes and fish. There is still debate about whether chocolate worsens acne, with a small study suggesting chocolate consumption may increase the inflammation that can contribute to acne (Netea et al. 2013). Different foods may have different effects on individuals, so it is worthwhile considering whether any foods act as triggers through the use of a food and symptom diary.

Changes to diet are more likely to influence mild cases of acne than severe cases, in which other factors such as genetics may have more influence. While stress does not cause acne, it can make it worse due to the effect on hormones increasing oil production.

SUMMARY AND KEY MESSAGES

Adolescence can be a challenging time in a person's life. There are many physical, cognitive, emotional and psychosocial changes occurring, with adolescents becoming much more aware of their surroundings and peers. Nutrient intake may not be the adolescent's priority. It is important for health professionals to understand the complex changes occurring throughout adolescence and how to best advocate for a healthy lifestyle to meet adolescent nutritional needs in this critical period of growth and development.

- Key nutrients to consider in adolescent diets include carbohydrate, fibre, protein, fat, iron, zinc, folate and calcium.
- The prevalence of overweight and obesity in adolescence is following an upward trend, with significant adverse health effects seen in adolescence and adulthood.
- Vegetarian diets in adolescence should be well planned in order to meet nutritional needs for growth and development and could be beneficial for maintaining a healthy weight and body.
- Eating disorders most commonly develop in adolescence and are often preceded by extreme dieting, low self-esteem and poor body image.
- Dietary intake can have a significant influence on energy levels and academic performance in adolescence.
- Sugar-sweetened beverages including energy drinks are replacing nutrient-dense foods and drinks in adolescent diets. Adolescents are building up a tolerance to high doses of caffeine, with potential adverse health effects.
- Although evidence linking diet and acne is limited, drinking plenty of water, choosing low-GI foods and increasing intake of omega-3 fatty acids may assist in reducing acne severity.

REFERENCES

ABS (Australian Bureau of Statistics) 1995. *National Nutrition Survey: Selected highlights, Australia, 1995*. Cat. no. 4802.0. Commonwealth of Australia. <www.abs.gov.au/ausstats/abs@.nsf/mf/4802.0>, accessed August 2013.

——2012. *Australian Health Survey: First results, 2011–12*. Cat. no. 4364.0.55.001. Commonwealth of Australia. <www.ausstats.abs.gov.au/ausstats/subscriber.nsf/0/1680ECA402368CCFCA257AC90015AA4E/$File/4364.0.55.001.pdf>, accessed 28 September 2014.

AIHW (Australian Institute of Health and Welfare) 2011. *Young Australians: Their health and wellbeing 2011*. Cat. no. PHE 140. Commonwealth of Australia. <www.aihw.gov.au/WorkArea/DownloadAsset.aspx?id=10737419259>, accessed 28 September 2014.

APA (American Psychiatric Association) 2013. *Diagnostic and Statistical Manual of Mental Disorders* 5th edn (*DSM-5*). APA, Washington, DC.

Bernstein, G.A., Carroll, M.E., Dean, N.W., Crosby, R.D. et al. 1998. Caffeine withdrawal in normal school-age children. *Journal of the American Academy of Child & Adolescent Psychiatry* 37: 858–65.

CSIRO (Commonwealth Scientific and Industrial Research Organisation), Preventative Health National Research Flagship & University of South Australia 2008. *2007 Australian National Children's Nutrition and Physical Activity Survey: Main findings.* Commonwealth of Australia, Canberra.

CTRU (Clinical Trials Research Unit) & Synovate 2010. *A National Survey of Children and Young People's Physical Activity and Dietary Behaviours in New Zealand: 2008/09; Key findings.* New Zealand Government. <www.health.govt.nz/system/files/documents/publications/cyp-physical-activity-dietary-behaviours-08-09-keyfindgs.pdf>, accessed 14 November 2014.

De Jesus Moreno Moreno, M. 2003. Cognitive improvement in mild to moderate Alzheimer's dementia after treatment with the acetylcholine precursor choline alfoscerate: A multicenter, double-blind, randomized, placebo-controlled trial. *Clinical Therapeutics* 25: 178–93.

Department of Health 2014. *Australia's Physical Activity and Sedentary Behaviour Guidelines.* Commonwealth of Australia. <www.health.gov.au/internet/main/publishing.nsf/content/health-pubhlth-strateg-phys-act-guidelines>, accessed 28 September 2014.

ESHRE Capri Workshop Group 2005. Noncontraceptive health benefits of combined oral contraception. *Human Reproduction Update* 11: 513–25.

Ferguson, E.L., Morison, I.M., Faed, J.M., Parnell, W.R. et al. 2001. Dietary iron intakes and biochemical iron status of 15–49 year old women in New Zealand: Is there a cause for concern? *New Zealand Medical Journal* 114: 134–8.

Garaulet, M., Ortega, F.B., Ruiz, J.R., Rey-López, J. et al. 2011. Short sleep duration is associated with increased obesity markers in European adolescents: Effect of physical activity and dietary habits; The HELENA study. *International Journal of Obesity* 35: 1308–17.

Garnett, S.P., Baur, L.A. & Cowell, C.T. 2008. Waist-to-height ratio: A simple option for determining excess central adiposity in young people. *International Journal of Obesity* 32: 1028–30.

Gibson, R.S., Heath, A.L. & Ferguson, E.L. 2002. Risk of suboptimal iron and zinc nutriture among adolescent girls in Australia and New Zealand: Causes, consequences, and solutions. *Asia Pacific Journal of Clinical Nutrition* 11: S543–S552.

Gómez-Pinilla, F. 2008. Brain foods: The effects of nutrients on brain function. Nature Reviews. *Neuroscience* 9: 568–78.

Grant, R., Bilgin, A., Zeuschner, C., Guy, T. et al. 2008. The relative impact of a vegetable-rich diet on key markers of health in a cohort of Australian adolescents. *Asia Pacific Journal of Clinical Nutrition* 17: 107–15.

Gunja, N. & Brown, J.A. 2012. Energy drinks: health risks and toxicity. *Medical Journal of Australia* 196: 46–9.

Hirano, Y., Obata, T., Kashikura, K. & Nonaka, H. 2008. Effects of chewing in working memory processing. *Neuroscience Letters* 436: 189–92.

Hoyland, A., Dye, L. & Lawton, C.L. 2009. A systematic review of the effect of breakfast on the cognitive performance of children and adolescents. *Nutrition Research Reviews* 22: 220–43.

Ingersoll, G.M. 1992. Psychological and social development. In McAnarney, E.R., Kreipe, R.E., Orr, D.P. & Comerci, G.D. (eds). *Textbook of Adolescent Medicine.* Elsevier, Philadelphia, PA.

Jakobsen, M.U., O'Reilly, E.J., Heitmann, B.L. & Pereira, M.A. 2009. Major types of dietary fat and risk of coronary heart disease: A pooled analysis of 11 cohort studies. *American Journal of Clinical Nutrition* 89: 1425–32.

Kuczmarski, R.J., Ogden, C.L., Grummer-Strawn, L.M., Flegal, K.M. et al. 2000. CDC Growth Charts: United States. *Advance Data* 314 (revised): 1–28.

Leger, D., Beck, F., Richard, J.-B. & Godeau, E. 2012. Total sleep time severely drops during adolescence. *PLoS One* 7(10): e45204.

Linkswiler, H.M., Zemel, M.B., Hegsted, M., Schuette, S. et al. 1981. Protein-induced hypercalciuria. *Federation Proceedings* 40: 2429–33.

McCann, J.C., Hudes, M. & Ames, B.N. 2006. An overview of evidence for a causal relationship between dietary availability of choline during development and cognitive function in offspring. *Neuroscience & Biobehavioral Reviews* 30: 696–712.

McLennan, W. & Podger, A. 1998. *National Nutrition Survey: Nutrient intakes and physical measurements, Australia, 1995* new issue. Cat. no. 4805.0. ABS, Commonwealth of Australia. <www.ausstats.abs.gov.au/ausstats/subscriber.nsf/0/CA25687100069892CA25688900268A6D/$File/48050_1995.pdf>, accessed 28 September 2014.

Mahan, K.L. & Escott-Stump, S. 2008. *Krause's Food and Nutrition Therapy*. Elsevier, St Louis, MO.

Ministry of Health 2012a. *The Health of New Zealand Adults 2011/12: Key findings of the New Zealand Health Survey*. New Zealand Government, Wellington.

——2012b. *Food and Nutrition Guidelines for Healthy Children and Young People (Aged 2–18 Years): A background paper*. New Zealand Government.

Morgan, K., Johnson, A.J. & Miles, C. 2013. Chewing gum moderates the vigilance decrement. *British Journal of Psychology* 105: 214–25.

Nambiar, S., Hughes, I. & Davies, P.S. 2010. Developing waist-to-height ratio cut-offs to define overweight and obesity in children and adolescents. *Public Health Nutrition* 13: 1566–74.

NEDC (National Eating Disorders Collaboration) 2010. *Eating Disorders—Prevention, Treatment and Management: An evidence review*. NEDC, Crows Nest, NSW.

Netea, S.A., Janssen, S.A., Jaegar, M., Jansen, T. et al. 2013. Chocolate consumption modulates cytokine production in healthy individuals. *Cytokine* 62: 40–3.

NHMRC (National Health and Medical Research Council) 2013. *Eat for Health: Australian Dietary Guidelines; Providing the scientific evidence for healthier Australian diets*. Commonwealth of Australia. <www.eatforhealth.gov.au/sites/default/files/files/the_guidelines/n55_australian_dietary_guidelines.pdf>, accessed 28 September 2014.

NHMRC (National Health and Medical Research Council) & Ministry of Health 2006. *Nutrient Reference Values for Australia and New Zealand: Including recommended dietary intakes*. Commonwealth of Australia and New Zealand Government. <www.nhmrc.gov.au/_files_nhmrc/publications/attachments/n35.pdf>, accessed 26 September 2014.

O'Dea, J.A. 2003. Consumption of nutritional supplements among adolescents: usage and perceived benefits. *Health Education Research* 18: 98–107.

O'Sullivan, T.A., Hafekost, K., Mitrou, F. & Lawrence, D. 2013. Food sources of saturated fat and the association with mortality: A meta-analysis. *American Journal of Public Health* 103: e31–e42.

O'Sullivan, T.A., Robinson, M., Kendall, G.E., Miller, M. et al. 2009. A good quality breakfast is associated with better mental health in adolescence. *Public Health Nutrition* 12: 249–58.

Rampersaud, G.C., Pereira, M.A., Girard, B.L., Adams, J. et al. 2005. Breakfast habits, nutritional status, body weight, and academic performance in children and adolescents. *Journal of the American Dietetic Association* 105: 743–60.

Rangan, A.M., Randall, D., Hector, D.J., Gill, T.P. et al. 2008. Consumption of 'extra' foods by Australian children: types, quantities and contribution to energy and nutrient intakes. *European Journal of Clinical Nutrition* 62: 356–64.

Roche, A.F. & Davilia, G.H. 1972. Late adolescent growth in stature. *Pediatrics* 50: 874–80.

Schlundt, D.G., Hill, J.O., Sbrocco, T., Pope-Cordle, J. et al. 1992. The role of breakfast in the treatment of obesity: A randomized clinical trial. *American Journal of Clinical Nutrition* 55: 645–51.

Smith, A. 2009. Effects of chewing gum on mood, learning, memory and performance of an intelligence test. *Nutritional Neuroscience* 12: 81–8.

Solomons, N.W. 1986. Competitive interactions of iron and zinc in the diet: Consequences for human nutrition. *Journal of Nutrition* 116: 927–35.

Tanner, J. 1962. *Growth at Adolescence*. Blackwell Scientific Publications, Oxford.

University of Otago & Ministry of Health 2011. *A Focus on Nutrition: Key findings of the 2008/09 New Zealand Adult Nutrition Survey*. New Zealand Government, Wellington.

Wainwright, P. 2000. Nutrition and behaviour: The role of n-3 fatty acids in cognitive function. *British Journal of Nutrition* 83: 337–9.

Weiss, A., Xu, F., Storfer-Isser, A., Thomas, A. et al. 2010. The association of sleep duration with adolescents' fat and carbohydrate consumption. *Sleep* 33: 1201–9.

Wood, A., Utter, J., Robinson, E., Ameratunga, S. et al. 2011. Body weight satisfaction among New Zealand adolescents: Findings from a national survey. *International Journal of Adolescent Medicine and Health* 24: 161–7.

ADDITIONAL READING

Blankson, K.L., Thompson, A.M., Ahrendt, D.M. & Patrick, V. 2013. Energy drinks: What teenagers (and their doctors) should know. *Pediatrics in Review* 34: 55–62.

Burris, J., Rietkerk, W. & Woolf, K. 2013. Acne: The role of medical nutrition therapy. *Journal of the Academy of Nutrition and Dietetics* 113: 416–30.

Drake, V.J. 2012. *Micronutrient Requirements of Adolescents Ages 14 to 18 Years*. Oregon State University. <http://lpi.oregonstate.edu/infocenter/lifestages/adolescents/index.html>, accessed 2 October 2014.

Kids Helpline (website) 2014. <www.kidshelp.com.au>, accessed 2 October 2014.

Ministry of Health 2003. *Food and Nutrition Guidelines for Healthy Adults: A background paper*. New Zealand Government. <www.health.govt.nz/system/files/documents/publications/foodandnutritionguidelines-adults.pdf>, accessed 16 March 2014.

National Eating Disorders Collaboration (website) 2014. <www.nedc.com.au/>, accessed 2 October 2014.

Stanton, R. 1997. *Healthy Vegetarian Eating*. Allen & Unwin, Sydney.

University of Sydney 2014. Glycemic Index (website). <www.glycemicindex.com/>, accessed 30 March 2014.

TEST YOUR UNDERSTANDING

Note: There is only one correct answer for each question.

1. Which of the following symptoms can result from excess energy drink consumption in adolescence?
   a. Heart palpitations
   b. Growth restriction
   c. Joint pain
   d. a and b
   e. b and c
2. Which of the following would be least appropriate for an exam day breakfast?
   a. Weet-bix with milk, canned apricots and sunflower seeds
   b. Porridge cooked with milk, served with sliced banana
   c. Baked beans on wholegrain toast with avocado
   d. White-bread toast with jam and a small can of energy drink
   e. Scrambled eggs on an English muffin
3. Which meal would be most likely to result in a high peak of glucose followed by a high insulin peak?
   a. Large serving of yoghurt
   b. Small banana with a kiwi fruit
   c. Small milkshake
   d. Handful of nuts and a hard-boiled egg
   e. Medium serving of rice bubbles with milk and honey
4. Which of the following statements about acne is false?
   a. More research is required to develop evidenced-based guidelines around diet for acne
   b. Reducing inflammation through increased consumption of water is helpful to decrease severity of acne

c. Decreasing insulin peaks through the use of low-GI foods can help to reduce acne
d. Stress does not cause acne directly but can make it worse by increasing sebum production
e. A Western-style diet high in processed foods may contribute to acne

5. Which of the below adolescent profiles is considered more at risk of developing an eating disorder?
   a. Female, high achiever, Tanner Stage 5 of puberty
   b. Male, high achiever, Tanner Stage 5 of puberty
   c. Male, anxious, low self-esteem, Tanner Stage 2 of puberty
   d. Female, anxious, low self-esteem, Tanner Stage 2 of puberty
   e. a and d
6. About which of the following nutrients do vegetarians need to be particularly concerned?
   a. Folate
   b. Vitamins A and C
   c. Iron and zinc
   d. Fibre
   e. None—same as for meat eaters
7. Which of the following is not a factor in MetS?
   a. Abdominal obesity
   b. Inflammation and acne
   c. High blood pressure
   d. Dyslipidaemia or high cholesterol
   e. Insulin resistance
8. Which of the following statements about early adolescence is false?
   a. The adolescent becomes increasingly independent
   b. The adolescent is aged eleven to fourteen years
   c. The adolescent is concerned about how their diet will affect them in adulthood
   d. Moral concepts are beginning to develop
   e. The adolescent is adjusting to a developing body and body image
9. Which of the following will result from excessive protein intakes, above requirements?
   a. Increased muscle mass and tone
   b. Increased calcium excretion
   c. Increased fat mass
   d. All of the above
   e. b and c
10. What is the cause of increased iron requirements during adolescence?
    a. Increased volume of red blood cells
    b. Menstruation commencing in females
    c. Increased physical activity
    d. a and b
    e. b and c

STUDY QUESTIONS

1. Describe three dietary and three practical factors to consider when advising on appropriate lunchbox foods for high schoolers.
2. Why is breakfast important for adolescents? What sorts of healthy breakfasts could you recommend for adolescents with the following issues?
   a. Diagnosed with coeliac disease
   b. Lactose intolerant
   c. Reports they don't have time to eat before school
3. List five physical signs of anorexia nervosa and five psychological or behavioural symptoms. What are three potential side effects of anorexia nervosa?
4. Identify three nutrients which may be more difficult to obtain from a vegetarian diet and explain why. List four vegetarian food options for each nutrient identified.
5. List and describe at least four factors contributing to overweight and obesity in adolescents.
6. Saturated fat intake in adolescents is higher than the recommended 10 per cent of total energy. Provide three examples of food changes that would reduce saturated fat and increase unsaturated fat intake in adolescents.

10

Young adult (19–25 years)

Regina Belski and Antonia Thodis

> **LEARNING OUTCOMES**
>
> Upon completion of this chapter you will be able to:
>
> - identify significant lifestyle changes that can occur in early adulthood
> - list the social and lifestyle factors that impact on nutrition and health
> - describe common nutrition-related health issues observed in young adults
> - distinguish the relevant key macronutrients and micronutrients
> - discuss the prevalence of mental health issues and their association with diet quality
> - identify support services available to young adults.

The young adult life stage, encompassing ages from 19 to 25, is the transition stage between adolescence and adulthood, where the rapid growth period of adolescence is ending and true independence in relation to food, health and other lifestyle choices begins. For Australian and New Zealand young adults this stage often involves further study and the commencement of professional employment as well as leaving the parental home, the formation of new relationships and in some instances the beginning of new families. All of these changes can play a significant role in the nutritional status of young adults. Within this chapter the nutritional needs of young adults are examined, as well as some of the nutrition-related health issues observed in this group, including obesity and overweight, alcohol and drug consumption, fad diets, living independently, commencing professional employment and starting a family.

LIFESTYLE CHANGES

Independent living

Young adulthood is a life stage during which significant transition in living arrangements and changes in type of accommodation occur due to lifestyle changes post-adolescence. Some young adults move out of home to live on campus at university; others move out with friends and partners. This change in circumstance can have a significant impact on food intake and nutritional status.

This can often be the first time that individuals become responsible for their own food selection and preparation, but nutrition may not be of primary concern in this age group, with other lifestyle factors taking higher priority. Cost, accessibility, ease of preparation and taste tend to top the list of priorities over high nutritional quality.

The cost of food is a factor that tends to be an issue for many young adults. Regardless of whether they are studying, working at entry-level rates or completing other training, income is often conservative, and after paying for accommodation and prioritising social life there may be limited amounts to spend on food.

Living arrangements, particularly in share accommodation, can also present challenges depending on the arrangements or rules of the share house when it comes to food. For example, 'hungry' roommates may find their way to fellow housemates' food provisions. This can leave a budget-conscious housemate with little to eat when they get home from work or study. The true problem in this situation is that some young adults live on extremely tight budgets and may simply not be able to replace the foods until their next pay, which can leave them with suboptimal and irregular eating habits.

Cooking skills and facilities can pose a challenge, particularly if limited, and if young adults lack confidence in food preparation. The result is often resorting to 'simple and quick' meals such as pasta with packet sauce, two-minute noodles, instant soup and toast or purchasing takeaway meals. Lack of appropriate cooking facilities and cooking utensils and equipment can also create an obstacle to healthy eating. The lack of a working oven and poor-quality pots, pans and utensils limit the types of foods that can be prepared and cooked.

Knowledge, attitudes and beliefs about food are influenced by culture, family and peers as well as knowledge around food labelling, nutrition

CASE STUDY 10.1 YOUNG ADULT (19–25 YEARS)

Profile and problem

Miss Z. is a 22-year-old female who has recently started a new full-time job as an assistant in a busy office. She shares a house with two friends. Twice a week she attends early morning meetings at the office and does not have time for breakfast. On these days she will buy a coffee and muffin mid-morning to get through until lunch, which she buys from a nearby cafe. She generally chooses a meat and salad sandwich or pasta. On the other mornings, breakfast is usually just a takeaway coffee and perhaps a piece of fruit. By the end of a long work day she reports feeling very hungry and too tired most nights to cook a meal so will stop off for a takeaway on her way home. She eats dinner out two to three times a week, especially after a run or the gym. Miss Z. is 165 centimetres tall and weighs 70 kilograms. She has always been active person but has noticed that she has gained weight since starting her new job and is concerned that she needs to 'go on a diet'.

Assessment

Miss Z.'s concern is that she is feeling tired more often and has gained weight. She is keen to make some adjustments. Her current food intake is lacking in certain areas, but her food choices are being heavily influenced by lifestyle factors that can be addressed to assist her manage her nutrition-related problems.

Treatment

Miss Z. has agreed to do the following:

- eat more substantial meals earlier in the day to avoid the afternoon and evening slump in energy
- plan morning meals ahead of time, including healthy options for breakfasts at home (a good cereal choice with reduced-fat milk to assist with meeting calcium requirements) and for breakfasts 'on the run'
- choose healthy options for mid-morning snacks
- prepare lunch ahead of time at least twice a week
- choose healthy options for eating out
- aim to eat out less and prepare meals at home more
- include high-protein foods such as meat, chicken, fish and legumes at lunch and dinner, which also assist with meeting requirements for iron
- include a handful of nuts daily as a snack to minimise feelings of hunger, which promotes suboptimal food choices.

Plan

Miss Z. will continue to see a dietitian who will offer her ongoing support to ensure the changes she has made are sustainable and that her weight management goals are realistic and met.

information, the media and advertising (AIHW 2012). Past eating preferences and experiences often continue to have an impact on the eating habits of young adults when they begin their independent life; however, the extent and degree of influence can differ from one individual to another. It is not uncommon for food choices to change considerably when a young adult moves out of the parental home. Greater freedom of food choice may present a conflict between 'healthy' and 'less healthy' food selection, given that a parent is no longer monitoring the intake and purchasing of the latter group. Foods that may have been restricted by parents, such as soft drinks, lollies or chips, are now able to be purchased and consumed in any desired quantity. These types of foods can take up a significant portion of the food budget, leaving little money for foods recommended by the Australian Dietary Guidelines and the New Zealand Food and Nutrition Guidelines, such as fish, lean meat, chicken, fruit, vegetables and low-fat dairy foods (Ministry of Health 2003; NHMRC 2013). The young adult's pantry can often look less healthy than is desirable.

Transportation can also be a challenge in regards to having an optimal nutritional intake. The lack of car ownership and reliance on public transport make the otherwise simple task of a weekly grocery shop, which may be more cost-effective on a tight budget, more complex. A young adult who relies on day-to-day shopping in order to buy amounts small enough to carry home by foot or on public transport has to address this additional obstacle. Bulk buys are often more cost-effective, and the consistent purchasing of small portions and packages can add further pressure to the already tight budgets of young adults.

While the rapid body composition changes seen in adolescence are well and truly over, this does not mean significant changes in body composition do not occur in young adults. This is often the age at which individual decisions around food and physical activity can have a big impact on body composition and shape. For example, at this age young men may decide to begin a gym routine and build muscle mass. Both men and women may be seeking weight loss, and this may have an impact on their food choices and body composition.

Young women are reportedly one of the most frequently dieting groups in today's society and are more likely to experiment with fad diets, weight-loss shakes and diet pills. The Young Women's Nutrition Survey, commissioned by the Dietitians Association of Australia in 2012–13, revealed that 68 per cent of 18–24-year-old women had resolved to eat more healthily in 2013, and 42 per cent were hoping to lose weight. Approximately two-thirds of the young women surveyed reported having tried to lose weight in the year prior to the survey, and just over a quarter (28 per cent) reported trying to lose weight was an ongoing issue for them (DAA 2013).

With more than 60 per cent of women reporting the desire to lose weight, it is not surprising that young adults, especially young females, are prone to trialling fad diets for weight management. The appeal of a 'quick fix' can drive a young female adult to try many undesirable and potentially dangerous diets. This particularly vulnerable group can often be taken in by strategic marketing and anecdotal stories of great success that are not based on quality evidence. The choice to spend a limited income on these weight-loss gimmicks rather than on a varied diet of healthy food-based choices may compromise nutritional status in both the short and long term.

A change in body composition from a 'healthy' weight-for-height in adolescence towards weight gain and overweight in young adulthood has been identified. In the United States this phenomenon is described as the freshman five (or freshman fifteen), when both male and female students gain a significant amount of weight (5–15 imperial pounds, or around 2–6 kilograms) in the first year of university. This has been attributed to the changes in lifestyle and food choices and significant increases in alcohol consumption when young adults move away from home to live at university. The phenomenon is not observed as clearly in Australia and New Zealand, as many young adults remain living in the family home while at university, which may impact less upon their lifestyle, eating and drinking behaviours. Longer time spent studying often results in more time spent managing on a low income. This can put additional strain on budget priorities and extend the time that an individual has suboptimal eating habits.

Young adults continuing to live at home well into their twenties and in some cases their thirties may have higher expectations placed on them to contribute to the household either financially or through doing chores. Therefore, they may need to become more involved in food purchasing and preparation and contribute financially to this household expense. This may have a direct impact on nutritional status, as the transition from having food prepared to having to prepare it themselves may lead to the consumption of the more 'easy' or 'convenient' foods, which may have limited nutritional value.

Starting professional employment

Nutritional status can be influenced by employment in a number of ways, including a person's economic status, a new peer group that may develop with commencing a new job, the type of employment and how much of it revolves around being physically active or sedentary. Gaining professional employment at this stage is very exciting and can improve the financial situation of a young adult. It may also impose time pressures, with long working hours leading to irregular eating and the consumption of 'convenient' over nutritional foods.

Some corporate office environments within Australia and New Zealand have made an effort to promote health and nutrition to improve the nutritional status of their employees by providing fresh fruit in the office once or twice per week; this is a common practice in many of the top five accounting firms. However, tea, coffee and biscuits are still the primary tea break options provided by many workplaces. The end of a long and busy work day may increase the temptation to save time and effort by purchasing a quick and simple takeaway dinner on the way home or rely on frozen meals.

The coffee culture has well and truly taken over Australia and New Zealand during recent decades. Having coffee with workmates has become the new smoke break, and unfortunately it is often paired with accompanying choices, such as a muffin or cake, that are often not ideal. Trying to fit in with new colleagues and peers often involves going out for lunches or after-work drinks, where less control over food choices impacts upon nutritional intake.

Habits that are formed early in working life can often be hard to change. Long work hours, reduced physical activity, relying on coffee to get through the day and fast and convenient food choices can quickly become the norm, and without a person's noticing it their nutritional status can often be detrimentally affected. Whether this habit leads to weight gain in the long term is debatable, but there is an association in females between skipping breakfast and having a higher body mass index (BMI). Poor choices are more likely, and overeating occurs due to feeling hungry later in the day. It is also well established that skipping breakfast negatively impacts upon cognitive function, academic performance and mood, all of which are undesirable.

Relationships

Another area of interest is how nutritional intake and status can be influenced by relationships, particularly in young adulthood, a time when new relationships are formed. Food can be central from the start, with many social arrangements being based around meals and food and likely to involve lunch, dinner, coffee or drinks. This, coupled with a busy social life, may result in many young adults eating out numerous times a week.

Settling down into a more comfortable relationship may mean that a partner's level of nutritional knowledge and subsequent food choices can significantly impact upon the other partner's food choices, either positively or negatively. Partners will introduce one another to new and different food options, and moving out of home and living together with a partner implies that the pantry will be accommodating the preferences of both individuals. The different food and dietary habits of one partner can influence the other, and, depending on whether the food choices are healthy or not, this can have an impact on the nutritional status of both parties.

The social aspect of eating may also lead to more food being consumed. This can also work in reverse, when a partner with healthy eating habits and good dietary choices encourages and teaches their partner with less optimal dietary choices about better food selection and healthy eating.

NUTRIENT REQUIREMENTS AND INTAKES

The following section provides a snapshot of young adults' requirements of macronutrients and micronutrients according to Australia and New Zealand's nutrient reference values (NRVs; see Chapter 1; NHMRC & Ministry of Health 2006). The NRVs promote and encourage healthy eating for the general population and aim to reduce the risk of chronic health problems such as obesity, heart disease, type 2 diabetes mellitus (T2DM) and some cancers. The onset of chronic diseases is associated with later adulthood (55 years and over), but the last few decades have seen the age of onset steadily decreasing. Therefore, the dietary intake and the foods chosen by young adults need to be high in nutritional quality within the context of the social and lifestyle choices that occur during this life stage. In the longer term, the diet and lifestyle habits of a young adult will very likely influence how well they age.

Energy

The estimated energy requirement (EER) for 19–30-year-old males with a reference body weight of 76 kilograms is between 10,800 kilojoules per day for sedentary activity to 13,800 kilojoules per day for moderate activity. For young adult females with a reference body weight of 61 kilograms, the EER ranges from 8100 kilojoules per day for those who are sedentary to 10,500 kilojoules per day for the moderately active (NHMRC & Ministry of Health 2006).

The 2011–12 Australian Health Survey reported that the mean daily energy intake for young adult males is approximately 11,000 kilojoules, the highest of all the age groups surveyed and much higher than the average energy intake for all Australian males (9655 kilojoules per day). For young adult females, the survey reported a mean daily energy intake of close to 8000 kilojoules per day, slightly higher than the average energy intake for Australian females (7402 kilojoules per day). These values for energy intake also included energy from dietary fibre (ABS 2014).

The desirable proportion of dietary energy from the energy yielding macronutrients is described as the acceptable macronutrient distribution range (AMDR; see Chapter 1). The survey found that for 19–30-year-olds the contribution of protein to energy was approximately 18 per cent, carbohydrate 45 per cent, total fat 31 per cent (all within the AMDR) and alcohol 4 per cent. The main contributors to the carbohydrate proportion of energy were foods such as sweet and savoury biscuits and pastries and pizza (20–23 per cent of energy), followed by bread, rice, noodles, pasta and breakfast cereal (15–17 per cent) (ABS 2014).

The survey results showed that total energy consumed from discretionary foods decreased in the age groups from 19–30 years and older when compared with the fourteen- to eighteen-year-old age group (ABS 2014). Nevertheless, for young adult males and females aged 19–30 years, approximately 37 per cent and 35 per cent of total energy, respectively, was derived from discretionary foods. Foods contributing the most to energy from discretionary foods reported were alcoholic beverages (almost 5 per cent) followed by cakes, muffins, scones and similar foods (3.4 per cent). Smaller proportions were derived from confectionery, biscuits, soft drinks, potato chips, snack foods, sugar, honey and syrups. In all older age groups, alcoholic drinks provided the largest source of energy from discretionary foods overall with 6 per cent of energy consumed by people aged nineteen years and over coming from alcoholic beverages (ABS 2014). The proportion of energy from discretionary foods is largely unchanged since the 1995 National Nutrition Survey (ABS 1995). Discretionary foods contributed 36 per cent of total energy intake for young adult males and females at that time. However, for the young adult age group in Australia, the contribution to energy intake from alcoholic beverages compared with other discretionary foods has changed. The 2011–12 survey suggests that energy derived from alcoholic beverages (including beers, wines, spirits, cider and other) has displaced some of the energy derived from other discretionary foods reported by young adults in 1995 (ABS 2014).

In 2008–09 in New Zealand the median energy intake for 19–30-year-olds was 11,817 kilojoules per day for males and 8245 kilojoules per day for females; as in Australia, this was higher than the national adult average (10,380 kilojoules for males and 7448 for females). The main foods contributing to energy intake for this age group were bread (8–10 per cent), grains and pasta (9 per cent) and bread-based dishes (9 per cent males; 5 per cent females). Non-alcoholic beverages and potatoes, kumara and taro both contributed around 7 per cent, and alcoholic beverages were 5 per cent for young females and 7 per cent for young males. Energy from fruit and vegetables, sugar and sweets was less than 5 per cent (University of Otago & Ministry of Health 2011).

Macronutrients

Protein

The recommended dietary intake (RDI) for young adults aged 19–30 years is 64 grams per day for males and 46 grams per day for females. In Australia, the mean daily intake of protein in this age group was 117 grams per day for males and 78 grams per day for females, which is higher than the RDI (ABS 2014). Food sources for protein are listed in Table 1.16.

In New Zealand, the median usual daily protein intake was 111 grams per day for males and 72 grams per day for females. Dietary sources of protein included bread as the largest contributor, followed by poultry, milk, beef, veal, grains, pasta, bread-based dishes, fish, seafood and pork (University of Otago & Ministry of Health 2011).

Carbohydrate

Limited data exist for estimated carbohydrate requirements, and these have not been set by the National Health and Medical Research Council for Australia and New Zealand (NHMRC 2013). The mean daily intake of carbohydrate for 19–30-year-old males was

297 grams, with most in the form of starch (159 grams) compared with sugar (131 grams). Females consumed less (213 grams), with the same pattern of a higher starch (112 grams) than sugar intake (100 grams). The major food groups contributing to total starch reported in the 2011–12 Australian Health Survey were breads, dishes where cereal is a major ingredient, flours, cereal grains (mainly rice), breakfast cereals and potatoes. Non-alcoholic beverages, including fruit and vegetable juices and soft drinks contributed 14–16 per cent of carbohydrates (ABS 2014).

The survey found 44 per cent of young adult males had consumed soft drink on the day prior to the 24-hour recall interview. This is higher than the percentage for the rest of the Australian population, which was under 30 per cent and more than for young adult females. The results indicated that rates of consumption of fruit and vegetables for young adults was low, with around 40 per cent of males and 50 per cent of females consuming fruit compared with 60 per cent for the whole population. Although approximately 75 per cent of young adults did report consuming vegetables on the day prior to the 24-hour recall, almost half of this consumption was from potatoes, including potato chips (ABS 2014).

The median usual daily intake of carbohydrate in New Zealand was 327 grams for males and 239 grams for females. The mean contribution to daily energy intake was 47 per cent for males and 49.3 per cent for females, which fell within the AMDR of 45–65 per cent of total energy from carbohydrate. For males and females respectively, bread (12 and 13 per cent), non-alcoholic beverages (14 and 12 per cent) and grains and pasta (12 and 11.3 per cent) were the three highest contributors to dietary carbohydrate. Non-alcoholic beverages also contributed the greatest proportion of total sugars in this age group compared with other age groups, followed by sugar and sweets and then fruit for both genders. For males, 28.8 per cent of total sugars were from non-alcoholic beverages, and for females the figure was 22.9 per cent. Males in this age group also showed the highest median intake of total sugars (140 grams per day) compared with all other age groups. For females the amount was 120 grams per day (University of Otago & Ministry of Health 2011).

Fat

The RDI for total fat has not been set, due to the essential nature of some fatty acids and the lack of data as to the exact requirements and dose-response relationships. The amounts for an adequate intake (AI) are shown in Table 1.10b, and are based on the highest median intakes of any of the gender-related age groups taken from the National Nutrition Survey of Australia (ABS 1995). National data for New Zealand was not available, so the same values were adopted. The AMDR aims to provide a suitable intake of fats, reduce chronic disease risk and allow for the sufficient intake of essential nutrients and the moderate intake of saturated fat (NHMRC & Ministry of Health 2006). Saturated and trans fatty acids combined should provide no more than 10 per cent of total energy. The World Health Organization (WHO) recommends that the contribution of trans fatty acids should not be more than 1 per cent of total dietary energy (WHO 2003).

High intakes of linoleic acid (omega 6 (n-6)) and alpha-linolenic acid (omega 3(n-3)) polyunsaturated fatty acids have been associated with a reduced risk of coronary heart disease (CHD), although individuals are encouraged to consume more omega 3 (n-3) rich foods than omega 6 (n-6). For total long-chain (omega 3 (n-3)) fatty acids, the AI was based on the median intakes of all adults of the relevant gender. Table 10.1 shows that young Australian adults exceed the AI for alpha-linolenic acid and long-chain omega 3 (n-3) fatty acids, women exceed the AI for linoleic while males consumed 1 gram less than recommended. Food sources of fats are described in Table 1.16.

The New Zealand median usual daily intake of total fat for males was 102 grams and for females 71 grams, which is higher than the national average. The proportion of energy from total fat for males was

Table 10.1 Fat intakes for 19–30-year-old Australian males and females and comparison to adult AIs for linoleic acid, alpha-linolenic acid and long-chain omega 3 (n-3) fatty acids

Fat	Daily fat intake gram (AI gram/day)	
	Males	**Females**
Total fat	92	70
Saturated	35	27
Monounsaturated	36	26
Polyunsaturated	14	11
Linoleic acid (n-6)	12 (13)	9 (8)
Alpha-linolenic acid (n-3)	1.4 (1.3)	1.4 (0.8)
Long-chain omega 3 (n-3) fatty acids	0.324 (0.16)	0.117 (0.09)

Note: Intake and AIs for long-chain fatty acids are typically given as milligrams.

Source: ABS (2014); NHMRC & Ministry of Health (2006).

33.6 per cent and for females 33 per cent, which was within the AMDR for fat (20–35 per cent of energy). The largest contributors of total dietary fat were butter and margarine, and within this age group bread-based dishes and potatoes, kumara and taro were significant contributors for both males and females.

The median usual daily intake of saturated fatty acids was 40.7 grams for males and 27.9 grams for females, and in terms of the proportion of energy coming from saturated fats this was higher than the recommended intake for saturated and trans fats together, which is no greater than 10 per cent of energy. For males it was 13.1 per cent and for females 12.9 per cent, both in excess of the recommendations (University of Otago & Ministry of Health 2011). Food sources of fat are described in Table 1.16.

Fibre

The AI for young adults for dietary fibre is 30 grams per day for males and 25 grams per day for females (NHMRC & Ministry of Health 2006). The AI is set at the median for dietary fibre intake in Australia and New Zealand based on the 1995 National Nutrition Survey of Australia (ABS 1995) and the 1997 National Nutrition Survey of New Zealand (Ministry of Health 1999). These nutrient reference values for dietary fibre have not been reviewed since these surveys were conducted.

An upper limit (UL) has been set for dietary fibre, given the potential adverse effects of very high intakes reported by several studies. However, the composition of dietary fibre is highly variable and it is difficult to associate a specific fibre type with a particular adverse outcome. In practical terms, any potential adverse effects of excessive amounts of dietary fibre consumption are unlikely at the level of the recommended intake, particularly if dietary fibre is derived from various sources.

Fibre intakes for both Australian and New Zealand 19–30-year-olds do not meet recommendations and are described in Table 1.15a. Food sources of fibre are described in Table 1.16.

Water

The AI of water from foods and fluids combined for adult males aged 19–30 years is approximately 3.4 litres per day and for females 2.8 litres per day. From fluids alone, such as plain water, milk and other drinks, the AI for 19–30-year-old males is 2.6 litres per day, or about 10 cups (each equivalent to 250 millilitres), and for females 2.1 litres per day, or about 8 cups.

The AIs for both genders were set at the level of the highest median intake from any of the four age categories (that is, from 19 to over 70 years of age). Adults who live or work in warm to hot climates are likely to require higher amounts than the AI for hydration, particularly if they are very active. A UL has not been set, but excess water intake can lead to **hyponatraemia**. Most reports of this condition have been linked to significant water overconsumption by athletes during marathon races.

The 2011–12 Australian Health Survey found the mean daily moisture intake for young adult males to be 3249 grams and for females to be 2618 grams. Moisture comes from both food and drink. Non-alcoholic beverages contributed the greatest proportion of moisture (64–68 per cent) with water contributing two-thirds of this, followed by soft drinks and fruit juices, which contributed less than one-tenth (ABS 2014). Similar data is not available for New Zealand.

Micronutrients

Folate

This B group vitamin is naturally occurring in foods such as green leafy vegetables, cereals, fruits and grains. Folic acid is the synthetic form of folate used in supplements or added to food. Folate is the commonly used name for folic acid and its derivatives and is reported as dietary folate equivalents. One microgram of dietary folate equivalents is equivalent to 1 microgram of food folate or 0.5 microgram of folic acid. An EAR of 320 micrograms per day and an RDI of 400 micrograms per day have been set for both men and women aged 19–30 years. A higher intake applies for women of childbearing age: an EAR of 520 micrograms per day and an RDI of 600 micrograms per day. For women who are planning pregnancy, it is recommended that an additional folate intake of 400 micrograms per day is consumed for at least one month before and three months after conception, in addition to consuming folate-containing foods aimed at the prevention of neural tube defects. Folic acid supplementation has a greater bioavailability compared with that of dietary folate sources.

In Australia, mandatory fortification of folate in wheat flour for bread-making has been enforced since 2009. In New Zealand, mandatory fortification of

Hyponatraemia is a rare condition in which the sodium level in the blood is abnormally low. In some cases, if untreated it may result in death.

folate of all bread was to commence in September 2012; however, following consultation voluntary fortification was continued. It is expected that fortification will result in an average of 120 micrograms of folic acid per 100 grams of bread in Australia. Organic flour and bread are exempt from mandatory fortification (AIHW 2011).

The 2011–12 Australian Health Survey found that the average daily amount of folate equivalents consumed from foods was 700 micrograms for young adult males and 536 micrograms for females (ABS 2014). Food sources of folate in Australian young adults are described in Table 1.17. The role of folate is recognised by the Ministry of Health in New Zealand and it has in place a comprehensive policy detailing folic acid supplementation. However, equivalent data describing the intake of folate by young adult males and females is not reported in the 2008–09 New Zealand Adult Nutrition Survey (University of Otago & Ministry of Health 2011).

Vitamin D

Studies have identified that in addition to vitamin D's role in bone health, a deficiency is also associated with chronic diseases such as diabetes, some cancers, chronic inflammatory diseases, autoimmune diseases and, of particular relevance to young adults, anxiety and depression. Appropriate exposure to sunlight and adequate dietary sources of vitamin D are essential to maintain vitamin D status. The amount of sun exposure received year-round appears to be adequate in meeting requirements for this age group, but with awareness and health promotion strategies to decrease the risk of skin cancer, as well as Australians spending more time indoors, vitamin D deficiency has become a real problem for Australians in most age groups.

The AI of 5 micrograms per day for young adult males and females aged 19–30 years is based on the amount of vitamin D required to maintain serum 25-hydroxy vitamin D, the form of vitamin D converted to active vitamin D in the kidney, at an average level with minimal exposure to sunlight. The reference level of serum 25-hydroxy vitamin D is variable in the literature. Young adults who follow a vegan diet may require fortified foods and/or supplementation if sun exposure is inadequate. Few foods contain significant amounts of vitamin D, and accurate estimates of dietary intakes in Australia and New Zealand are not available.

In Australia, the major dietary sources of vitamin D are fortified margarine spreads, skim milk, powdered milk, yoghurts and cheese. Most of these foods have been mandated for fortification in Australia, but in New Zealand this is not the case. Voluntary fortification of margarine, fat spreads, dried milk and skim milk, legume beverages and 'food' drinks has been permitted in New Zealand since 1996. Other dietary sources naturally high in vitamin D include egg yolks, liver, fatty fish and their oils, such as salmon and herrings; however, these foods are not commonly consumed by young adults in Australia and New Zealand.

Calcium

The EAR and RDI for males and females aged 19–30 years are 840 milligrams per day and 1000 milligrams per day respectively and remain the same for young adult women who are pregnant or lactating. In 2011–12 the mean daily calcium intake for young adult males is 954 milligrams and for females it is 765 milligrams, both lower than the RDI (ABS 2014). The main sources of dietary calcium for males are milk (20 per cent), cheese (9 per cent), yoghurt (4 per cent), flavoured milks and milkshakes (14 per cent). For females the figures are milk (17 per cent), cheese (8 per cent), yoghurt (5 per cent) and flavoured milks and milkshakes (4 per cent) (ABS 2014).

In New Zealand, the median calcium intake for males aged 19–30 years was 967 milligrams per day and for females 704 milligrams per day. Milk was the greatest contributor to dietary calcium for males and females (both at 21 per cent), followed by non-alcoholic beverages (11 per cent) and cheese for males (9 per cent) but dairy products for females (7.4 per cent). Tables 1.15a and 1.17 describe intakes and food sources in more detail.

Iron

Iron deficiency is a worldwide issue, but the prevalence in Australia and New Zealand is largely unknown. It is clear that the depletion of iron stores is more common in 'at risk' groups, including young adult women. For men and women aged 19–30 years, the EAR is 6 milligrams and 8 milligrams per day, and the RDI is 8 milligrams and 18 milligrams per day, respectively. For pregnant young adult women the EAR is 22 milligrams per day and the RDI 27 milligrams per day. Recommended intakes are lower for lactating women: the EAR is 6.5 milligrams per day and the RDI 9 milligrams per day, based on estimates which assume that for lactating women menstruation does not resume until six months of exclusive breastfeeding postpartum. The most bioavailable sources of iron are red meat, fish, poultry and wholegrain cereals.

CASE STUDY 10.2 ***SPORT***

Adrienne Forsyth

Profile and problem

Mr G., a 23-year-old male, plays football for a local Australian Football League team. He has signed a contract for $120,000 over three years and is hoping to outperform his older brother, who plays for a rival team. Mr G. is 190 centimetres tall and weighs 78 kilograms. He has no health problems and has never worried about his diet in the past. He would like to have his diet assessed to determine whether there is anything he could change to improve his performance.

Assessment

Mr G. consumes in excess of 16,000 kilojoules per day. He trains five days per week for at least two hours per day and plays one game per week. Games are 120 minutes, with breaks every 30 minutes. Mr G. eats breakfast at the club: usually toast with eggs and a bowl of cereal. He has a protein shake after training and snacks on protein bars through the afternoon. For dinner he is often in a rush to see friends, so he picks up takeaway: either chicken or a burger and fries. He avoids soft drinks because he has heard that sugar is fattening.

Treatment

Mr G. has agreed to the do following:

- continue eating breakfast at the club
- pack a lunch with two chicken salad sandwiches, two pieces of fruit and a yoghurt
- pack snacks that are higher in carbohydrate, such as muesli or cereal bars
- plan meals for the week in advance; cook once or twice and freeze meals for the week
- have a bowl of cereal or a smoothie made with milk, yoghurt and fruit for a snack in the evening.

Plan

Mr G. will continue to meet with his team dietitian for regular dietary assessment. Ongoing anthropometric measurements including weight, circumferences and skin folds will determine whether weight is being maintained with his current dietary plan.

Mean iron intakes from food and beverages reported in the 2011–12 Australian Health Survey for young adults were 13.5 milligrams for males and 9.7 milligrams for females. The main sources of iron were breakfast cereals (contributing 20–25 per cent), cereal-based dishes (18–19 per cent), and meat, poultry and dishes containing meat and poultry (14–18 per cent) (ABS 2014).

The median intake of iron for adults aged 19–30 years in New Zealand was 13.9 milligrams per day for males and 10.2 milligrams per day for females. The main sources of dietary iron for males and females were bread, bread-based dishes, grains and pasta, as well as breakfast cereals (contributing 8–10 per cent). Meat such as beef and veal, and poultry and vegetables were lower contributors of dietary iron, at 4–7 per cent for both males and females (University of Otago & Ministry of Health 2011). Tables 1.15a and 1.17 describe intakes and food sources in more detail.

Zinc

The zinc requirement is based on estimates of the minimal amount of absorbed zinc required to equal total daily excretion of endogenous zinc (FNB:IOM 2001). The EAR and RDI for men are 12 milligrams and 14 milligrams per day and for women 6.5 milligrams and 8 milligrams per day, respectively. Increased requirements for pregnancy and lactation are estimated for women aged 19–30 years. For pregnancy the EAR and RDI are 9 milligrams and 11 milligrams per day and for lactation 10 milligrams and 12 milligrams per day, respectively. Mean daily intake of zinc by the Australian young adult male population was 9.2 milligrams and for females it was 11.4 milligrams. For this group the main dietary sources of zinc were meat, poultry and dishes containing meat and poultry, and cereal-based products and dishes. Milk products and dishes provided 10 per cent of dietary zinc (ABS 2014).

The New Zealand median intake of zinc for adults aged 19–30 years was 14 milligrams for males and 8.9 milligrams for females. The main sources of dietary zinc for males and females were grains, pasta, bread and bread-based dishes. The contribution of dietary zinc from beef and veal was greater for males (8 per cent) than for females (6 per cent). Tables 1.15a and 1.17 describe intakes and food sources in more detail.

PHYSICAL ACTIVITY

Australia's Physical Activity and Sedentary Behaviour Guidelines for Adults from the Department of Health and the New Zealand Ministry of Health recommendations outline the minimum levels of physical activity required to gain a health benefit and ways to incorporate incidental physical activity into everyday life. Brown and colleagues (2012) reviewed the existing evidence, which finds that physical activity is associated with psychosocial health benefits in otherwise healthy adults. Therefore, to encourage being physically active and limit daily sedentary behaviour, Australia's Department of Health followed up this review with the 2014 release of guidelines for adults aged 18–64 years, irrespective of cultural background, gender or ability as shown in the box below.

Australian data from the 2011–12 Australian Health Survey indicated that 46 per cent of people aged 18–24 years were inactive or insufficiently inactive (ABS 2013a). Sixty per cent of New Zealand males and 50 per cent of females aged 15–24 and 25–34 years were classified as physically active. The 2012 New Zealand Health Survey does not report specific recommendations for young adults. The recommendations are for all adults to do at least 30 minutes of moderate-intensity physical activity (such as brisk walking or equivalent vigorous activity) at least five days a week. Doing more than this amount of daily activity (or at a higher intensity) can give additional health benefits and help people lose weight (Ministry of Health 2012).

Australia's Physical Activity and Sedentary Behaviour Guidelines for Adults (18–64 years)

Physical Activity Guidelines

- Doing any physical activity is better than doing none. If you currently do no physical activity, start by doing some, and gradually build up to the recommended amount.
- Be active on most, preferably all, days every week.
- Accumulate 150 to 300 minutes (2½ to 5 hours) of moderate intensity physical activity or 75 to 150 minutes (1¼ to 2½ hours) of vigorous intensity physical activity, or an equivalent combination of both moderate and vigorous activities, each week.
- Do muscle strengthening activities on at least 2 days each week.

Sedentary Behaviour Guidelines

- Minimise the amount of time spent in prolonged sitting.
- Break up long periods of sitting as often as possible. (Department of Health 2014)

COMMON CONCERNS AND HEALTH CONDITIONS

Mental health

Mental health disorders are more often diagnosed in adolescence and young adulthood compared with other life stages and can affect young adults' psychological development, and their perceptions, emotions, behaviour and social wellbeing. Issues around mental health can lead to sleep disturbance and drug abuse and can increase the difficulty in attaining education, employment and job security, all of which negatively impacts upon dietary intake and subsequently health, and influences mental health in later life.

The Australian Institute of Health and Welfare's report on the health and wellbeing of young Australians indicated that the prevalence of mental disorders was 26 per cent in young people aged 16–24 years (AIHW 2007). The most prevalent conditions reported were depression followed by anxiety-related problems, and they were more common among women than men. The New Zealand Health Survey of 2011–12 reported that among 15–24-year-olds, 7 per cent of males had been diagnosed with a mental disorder compared with 14 per cent of females. The prevalence increased with the next age group, 25–34 years old. Overall, women were more likely to have been diagnosed with a common mental disorder in their lifetime than men (Ministry of Health 2012). Psychological or mental distress was assessed separately and refers to an individual's experience of symptoms such as anxiety, confused emotions, depression or rage, which was higher for young women aged 15–24 years than for any other group (see Chapter 19).

Treatment of anxiety and depression has generally been pharmacological, but recent findings have linked the importance of diet quality and adequate sleep with reducing the symptoms and incidence of depression across all age groups. It is unclear whether the depressive symptoms precede the poor dietary intake and lead to a person making poorer dietary choices or whether poor dietary quality over time predisposes a person to symptoms of anxiety and depression. The emerging picture is that our understanding of the importance of food choices and overall diet quality, which have been extensively researched for other chronic lifestyle diseases, is still in its infancy when it comes to mental health. However, high dietary quality and the consumption of a Mediterranean-style diet in particular has been linked to positive effects on reducing anxiety and depression, increasing feelings of wellbeing and the strengthening the immune system. It is not surprising that enhanced diet quality assists with meeting requirements for most nutrients. Alongside diet quality and interrelated with its effect on mood is physical activity.

Homelessness

Levels of mental health problems have been reported to be higher in young adults who are homeless. Regardless of whether homelessness arises when a young adult is part of a family unit or living on their own, the negative social and health consequences are great and are amplified when mental health problems are added. During periods of homelessness, young adults are at greater risk of suboptimal nutrition, illnesses, sexually transmitted diseases, physical and sexual assault and social isolation. About 31 per cent of young people were homeless at some stage in 2006, although accurate estimates are difficult to obtain (AIHW 2011). The cycle of issues can have long-term health consequences for a significant proportion of the population and have

Homelessness services

Some key services and websites that professionals should have knowledge of and be able to direct young adults to are listed below.

Australia

- *Centrelink* (Department of Human Services) <www.humanservices.gov.au/customer/dhs/centrelink>
- *Supported accommodation* <www.aihw.gov.au/supported-accommodation-assistance-program/>
- *Youth allowance* <www.humanservices.gov.au/customer/enablers/centrelink/youth-allowance/eligibility-for-youth-allowance>
- *Newstart*—Allowance eligible to people from 22 years of age <www.humanservices.gov.au/customer/services/centrelink/newstart-allowance>
- *Community Health Services*—Available to all residents in all states and territories of Australia, offering healthcare services and health promotion strategies, particularly for people on low incomes and Health Care Card holders. Services are for all residents of the local area; fees are charged for services according to the client's ability to pay
- *Dietitians Association of Australia* <http://daa.asn.au>
- *Headspace* <www.headspace.org.au>
- *Lifeline* <www.lifeline.org.au>

New Zealand

- *Depression.org.nz* <www.depression.org.nz/>
- *Family Services Directory*—Lists organisations, services and providers in all areas of New Zealand that can help to provide support to young families <www.familyservices.govt.nz/directory/>
- *Feeding Our Families* <www.feedingourfamilies.org.nz/>
- *Health Promotion Agency*—Delivers cost-effective programs to promote health, wellbeing and healthy lifestyles aimed at disease, illness and injury prevention <www.hpa.org.nz/>
- *TheLowDown.co.nz* <www.thelowdown.co.nz/>
- *Youth Service*—Provides information regarding the best options for education, training or work-based learning to build skills and attain employment for youth aged seventeen years and over <www.youthservice.govt.nz/>
- *Youthline*—A collaboration of youth development organisations across New Zealand that work with young people and their families to enable access to help and support services <www.youthline.co.nz>

been identified as areas of high priority to be addressed in both Australia and New Zealand.

Alcohol and drug consumption

Substance use and in some cases abuse can be a problem in young adults. Alcohol consumption is high in young adults in both Australia and New Zealand, with **binge drinking** being a significant concern. People in their twenties are one and a half times more likely to binge drink than teenagers. The National Drugs Strategy Household Survey is commissioned by the Department of Health and conducted every three years in Australia (AIHW 2013). Approximately 37 per cent of people aged fourteen years and older drank alcohol weekly and 6.5 per cent daily. The frequency appears to have decreased since the 2010 survey. The proportion of males over fourteen years drinking daily was 8.5 per cent, which was higher than for females at 4.6 per cent. 'Risky' alcohol consumption (more than two standard drinks per day) in males aged 18–24 years was 27 per cent compared with females at 14.6 per cent. For 25–29-year-old males the proportion was higher at 31.9 per cent) than the 18–24-year-old males. Approximately 10 per cent of females aged 25–29 years engaged in risky-type drinking. However, there was a larger proportion of 18–24-year-old males and females (56 per cent and 67.5 per cent) who engaged in low-risk alcohol consumption (no more than two standard drinks per day). This was consistent for the 24–29-year-old males and females (54.5 per cent and 70 per cent) (AIHW 2013).

Illicit drugs

The use of illicit drugs such as amphetamines and inappropriate use of legal drugs such as painkillers can have harmful effects. Drug-related health problems vary greatly, depending on the type of drug, amount used and the duration of use. Common problems range from psychological and behavioural effects such as delusions, hallucinations and aggressive or erratic behaviour to physiological effects like high blood pressure, respiratory problems and kidney, liver and brain damage (AIHW 2007). The 2013 National Drugs Strategy Household Survey (AIHW 2013) found that 42 per cent of people in Australia aged fourteen years and over had used illicit drugs at least once, with males more likely than females to use illicit drugs (18 per cent compared with 12 per cent). In the 20–29 year age group, approximately 32 per cent of males and 23 per cent of females had used illicit drugs within the preceding twelve months. The most commonly reported drug used by people aged fourteen years and over was marijuana (cannabis) (35 per cent), followed by ecstasy (11 per cent) and methamphetamines (10 per cent).

> **Binge drinking** is drinking too much alcohol on a single occasion. Usually, it means drinking more than the recommended amount of alcohol, often with the intention of 'getting drunk'. It has also been described as a mean level of drinking that puts one at risk or high risk for short-term harm (injury or death).

Overweight and obesity

Young adults are not immune to the obesity epidemic that is sweeping Australia, New Zealand and most of the developed world. Between 1995 and 2012, there were increases in the rate of overweight and obesity in a number of male and female age groups. In the young adult age group this included females aged 18–24 years, with a rise of 9 per cent (ABS 2013b).

According to the Australian Bureau of Statistics, in 2011–12 young adults had the lowest rates of overweight and obesity compared to any other stage of adulthood, with 42 per cent of males and 35 per cent of females falling into the category (ABS 2013b). This is not surprising, as the older people are, the more time they have had to develop overweight and obesity, especially if weight is slowly increasing annually. However, even though these rates are lower than for older adults, they are not low and suggest that more than a third of young adults are overweight or obese. Statistics from New Zealand appear to be in line with available Australian data. For adults aged 19–30 years, approximately 53 per cent of males and 47 per cent of females fit into the overweight and obese category (University of Otago & Ministry of Health 2011).

The high rates of overweight and obesity are of concern for this age group. The higher prevalence in New Zealand compared with Australia may be attributable to the age group being larger in the New Zealand data or to physical differences between the ethnic groups represented within the populations.

The literature strongly suggests that being overweight during adolescence has important social and economic consequences in later life and that being overweight as a young adult increases the risk of this continuing into later life (Gortmaker et al. 1993). This increases the risk of metabolic health complications, such as development of T2DM and cardiovascular disease (CVD).

SUMMARY AND KEY MESSAGES

Young adulthood is a life stage in which lifestyle and social factors are dominant and can influence health, depending on choices made by individuals. It is a time when foundations for ageing well are being laid; thus, it is important to provide this age group with accurate public health messages and address nutrient requirements for both short-term and long-term health. Some of the nutrition-related challenges described in this chapter are also relevant to individuals over the age of 25 years. With the thirties now being referred to as the new twenties, individuals are spending more time in studies, living in the family home for longer and waiting longer to start their family.

- Young adults face numerous challenges, many of which may impact on their nutritional intake and status, including further study, the commencement of professional employment and the formation of new relationships.
- Young adults are at risk of numerous nutrition-related health problems, including obesity and overweight, alcohol and drug consumption and fad dieting.
- Current dietary practices of many young adults are suboptimal, particularly with regard to poor intakes of fibre and calcium for both genders.
- Young females need to improve iron consumption to avoid the development of iron deficiency anaemia.
- Binge drinking is a significant problem in the young adult age group and needs to be addressed.

REFERENCES

ABS (Australian Bureau of Statistics) 1995. *National Nutrition Survey: Selected highlights, Australia, 1995.* Cat. no. 4802.0. Commonwealth of Australia. <www.abs.gov.au/ausstats/abs@.nsf/mf/4802.0>, accessed August 2013.

——2013a. *Australian Health Survey: Physical activity*, 2011–12. Cat. no. 4364.0.55.004. Commonwealth of Australia. <www.abs.gov.au/AUSSTATS/abs@.nsf/DetailsPage/4364.0.55.004 2011-12?OpenDocument>, accessed 30 November 2014.

——2013b. Overweight/obesity. *Gender Indicators, Australia, Jan 2013.* Cat. no. 4125.0. Commonwealth of Australia. <www.abs.gov.au/ausstats/abs@.nsf/Lookup/4125.0main+features3330 Jan%202013>, accessed 27 October 2013.

——2014. *Australian Health Survey: Nutrition First Results—Food and Nutrients, 2011–12.* Cat. no. 4364.0.55.007. Commonwealth of Australia. <www.abs.gov.au/AUSSTATS/abs@.nsf/DetailsPage/4364.0.55.0072011-12?OpenDocument>, accessed 14 November 2014.

AIHW (Australian Institute of Health and Welfare) 2007. *Young Australians: Their health and wellbeing 2007.* Cat. no. PHE 87. Commonwealth of Australia. <www.aihw.gov.au/WorkArea/DownloadAsset.aspx?id=6442459812>, accessed 3 October 2014.

——2011. *Mandatory Folic Acid and Iodine Fortification in Australia and New Zealand: Baseline report for monitoring.* Cat. no. PHE 139. Commonwealth of Australia. <www.aihw.gov.au/WorkArea/DownloadAsset.aspx?id=10737418918&libID=10737418917>, accessed 3 October 2014.

——2012. *Australia's Food and Nutrition 2012.* Cat. no. PHE163. Commonwealth of Australia. <www.aihw.gov.au/WorkArea/DownloadAsset.aspx?id=10737422837>, accessed 14 November 2014.

——2013. *National Drugs Strategy Household Surveys (NDSHS).* Highlights from the 2013 survey. Commonwealth of Australia. <www.aihw.gov.au/alcohol-and-other-drugs/ndshs/>, accessed 14 November 2014.

Brown, W.J., Bauman, A.E., Bull, F.C. & Burton, N.W. 2012. *Development of Evidence-based Physical Activity Recommendations for Adults (18–64 years)*. Commonwealth of Australia. <www.health.gov.au/internet/main/publishing.nsf/Content/health-pubhlth-strateg-phys-act-guidelines/$File/DEB-PAR-Adults-18-64years.pdf>, accessed 14 November 2014.

DAA (Dietitians Association of Australia) 2013. *Study: Young women resolve (again) to eat better in the New Year.* Media release. 4 January. Expertguide.com au. <www.expertguide.com.au/news/article.aspx?ID=1497>, accessed 3 October 2014.

Department of Health 2014. *Australia's Physical Activity and Sedentary Behaviour Guidelines.* Commonwealth of Australia. <www.health.gov.au/internet/main/publishing.nsf/content/health-pubhlth-strateg-phys-act-guidelines>, accessed 3 October 2014.

FNB:IOM (Food and Nutrition Board: Institute of Medicine) 2001. *Dietary Reference Intakes for Vitamin A, Vitamin K, Arsenic, Boron, Chromium, Copper, Iodine, Iron, Manganese, Molybdenum, Nickel, Silicon, Vanadium and Zinc.* National Academy Press, Washington.

Gortmaker, S.L., Must, A., Perrin, J.M., Sobol, A.M. et al. 1993. Social and economic consequences of overweight in adolescence and young adulthood. *New England Journal of Medicine* 329: 1008–12.

Ministry of Health 1999. *NZ Food: NZ People; Key results of the 1997 National Nutrition Survey.* New Zealand Government. <www.moh.govt.nz/notebook/nbbooks.nsf/0/62c5d9d4c418c4e74c2567d9007186c2/$FILE/nns.pdf>, accessed 28 September 2014.

——2003. *Food and Nutrition Guidelines for Healthy Adults: A background paper.* New Zealand Government. <www.health.govt.nz/system/files/documents/publications/foodandnutritionguidelines-adults.pdf>, accessed 16 March 2014.

——2012. *The Health of New Zealand Adults 2011/12: Key findings of the New Zealand Health Survey.* New Zealand Government. <www.health.govt.nz/system/files/documents/publications/health-of-new-zealand-adults-2011-12-v2.pdf>, accessed 3 October 2014

NHMRC (National Health and Medical Research Council) & Ministry of Health 2006. *Nutrient Reference Values for Australia and New Zealand: Including recommended dietary intakes.* Commonwealth of Australia & New Zealand Government. <www.nhmrc.gov.au/_files_nhmrc/publications/attachments/n35.pdf>, accessed 26 September 2014.

——2013. *Eat for Health: Australian Dietary Guidelines; Providing the scientific evidence for healthier Australian diets.* Commonwealth of Australia. <www.eatforhealth.gov.au/sites/default/files/files/the_guidelines/n55_australian_dietary_guidelines.pdf>, accessed 16 March 2014.

University of Otago & Ministry of Health 2011. *A Focus on Nutrition: Key findings of the 2008/09 New Zealand Adult Nutrition Survey.* New Zealand Government. <www.health.govt.nz/system/files/documents/publications/a-focus-on-nutrition-v2.pdf>, accessed 2 October 2014.

WHO (World Health Organization) 2003. *Diet, Nutrition and the Prevention of Chronic Diseases.* WHO Technical Report Series 916. <www.who.int/hpr/NPH/docs/who_fao_expert_report.pdf>, accessed 14 November 2014.

ADDITIONAL READING

Australian Bureau of Statistics 2008. Risk taking by young people. *Australian Social Trends, 2008.* Cat. no. 4102.0. Commonwealth of Australia. <www.abs.gov.au/AUSSTATS/abs@.nsf/Lookup/4102.0Chapter5002008>, accessed 27 October 2013.

Health Promotion Agency (website) 2014. <www.hpa.org.nz/>, accessed 3 October 2014.

TEST YOUR UNDERSTANDING

Note: There is only one correct answer for each question.

1. Which of the following social and lifestyle factors may impact on the nutrition and health of young adults?
   a. Living arrangements
   b. Cooking skills and available facilities
   c. Transport
   d. Personal preference
   e. All of the above
2. What is the prevalence of mental health issues in young adults in Australia?
   a. 5 per cent
   b. 10 per cent
   c. 17 per cent
   d. 26 per cent
   e. 38 per cent
3. Which of the following nutrients have been reported to be consumed in amounts below the RDI by young adult females?
   a. Calcium
   b. Iron
   c. Fibre
   d. All of the above
   e. b and c only
4. Which of the following nutrients have been reported to be consumed in amounts below the RDI or AI by young adult males?
   a. Zinc
   b. Fibre
   c. Iron
   d. Protein
   e. a and b
5. What was the total reported saturated fat intake of young adults?
   a. As per the recommendations
   b. Below the recommendations
   c. Above the recommendations
   d. The same as monounsaturated fat
   e. None of the above
6. Approximately what percentage of young adult females in Australia and New Zealand fitted into the overweight and obese category?
   a. 16–27 per cent
   b. 35–47 per cent
   c. 51–8 per cent
   d. 61–5 per cent
   e. None of the above
7. Approximately what percentage of young adult males in Australia and New Zealand fitted into the overweight and obese category?
   a. 10–12 per cent
   b. 22–9 per cent
   c. 31–40 per cent
   d. 42–53 per cent
   e. 54–9 per cent
8. Approximately what percentage of young Australian males aged 20–29 years had recently used illicit drugs (in the 12 months preceding the 2013 National Drugs Strategy Household Survey)?
   a. 7 per cent
   b. 14 per cent
   c. 21 per cent
   d. 32 per cent
   e. 44 per cent
9. What is the recommend physical activity requirement for young adults?
   a. 30 minutes at least four days a week
   b. 30 minutes at least five days a week
   c. 45 minutes at least four days a week
   d. 45 minutes at least five days a week
   e. 60 minutes at least five days a week
10. Which of the following issues affect the young adult age group in Australia and New Zealand?
   a. Homelessness
   b. Binge drinking
   c. Mental illness
   d. Drug use
   e. All of the above

STUDY QUESTIONS

1. Describe the most common nutrition-related challenges young adults face.
2. Describe the current macronutrient consumption trends of young adolescents in Australia and New Zealand.
3. What nutrients are consumed in insufficient amounts by young adult males?
4. What nutrients are consumed in insufficient amounts by young adult females?
5. What is binge drinking, and why is it a problem?
6. Outline how the physical activity recommendations for young adults can be put into practice by young adults.

11

Early and middle adult (26-60 years)

Catherine Itsiopoulos and Audrey Tierney

LEARNING OUTCOMES

Upon completion of this chapter you will be able to:

- describe how physiological changes during adulthood impact upon nutrient requirements and nutritional status
- list the nutrient requirements of adults
- identify the key diet-related risk factors for chronic disease in adults
- recommend diet and lifestyle strategies for prevention of chronic disease in adults.

The adult years encompassing ages 26–60 are considered the most productive years in a person's life. In the early adult years individuals are gaining independence by leaving the parental home, graduating from tertiary or vocational programs or university and starting regular employment and a career. This period has a major impact on nutritional status because it involves planning, shopping and preparing food for oneself and making food choices when eating out. There are many factors that impact on the quality of the diet, such as eating out and partying with friends, where meals may be irregular and not well balanced with respect to nutrient density (for example, high in fat, processed carbohydrate and salt and low in fibre and nutrients). There may also be a sharp increase in consumption of alcohol associated with regular socialising. The majority of growth and development has occurred, and the adult years, in particular the early adult years, are critical for the establishment of healthy diet and lifestyle patterns which will in turn impact the prevention of chronic disease.

PHYSIOLOGICAL CHANGES

For the majority of individuals, the growth process is completed by the mid-twenties. Body functions are fully developed and senses are sharp. For this period of adulthood and until the age of 30, bone density continues to accrue and muscular strength increases, although this is dependent on the type and amount of physical activity in which the individual participates.

The physiological signs characteristic of middle age in adulthood include a decline in lean body mass or muscle mass and an increase in fat mass. This coincides with a decline in metabolic rate and resting energy expenditure in early adulthood. The loss of muscle mass is often accompanied by a decrease in fitness and strength of the body. Distribution of body fat can increase a person's risk of developing metabolic disease, with abdominal fat more metabolically active and excess associated with increased risk of developing insulin resistance, type 2 diabetes mellitus (T2DM) and cardiovascular disease (CVD), as opposed to glutaeal, or buttock, fat which is more benign metabolically and is usually not associated with increased risk of chronic disease. The presence of disease can also lead to a decline in energy expenditure and therefore can further affect energy needs and metabolic rate.

As we age, our skin becomes less supple and more prone to age and sun spots, which increase with more sun exposure. With respect to the reproductive system, the **climacteric change** precedes actual menopause for women and occurs gradually over a ten-year period while the production of oestrogen drops. Fertility declines, leading to an end in reproductive capacity. In women, the monthly cycle shortens and becomes more irregular. Climacteric ends with menopause, which is the end of menstruation. This usually occurs from 45 to 55 years and is also associated with other physical and psychological symptoms such as headaches, sleep loss and cold and hot sweats. In men in late adulthood, fertility diminishes but is retained. Quantity of semen and sperm decrease after 40 years, but sperm production continues throughout life. Testosterone production declines gradually with age. Stress, alcohol and heart and other diseases increase the decline.

NUTRIENT REQUIREMENTS AND INTAKES

The nutrient requirements of people throughout all stages of life, including adults, are detailed in Chapter 1. A summary of selected macronutrient and micronutrient intakes from the 2011–12 Australian Health Survey (ABS 2014) and the 2008–09 New Zealand Adult Nutrition Survey (University of Otago & Ministry of Health 2011) are also detailed in Table 1.15. Food sources of nutrient are described in Tables 1.16 and 1.17.

As described in Chapter 1, energy requirements of individuals are dependent on age, height, muscularity, physical activity, stage of life and state of health. As adults have achieved their full growth potential by their mid-twenties, there are no additional energy requirements for growth, and therefore energy requirements are lower than the adolescent and young adult years and continue to decline throughout the adult years as lean body mass reduces. As for other life stages, energy requirements of adults are predicted using equations such as Schofield's (1985), based on gender, age, weight, height and level of physical activity, to maintain energy balance and hence maintain weight in healthy individuals.

> **Climacteric change** occurs prior to menopause in women and exhibits a range of symptoms from hot flushes to changes in body fat and painful menstruation.

Basal requirements (basal metabolic rate [BMR] calculated using the Schofield (1985) equation (see Chapter 1) are multiplied by a physical activity level (PAL) to determine total energy requirements. PALs vary between 1.2 (bed rest) and 2.2 (very active or heavy occupational work). During the adult years a person's PAL can change dramatically, depending on their occupation and activities outside work. For a person who maintains heavy occupational work (for example, a builder, gardener, landscaper or industrial cleaner), physical activity remains high, and therefore energy requirements remain high and weight gain is less likely. However, for the majority who undertake sedentary employment (for example, office work), physical activity is low, and dietary energy intake should be adjusted to prevent weight gain.

For an average woman (65 kilograms, aged 30–40, office worker), daily energy needs are 8000–8500 kilojoules (for weight maintenance). For an average man (80 kilograms, aged 30–40, office worker) daily energy needs are 10,150–10,500 kilojoules (for weight maintenance). Energy requirements are adjusted based on weight status, and as a general rule 2100 kilojoules per day are subtracted from the estimated total energy requirements for weight loss, and a similar amount is added per day for weight gain.

During the adult years energy intake may often exceed requirements, leading to weight gain, as adults continue to eat as they did during the early adult years, consuming large portions of food and energy-dense foods and not adjusting intake according to a reduction in energy requirements.

The 2011–12 Australian Health Survey reported that the mean daily energy intake for adult males 31–50 years is approximately 10,220 kilojoules, which reduces to 9345 kilojoules when 51–70 years. For females, the survey reported a mean daily energy intake of 7540 kilojoules per day at 31–50 years, decreasing to 7268 kilojoules when 51–70 years. These values for energy intake also include energy from dietary fibre (ABS 2014). In the 2008–09 New Zealand Adult Nutrition Survey, males aged 31–50 years were consuming 11,376 kilojoules reducing to 9158 when 51–70 years. Energy intake for women aged 31–50 years was 7821 kilojoules decreasing to 7071 kilojoules when 51–70 years (University of Otago & Ministry of Health 2011).

The Australian Dietary Guidelines recommend an acceptable macronutrient distribution range (AMDR) for energy from protein, fat and carbohydrate (see Chapter 1; NHMRC 2013a). Across male and female adults (31–70 years) the contribution to energy was approximately 18–19 per cent for protein, 42–45 per cent for carbohydrate and 30–31 per cent for total fat. Both males and females in the 51–70 age group were consuming carbohydrate at lower levels than recommended by the AMDR. While total fat is within the AMDR, for all adults saturated fat is outside the range at 11.5–11.6 per cent for Australian males and females, respectively, and 13.3–13.6 per cent for New Zealand males and females, respectively; intake of saturated and trans fats should take up no more than 10 per cent of energy (ABS 2014; University of Otago & Ministry of Health 2011). Details of food sources are described in Tables 1.16 and 1.17.

HEALTH CONDITIONS

Overweight and obesity

Prevalence

Sixty-three per cent of Australian adults had a body mass index (BMI) in either the overweight or the obese range in 2011–12. Obesity prevalence was similar among males and females (28 per cent). The prevalence of overweight and obesity has increased since the National Nutrition Survey in 1995, rising from 64 per cent to 70 per cent among males and from 49 per cent to 56 per cent among females (ABS 1995, 2012).

Causes

Obesity has transpired as a complex metabolic disease with vast environmental and genetic influences. On the whole, the environment in Australia encourages energy imbalance and is seen as obesogenic (Egger & Swinburn 1997). With abundance of food and declining physical activity, energy consumption is favoured over energy expenditure, and thus weight gain inevitably occurs. Aside from physical activity and excess energy intake, other environmental factors, discussed below, have been implicated in the upsurge of obesity and are likely to interact with a person's genes.

Changes to the food supply have led to a wide availability of cheap processed foods that provide excess kilojoules and have levels of saturated fats, salt and sugar well above the recommended intakes for good health and weight control. The portion size of many packaged foods, restaurant meals and takeaway snacks has increased, with the cost decreasing. Meanwhile, the relative cost of fresh produce has increased.

Cultural and social aspects of eating have changed. It is commonplace that foods high in fat and sugar and increased consumption of alcohol are more available in the workplace, cultural activities and sporting and family events. Town planning and the built environment discourage physical activity and active travel (for example, walking or cycling). Changes to work environments have led to physically active workplaces being replaced with more sedentary spaces. Increased work and life pressures mean longer working hours leave less time for food preparation, family time and physical activity. Disrupted sleep, or too long or too short periods of sleep, can disturb metabolic processes and interfere with systems for appetite control.

National surveys have identified factors contributing to the increasing prevalence of overweight and obesity among adults. Comparison of the results of the 1995 National Nutrition Survey (McLennan & Podger 1998) with those of the 1983 National Dietary Survey of Adults (AIHW 2001) showed a significant increase in energy intake. The 2007–08 National Health Survey showed that 37 per cent of adults exercised sufficiently to obtain benefits to their health (AIHW 2010).

Components of energy balance (energy intake and energy expenditure) interact with each other to affect body weight. The body attempts to maintain energy balance and protect existing body weight through a complex negative feedback system involving hormones such as **ghrelin**, which increases hunger; cholecystokinin, peptide YY and glucagon-like peptide 1, which inhibit food uptake in the short term; and leptin and insulin, which inhibit food intake in the long

Ghrelin is a peptide hormone that is secreted into the circulatory system in response to extended periods of fasting and stimulates appetite. It is known to play a role in cell proliferation, migration and invasion, and programmed cell death in a number of cancer cell lines, though evidence conflicts as to whether it plays a role in promoting or inhibiting cancer growth.

CASE STUDY 11.1 PREVENTION OF TYPE 2 DIABETES MELLITUS

Profile and problem

Mrs R. is a 46-year-old married primary teacher who lives with her husband and two teenage children. She is 164 centimetres tall and weighs 92.8 kilograms. She is experiencing some hot flushes and some period irregularity; otherwise, her general health is quite good, and she has rarely needed to see her general practitioner. She is also worried about her teenage son, who is overweight and has been diagnosed with young adult T2DM. Both of her parents are in their early seventies and are overweight, and her father has T2DM. She has been trying to lose weight since she was 30 years old. She has tried lots of diets, including the fasting diet, the soup diet and several detox diets, but none has helped.

Assessment

Mrs R. is concerned about her weight and the high chance that she may too develop T2DM. She is prepared to make necessary changes to her diet and lifestyle to prevent diabetes but also to be a good role model to her two children. Her current diet shows that she eats only fruit for breakfast, often fasts for most of the day and snacks on cheese and biscuits while making dinner. She eats vegetables and some meat most nights, together with a glass of wine. She snacks on nuts, cheese and chips late into the evening. She wants help to lose weight over the next six months, before she goes on holiday.

Treatment

Mrs R. has agreed to do the following:

- get a full check-up with her general practitioner for blood glucose, lipids and blood pressure readings
- keep a food record diary (either paper based or using a diet and fitness app on her phone)
- aim for weight loss of 5 per cent of her total body weight for her next review consultation in six weeks' time
- join a local gym and participate in group classes three times a week plus a yoga class once a week
- establish a more regular eating pattern, with three small meals and three snacks throughout the day, avoiding long periods of fasting, including two to three low-fat dairy serves and some alcohol-free days
- limit snacking in the evening and keep sweet treats to once a week.

Plan

Mrs R. will be monitored by a dietitian in the community every six weeks for the next six months. She will bring the results of her general practitioner check-up to the next appointment, and anthropometric measurements including waist and hip circumference will be taken. Diet records will be assessed for ongoing individualised and tailored advice to help Mrs R. lose weight.

term. The body will normally try to defend against weight changes, either losses or gains. However, in the case of continuous positive energy balance, weight gain will occur until a new set point or weight is established. The same physiological mechanisms then aim to maintain energy balance at the higher weight and will defend against weight loss by increasing appetite and reducing energy expenditure (Rosenbaum et al. 2008; Sumithran et al. 2011).

Restrained, emotional (that is, comfort eating) and external eating behaviours are related to unconscious eating that leads to weight gain. Particular behaviours can become habits if they are repeated often over time so that they become almost an automatic response to certain stimulants or situations. Psychological factors such as stress and underlying personal issues can lead to a lack of energy and motivation and increased food consumption (for example, emotional eating), which may indirectly contribute to weight gain. There is a strong association between mood disorders and obesity; people with obesity are more likely to become depressed over time, and people with depression are more likely to become obese. Obesity may increase risk factors for depression, such as body dissatisfaction and low self-esteem. In turn, depression and poor body image can affect people's ability and willingness to eat healthily and exercise regularly. Disturbed eating patterns and eating disorders are also associated with increased risk of both obesity and depression (Luppino et al. 2010).

Physical and developmental factors such as impaired mobility (for example, due to physical disability, ageing or obesity) can affect an individual's capacity to adopt a healthy lifestyle and undertake physical activity.

Prevention and treatment

Health professionals in primary health care are the first line of intervention for weight management and an important and trusted source for information surrounding weight loss and health benefits. Consistent messages across clinical and public health disciplines are fundamental to addressing overweight and obesity in Australia. A range of tools are available for healthcare professionals' use to provide individualised advice for people to address overweight and obesity, along with other lifestyle risk factors for chronic disease.

Weight management programs may be effective if they are tailored to the local and cultural context and involve the expertise of other primary healthcare professionals and specialists as indicated for that individual. When counselling or intervening with clients who are obese, the Five As approach is a useful tool:

Ask and *assess*—Discover current lifestyle behaviours and BMI, comorbidities and other factors related to health risks.

Advise—Promote the benefits of a healthy lifestyle and explain the benefits of weight management.

Assist—Develop a weight management program that includes lifestyle interventions tailored to the individual (e.g. based on severity of obesity, risk factors, comorbidities) and plan for review and monitoring.

Arrange—Make regular follow-up visits and referrals as required (e.g. to a dietitian, exercise physiologist) (NHMRC 2013b).

A US national database of self-reported long-term weight management identified the following weight management strategies as being successful (Wing & Phelan 2005):

- maintaining high levels of physical activity and limiting sedentary activities (e.g. television viewing)
- eating a diet low in kilojoules
- regularly eating breakfast
- maintaining a consistent eating pattern throughout the week and year
- identifying triggers of emotional eating and developing alternative strategies for regulating mood
- frequently monitoring weight
- catching lapses before they become large-scale weight gains.

The greatest health risks for individuals who are overweight or obese is CVD with its associated risk factors (elevated blood pressure and lipids), T2DM and some cancers. When comorbidities are present, the need for weight management is heightened.

Metabolic syndrome

Metabolic syndrome (MetS) is characterised by hyperinsulinaemia, impaired glucose tolerance, dyslipidaemia, hypertension and obesity. This cluster of components has been recognised for many years and was formally labelled in 1988 by endocrinologist Gerald Reaven, who suggested that insulin resistance was the key underlying factor (Reaven 1988). The cause of MetS is largely unknown, but factors such as ageing, obesity, sedentary lifestyle and genetics are implicated. The syndrome predicts CVD and T2DM. Waist circumference has been used as the optimum variable to predict those who are likely to develop MetS and has been shown to be a better indicator than BMI.

Non-alcoholic fatty liver disease (NAFLD) and non-alcoholic steatohepatitis are also considered manifestations of MetS. Reproductive disorders (such as polycystic ovary syndrome—PCOS) in woman and erectile dysfunction or decreased total testosterone (low testosterone-binding globulin) in men can be attributed to MetS.

There are several terms in existence to describe this clustering of components, including syndrome X, deadly quartet, insulin resistance syndrome and cardiometabolic syndrome, as well as numerous ways of defining it, with seven definitions proposed between 1998 and 2005 (see box below). Differences between the

Non-alcoholic fatty liver disease is a significant cause of liver disease. It occurs where there are fat deposits in the liver (steatosis) not due to excessive alcohol use. The pathogenesis of NAFLD is not completely understood, but insulin resistance is strongly implicated. Individuals with MetS are twice as likely to die from and have a three-fold higher risk of developing heart attack or stroke compared with people without the syndrome, and a five-fold higher risk of developing T2DM (Isomaa et al. 2001).

Timeline of the most popular definitions of the metabolic syndrome

Year	Organization
1998	WHO (World Health Organization)
1999	EGIR (European Group for the Study of Insulin Resistance)
2001	NCEP ATP III (National Cholesterol Education Program/Adult Treatment Panel)
2002	ACE (American College of Endocrinology)
2004	AHA/NHLBI (National Heart, Lung, and Blood Institute/American Heart Association)
2005	IDF (International Diabetes Federation)
2005	AHA/NHLBI (National Heart, Lung, and Blood Institute/American Heart Association)

definitions include the essential criteria, the emphasis on adiposity and the priority of assignment. In 2009, several major organisations met in an attempt to unify criteria. It was agreed that 'there should not be an obligatory component, but that waist measurement would continue to be a useful preliminary screening tool' (Alberti et al. 2009). As per the NCEP ATP III criteria, three abnormal findings out of five would qualify a person for the metabolic syndrome (Alberti et al. 2009). The various definitions share the same goal, however: to identify high-risk individuals who would most benefit from early and more intensive disease prevention.

Prevalence

The Australian Diabetes, Obesity and Lifestyle study in 2000 found that 19 per cent of Australians aged 25 years and over met the criteria for a diagnosis of MetS (Zimmet et al. 2005). The prevalence of MetS increases with age, rising steeply after the third decade and reaching a peak in men aged 50–70 years and in women aged 60–80 years. Although the high prevalence of T2DM and glucose intolerance has been previously attributed to ageing itself, data suggest that the age-related decline in insulin sensitivity is associated with abdominal obesity and inactivity, such that older persons who are physically active and do not have increased abdominal girth are much less likely to develop T2DM (Cefalu et al. 1995). The prevalence of MetS increases with obesity as assessed by BMI or waist circumference.

Causes

Insulin resistance is recognised as a characteristic trait of T2DM and contributes to abnormalities in liver, fat, muscle tissue and pancreatic beta-cells. Insulin resistance is associated with ageing and a sedentary lifestyle. Higher rates of insulin resistance were found in T2DM and hypertriglyceridaemia and in the low high-density lipoprotein cholesterol states, with the latter two almost never occurring as isolated disorders and nearly always associated with insulin resistance. This suggests that insulin resistance is almost ever-present when several metabolic disorders cluster together, and more so than when metabolic disorders are isolated (Bonora et al. 1998).

A relationship between obesity and insulin resistance is seen across all ethnic groups and body weights. Central obesity has become a more defining factor in the development of MetS than peripheral body fat distribution. Adipose tissue is not purely a storage tissue but also acts as an endocrine organ with adipokines (hormones and inflammatory cytokines derived from adipose tissue), playing an essential role in overall insulin sensitivity and insulin resistance (Ruan & Lodish 2004). The importance of being overweight in relation to risk of developing MetS and T2DM is supported by intervention studies demonstrating an improvement in all abnormalities clustering in MetS in individuals who reduce their body weight.

Adverse lifestyle factors independently associated with greater risk of having MetS include lack of physical activity, high carbohydrate intake and smoking. Insulin action has been shown to be greater in physically active individuals when compared to sedentary individuals. Physical activity could exert protective effects on MetS through improving the individual components. Zhu et al. (2004) also observed a lower risk of having MetS in light to moderate drinkers. With heavy drinkers, men had an increased risk of having MetS consistent with a J-shaped effect of alcohol on CVD risk (whereby a low-to-moderate intake seems to be associated with a reduction, whereas higher consumption is associated with increased risk). Socio-economic deprivation is also associated with clinical features of MetS.

Many facets of the composition of the diet (carbohydrate, fat, fibre, vitamins, minerals, alcohol) have been considered in the modulation of insulin resistance. However, attention has been focused recently on the quality and quantity of dietary fat and its influence on insulin sensitivity. Both impaired glucose tolerance and T2DM are increased in persons with a high amount of fat, especially saturated fat. Public health recommendations aim to reduce the intake of dietary fat in the adult population, with particular emphasis on the reduction of saturated fat. However, the alternative low-fat diet is not without its negative attributes. Results of the National Health and Nutrition Examination Survey in the United States indicated that the replacement of dietary fat with dietary carbohydrate failed to reverse the trend of increasing obesity rates, as carbohydrate increases blood glucose concentrations, stimulating the release of insulin, which in turn promotes the growth of fat tissue and thus can cause weight gain (German & Dillard 2004).

Considering dietary factors in more detail, it is important to highlight the beneficial role of the Mediterranean dietary pattern, characterised by the daily consumption of olive oil, fruits, vegetables, tree nuts, legumes and whole grains and the weekly consumption of fish and poultry; a relatively low consumption of red meat; and a moderate consumption of alcohol, normally with meals. In a systematic review in 2013, evidence from prospective cohort studies, cross-sectional studies and clinical trials supported the beneficial role of adherence to the Mediterranean dietary pattern regarding MetS presence and progression (Esposito et al. 2013).

The optimal diet to ameliorate insulin sensitivity and prevent the development of MetS is under debate and scrutiny. The diet for prevention and treatment should constitute low intakes of saturated fat and cholesterol (for unfavourable effects on insulin sensitivity, blood pressure and plasma lipids), while monounsaturated fat in moderate quantities should be permitted. Total fat does not need to be drastically reduced, as within certain limits (35–40 per cent) it is the quality and not the overall quantity that has untoward effects. Low-fat diets as promoted in the past led to high carbohydrate intakes, which are not without blame in relation to exacerbation of some metabolic features. The optimal range for the percentage of dietary energy gained from macronutrients, and the recommendations for cholesterol and dietary fibre, in order to achieve weight loss and maintenance in insulin-resistant states is shown below:

Carbohydrate	45–60 per cent
Sugars	less than 10 per cent
Protein	15–25 per cent
Total fat	25–30 per cent
Saturated fat	less than 8 per cent
Monounsaturated fat	10–20 per cent
Polyunsaturated fat	5 per cent
Cholesterol	less than 200 milligrams per day
Dietary fibre	30–40 grams per day (half should be soluble fibre). (McAuley & Mann 2006)

The preferred sources of carbohydrate include vegetables, legumes, intact fruit and wholegrain cereals that are high in dietary fibre. Trans fatty acids should be avoided.

Trends in diet and physical activity coupled with genetic susceptibility must account for the recent and dramatic rise in the incidence of MetS and T2DM, demonstrating the important impact of gene–environment interactions. Evidence for a genetic basis for MetS and T2DM has been derived from studies of families, twins and populations with genetic admixture. High heritability estimates have been reported for fasting glucose, insulin, triglycerides and high-density lipoprotein cholesterol concentrations. The heritability estimates for hyperinsulinaemia, hypertension and hypertriglyceridaemia are low, indicating a more important environmental influence on these components of MetS.

Prevention and treatment

Prevention and treatment should focus on increasing physical activity and reducing weight. Associated conditions such as PCOS, sleep apnoea and NAFLD need to be identified and managed and the use of medications associated with weight gain modified where possible. There is no specific pharmacotherapy. Treatment should focus on management of individual risk factors, such as lipids, blood pressure and glucose.

Diagnosis of MetS is useful in focusing attention on central adiposity and insulin resistance as risk factors both for the syndrome and for cardiovascular and diabetes morbidity and mortality. Its assessment requires measurement of waist circumference—a simple but seldom-performed procedure in general practice. The most essential components for the prevention and management of MetS are measures to change diet and physical activity in order to achieve and sustain weight loss.

Diabetes

Diabetes mellitus is a disorder of the endocrine system. The disease is caused by inadequate production of insulin or insulin resistance, or both. It results in high blood sugar levels.

Prevalence

The 2005 follow-up study to the Australian Diabetes, Obesity and Lifestyle study showed that 1.7 million Australians had diabetes but that up to half of the cases of T2DM remained undiagnosed (Barr et al. 2006), making it one of the most challenging nd concerning public health issues. It is the fastest rowing chronic condition in Australia. This results in ıbstantial morbidity and mortality, particularly from cardiovascular complications, eye and kidney diseases and limb amputations.

There are two main types of diabetes. Type 1 diabetes mellitus (T1DM) is discussed in detail in Chapter 8 and type 2 diabetes mellitus (T2DM) is discussed below. Genetic risk factors for T1DM (including a family history of diabetes) are also important for T2DM but the major risk factors are being overweight or obese, undertaking little physical activity and poor dietary intake. Aboriginals and Torres Strait Islanders carry a higher risk of developing diabetes than non-Indigenous Australians. Typically, T2DM is diagnosed in middle to late adulthood, although it is increasingly diagnosed in younger adults and adolescents. Diet, exercise, medication or combinations of these are the principal management methods for achieving optimal blood glucose levels in T2DM. T2DM remains undetected in many people until complications develop and are diagnosed. The prevalence of T2DM rises with age and is higher in men than in women.

There is a strong genetic predisposition to T2DM. A 2006 study found a strong association between T2DM and particular single nucleotide polymorphisms of the transcription factor 7-like 2 gene in Icelandic, Danish and US white cohorts (Grant et al. 2006). This finding has been widely reproduced in several populations and also been shown to predict the conversion from impaired glucose tolerance to overt diabetes (Florez et al. 2006).

Impaired fasting glucose (IFG) and impaired glucose tolerance (IGT) are conditions where the blood sugar levels are higher than normal. They are often referred to as the prediabetic state. Individuals with prediabetes are at an increased risk of developing T2DM and have increased risk of CVD (Magliano et al. 2008). Lifestyle changes such as increasing exercise and modifications to diet are recommended for people who are diagnosed with IFG and IGT.

Gestational diabetes mellitus (GDM) is a form of diabetes that occurs early in the first trimester of pregnancy and may be associated with birth defects or spontaneous abortion. Being overweight or obese increases the risk of developing GDM. A family history of diabetes or previous history of prediabetes or having GDM with a prior pregnancy and being over the age of 25 years also increases the risk.

Diabetes may cause multiple complications, including:

- increased likelihood of CVD
- damage to the eyes (diabetic retinopathy)
- damage to the kidneys
- nerve damage to the feet and poor circulation to the legs and feet
- increased likelihood of infection.

Healthy diet patterns, weight loss for obese and overweight individuals and increasing physical activities are encouraged to prevent and treat diabetes prior to the implementation of medication and insulin, which may be required to stabilise blood sugar levels and prevent microvascular complications. These modifiable risk factors are largely preventable and are targeted in many public health campaigns and diet and lifestyle interventions.

Prevention

The main primary preventative strategies for preventing T2DM are adopting a healthy lifestyle, including a well-balanced healthy diet and moderate physical activity levels.

Complications of diabetes, including heart attack, stroke and vision impairment, can be reduced with monitoring and strict control of blood glucose levels, blood pressure and lipids, as shown in the United Kingdom Prospective Diabetes Study (Stratton et al. 2000).

Cardiovascular disease

CVD is a group of disorders of the heart and blood vessels that includes the following:

- coronary heart disease (CHD)
- cerebrovascular disease
- peripheral arterial disease
- rheumatic heart disease
- congenital heart disease
- deep vein thrombosis.

CVD is a major cause of disease burden and death in Australia and New Zealand (AIHW 2014; Ministry of Health 2014). Furthermore, with the increasing prevalence of T2DM (secondary to increasing overweight and obesity), premature CVD as a complication of T2DM is rising. The costs of CVD are considerable. Of $121 billion spent in 2009–10 on health in Australia, 96 per cent was recurrent spending, with the greatest share being spent on hospital services, overwhelmingly public hospital services. Spending on CVD was $8 billion, higher than for any other disease.

Heart disease

Prevalence

In Australia, CHD, also known as ischaemic heart disease, is the most common form of CVD, with two major clinical presentations: acute **myocardial infarction** and angina. CHD is the major cause of death for New Zealanders. Despite the advances in cardiovascular care and treatments, in 2009 CHD accounted for 16 per cent of all deaths in Australia (which was 49 per cent of cardiovascular deaths) and 17.5 per cent of all deaths in New Zealand (AIHW 2014; Ministry of Health 2014).

Causes

Heart disease is a progressive narrowing of the blood vessels that supply the heart muscle (the coronary arteries) with blood and oxygen, due to the build-up of fatty deposits. This process is called atherosclerosis and can begin as early as childhood and progresses through to adult years until the narrowing becomes so severe that a clot forms which results in a heart attack.

Atherosclerosis can also occur in the small vessels in the brain, leading to stroke, or other vessels supplying vital organs such as the kidneys. The process of atherosclerosis was once thought to be caused by some injury to the inner wall of the blood vessels (such as high blood pressure causing an initial defect) followed by accumulation of fats. Now, however, it is thought that atherosclerosis begins by a process of inflammation.

The initial damage to the vessel wall that triggers inflammation can be caused by many factors, such as smoking, excessive alcohol, poor diet and stress through a process of free radical damage. During inflammation white blood cells attack the inner walls of the blood vessels, which forms a plaque. This plaque begins to fill up with fats and other particles such as collagen and elastin. The fats include low-density lipoprotein ('bad' cholesterol) that has been oxidised (chemically modified by the action of free radicals). The process continues until the plaques are large, blocking the flow of blood to the heart muscle. The plaques eventually rupture, which causes a blood clot, and this totally blocks the artery, causing a heart attack.

> **Myocardial infarction** is commonly known as a heart attack. It is caused by a disruption in the blood supply, and thus oxygen, to the heart resulting in damage to the heart. An infarction is usually caused by a blockage of the coronary (heart) arteries.

The well-known risk factors for heart disease include family history, high blood pressure, smoking, high cholesterol levels, diabetes, and abdominal obesity. The majority of these risk factors, with the exception of family history, are considered to be modifiable; that is, we have some level of control over them and can take positive steps to improve them.

It is well known that a Western diet consisting of highly processed foods that are high in starch, sugar, salt and animal (saturated) fat is linked to heart disease, as it impacts on many of the risk factors. For example, a diet high in salt is linked to high blood pressure, particularly in susceptible people. In a typical Western diet the major sources of salt include pre-prepared or takeaway meals, commercial breads and some cereals, canned foods, sauces and pickles, snack foods (for example, crisps) and added salt.

A diet high in animal (saturated) fats is linked to high blood cholesterol, specifically the atherogenic low-density lipoprotein cholesterol. In a typical Western diet the major sources of saturated fats include fatty meats, processed meats (for example, salami and other luncheon meats), deep-fried takeaway foods, cakes, pastries, spreads and chocolates.

A diet high in sugars and highly processed carbohydrates and which is low in fibre is linked to obesity (especially in children). In a typical Western diet the major sources of sugar include soft drinks, juices, energy drinks, cakes, pastries, lollies and chocolates. A diet that is highly processed, with a short supply of fresh fruits and vegetables, will be low in natural antioxidants that are known to be critically important in preventing the early steps in the process of atherosclerosis.

Prevention

There are many factors that are linked to atherosclerosis that can be prevented by diet and lifestyle. An analysis of studies from over 50 countries around the world investigating factors that are linked to heart disease showed that over 90 per cent of all heart disease can be prevented with a healthy diet (particularly fruits and vegetables) and regular physical activity (Yusuf et al. 2004).

A systematic review and meta-analysis of 72 published observational studies and **randomised controlled trials** of 659,258 people from eighteen countries evaluated associations between dietary fatty acid intake, fatty acid biomarkers (for example, plasma fatty acids) and fatty acid supplementation trials and

Randomised controlled trials are the 'gold standard' in clinical trials. Subjects are randomly allocated to the intervention or placebo or other control arm of the study.

coronary disease risk (Chowdhury et al. 2014). The results did not support the well-known association between saturated fats and heart disease risk, which caused significant worldwide debate on the well-known diet–heart disease hypothesis.

The results clearly demonstrated that trans fats derived from highly processed foods were positively associated with increase coronary risk; however, saturated fats were not. Furthermore, monounsaturated and polyunsaturated fats were not found to be protective for coronary risk. It is important to note, however, that the review focused on fatty acids intakes independent of sources of fatty acids and dietary patterns (Chowdhury et al. 2014). For example, in a Mediterranean-style diet monounsaturated fats are

CASE STUDY 11.2 DISADVANTAGE, FOOD INSECURITY AND HYPERTENSION

Sharon Croxford

Profile and problem

Mr D., a 55-year-old man, presents at a service for people experiencing homelessness, as he is feeling physically unwell. Upon interview Mr D. discloses that he has been living on the streets for the past nine months. He had one short spell in a rooming house but found that he could not cope with the other residents and the state of the house, so he left. In the year before becoming homeless, Mr D. had lost his job as an IT specialist, separated from his partner and lost his house. Upon finding himself without a home, he drifted into the central business district, sleeping in parks and doorways. Mr D. tends to keep to himself and has not made many connections since losing his home. His mental health, along with his physical health, has deteriorated recently, and this has prompted him to access the current service. Mr D. has been diagnosed with hypertension.

Assessment

Mr D. is a thin, tall man who has lost 10 kilograms in the past nine months. Despite this weight loss his blood pressure is 160/90. He is unfamiliar with the services available that could have supported him and has lived on tinned baked beans alone for six months. He is working with an agency to find accommodation; however, based on his previous experience he is wary of what is available. It is likely that housing may take up to six months to secure. In the meantime, Mr D. will not have access to any storage or cooking facilities, although he has expressed an interest in cooking.

Treatment

Mr D. has agreed to do the following:

- trial the Café Meals Program, through which he will pay $2 for a meal up to the value of $12 at one of three cafes within the central business district
- visit one of two free breakfast options within the central business district
- attend and participate in the men's barbecue held at a garden with barbecue facilities within the central business district each week.

Mr D. does not feel he will be able to take any medication for his hypertension until he has worked on securing housing and accessing more and better food.

After a four-week trial of the Café Meals Program, Mr D. has purchased an average of five meals per week. He has been able to access breakfast at only one location within the central business district, as he felt intimidated at the other venue. He has attended the men's barbecue and has participated in preparing, cooking and eating a meal. His choices at all meals were healthy. It appears that Mr D. is to be fast-tracked into an over-55s accommodation facility, where he will live in a self-contained one-bedroom flat outside the central business district.

Plan

Mr D. has been referred to a local health service for follow-up. He will be linked in with the dietitian where they will monitor his transition back into housing and managing his own shopping and meal preparation. Mr D. will also see the general practitioner for management of his hypertension.

usually derived from olive oil and nuts, whereas in a Westernised dietary pattern monounsaturated fats may be derived primarily from fatty meats and processed foods (for example, chicken nuggets). In the case of saturated fats, a traditional Polynesian diet may be rich in saturated fats from coconut (the oil, milk and flesh), whereas in a Western dietary pattern saturated fats are usually derived from fatty meats and highly processed foods (for example, fried chicken, meat pies, pastries, fish and chips and other fast foods). These dietary patterns are vastly different and will have a different impact on coronary risk.

The supplementation trials, such as those for long-chain omega-3 fats (eicosapentaenoic acid [EPA] and docosahexaenoic acid [DHA]), reviewed in the systematic review and meta-analysis did not demonstrate overall protective effects for coronary risk. However, increased plasma levels of long-chain omega-3 fatty acids were shown to be protective (Chowdhury et al. 2014). This shows that long-chain omega-3 fats are important in protection from CHD but most likely when derived from dietary sources (such as fatty fish), not supplements.

It is clear that saturated fats (primarily from anima products) increase total cholesterol by increasing low-density lipoprotein cholesterol, which leads to atherosclerosis and heart disease. Replacing saturated fats with polyunsaturated fats (for example, seed and vegetable oils, nuts and fish) or monounsaturated fats (for example, olive oil, canola, peanut or macadamia oil, nuts and avocado) reduces low-density lipoprotein cholesterol and increases high-density lipoprotein cholesterol and therefore protects from heart disease.

Trans fats are mainly found in high-fat processed foods and some types of spreads (for example, table margarine) that are linked to increased low-density lipoprotein cholesterol and reduced high-density lipoprotein cholesterol. Dietary cholesterol (found in eggs, shellfish and offal) is not as harmful as originally thought, and in most people a regular intake of these foods does not raise blood cholesterol. By far, animal (saturated) fats have a much stronger effect in raising blood cholesterol. Furthermore, eggs are a good source of protein, omega-3 fats and iron, so there is no need to specifically avoid them. Offal is a rich source of

Secondary prevention of coronary heart disease

The National Heart Foundation in collaboration with the Cardiac Society of Australia and New Zealand has developed the following evidence-based practice guidelines for the management of patients with CHD. Management of the lifestyle and behavioural risk factors are noted here. Discussion of pharmacotherapy is beyond the scope of this chapter.

Smoking

Patients with CHD completely stop smoking and avoid second-hand smoke.

Nutrition

Patients with CHD establish and maintain healthy eating. This includes:

- limiting saturated fatty acid intake to less than 7 per cent and trans fatty acid intake to less than 1 per cent of total energy intake
- consuming 1 gram of EPA plus DHA and more than 2 grams of alpha-linolenic acid daily
- limiting salt intake to 4 grams or less per day (1550 milligrams sodium).

Encourage patients with CHD to adopt a healthy eating pattern that includes:

- mainly plant-based foods (e.g. fruits, vegetables, pulses and a wide selection of wholegrain foods)
- moderate amounts of reduced-, low- or no-fat dairy products
- moderate amounts of lean unprocessed meats, poultry and fish
- moderate amounts of polyunsaturated and mono-unsaturated fats (e.g. olive oil, canola oil, reduced-salt margarines).

Alcohol

Patients with CHD consume a low-risk amount of alcohol. The National Health and Medical Research Council's recommendations for safe alcohol consumption advise drinking no more than one to two standard alcoholic drinks per day, for men and women (NHMRC 2009).

Physical activity

Patients with CHD do at least 30 minutes of moderate-intensity physical activity on most, if not all, days of the week (i.e., 150 minutes per week minimum). (Briffa et al. 2006; WHO 2010)

Why is a Mediterranean diet good for your heart?

'Adopting a Mediterranean diet after a heart attack is almost three times as powerful in reducing mortality as taking a statin' (Malhotra 2013). The popularity of the traditional Mediterranean diet in the primary and secondary prevention of CHD has continued to grow since the findings of the Seven Countries Study of the 1960s (a prospective cohort study investigating the association between diet and lifestyle and mortality from CHD in seven countries), which showed that consumption of a high-fat Mediterranean diet was associated with lower risk of CHD and all-cause deaths (Keys et al. 1986). The most impressive findings in support of the Mediterranean diet were from the PREDIMED study (a primary prevention trial of 7500 people across multiple centres in Spain), which showed that after five years on a Mediterranean diet (supplemented with nuts or olive oil) there was a 30 per cent reduction in risk of death from CHD and a 50 per cent reduced risk of T2DM (Estruch et al. 2013). Many practice guidelines around the world are gradually being modified to include elements of the traditional Mediterranean diet in the management of CHD.

The heart health benefits of the traditional Cretan Mediterranean diet have been attributed in part to the high content of plant-derived bioactive phytochemicals (such as carotenoids, flavonoids and polyphenols) from fruits, vegetables, olive oil, nuts and wine, which could play a significant role in heart disease protection by acting at the first step of preventing inflammation of the coronary blood vessels caused by free radical damage. For details of the diet see Chapter 2.

The nutritional benefits of the traditional Cretan Mediterranean diet, with a particular focus on nutrients that are protective for heart disease, stroke and cancer, include a healthy balance of fats, with low quantities of saturated (animal) fats and a good source of monounsaturated fats (olive oil, nuts) and omega-3 fats (EPA and DHA from fish). It is also rich in bioactive phytochemicals with high antioxidant potential: there are high amounts of vitamin C, vitamin E, beta-carotene, glutathione, phytoestrogens and phytochemicals from green leafy vegetables; phenolic compounds from wine and olive oil; high intakes of tomatoes, onions, garlic and herbs, especially oregano, mint, rosemary, parsley and dill, which contain lycopene, allylthiosulfinates, salicylates, carotenoids, indoles, monoterpenes, polyphenols and flavonoids; and other phytochemicals used in cooking vegetables, meat and fish.

most nutrients, especially iron, and shellfish are high in important minerals such as zinc.

Whole fish and long-chain omega-3 fatty acids, such as EPA and DHA, are associated with a lower risk of mortality from CHD and stroke. The mechanisms of action are thought to be related to improvements in dyslipidaemia (high triglycerides and low high-density lipoprotein cholesterol), which is associated with increased diabetes risk. Long-chain omega-3 fats also act by inhibiting platelet formation and hence prevention of clot formation (or thrombus) in the arteries, which is the last (often fatal) step in heart disease and stroke.

Folate is important in DNA synthesis and repair, and its role is well known in the prevention of neurological abnormalities in newborns, such spina bifida. However, it is also important in the prevention of cancer and heart disease. Folate deficiency has been linked to high levels of an inflammatory blood hormone called homocysteine, which is linked to increased heart disease risk. Folate is mainly found in green leafy vegetables such as spinach, broccoli, asparagus and legumes and some fruits such as citrus. Studies using folate supplements to prevent heart disease have not been successful; therefore, diet is the best approach.

Many foods naturally contain antioxidants (such as vitamin C in oranges) to prevent them from spoiling. The antioxidants form part of the food's natural defence system. There are many different types of antioxidants in the diet that can help to boost the immune system and prevent free radical damage, such as vitamin C, vitamin E, carotenoids, phenolics, selenium and zinc.

Stroke

There are two main types of **stroke**: haemorrhagic stroke, which occurs when blood vessels in the brain

Stroke occurs when a blood vessel to the brain is suddenly blocked or bleeds. As a result, brain function may be lost, and activities such as movement, thinking and communication may be impaired.

rupture and bleed causing damage to the nearby brain tissue, and thrombotic stroke, which occurs when blood vessels in the brain become blocked and a thrombus forms, causing a lack of blood flow to the part of the brain supplied by the vessels. The latter type is similar to atherosclerosis and thrombus that occurs with CHD and myocardial infarction.

Prevalence

Stroke is the second-biggest killer after CHD in Australia and the third-biggest killer in New Zealand. It is the major cause of serious adult disability. In 2012, about 50,000 Australians and 9000 New Zealanders suffered a stroke. There are an estimated 60,000 people in New Zealand and 420,000 in Australia (30 per cent under the age of 65) living with the effects of stroke. The estimated burden of disease costs for stroke in Australia in 2010 was $49.3 billion, similar to anxiety and depression (AIHW 2013).

Causes

Stroke can affect all ages; however, over 80 per cent of all deaths from stroke occur in people aged 75 and over. Obesity, hypertension, hyperlipidaemia and impaired glucose tolerance or insulin resistance are biological risk factors associated with increased risk of CVD (stroke and CHD). They are impacted by the behavioural, or modifiable, risk factors noted below:

- *Tobacco smoking*—Of all behavioural risk factors, tobacco smoking is responsible for the greatest burden on the health of Australians, accounting for 9.7 per cent of the total burden of disease. Tobacco smoking increases risk of all vascular disease (stroke and CHD) through multiple mechanisms, including free radical damage to the vascular endothelium from the toxic chemicals in smoke, the carbon monoxide effects on the endothelium and the effects of nicotine on elevating blood pressure.
- *Excessive alcohol consumption*—This increases blood pressure, causes vascular inflammation and has direct toxic effects on the brain.
- *Poor diet*—No one single nutrient is responsible. As for CHD, a Western diet consisting of highly processed foods that are high in starch, sugar, salt and animal (saturated) fat is linked to increased stroke risk, by impacting on plasma low-density lipoprotein cholesterol, blood pressure and obesity.
- *Physical inactivity*—This leads to obesity, which in turn impacts blood pressure and blood lipids. Increasing physical activity can reduce stroke risk by reducing weight and blood pressure and improving blood lipid profile (reducing low-density lipoprotein and increasing high-density lipoprotein cholesterol).

Table 11.1 shows the relationships that exist between the different risk factors.

Cancer

Prevalence

Cancer is a leading cause of death in Australia and New Zealand, accounting for nearly one-third of all deaths in New Zealand in 2010 and three in ten deaths in Australia in 2011 (AIHW 2014; Ministry of Health 2014). One in two Australian men and one in three Australian women will be diagnosed with cancer by the age of 85. Cancer diagnosis rates are on the rise in Australia and New Zealand; however, survival rates for common cancers are increasing. In 2011 approximately 128,000 new cases of cancer were diagnosed in Australia, with 21,235 new diagnoses in New Zealand in 2010. The direct health system costs of cancer in 2011 in Australia were $3.8 billion, representing 7.2 per cent of all health system costs.

In 2011, the most common cancers in Australia (excluding non-melanoma skin cancer) were prostate, colorectal, breast, melanoma and lung cancer, accounting for 60 per cent of all cancers. Similarly, in 2010, the

Table 11.1 Relationships between risk factors for stroke

Behavioural risk factors	Biological risk factors			
	Obesity	Hypertension	Hyperlipidaemia	Insulin resistance
Smoking	–	✓	–	–
Excess alcohol	✓	✓	–	–
Poor diet	✓	✓	✓	✓
Inactivity	✓	✓	–	–

most common cancers in New Zealand were prostate and colorectal cancer, followed by breast cancer and melanoma. Lung cancer was the most common cause of cancer death in both countries, followed by prostate cancer (in men) and breast cancer (in women).

Causes

All of the risks and causes of cancer are not known. However, there are a number of chemical, physical and biological agents that have been shown to trigger the mutations in the cell DNA that lead to cancer. These are called carcinogens and include tobacco, ultraviolet radiation and asbestos. Common modifiable risk factors of different cancers include the following:

- Smoking is responsible for one in five cancer deaths (lung, oral cavity, larynx, pharynx, oesophagus).
- Alcohol consumption is related to 3 per cent of cancers (oral cavity, larynx, pharynx, oesophagus, liver).
- Skin cancers are caused by excessive exposure to ultraviolet radiation (sunlight).
- Diet-related factors are associated with about 30 per cent of all cancers.
- Obesity increases the risk of cancer (breast, endometrium, colorectum, kidney, oesophagus). Conversely, physical activity reduces risk of colorectal, breast and endometrial cancer.

It is important to note that cancer can sometimes develop without any specific causes.

Most evidence of the link between diet and cancer prevention comes from large population studies. Populations that have a high risk or a low risk for specific cancers can be examined with regard to their exposures, such as diet and lifestyle factors (including occupational hazards), and inferences can be made about protective or causative effects. Countries with high rates of stomach cancer, such as Japan and China, have a high intake of pickled foods; therefore, it is hypothesised that long-term consumption of pickled foods increases the risk of stomach cancer.

A report by the World Cancer Research Fund (WCRF/AICR 2007) provided a detailed analysis of all studies that had examined the link between diet and cancer and concluded that processed meats, red meat, salty foods, pickled foods and food high in calcium are associated with a probable or convincing increased risk of cancer. This report also reviewed all studies that had examined protective effects of diet on specific cancers and found that protective foods include leafy green vegetables, tomatoes, fresh fruits, garlic, onions, high-fibre foods and food containing selenium (such as shellfish, octopus, beans and nuts).

Most of the available evidence on diet and cancer is centred on individual foods or food groups and isolated nutrients. But people consume dietary patterns, not individual foods or nutrients. It is well recognised that dietary patterns rather than individual foods or nutrients are more predictive of disease risk, because the overall diet is considered and there is less risk of excluding important components of the diet, particularly in diseases with multiple dietary associations such as cancer.

The World Cancer Research Fund reported that a Western-type diet defined by a high consumption of dairy foods, meat and excess energy is linked to a higher risk of cancer, and an eating pattern consisting mainly of plant foods, such as fresh fruit and vegetables, wholegrain cereals, tea and nuts, along with fish, is protective for cancer. Furthermore, the expert panel concluded that whole diets are more important, and healthy diet patterns should ideally be followed throughout life to be most effective in preventing cancer.

SUMMARY AND KEY MESSAGES

This age group should aim to be in the best physical, cognitive and emotional condition as possible. It is a time in life during which people have multiple roles and responsibilities. Lifestyle determinants of health, such as adhering to a healthy diet and participating in regular physical activity, are control measures that should be undertaken by this age group to avoid the untoward effects of non-communicable diseases.

- Overweight and obesity are associated with a wide range of other conditions, particularly CVD, T2DM and some cancers.
- The risk of comorbidity appears to rise with increasing BMI.
- Even small amounts of weight loss bring health benefits, including lowered cardiovascular risk; prevention, delayed progression or improved control of T2DM; and improvements in other health conditions.
- Lifestyle changes that include reduced energy intake and increased physical activity have health benefits that are independent of weight loss.
- The health benefits of the traditional Mediterranean diet have been attributed in part to the high content of plant-derived bioactive phytochemicals (such as carotenoids, flavonoids and polyphenols) from fruits, vegetables, olive oil and wine, and could play a significant role in reducing chronic disease, particularly CVD, diabetes and a number of cancers, such as breast, prostate and bowel.

REFERENCES

ABS (Australian Bureau of Statistics) 1995. *National Nutrition Survey: Selected highlights, Australia, 1995*. Cat. no. 4802.0. Commonwealth of Australia. <www.abs.gov.au/ausstats/abs@.nsf/mf/4802.0>, accessed 14 August 2013.

——2012. *Australian Health Survey: First results, 2011–12*. Cat. no. 4364.0.55.001. Commonwealth of Australia. <www.ausstats.abs.gov.au/ausstats/subscriber.nsf/0/1680ECA402368CCFCA257AC90015AA4E/$File/4364.0.55.001.pdf>, accessed 28 September 2014.

——2014. *Australian Health Survey: Nutrition first results—food and nutrients, 2011–12*. Cat. no. 4364.0.55.007. Commonwealth of Australia. <www.abs.gov.au/AUSSTATS/abs@.nsf/DetailsPage/4364.0.55.0072011-12?OpenDocument>, accessed 14 November 2014.

AIHW (Australian Institute of Health and Welfare) 2001. *Australia's Welfare, 2001: The fifth biennial report of the Australian Institute of Health and Welfare*. Cat. no. AUS 24. Commonwealth of Australia. <www.aihw.gov.au/WorkArea/DownloadAsset.aspx?id=6442453123>, accessed 28 September 2014.

——2010. *Australia's Health 2010: The twelfth biennial health report of the Australia Institute of Health and Welfare*. Australia's Health Series 12, cat. no. AUS 122. Commonwealth of Australia. <www.aihw.gov.au/WorkArea/DownloadAsset.aspx?id=6442452962>, accessed 28 September 2014.

——2013. *Stroke and Its Management in Australia: An update*. Cat. no. CVD 61. Commonwealth of Australia. <www.aihw.gov.au/WorkArea/DownloadAsset.aspx?id=60129543611>, accessed 14 November 2014.

——2014. *Australia's Health 2014*. Cat. no. AUS 178. Commonwealth of Australia. <www.aihw.gov.au/WorkArea/DownloadAsset.aspx?id=60129548150>, accessed 14 November 2014.

Alberti, K., Eckel, R.H., Grundy, S.M., Zimmet, P.Z. et al. 2009. Harmonizing the Metabolic Syndrome: A joint interim statement of the International Diabetes Federation Task Force on Epidemiology and Prevention; National Heart, Lung, and Blood Institute; American Heart Association; World Heart Federation; International Atherosclerosis Society; and International Association for the Study of Obesity. *Circulation* 120: 1640–45.

Barr, E.L.M., Magliano, D.J., Zimmet, P.Z., Polkinghorne, K.R. et al. 2006. *AusDiab 2005: The Australian Diabetes, Obesity and Lifestyle Study; Tracking the accelerating epidemic: Its causes and outcomes.* Baker IDI Heart & Diabetes Institute. <www.bakeridi.edu.au/Assets/Files/AUSDIAB_Report_2005.pdf>, accessed 28 September 2014.

Bonora, E., Kiechl, S., Willeit, J., Oberhollenzer, F. et al. 1998. Prevalence of insulin resistance in metabolic disorders: The Bruneck Study. *Diabetes* 47: 1643–9.

Briffa, T., Maiorana, A., Allan, R., Oldenburg, B. et al. 2006. *National Heart Foundation of Australia Physical Activity Recommendations for People with Cardiovascular Disease.* <www.heartfoundation.org.au/SiteCollectionDocuments/PAR4CVD.pdf>, accessed 14 November 2014.

Cefalu, W.T., Wang, Z.Q., Werbel, S., Bell-Farrow, A. et al. 1995. Contribution of visceral fat mass to the insulin resistance of aging. *Metabolism* 44: 954–9.

Chowdhury, R., Warnakula, S., Kunutsor, S., Crowe, F. et al. 2014. Association of dietary, circulating, and supplement fatty acids with coronary risk: A systematic review and meta-analysis. *Annals of Internal Medicine* 160: 398–406.

Egger, G. & Swinburn, B. 1997. An 'ecological' approach to the obesity pandemic. *British Medical Journal* 315: 477–80.

Esposito, K., Kastorini, C.M., Panagiotakos, D.B. & Giugliano, D. 2013. Mediterranean diet and metabolic syndrome: An updated systematic review. *Reviews in Endocrine and Metabolic Disorders* 14: 255–63.

Estruch, R., Ros, E., Salas-Salvado, J., Covas, M. et al. The PREDIMED Study Investigators 2013. Primary prevention of cardiovascular disease with a Mediterranean diet. *New England Journal of Medicine* 368: 1279–90.

Florez, J.C., Jablonski, K.A., Bayley, N., Pollin, T.I. et al. 2006. TCF7L2 polymorphisms and progression to diabetes in the Diabetes Prevention Program. *New England Journal of Medicine* 355: 241–50.

German, J.B. & Dillard, C.J. 2004. Saturated fats: What dietary intake? *American Journal of Clinical Nutrition* 80: 550–9.

Grant, S.F., Thorleifsson, G., Reynisdottir, I., Benediktsson, R. et al. 2006. Variant of transcription factor 7-like 2 (TCF7L2) gene confers risk of type 2 diabetes. *Nature Genetics* 38: 320–3.

Isomaa, B.P., Almgren, T., Tuomi, B., Forsen, K. et al. 2001. Cardiovascular morbidity and mortality associated with the metabolic syndrome. *Diabetes Care* 24: 683–9.

Keys, A., Menotti, A., Karvonen, M.J., Aravanis, C. et al. 1986. The diet and 15-year death rate in the seven countries study. *American Journal of Epidemiology* 124: 903–15.

Luppino, F.S., de Wit, L.M., Bouvy, P.F., Stijnen, T. et al. 2010. Overweight, obesity, and depression: A systematic review and meta-analysis of longitudinal studies. *Archives of General Psychiatry* (now called *JAMA Psychiatry*) 67: 220–9.

McAuley, K. & Mann, J. 2006. Thematic review series: Patient-oriented research; Nutritional determinants of insulin resistance. *Journal of Lipid Research* 47: 1668–76.

McLennan, W. & Podger, A. 1998. *National Nutrition Survey Users' Guide 1995* new issue. Cat. no. 4801.0. ABS, Commonwealth of Australia. <www.ausstats.abs.gov.au/ausstats/subscriber.nsf/0/CA25687100069892CA256889002102FD/$File/48010_1995.pdf>, accessed 3 October 2014.

Magliano, D.J., Barr, E.L., Zimmet, P.Z., Cameron, A.J. et al. 2008. Glucose indices, health behaviors, and incidence of diabetes in Australia: The Australian Diabetes, Obesity and Lifestyle Study. *Diabetes Care* 31: 267–72.

Malhotra, A. 2013. Saturated fat is not the major issue. *British Medical Journal* 347: f6340.

Ministry of Health 2014. *New Zealand Burden of Diseases, Injuries and Risk Factors Study*, 2006-2016. New Zealand Government. <www.health.govt.nz/nz-health-statistics/health-statistics-and-data-sets/new-zealand-burden-diseases-injuries-and-risk-factors-study-2006-2016>, accessed 14 November 2014.

NHMRC (National Health and Medical Research Council) 2009. *Australian Guidelines to Reduce Health Risks from Drinking Alcohol.* Commonwealth of Australia. <www.nhmrc.gov.au/_files_nhmrc/publications/attachments/ds10-alcohol.pdf>, accessed 14 November 2014.

——2013a. *Clinical Practice Guidelines for the Management of Overweight and Obesity in Adults, Adolescents and Children in Australia 2013*. Commonwealth of Australia. <www.nhmrc.gov.au/_files_nhmrc/publications/attachments/n57_obesity_guidelines_131204_0.pdf>, accessed 28 September 2014.

——2013b. *Eat for Health: Australian Dietary Guidelines; Providing the scientific evidence for healthier Australian diets*. Commonwealth of Australia. <www.eatforhealth.gov.au/sites/default/files/files/the_guidelines/n55_australian_dietary_guidelines.pdf>, accessed 16 March 2014.

Reaven, G.M. 1988. Role of insulin resistance in human disease. *Diabetes* 37: 1595–1607.

Rosenbaum, M., Hirsch, J., Gallagher, D.A. & Leibel, R.L. 2008. Long-term persistence of adaptive thermogenesis in subjects who have maintained a reduced body weight. *American Journal of Clinical Nutrition* 88: 906–12.

Ruan, H. & Lodish, H.F. 2004. Regulation of insulin sensitivity by adipose tissue-derived hormones and inflammatory cytokines. *Current Opinion in Lipidology* 15: 297–302.

Schofield, W.N. 1985. Predicting basal metabolic rate, new standards and review of previous work. *Human Nutrition. Clinical Nutrition* 39: 5–41.

Stratton, I.M., Adler, A.I., Neil, H.A., Matthews, D.R. et al. 2000. Association of glycaemia with macrovascular and microvascular complications of type 2 diabetes (UKPDS 35): Prospective observational study. *British Medical Journal* 321: 405–12.

Sumithran, P., Prendergast, L.A., Delbridge, E., Purcell, K. et al. 2011. Long-term persistence of hormonal adaptations to weight loss. *New England Journal of Medicine* 365: 1597–604.

University of Otago & Ministry of Health 2011. *A Focus on Nutrition: Key findings of the 2008/09 New Zealand Adult Nutrition Survey*. New Zealand Government. <www.health.govt.nz/system/files/documents/publications/a-focus-on-nutrition-v2.pdf>, accessed 2 October 2014.

WCRF/AICR (World Cancer Research Fund/American Institute for Cancer) 2007. *Summary: Food, Nutrition, Physical Activity and the Prevention of Cancer: A global perspective*. Washington, DC. <www.dietandcancerreport.org/cancer_resource_center/downloads/summary/english.pdf>, accessed 14 November 2014.

WHO (World Health Organization) 2010. *Global Recommendations on Physical Activity for Health*. WHO. <http://whqlibdoc.who.int/publications/2010/9789241599979_eng.pdf?ua=1>, accessed 14 November 2014.

Wing, R.R. & Phelan, S. 2005. Long-term weight loss maintenance. *American Journal of Clinical Nutrition* 82(1 Suppl): 222S–225S

Yusuf, S.S., Hawken, S., Ounpuu, S., Dans, T. et al. 2004. The INTERHEART Study Investigators 2004. Effect of potentially modifiable risk factors associated with myocardial infarction in 52 countries (the INTERHEART study): Case-control study. *Lancet* 364: 937–52.

Zhu, S., St-Onge, M.P., Heshka, S. & Heymsfield, S.B. 2004. Lifestyle behaviors associated with lower risk of having the metabolic syndrome. *Metabolism* 53: 1503–11.

Zimmet, P.Z., Alberti, K.G. & Shaw, J.E. 2005. Mainstreaming the metabolic syndrome: A definitive definition. *Medical Journal of Australia* 183: 175–6.

ADDITIONAL READING

Baker IDI Heart & Diabetes Institute 2012. *Diabetes: The silent pandemic and its impact on Australia*. Diabetes Australia. <www.diabetesaustralia.com.au/Documents/DA/What's%20New/12.03.14%20Diabetes%20management%20booklet%20FINAL.pdf>, accessed 6 October 2014.

NHFA (National Heart Foundation of Australia) & CSANZ (Cardiac Society of Australia and New Zealand) 2012. *Reducing Risk in Heart Disease: An expert guide to clinical practice for secondary prevention of coronary heart disease*. NHFA. <www.heartfoundation.org.au/SiteCollectionDocuments/Reducing-risk-in-heart-disease.pdf>, accessed August 2013.

World Stroke Campaign (website) 2014. World Stroke Organization. <www.worldstrokecampaign.org>, accessed 6 October 2014.

TEST YOUR UNDERSTANDING

Note: There is only one correct answer for each question.

1. The physiological signs characteristic of adulthood typically include which of the following?
   a. Decline in lean body mass and increase in fat mass
   b. Loss of bone density and muscular strength
   c. Increase in resting metabolic rate
   d. Increase in fitness
   e. Decline in fat mass and increase in muscle mass
2. Which environmental factor is implicated in the obesity epidemic?
   a. Increase in abdominal fat and loss of lean body mass
   b. Family history of obesity and diabetes
   c. Disruption in the feedback mechanisms of hormones that increase hunger
   d. Availability of easily accessible processed foods high in sugar and saturated fat
   e. Bad mood
3. Which of the following is a successful weight management strategy?
   a. Monitoring weight every so often
   b. Having two meals a day
   c. Identifying triggers of emotional eating and eating a diet low in kilojoules
   d. Consistent low levels of physical activity
   e. Monitoring blood pressure and lipid levels
4. What is the underlying factor of MetS?
   a. High blood pressure
   b. Low levels of physical activity
   c. Increased waist circumference
   d. Insulin resistance
   e. High cholesterol
5. What is the optimum diet for weight loss in insulin-resistant patients?
   a. Low fat and high carbohydrate
   b. Moderate carbohydrate, low saturated fat and high monounsaturated fat
   c. High monounsaturated and polyunsaturated fat (20 per cent and 10 per cent, respectively)
   d. High protein, low fat and low carbohydrate
   e. Equal amounts of carbohydrate, fat and protein
6. Which of the following statements is false?
   a. Physical activity increases muscle mass and decreases fat mass
   b. Bone mass continues to accrue in early adulthood
   c. Emotional, restrained and external eating may increase energy intake
   d. Leptin increases appetite
   e. Insulin resistance is thought to be the key underlying factor of MetS
7. Which of the following dietary patterns is associated with probable risks of cancer?
   a. Pickled foods, high-fibre foods, soft drinks
   b. Alcohol, leafy greens, tomatoes
   c. Foods high in selenium, garlic, onions
   d. Red meat, processed meat, foods high in calcium
   e. All of the above
8. Which of the following strategies is not part of the management of CHD?
   a. Do at least 30 minutes of moderate-intensity exercise per day
   b. Reduce saturated fatty acid intake to below 7 per cent of total energy intake
   c. Replace all sugar in the diet with low-sugar or artificially sweetened products
   d. Limit salt intake to 4 grams or less per day
   e. a and c
9. Which of the following is not an established risk factor for stroke?
   a. High blood pressure
   b. Inactivity
   c. Obesity
   d. Moderate alcohol consumption
   e. None of the above
10. Which statement is correct in relation to energy expenditure and adults?
    a. Energy expenditure continues to rise during the adult years due to an increase in metabolic rate
    b. Energy expenditure starts to decline during the adult years due to loss of lean body mass
    c. Energy expenditure starts to decline during the adult years due to inability to exercise
    d. Energy expenditure remains stable throughout the adult years and declines only after the age of 65 years
    e. Energy expenditure remains stable throughout the adult years

STUDY QUESTIONS

1. Compare and contrast three definitions of MetS (e.g. International Diabetes Federation, National Heart, Lung, and Blood Institute, World Health Organization [WHO]) in their essential criteria, emphasis on adiposity and priority of individual factors. Comment on how using one definition over another affects prevalence rates between studies.
2. Discuss what societal, cultural and genetic factors are attributing to the growing prevalence of T2DM in younger age groups.
3. As a health professional, what is the ideal way to intervene and counsel a 45-year-old male truck driver who attends your clinic and who is obese and has a history of high blood pressure? Design a weight management plan that would suit his individual needs.
4. Describe the modifiable risk factors that are common for CVD, CHD and stroke. Design a meal plan that would be suitable for the secondary prevention of CHD.
5. Define the dietary factors that have a probable or convincing positive association with cancer risk and the dietary factors that are protective for cancer. What dietary advice would you provide to someone who has a strong family history of bowel cancer?

12

Older adult (61–84 years)

Michelle Miller and Alison Yaxley

> **LEARNING OUTCOMES**
>
> Upon completion of this chapter you will be able to:
>
> - appreciate the role of genetics in ageing successfully
> - discuss the nutrient requirements of older adults
> - identify the key risk factors and areas of concern for potential nutritional deficiencies
> - describe the impact that transition to retirement can have on nutritional health
> - list the key diet-responsive chronic diseases in older adults and describe the role of nutrition in optimising management and avoidance of long-term complications.

As with the rest of the world, the populations of Australia and New Zealand are ageing, due to a decrease in the fertility rate and number of births and an increase in life expectancy. Life expectancy in Australia has continued to rise steadily. Those born in 2010–12 are expected to live to 79.9 years for males and 84.3 years for females, an increase from 67.6 years and 74.2 years, respectively, 50 years ago (ABS 2011, 2013a). The median age of the Australian population has increased from 33 years to 37.3 years in the past two decades. There was a 3.7 per cent increase in the number of people aged 65 years and over between 2012 and 2013, and figures indicate that 14.4 per cent of the population was included in that age group in 2013, an increase from 11.1 per cent twenty years ago (ABS 2013b).

There is no indication that growth trends will slow down in the near future. Australian population projections indicate that by 2061 the 65 years and over age group will make up approximately 22 per cent of the population, based on the current rate of increase (ABS 2013c). The picture in New Zealand is similar. Life expectancy at birth in New Zealand in 2010–12 was reported to be 79.3 years for males and 83 years for females (Statistics New Zealand 2013). The median age in New Zealand increased to 37 years in 2012, from 26 years four decades earlier. The representation of those aged 65 years and over in the New Zealand population has doubled since 1980 and is expected to double again by 2036, to at least 21 per cent. While figures reported in 2012 indicate that there are around two-thirds as many older New Zealanders as there are children, these figures are expected to rise as the baby boomers age over the next 20 years (Statistics New Zealand 2012).

The increase in years of life brings with it challenges for maintaining quality of life. While there is evidence to suggest that genetics play an important role in both, environmental factors also make a contribution. This chapter outlines current Australian and New Zealand dietary recommendations for older adults and highlights some of the pivotal stages from a nutrition perspective as age moves beyond the sixth decade of life. As well as presenting the challenges posed by the increased prevalence of chronic disease in older age, the chapter discusses the role that nutrition can play in the prevention and management of these potentially debilitating conditions that can impact on both quality and quantity of life.

PHYSIOLOGICAL AND LIFESTYLE CHANGES

Genetics of ageing and the link to nutrition

Numerous theories have been posed on the role of genetics in extending life, many of which have their origins in genetic alterations of animals. While it was once thought that extending life would correlate highly with extending the time of suffering from chronic disease, it is emerging that a mutation that impacts to slow ageing also reduces the risk or postpones the development of chronic disease. Many of these mutations are known to be influenced by nutrient sensors in addition to stress sensors; hence, diet is emerging as a powerful factor in the promotion of healthy ageing (Kenyon 2010). One of the earliest examples of a dietary factor that has been demonstrated to extend life is **caloric restriction**.

It is becoming increasingly evident, however, that the longevity response to caloric restriction is not simple; it is dependent on the nutrient-sensing pathways activated and the way in which the caloric restriction occurs. In 2012 the journal *Nature* published a study highlighting that in rhesus monkeys, genetics and healthy diets matter more than caloric restriction for longevity (Mattison et al. 2012).

There is a range of other suspected determinants of ageing, many of which also have potential nutrient-mediating influences. The relationship between telomeres and diet is one area receiving increasing attention. Telomeres are located at the ends of human chromosomes and consist of TTAGGG (T is thymine, A is adenine, G is guanine nucleotides) DNA sequence repeats. Their main functions are to maintain **genomic stability** and help protect cells' chromosomes from degradation. Cells are generally unaffected by a small amount of telomere erosion; however, once they reach a critically short length, the ends become uncapped, causing **cell senescence**, signalling the arrest of cell proliferation and subsequently leading to **apoptosis**. Erosion of telomere length has been postulated to be a causal factor in cellular ageing, with older adults reported to have shorter telomeres than younger adults (Steenstrup et al. 2013). Progressive telomere erosion has also been associated with chronic diseases in humans including cardiovascular disease (CVD), type 2 diabetes mellitus (T2DM) and vascular dementia.

The role of micronutrients in maintenance of genomic stability has been extensively reviewed, and omega-3 fatty acids, as well as the antioxidants vitamins C and E, selenium, carotenoids and polyphenols, have been shown to play a role in DNA oxidation. Furthermore, niacin, zinc and folate have been shown to assist with DNA repair mechanisms (Fenech 2010). Few experimental studies have been undertaken specifically to determine the effect of diet on telomere length. A randomised controlled trial found that the proportion of cells with telomere erosion was lower in those consuming a Mediterranean-style diet (Marin et al. 2012). In a secondary analysis of a randomised controlled trial in healthy middle-aged and older adults, a difference in telomere length of 29 base pairs was demonstrated after four months of a high-dose omega-3 fatty acid supplement (2.5 grams) versus a low dose (1.25 grams) (Kiecolt-Glaser et al. 2012). This area of investigation is emerging, and it is expected that future evidence will demonstrate convincingly that selected dietary patterns can delay the shortening of telomeres, impact positively on ageing and delay or prevent the onset of chronic disease.

Many physiological changes in this life stage also affect the body structure. Gastrointestinal changes slow down functions of digestion and absorption of nutrients and subsequently increase the risk of nutrient

Caloric restriction is the limitation of dietary energy intake (kilojoules or calories).

Genomic stability is maintenance of the genome to minimise gene mutations.

Cell senescence is the phase in a cell's life when it is no longer able to divide.

Apoptosis describes the death of cells that is deliberate, that is programmed, rather than by necrosis.

deficiencies. Bone density, skeletal muscle, smooth muscle and muscle from organs are often reduced. The ability to regenerate muscle is impaired with age. Alterations in the senses of sight, hearing, smell and taste are also common with increasing age and can impact on the ability of an older adult to meet their dietary requirements.

Transition to retirement

Maintaining healthy body weight into older age is important for reducing risk of chronic disease and disability. Retirement, a key milestone that often coincides with entering older age, represents a significant life transition in which alterations in lifestyle are known to occur, including changes in dietary intake. The magnitude and direction of the dietary changes and the implications these changes have in terms of weight status are important considerations for public health planning and policy, as the proportion of the populace retiring in Australia is expected to continue to grow into the foreseeable future as the population ages.

While there is some inconsistency in the literature when it comes to summarising the impact of retirement on weight status, the majority suggests that weight gain is likely. The key predictor of weight gain appears to be the type of employment that was undertaken prior to retirement. Nooyens and colleagues (2005) reported that men from active employment gained more weight and had larger increases in waist circumference compared to men retiring from sedentary employment.

CASE STUDY 12.1 ***MALNUTRITION***

Carol Wham

This case study is included in the older adult section; however, the problem, assessment and treatment would equally be observed in later stage older adults. It is important to point out that those who live with disadvantage often experience poor health at an earlier age.

Profile and problem

Mrs S. is a 68-year-old Māori woman recently bereaved from her husband. She lives independently in the community. In the past she has been a heavy smoker and was diagnosed with chronic obstructive pulmonary disease eight years ago. Mrs S. has three adult children with families and despite her pension being her only income likes to provide financial support to her *mokopuna* (grandchildren). In spite of experiencing constant fatigue, recent severe weight loss (15 per cent in six months) and a poor appetite she doesn't want to burden her son who lives nearby. Her mobility is severely compromised by a lack of energy, and she has difficulty getting to the shops. Mrs S. has a BMI of 17.

Assessment

Mrs S. has a poor appetite and reports she is no longer motivated to cook after the loss of her husband. While she manages tea and toast for breakfast and appears to have an adequate fluid intake from hot drinks, her diet is low in total energy, dietary variety and protein-rich foods. By mid-afternoon Mrs S. is severely fatigued and frequently has nil per mouth apart from black coffee until the following day.

Treatment

Mrs S. has agreed to take part in a six-week meals for rehabilitation program. To assist her to gain weight, nutrition support will include the following:

- Meals on Wheels: two meals a day, seven days a week
- intensive energy-dense snacks throughout the day
- oral supplements providing 4.2 kilojoules per millilitres to be taken between meals
- maintenance of regular fluid intake.

Mrs S. is keen to provide after-school care for her grandchild who has moved to a new school nearby. She is motivated to regain a healthy weight and prepare an early dinner for her grandchild during the weeknights to relieve the working parents. Mrs S. will be able to teach her granddaughter to cook a 'boil up' using *puha* (sow-thistle) and spinach from her garden.

Plan

Mrs S. will be monitored by a registered community dietitian on a weekly basis to ensure she has an adequate weight gain, restores her appetite and is independent in procuring and preparing meals. She will be assisted with budgetary advice and meal plan ideas incorporating traditional Māori foods.

Weight gain was also demonstrated to be likely post-retirement by Forman-Hoffman and colleagues (2008), who reported weight gain among women retirees, specifically those who were of normal weight on retirement and previously working in a blue-collar job, compared to women who continued to work.

The key dietary factors that have been explored during or after retirement include energy, protein, fat, carbohydrate, fibre, alcohol and location of meals. The evidence suggests that while frequency of vegetable consumption among retirees compared to those who were employed increased (Nooyens et al. 2005), there was no significant difference in fibre intake (Davies et al. 1986) or in distribution of nutrients (Lauque et al. 1998). Alcohol consumption has been shown to increase more in men retiring from sedentary jobs compared to those in continued sedentary employment (Nooyens et al. 2005). There also appears to be an increase in the frequency of retirees eating at a restaurant or friend's house (Lauque et al. 1998), but overall income spent on eating out decreases (Chung et al. 2007).

There is some evidence that health promotion is effective at improving health outcomes among retirees; however, there appear to be important predictors of uptake (Wilson & Palha 2007). These include being male, relatively young (under 70 years) and having a low-risk health status. The last of these is concerning, given those at high risk have greater needs and can achieve better outcomes from a clinical and cost perspective. There is some way to go in identifying the most successful health promotion programs for retirees. One high-quality randomised controlled trial of an individually tailored energy balance program delivered for twelve months was unable to demonstrate significant differences between groups in body weight, body composition, fruit and vegetable consumption or consumption of sliced meat, meat, sugar added to tea, fat or total energy (Werkman et al. 2010).

NUTRIENT REQUIREMENTS AND INTAKES

The well-established changes in physiology experienced by people aged 61–84 years have been considered in determining dietary recommendations for older adults in combination with the best available scientific evidence. The nutrient reference values (NRVs) for Australia and New Zealand outline recommendations for healthy people, with a primary focus being the achievement of physiological needs (NHMRC & Ministry of Health 2006). Guideline 1 specifically highlights the need for older adults to consume nutritious foods and keep physically active to help maintain muscle strength and a healthy weight. Other guidelines provide minimum recommended serves of fruits, vegetables, lean meats and alternatives and dairy and alternatives for older adults and refer to tips for older adults to achieve the guidelines. They also highlight the need to prepare and store food safely, which is particularly important for older adults, as the complications of food poisoning can be severe in those with a weakened immune system. The New Zealand Ministry of Health (2013a) Food and Nutrition Guidelines for Healthy Older People give guidelines of a similar nature to the Australian guidelines, with a specific focus on those aged 65 years and over. A summary of selected macronutrient and micronutrient intakes from the 2011–12 Australian Health Survey (ABS 2014) and the 2008–09 New Zealand Adult Nutrition Survey (University of Otago & Ministry of Health 2011) are detailed in Table 1.15. Food sources of nutrients are described in Tables 1.16 and 1.17.

Energy

The recommendations for dietary energy are prescribed according to age and gender, with the goal of maintaining a body mass index (BMI) of 22, consistent with the midpoint of the healthy weight range. Estimates of total energy requirements for older adults are complicated, given the evidence to suggest that the desirable healthy weight range should be set higher for improved health outcomes (Rejeski et al. 2010) and the potential that the estimates are based on predictive equations that have not been validated in this age group and may therefore overestimate requirements as a result of a decline in muscle mass with age (NHMRC & Ministry of Health 2006). The estimated energy requirements (EERs) for older adults are outlined in Table 1.7.

Data from the 2011–12 Australian Health Survey highlight that older women are consuming 7268 kilojoules for 51–70-year-olds and 6570 kilojoules for greater than 71 years, with males consuming approximately 2100 and 1600 kilojoules more than females in the respective age groups (ABS 2014). The 2008–09 New Zealand Adult Nutrition Survey found that 51–70-year-old males consumed more energy than their Australian counterparts at 9371 kilojoules. Females (51–70 years) consumed 7205 kilojoules. As New Zealanders move to the over 70 years groups they consume approximately 1000 kilojoules less than when 51–70 years (University of Otago & Ministry of Health 2011).

Macronutrients

Protein

The recommendations for dietary protein across the lifespan are consistently higher for males than females, and at age 51–70 years the recommendations are estimated to be the same as those for younger adults. Recommendations increase by about 25 per cent for those aged over 70 years, although the evidence to support this increase in requirements is limited. These levels equate to 15–25 per cent of total energy. From the 24-hour recalls used to derive the nutrient intake data in the 2011–12 Australian Health Survey, older adults were consuming approximately 18.5 per cent of their total energy intake as protein (ABS 2014). In 2008–09, New Zealand older adults were getting 17.6 per cent of their energy from protein (University of Otago & Ministry of Health 2011).

Carbohydrate

While there are no dietary recommendations in the form of EAR and RDI for carbohydrate, as there is insufficient evidence to support these, the acceptable intake of carbohydrate is implicit within the acceptable macronutrient distribution range (AMDR) for reducing chronic disease risk. The recommendation is set at 45–65 per cent of total energy from predominantly low-energy density and/or low-glycaemic index (GI) foods. This recommendation is consistent for adults of all ages. For 51–70-year-olds, carbohydrates provided 42 per cent of their total energy intakes, with persons 71 years and older consuming 44 per cent (ABS 2014). Older New Zealanders consumed 45–49 per cent of energy from carbohydrates, with older males and females at the higher end of the range (University of Otago & Ministry of Health 2011).

Fat

The recommendations for dietary fat are not provided in the form of an EAR or RDI, as only some fats are required from dietary sources. Rather, recommendations are based on an AMDR, reported as a proportion of energy from fat. The recommendation for dietary fat is 20–35 per cent of total energy for all adults, in an effort to prevent the onset of chronic disease, specifically obesity and CVD, which is especially important for this stage of the lifespan.

Due to the established and emerging protective effects of omega-3 fatty acids against the development of chronic disease, there have been some additional recommendations provided in the form of an adequate intake (AI), which is 0.4–1 per cent of total energy. The recommended AI is consistent for all adults, with the suggestion that men should consume 160 milligrams per day and women 90 milligrams per day of total long-chain omega-3 (n-3) fatty acids. For the older age groups (51–70 years and 71 years and older), in the 2011–12 Australian Health Survey, fat contributed an average 30.7 per cent and 30.2 per cent of the population's dietary energy intake, respectively. Saturated fat (including trans fatty acids) contributed an average 11.8 per cent and 12.3 per cent of energy, respectively, while monounsaturated fat also contributed 11.8 per cent and 11.1 per cent, respectively. Polyunsaturated fat contributed 4.8 per cent and 4.6 per cent, respectively. Linoleic acid contributed 3.9 per cent of energy, just below the lower boundary of the AMDR (4-10 per cent). Alpha-linolenic acid, a plant-based omega-3 (n-3) fatty acid found in vegetable oils such as canola and linseed or flaxseed, nuts and seeds, contributed on average 0.6 per cent to total dietary energy (ABS 2014).

The proportion of energy from fat declines as New Zealanders age, at 33–34 per cent for 51–70 years and 32 per cent for over 70 years, according to the 2008–09 New Zealand Adult Nutrition Survey. Saturated fat contributed 13 and 12 per cent for the 51–70 years and over 70 years groups respectively. Monounsaturated fat varied between males and females in both age groups, with 29.5 per cent and 23.5 per cent for 51–70-year-old males and females respectively, decreasing by 5 per cent for both males and females at over 70 years. A similar pattern was seen for polyunsaturated fat, yet overall percentage energy was 8–11 per cent across sexes and age groups (University of Otago & Ministry of Health 2011).

Fibre

Recommendations for dietary fibre and its components (cellulose, hemicellulose, lignin, pectin, resistant starch) were established with consideration given to gastro-intestinal function, adequate laxation and the median fibre intake of the Australian and New Zealand populations. The AI is higher for older males (30 grams) than older females (25 grams), but there is no difference within gender for those aged 51–70 years compared to those aged over 70 years. Of note, both males and females in the older age categories (51–70 years and 71 years and older) did not meet the AI in the 2011–12 Australian Health Survey. Dietary fibre intakes for males averaged 24.8 grams and 25.1 grams, respectively, and females 22.2 grams and 21 grams, respectively

(ABS 2014). Females consumed more fibre than males in New Zealand in 2008-09 (20–21 grams compared with 17–18 grams), again less than the AI for both sexes and all age groups (University of Otago & Ministry of Health 2011).

Water

Water is particularly important in older age, due to the decline in kidney function, impact on oral health, use of medications such as diuretics and subsequent consequences of dehydration including confusion, bladder infections, constipation and functional decline. It can, however, be challenging to meet fluid requirements, as a result of a reduction in the thirst mechanism in older age. Due to the significant variability in requirements, an AI has been set which is consistent for all adults and varies by gender: males have a daily AI to be achieved from all foods and fluids set higher (3.4 litres) than females (2.8 litres). In the 2011–12 Australian Health Survey, persons aged 51–70 years consumed 2.8 litres of water and persons greater than 71 years consumed 2.3 litres, which included plain drinking water (ABS 2014).

Micronutrients

Vitamins

The extent to which older adults are increasingly turning to the use of multivitamin supplements should not be underestimated. A census conducted in 2009–10 reported that 27 per cent of Australians aged 65 years and over used multivitamin and mineral supplements (Morgan et al. 2012). Although lower than the Australian prevalence figures, data from the 2008–09 New Zealand Adult Nutrition Survey indicate that a significant number of adults aged 51 years and over (17.4 per cent) used multivitamin and mineral supplements (University of Otago & Ministry of Health 2011). While there is certainly a need for supplementation in situations where diet is unable to meet increased requirements, where overt deficiency is confirmed or where drug–nutrient interactions are evident, for the most part, a healthy diet consistent with national dietary guidelines should ensure that an older adult achieves all of their requirements for the vast array of essential vitamins.

For some vitamins, the recommended requirements for older adults are no different from those for younger adults. These include vitamin B12, folate and vitamin C. Interestingly, for these vitamins there is also no difference in recommendations according to gender. In the 2011–12 Australian Health Survey, adults aged 71 years and older consumed on average 4.1 micrograms of vitamin B12, 612.8 micrograms of folate equivalents and 93.6 milligrams of vitamin C (ABS 2014). In 2008–09 older New Zealand women were consuming 3.2 micrograms, and men 5.7 micrograms, of vitamin B12. Women were consuming 98 milligrams of vitamin C, while men consumed 103 milligrams. Folate was not reported (University of Otago & Ministry of Health 2011). Megaloblastic anaemia can be caused by both folate and vitamin B12 deficiency; however, while high-dose folic acid alone can correct the anaemia it will not correct the neurological damage arising from inadequate vitamin B12. While the evidence is that folic acid fortification is safe in the general population, the compromised vitamin B12 status that is common in older adults may be masked in the presence of large amounts of folic acid from the food supply.

Similarly, vitamin A, thiamin, niacin, pantothenic acid, biotin, choline, vitamin E and vitamin K are recommended at the same amount for all adult age groups, but there is an increased recommendation for males compared to females. Dietary intake data revealed that persons aged 71 years and older had on average 957.5 micrograms for males and 853.6 micrograms for females of dietary vitamin A; 1.7 milligrams and 1.3 milligrams of thiamin for older males and females, respectively; 37.5 milligrams and 31.6 milligrams of niacin for older males and females, respectively; and 9.3 milligrams and 8.3 milligrams of Vitamin E for males and females aged 71 years and older, respectively (ABS 2014). In New Zealand in 2008–09 the same age group were consuming 1236 micrograms (males) and 844 micrograms (females) of vitamin A; 1.6 milligrams (males) and 1.2 milligrams (females) of thiamin; 32 milligrams (males) and 24.2 milligrams (females) of niacin and 10.5 milligrams (males) and 8.7 milligrams (females) of vitamin E (University of Otago & Ministry of Health 2011).

For riboflavin, the same increased recommendation for males compared to females is observed for older adults, but it is not until an adult reaches over 70 years that there is a further increase in recommendation for this vitamin (see Table 12.1). Older males were found to have on average 0.2 milligrams more dietary riboflavin than older females in the 2011–12 Australian Health Survey (1.8 milligrams and 1.6 milligrams, respectively) (ABS 2014). The same was found for older New Zealanders in 2008–09, with men consuming 1.6 milligrams and women 1.2 milligrams

(University of Otago & Ministry of Health 2011). For vitamin B6 the recommendations are greater for older adults compared with younger adults, and males also have greater requirements (see Table 12.1). Again, males aged 71 years and older had higher dietary intakes of Vitamin B6 than their female counterparts (1.3 milligrams versus 1.1 milligrams) (ABS 2014) as was the case in New Zealand in 2008–09, with males consuming 1.6 milligrams and females 1.3 milligrams (University of Otago & Ministry of Health 2011).

Vitamin D is recommended in similar amounts for both older males and older females; however, the AI for 51–70 years (10 micrograms) is deemed to be twice that of younger adults (5 micrograms) and three times that for over 70 years (15 micrograms per day). Normal Vitamin D status is difficult to achieve in older adults due to reduced exposure to the sun and therefore there is an increased requirement to meet needs through the diet.

Summaries of food sources of nutrients are in Table 1.17.

Minerals

Similar to the case of multivitamin supplements, there has been an increase in the use of over-the-counter mineral preparations in recent years. Again, while these preparations have their place under some circumstances, the majority of healthy older people should be able to achieve the recommended intake of the various minerals required for good health from a diverse diet consistent with national dietary guidelines. The minerals iodine, molybdenum and phosphorus are not deemed to be required in any greater amount by older adults compared to younger adults, and the recommendations are for both males and females. This also applies to sodium, although sodium has its recommendations stated in terms of AI rather than EAR or RDI.

Males do have increased requirements for the minerals chromium, copper, fluoride, selenium and zinc; however, this does not alter with increasing adult age. This also applies to manganese and potassium, although recommendations for these minerals are stated in terms of AI as opposed to EAR or RDI. Recommendations for magnesium increase at 31 years but then remain stable through older age, with males consistently having recommendations greater than females. For iron the recommendations remain stable across the adult years for males but decrease for females from 51 years, as a reflection of reduced requirements post-menopause. In contrast, females have increased requirement for calcium at 51 years, and this remains stable thereafter, whereas males have a recommendation that is consistently lower than females up to 70 years, when it then increases to the same level of requirement (see Table 12.2).

In the 2011–12 Australian Health Survey intake data, for males in the older age group (71 years and older) intakes were 726.4 milligrams of calcium, 11.6 milligrams of iron, 2216.7 milligrams of sodium and 10.5 milligrams of zinc. For females in the same age group, intakes were less for all minerals at 674.2 milligrams of calcium, 9.2 milligrams of iron, 1772.5 milligrams of sodium and 9 milligrams of zinc (ABS 2014). Calcium, iron and zinc were reported in the 2008–09 New Zealand Adult Nutrition Survey. Older men consumed 785 milligrams of calcium, 11.8 milligrams of iron and 9.8 milligrams of zinc. Older women consumed 710 milligrams of calcium, 10.2 milligrams of iron and 7.9 milligrams of zinc (University of Otago & Ministry of Health 2011). Sodium was not reported in this survey.

Table 12.1 Daily EARs (RDIs) for riboflavin and vitamin B6 for older adults (51+ years) in Australia and New Zealand

Gender	Age (years)	Amount of nutrients (mg)	
		Riboflavin	Vitamin B6
Male	51–70	1.1 (1.3)	1.4 (1.7)
	> 70	1.3 (1.6)	1.4 (1.7)
Female	51–70	0.9 (1.1)	1.3 (1.5)
	> 70	1.1 (1.3)	1.3 (1.5)

Notes: EAR: estimated average requirement; RDI: recommended dietary intake.

Source: NHMRC & Ministry of Health (2006).

Table 12.2 Daily EARs (RDIs) for calcium for older adults (51+ years) in Australia and New Zealand

Gender	Age (years)	Amount of calcium (mg)
Male	51–70	840 (1000)
	>70	1100 (1300)
Female	51+	1100 (1300)

Notes: EAR: estimated average requirement; RDI: recommended dietary intake.

Source: NHMRC & Ministry of Health (2006).

CASE STUDY 12.2 CARDIOVASCULAR DISEASE

Profile and problem

Mr V. is a 66-year-old retired Australian man who lives with his wife of 45 years in their own home. He has struggled with his weight for a number of years and is currently obese, with a BMI of 35.7. Mr V. has a family history of CVD, with both parents passing away from stroke, and suffers hypertension, hypercholesterolaemia and chronic leg pain on walking. Two weeks ago he was diagnosed with peripheral arterial disease, and his leg pain was identified as intermittent claudication. He is an ex-smoker, engages in very little physical activity and enjoys his food, stating that his wife is a good traditional cook.

Assessment

Mr V. wants to reduce the progression of his peripheral arterial disease and is motivated to lose some weight to avoid surgery in the future. His wife currently cooks traditional meals of meat and three vegetables or fried foods such as fish and chips, so he is aware that she will need to be included in the changes. Mrs V. fries much of their food and roasts vegetables with roast meat; therefore, Mr V.'s diet is likely high in fat, including saturated fat. They consume full-fat dairy products and are fond of biscuits and cakes at afternoon tea and supper. Mr V. is also a social drinker, regularly consuming four to six full-strength beers at the local football club on a Friday evening.

Treatment

Mr V. has agreed to take part in a group weight management program for older men run by his local community health centre, with a view to losing 5 kilograms in the next three months. He will start to visit the local swimming pool and do some gentle walking in the water to build up his tolerance to exercise and assist weight loss. His general practitioner has advised that he should also commence a treadmill walking program, which should increase his pain-free walking distance. With Mrs V.'s assistance, Mr V. has agreed to do the following:

- reduce the portion size of his meals, particularly meat
- reduce snacking, and make any snacks a healthy choice
- include salads, nuts and legumes in his diet
- eat grilled foods instead of fried foods
- switch to low-fat dairy products
- reduce his alcohol intake.

Plan

Mr V. will need to follow a weight management program for some time in order to reach his healthy weight range. He will need to be monitored by a dietitian on an ongoing basis to ensure that his diet is appropriate for each stage of his weight change. Unless his weight reduces there is a risk that he may require bypass surgery on his symptomatic leg. In addition, weight reduction will assist in reducing both hypertension and hypercholesterolaemia, both risk factors for peripheral arterial disease, and in increasing his exercise capacity.

HEALTH CONDITIONS

Evidence indicates that the prevalence of chronic disease in Australia ranges from 50 to 80 per cent of the adult population and is approximately 50 per cent in the New Zealand population. There tends to be a higher prevalence of chronic conditions among the Indigenous peoples of both Australia and New Zealand. With increasing age comes greater incidence of chronic disease, with the prevalence of multi-morbidity in those aged 65 years and over reported to be 65 per cent, increasing to 80 per cent in those aged 75 years and over (Erny-Albrecht & McIntyre 2013). Diet plays a role in the prevention and management of many of these conditions. This section describes the most common conditions and associated nutrients.

Osteoporosis

Osteoporosis is a condition in which bone becomes more porous, therefore more fragile. It arises when bone loss is greater than bone remodelling, resulting in low bone density. Osteoporosis leads to an increased risk of fractures, which can occur with relatively minor trauma, most commonly in the hip, pelvis, wrist and spine. Fractures can cause pain, loss of function and independence and often premature mortality. In Australia over one million people have osteoporosis and 6.3 million people have osteopenia, a less severe condition of low bone density which may progress to osteoporosis over time (Ebeling et al. 2013). It is estimated that approximately 30 per cent of male and 50 per cent of female New Zealanders aged over 60 years will develop osteoporosis (Brown et al. 2007).

Risk factors for osteoporosis include both modifiable and non-modifiable components. There is some genetic component to bone health, such that family history of osteoporosis may increase the risk of incidence of the condition. Age, gender and menopausal status are also non-modifiable risk factors for osteoporosis. Being female is a major contributor to risk of osteoporosis, with women four times more likely to develop the condition than men. Peak bone mass in adults is usually reached by the age of 30 years and remains relatively stable in women until menopause, at the age of 40–50 years. At that time oestrogen levels decline suddenly, resulting in rapid loss of bone density of around 10 per cent of bone mass over the five years following menopause. A more gradual decrease of approximately 1 per cent per year is more normal after that initial decline. Men develop osteoporosis later in life, as their testosterone levels drop more gradually from mid-life until around 70 years of age, when their risk is similar to that of women of the same age.

Conversely, based on age-standardised rates, in Aboriginals and Torres Strait Islanders, a diagnosis of osteoporosis is more common in men than in women. Furthermore, Indigenous Australian men are twice as likely to be diagnosed as non-Indigenous men, and Indigenous Australian women are half as likely to be diagnosed as non-Indigenous women (AIHW 2011a). This is not the case for Māori New Zealanders, who are less likely to be diagnosed with osteoporosis than their counterparts of European descent. This is thought to be due to their higher body weight and therefore higher bone density.

Major modifiable risk factors for osteoporosis are inadequate calcium and vitamin D status and physical inactivity; however, the list also includes smoking, alcohol misuse, use of corticosteroids and low socio-economic status. Aboriginals and Torres Strait Islanders, particularly males, are more likely to report higher prevalence of modifiable risk factors such as smoking, alcohol misuse, physical inactivity and low socio-economic status.

As the main mineral in the human skeleton, calcium is key for the prevention and management of osteoporosis. Evidence in older adults suggests increased calcium requirements; however, nutrient intake in this group can be suboptimal. Calcium status is affected by parathyroid hormone levels, which increase as a part of normal ageing, resulting in greater release of calcium from bone to maintain blood calcium. Low vitamin D levels increase parathyroid hormone secretion as well as impairing intestinal absorption of calcium and phosphorus, another key bone mineral. Vitamin D deficiency is common in Australia, particularly among frail institutionalised elderly people, who may not have adequate exposure to sunlight. Supplementation of the diet with calcium and vitamin D is a key strategy to maximise bone density in this group.

Exercise is an important aspect of attaining peak bone mass in younger years; however, it can also be effective in maximising bone health in older adults. Weight-bearing exercise can be useful in maintaining bone density in the hips, and obesity can be protective due to the weight-bearing nature of the body composition in that group. In general, however, resistance training is more appropriate for older adults, even frail older adults, and can help to improve balance and coordination, which reduces the risk of falls and fractures.

Overweight and obesity

Tackling obesity is one of the nine National Health Priority Areas in Australia, selected because of the burden inflicted on the community by this condition. Figures from 2013 indicate that Australia is among the most overweight countries in the world, and five million Australians are identified as obese, with a BMI greater than 30 (Monash Obesity and Diabetes Institute 2013). Over 30 per cent of New Zealand adults, or just over 1.1 million people, are reported to be obese, with obesity over-represented in the Māori and Pacific populations (Parliamentary Library 2014). Overweight and obesity are associated with increased risk of many of the chronic diseases which are prevalent in the Australian and New Zealand populations.

While overweight and obesity affect all age groups, the indication is that rates increase with age and peak in late middle age and early old age. Figures from the 2011–12 Australian Health Survey showed that rates peaked for males at around 80 per cent between 45 and 74 years and for females at around 70 per cent between 55 and 74 years, before declining into older age. The figures are similar for Aboriginals and Torres Strait Islanders, with around 75 per cent of males and females aged 55 years and over classified as overweight or obese. However, the Indigenous Australian population is almost twice as likely to be obese as the non-Indigenous population (AIHW 2011b).

The 2012–13 New Zealand Health Survey found that obesity rates peaked at 39 per cent in the 65–74-year age group then declined to less than 30 per cent in those aged 75 years and over. Obesity rates were higher in Māori and Pacific peoples than in the non-Indigenous adult population, with 68 per cent and 48 per cent, respectively, found to have a BMI greater than 30 (Marriott & Sim 2014).

Weight gain is a common consequence of normal ageing, due to physiological and lifestyle changes in the later stages of life. Body composition changes lead to a decrease in lean mass and an increase in, and redistribution of, fat mass, with consequent reduction in total body water and energy expenditure for older adults. Loss of lean mass can make it more difficult for older people to remain active, and this, combined with the reduction in activity which commonly occurs with retirement, creates ideal conditions for energy imbalance and weight gain.

The loss of muscle mass with ageing is termed sarcopenia (see Chapter 13); however, this is often masked by obesity, in the condition called sarcopenic obesity, defined as low muscle mass with excess fat mass, particularly centrally. Strategies for weight management in older adults should always include physical activity, particularly **resistance training**, to maximise maintenance of muscle mass. Dietary strategies should focus on nutrient-dense foods to enable requirements to be met without increasing the intake of energy-dense foods. Strategies should also consider potential reduced intake as a result of early satiety, which is common in older adults. Goals should include maximising function to reduce loss of independence and increase quality of life as ageing continues.

Cardiovascular disease

CVD is a term covering a range of conditions of the heart and blood vessels both centrally (coronary heart disease [CHD], cerebrovascular disease) and peripherally (peripheral vascular disease) and includes hypertension and hyperlipidaemia. In acknowledgement of the impact of CVD on the Australian population, it is also one of the nine National Health Priority Areas identified by the Commonwealth government. The New Zealand Health Strategy report on progress for 2012 included improved cardiovascular risk assessment among the general population as one of the government's health targets (Ministry of Health 2012). CVD is the leading cause of death in Australia and New Zealand. Over 15 per cent of Australians are affected by CVD, and there were 45,000 CVD-related deaths nationally in 2011 (Heart Foundation n.d.). In New Zealand, CVD is responsible for 40 per cent of all deaths each year (Heart Research Institute n.d.).

Although the overall prevalence of CVD in Australia and New Zealand has decreased steadily since the late 1960s (more so in Australia than in New Zealand), the proportion of those with CVD increases with age. The evidence indicates that almost 30 per cent of Australians aged 75 years and over had CVD in 2011–12, rising to 47 per cent for men aged 85 years and over. It is more common in males than in females at any age (5.5 versus 4.5 per cent in Australia). In New Zealand, the prevalence of CVD is higher in the Māori population than in any other ethnic group, consistent with the increased prevalence among Indigenous Australians (Ministry of Health 2013b).

CHD is the predominant condition in CVD, accounting for almost half of the deaths attributed to CVD in Australia and one-quarter of all deaths in New Zealand. As with CVD overall, prevalence of CHD increases with age, with 19 per cent of Australians aged 75 years and over reporting a long-term condition in 2011–12 (ABS 2013c). It is estimated that one in eighteen New Zealand adults has CHD, with the highest prevalence among those aged 75 years and over, including males at 28.4 per cent and females at 24.2 per cent (Ministry of Health 2013b).

Risk factors for CVD remain the same in older age as they are for younger people. However, hypertension, hyperlipidaemia, obesity and physical inactivity are common in older age groups and may account for some of the increased prevalence of CVD in these groups. Recommendations for a diet in line with healthy eating guidelines apply; however, they may need to be adjusted for weight management, within the smaller appetite of older adults and balanced with competing requirements of multiple comorbidities.

Type 2 diabetes mellitus

Figures indicate that around 800,000 Australians and 200,000 New Zealanders suffer from T2DM. Similarly to many other chronic conditions, diabetes is more common in males (56 per cent) than in females (44 per

Resistance training is any form of physical training that includes the use of weights.

cent), and in the Indigenous populations of both Australia and New Zealand than in their Caucasian populations. The prevalence of T2DM increases with age, with the majority of those with T2DM being over 64 years.

Although there is a strong genetic link in development of T2DM, the lifestyle risk factors include obesity and physical inactivity. Health outcomes are generally related to vascular damage resulting from elevated blood glucose levels. People with diabetes are three times more likely to have hypertension, are more likely to develop atherosclerosis and are at increased risk of major CVD incidents. Older adults are already at increased risk of many of the poor health outcomes associated with T2DM simply by virtue of age. Management of T2DM should include increased physical activity and dietary intervention for weight management, as well as diet planning to moderate blood glucose levels and minimise the incidence of vascular disease.

SUMMARY AND KEY MESSAGES

Globally, the population is ageing and Australia and New Zealand are no exceptions to that trend. With increased life expectancy it is important to maintain adequate quality of life for as long as possible. Recent evidence indicates that genetics play an important role in successful ageing, with contributing roles also for lifestyle and diet. Maintaining an active lifestyle and consuming a diet in line with recommendations of the Australian Dietary Guidelines (NHMRC 2013) or the New Zealand Ministry of Health's (2013a) Food and Nutrition Guidelines for Healthy Older People are key to reduction in incidence and severity of chronic disease and to successful ageing.

- Many genetic mutations are known to be influenced by nutrition as well as stress; therefore, diet is emerging as an important factor in the promotion of healthy ageing.
- Shortened telomeres are associated with older age and with chronic disease at any age so contribute to the increased prevalence of chronic disease with ageing.
- Numerous physiological changes occur in older age which impact on the ability of older adults to meet nutrient requirements.
- Energy requirements are reduced in older adults due to loss of lean mass occurring as a result of ageing, to decrease in physical activity secondary to retirement and to reduced ability to exercise as a result of loss of lean mass.
- In older adults, nutrient requirements may vary between males and females and may be different from those in younger adults.
- Changes in lifestyle following retirement often result in weight gain, changes in dietary intake and increased alcohol consumption.
- Chronic disease, such as osteoporosis, overweight and obesity, CVD and T2DM, is prevalent in older adults, with prevalence increasing with age and multi-morbidity being common.

REFERENCES

ABS (Australian Bureau of Statistics) 2011. *Australian Social Trends, Mar 2011.* Cat. no. 4102.0. Commonwealth of Australia. <www.abs.gov.au/AUSSTATS/abs@.nsf/Lookup/4102.0Main+Features10Mar+2011>, accessed 23 November 2014.

——2013a. *Gender Indicators, Australia, Feb 2014.* Cat. no. 4125.0. Commonwealth of Australia. <www.abs.gov.au/ausstats/abs@.nsf/Lookup/4125.0main+features3110Feb%202014>, accessed 23 November 2014.

——2013b. *Australian Demographic Statistics, Jun 2013.* Cat. no. 3101.0. Commonwealth of Australia. <www.abs.gov.au/ausstats/abs@.nsf/0/1CD2B1952AFC5E7ACA257298000F2E76?OpenDocument>, accessed 23 November 2014.

——2013c. *Population Projections, Australia, 2012 (base) to 2101.* Cat. no. 3222.0. Commonwealth of Australia. <www.abs.gov.au/AUSSTATS/abs@.nsf/DetailsPage/3222.02012%20(base)%20to%202101?OpenDocument>, accessed 23 November 2014.

——2014. *Australian Health Survey: Nutrition First Results—Food and Nutrients, 2011–12*. Cat. no. 4364.0.55.007. Commonwealth of Australia. <www.abs.gov.au/AUSSTATS/abs@.nsf/DetailsPage/4364.0.55.0072011-12?OpenDocument>, accessed 14 November 2014.

AIHW (Australian Institute of Health and Welfare) 2011a. *A Snapshot of Osteoporosis in Australia 2011*. Arthritis Series 15, cat. no. PHE 137. Commonwealth of Australia. <www.aihw.gov.au/WorkArea/DownloadAsset.aspx?id=10737418747&libID=10737418746>, accessed 5 October 2014.

——2011b. *The Health and Welfare of Australia's Aboriginal and Torres Strait Islander People: An overview*. Cat. no. IHW 42. Commonwealth of Australia. <www.aihw.gov.au/WorkArea/DownloadAsset.aspx?id=10737418955>, accessed 24 November 2014.

Brown, P., McNeill, R., Radwan, E. & Willingale, J. 2007. *The Burden of Osteoporosis in New Zealand: 2007–2020*. Osteoporosis New Zealand. <www.iofbonehealth.org/sites/default/files/PDFs/white_paper_new_zealand_2007.pdf>, accessed 24 November 2014.

Chung, S., Popkin, B.M., Domino, M.E. & Stearns, S.C. 2007. Effect of retirement on eating out and weight change: An analysis of gender differences. *Obesity* 15: 1053–60.

Davies, L., Holdsworth, M. & MacFarlane, D. 1986. Dietary fibre intakes in the United Kingdom before and after retirement from work. *Human Nutrition. Applied Nutrition* 40: 431–9.

Ebeling, P.R., Daly, R.M., Kerr, D.A. & Kimlin, M.G. 2013. Building healthy bones throughout life: An evidence-informed strategy to prevent osteoporosis in Australia. *MJA Open* 2(1 Suppl.). <www.mja.com.au/open/2013/2/1/building-healthy-bones-throughout-life-evidence-informed-strategy-prevent-osteoporosis>, accessed 5 October 2014.

Erny-Albrecht, K. & McIntyre, E. 2013. The growing burden of multimorbidity. *Public Health Care Research & Information Service Research RoundUp* 31; August 2013.

Fenech, M.F. 2010. Dietary reference values of individual micronutrients and nutriomes for genome damage prevention: Current status and a road map to the future. *American Journal of Clinical Nutrition* 91: 1438S–54.

Forman-Hoffman, V.L., Richardson, K.K., Yankey, J.W., Hillis, S.L. et al. 2008. Retirement and weight changes among men and women in the health and retirement study. *Journals of Gerontology Series B: Psychological Sciences and Social Sciences* 63: S146–53.

Heart Foundation n.d. Data and statistics (webpage). <www.heartfoundation.org.au/information-for-professionals/data-and-statistics/Pages/default.aspx>, accessed 24 November 2014.

Heart Research Institute n.d. *Heart Disease in New Zealand*. Heart Research Institute. <www.hri.org.nz/about-heart-disease/heart-facts/>, accessed 24 November 2014.

Kenyon, C.J. 2010. The genetics of ageing. *Nature* 464: 504–12.

Kiecolt-Glaser, J.K., Belury, M.A., Andridge, R., Malarkey, W.B. et al. 2012. Omega-3 supplementation lowers inflammation in healthy middle-aged and older adults: A randomized controlled trial. *Brain, Behavior, and Immunity* 26: 988–95.

Lauque, S., Nourashemi, F., Soleilhavoup, C., Guyonnet, S. et al. 1998. A prospective study of changes on nutritional patterns 6 months before and 18 months after retirement. *Journal of Nutrition, Health and Aging* 2: 88–91.

Marin, C., Delgado-Lista, J., Ramirez, R., Carracedo, J. et al. 2012. Mediterranean diet reduces senescence-associated stress in endothelial cells. *Age* 34: 1309–16.

Marriott, L. & Sim, D. 2014. *Indicators of Inequality for Māori and Pacific People*. Victoria University of Wellington. <www.victoria.ac.nz/sacl/centres-and-institutes/cpf/publications/pdfs/2015/WP09_2014_Indicators-of-Inequality.pdf>, accessed 24 November 2014.

Mattison, J.A., Roth, G.S., Beasley, T.M., Tilmont, E.M. et al. 2012. Impact of caloric restriction on health and survival in rhesus monkeys from the NIA study. *Nature* 489: 318–21.

Ministry of Health 2012. *Implementing the New Zealand Health Strategy 2012*. New Zealand Government. <www.health.govt.nz/system/files/documents/publications/implementing-new-zealand-health-strategy-2012.pdf>, accessed 7 October 2014.

——2013a. *Food and Nutrition Guidelines for Healthy Older People: A background paper.* New Zealand Government. <www.health.govt.nz/system/files/documents/publications/food-and-nutrition-guidelines-healthy-older-people-background-paper-jan2013.pdf>, accessed 16 March 2014.

——2013b. *New Zealand Health Survey: Annual update of key findings 2012/13.* New Zealand Government.<www.health.govt.nz/system/files/documents/publications/new-zealand-health-survey-annual-update-2012-13-dec13-v2.pdf>, accessed 24 November 2014.

Monash Obesity and Diabetes Institute 2013. Obesity in Australia (webpage). Monash University. <www.modi.monash.edu.au/obesity-facts-figures/obesity-in-australia/>, accessed 24 November 2014.

Morgan, T., Williamson, M., Pirotta, M., Stewart, K. et al. 2012. A national census of medicines use: A 24-hour snapshot of Australians aged 50 years and older. *Medical Journal of Australia* 196: 50–3.

NHMRC (National Health and Medical Research Council) 2013. *Australian Dietary Guidelines; Providing the scientific evidence for healthier Australian diets.* Commonwealth of Australia. <www.eatforhealth.gov.au/sites/default/files/files/the_guidelines/n55_australian_dietary_guidelines.pdf>, accessed 16 March 2014.

NHMRC (National Health and Medical Research Council) & Ministry of Health 2006. *Nutrient Reference Values for Australia and New Zealand: Including recommended dietary intakes.* Commonwealth of Australia & New Zealand Government. <www.nhmrc.gov.au/_files_nhmrc/publications/attachments/n35.pdf>, accessed 26 September 2014.

Nooyens, A.C., Visscher, T.L., Schuit, A.J., van Rossum, C.T. et al. 2005. Effects of retirement on lifestyle in relation to changes in weight and waist circumference in Dutch men: A prospective study. *Public Health Nutrition* 8: 1266–74.

Parliamentary Library 2014. *Obesity and Diabetes in New Zealand.* New Zealand Government. <http://apo.org.au/files/Resource/nzparliamentarylibrary_obesityanddiabetesinnewzealand_oct_2014.pdf>, accessed 24 November 2014.

Rejeski, W.J., Marsh, A.P., Chmelo, E. & Rejeski, J.J. 2010. Obesity, intentional weight loss and physical disability in older adults. *Obesity Reviews* 11: 671–85.

Statistics New Zealand 2012. *Demographic Projections from Statistics New Zealand: Aims, methods, and results.* New Zealand Government. <www.treasury.govt.nz/government/longterm/externalpanel/pdfs/ltfep-s1-06.pdf>, accessed 24 November 2014.

——2013. *New Zealand Period Life Tables: 2010–12.* New Zealand Government. <www.stats.govt.nz/browse_for_stats/health/life_expectancy/NZLifeTables_HOTP10-12.aspx>, accessed 23 November 2014.

Steenstrup, T., Hjelmborg, J.B., Mortensen, L., Kimura, M. et al. 2013. Leukocyte telomere dynamics in the elderly. *European Journal of Epidemiology* 28: 181–7.

University of Otago & Ministry of Health 2011. *A Focus on Nutrition: Key findings of the 2008/09 New Zealand Adult Nutrition Survey.* New Zealand Government. <www.health.govt.nz/system/files/documents/publications/a-focus-on-nutrition-v2.pdf>, accessed 2 October 2014.

Werkman, A., Hulshof, P.J., Stafleu, A., Kremers, S.P. et al. 2010. Effect of an individually tailored one-year energy balance programme on body weight, body composition and lifestyle in recent retirees: A cluster randomised controlled trial. *BMC Public Health* 10. <www.ncbi.nlm.nih.gov/pmc/articles/PMC2845102/>, accessed 5 October 2014.

Wilson, D.M. & Palha, P. 2007. A systematic review of published research articles on health promotion at retirement. *Journal of Nursing Scholarship* 39: 330–7.

ADDITIONAL READING

Vijg, J. & Suh, Y. 2005. Genetics of longevity and aging. *Annual Review of Medicine* 56: 193–212.

Wheeler, H.E. & Kim, S.K. 2011. Genetics and genomics of human ageing. *Philosophical Transactions of the Royal Society B: Biological Sciences* 366: 43–50.

TEST YOUR UNDERSTANDING

Note: There is only one correct answer for each question.

1. Telomeres are located at the terminal portion of chromosomes. They consist of which of the following repeats?
   a. TTTAGG
   b. TTAAGG
   c. ATTGGG
   d. TTAGGG
   e. GGAATT
2. Which of the senses is not commonly affected in normal ageing?
   a. Sight
   b. Touch
   c. Hearing
   d. Smell
   e. Taste
3. Which of the following micronutrients is required in larger amounts by women aged 51–70 years?
   a. Iron
   b. Selenium
   c. Zinc
   d. Calcium
   e. Riboflavin
4. Which of the following micronutrients is required in larger amounts by all adults aged over 70 years compared with those aged 51–70 years?
   a. Vitamin B12
   b. Iron
   c. Choline
   d. Vitamin D
   e. Iodine
5. Which of the following is true when considering life after retirement?
   a. Weight gain is likely
   b. There is increased frequency of eating meals outside the home
   c. Physical activity is likely to decrease
   d. a and c
   e. a, b and c
6. How many Australians currently have conditions associated with low bone mineral density?
   a. One million
   b. Six million
   c. Over seven million
   d. 500,000
   e. Three million
7. Which of the following is a risk factor for osteoporosis?
   a. Family history
   b. Gender
   c. Inadequate vitamin D status
   d. a and b
   e. a, b and c
8. Which of the following is not a body composition change normally associated with the ageing process?
   a. Increased fat mass
   b. Redistribution of fat mass
   c. Increase in lean mass
   d. Reduction in total body water
   e. Reduction in lean mass
9. At what age is CVD more common in males than in females?
   a. At any age
   b. 51 years and over
   c. In the eighth decade of life
   d. Over 21 years
   e. In middle age
10. T2DM is common in older adults and increases the risk factors for other chronic conditions, including which of the following?
    a. High blood pressure
    b. Stroke
    c. Peripheral vascular disease
    d. Atherosclerosis
    e. All of the above

STUDY QUESTIONS

1. Discuss the mechanisms postulated to be involved in the genetics of ageing.
2. What are the major changes which occur in transition to retirement?
3. What are the key physiological changes which occur in ageing?
4. What effect does ageing have on the energy requirements of humans?
5. Describe the nutrient requirements for older adults.
6. What are the main chronic diseases common in older adults?

13

Later stage older adult (85 years and over)

Michelle Miller and Alison Yaxley

LEARNING OUTCOMES

Upon completion of this chapter you will be able to:

- describe the common physiological changes associated with older age
- list the risk factors for nutritional deficiency in the oldest-old
- identify the nutrient requirements of the oldest-old
- discuss wasting conditions common in this age group and methods of their identification and treatment.

Of all the age groups in Australia, the fastest growing population sector is older adults aged 85 years and over, also known as the oldest-old. Over the period from June 2010 to June 2011 there was a 5.3 per cent increase in Australians aged 85 years and over, bringing this population group to 415,400, with an even larger 18.2 per cent increase in the 100 years and over age group, bringing that group to 4252 people. Since 1995, there has been a 169 per cent explosion in numbers of the oldest-old, far exceeding the total population growth of only 31 per cent (ABS 2008). Projections indicate that this group will make up 5.6–6 per cent of the population by 2061, up from 2.3–2.7 per cent in 2007. With the Australian population forecast to reach between 36.8 and 48.3 million by 2061, there is the prospect of approximately 1.9 million adults aged 85 years or over (ABS 2013). The situation is not dissimilar in New Zealand, where, although not the fastest growing group, there is considerable growth in numbers of the oldest-old, outpaced only by growth in the 60–69 years age group. The 2013 New Zealand census figures indicate approximately 73,000 people aged 85 years and over, a 29.4 per cent increase since 2006. Projections forecast that 7 per cent of the New Zealand population will be aged 85 years or over by 2061, compared with 3.5 per cent in 2012 (Statistics New Zealand 2012).

An increase in quantity of life provides challenges for maintaining quality of life. Further to the issues of ageing outlined in Chapter 12, there are other key issues in those aged 85 years and over which arise as a result of living to an extended age. This chapter outlines some of these key issues, specifically risk factors for poor nutritional health, including frequent illness and subsequent hospital admissions, residential accommodation and a range of other physiological changes that are more common in this age group. It also describes nutrient requirements particular to this stage of the lifespan and wasting conditions, including identification methods and potential treatment strategies.

PHYSIOLOGICAL CHANGES

Body composition

In terms of body composition, the consequences of ageing are a decrease in fat-free mass, mainly appendicular skeletal muscle mass, and an increase in and redistribution of fat mass without necessarily demonstrating weight change. In general, body weight and obesity level off around the age of 80 years before beginning a gentle decline.

Skeletal muscle has considerable metabolic activity; any loss may exacerbate the decrease in resting energy expenditure that occurs with age, independent of body composition. Sarcopenia, the loss of muscle mass, can be limiting in the older adult, particularly in terms of maintaining good movement and balance, and potentially increasing the risk of fall. The risk factors for sarcopenia include weight loss (unintentional or intentional) and being mostly sedentary. This reduction to skeletal muscle is thought to promote increased fat mass and reduction in fat-free mass. Furthermore, this can be accompanied by a decrease in appetite and increased systemic inflammation, both of which can impact on maintenance of optimal nutritional status.

Some extra weight has been shown to be protective for older adults when compared to younger people; this is termed the obesity paradox. The complications associated with obesity in younger adults do not tend to apply in the latest stage of the lifespan, with higher body weight protecting the older adult from functional decline and low body weight associated with increased risk of mortality. Furthermore, there is evidence that a body mass index (BMI) in the overweight range is associated with a reduced risk of mortality in older adults (Janssen & Mark 2007). Recommendations for a healthy weight range for older adults are therefore less restrictive than for the general population. While there are no evidence-based guidelines on BMI for older adults, the healthy range 22–27 is often applied in practice and is acknowledged as potentially appropriate in the nutrient reference values (NRVs) for Australia and New Zealand (NHMRC & Ministry of Health 2006).

Sensory

Taste and smell both play a role in the regulation of appetite, and the sight of food increases appetite such that the food is more likely to be eaten if it is seen. Even hearing affects dietary intake, as consumption can be stimulated by the sound of food and drink such as the crunch of a crisp apple or the fizz upon opening a bottle of carbonated drink. However, the ageing process negatively affects all senses (except touch), including those associated with food consumption, and such changes can be more severe in conjunction with certain health conditions, polypharmacy and therapeutic interventions that are more common in older adults. These changes can all negatively impact on the amount and quality of food consumed by the oldest-old.

Taste can be altered by degrees, ranging from distorted taste perception to complete loss of taste. Research shows that taste perception changes so that some compounds such as fat and salt must be almost five times stronger to be perceived by older adults (Schiffman 2009). Changes in taste can also increase the possibility of food-borne illness in this already vulnerable group, through the reduced ability to identify foods as inappropriate to eat. Diminished sight can contribute to this danger.

The sense of smell deteriorates in older age, and the ability to distinguish between aromas is also diminished, with research indicating that over 60 per cent of those aged 80 years or over have some degree of olfactory impairment (Boyce & Shone 2006). It may also be that this type of impairment is responsible for some of the deterioration in taste experienced in this age group. While reduced ability to smell foods contributes to the risk of food-borne illness, the fact that taste and smell are physiologically linked to the regulation of appetite means that this kind of sensory impairment has implications in terms of nutritional risk.

Gastrointestinal

Normal ageing brings about the decline of some key functions in the gastrointestinal tract, with some particularly concerning in the oldest-old. The mouth undergoes a number of important changes that impact upon the ability of this vulnerable group to consume adequate food to meet their needs. It is common for older adults to have poor dentition. This can be due to wear and tear on teeth over a number of years, gum disease resulting in pain or loose teeth, or ill-fitting dentures. Regardless of the cause, this can result in an inability or lack of desire to eat due to discomfort, thus increasing nutritional risk. Saliva production is reduced with older age and is often compounded by medications. This can increase difficulty in chewing and swallowing, due to lack of lubrication and leads to inadequate production of **salivary amylase**. This enzyme together with chewing food is necessary to signal the start of the digestion process. Swallowing difficulties, or **dysphagia**, are also common and may manifest acutely or gradually. Complaints of difficulty in swallowing food or medication should always be investigated thoroughly by a speech pathologist and may require a texture-modified diet to avoid choking risk or **aspiration pneumonia**.

Oesophageal function may also be reduced and can cause gastro-oesophageal reflux in this population. Gastro-oesophageal reflux disease, principally heartburn, is reported in 15 per cent of the population globally; however, in older adults this is more likely to manifest as dysphagia, regurgitation or vomiting (D'Souza 2007). Although gut motility and emptying and small bowel motility are usually normal in older adults, the motility of the colon may decrease and can contribute to the constipation that often manifests in older age. Changes in the environment of the stomach can lead to gastric ulcer, and reduction in exocrine pancreatic secretions can lead to maldigestion and malabsorption. In particular, there is a reduction in vitamin B12 absorption with increasing age.

Cognition

A number of physical changes in the brain that are a part of normal ageing manifest in cognitive changes that affect all older adults to some degree. Brain mass declines after the early twenties, and for most people the cortex begins to shrink during their forties, at which time cognitive decline tends to accelerate. Neurons shrink or atrophy, and there is less consecutiveness among neurons. The ageing brain has reduced blood flow and generally becomes less efficient. However, the degree to which these changes occur varies across the population, as does the degree to which the changes affect cognitive decline.

Salivary amylase is an enzyme in saliva which breaks down starches into sugars.

Dysphagia is difficulty or discomfort in swallowing.

Aspiration pneumonia is an acute respiratory infection which occurs when food or fluid is inhaled into the lungs.

Expected changes in older age include short-term memory loss, which may manifest as an inability to remember where you put the car keys or to remember certain words or phrases; however, long-term memory is usually preserved. Learned knowledge and behaviours and single-focus attention tend to be preserved, but the learning of new things and the ability to multitask can be impaired with ageing. While vocabulary usually survives intact into older age, the process of speaking can become more taxing, as it may be difficult to find the correct words or to remember names. Problem-solving skills are maintained but solving new issues can take longer. Finally, the speed at which older people process information slows, so that doing things takes longer. These changes are also affected by environmental factors such as medications, general health, depression and physical changes such as loss of hearing. Good nutrition also has a key role to play to maximise normal brain function. There are adults in their nineties and beyond who are highly functioning—they live independently, prepare meals and care for grandchildren and great-grandchildren. The Blue Zones (Buettner 2012) project has identified characteristics of longevity in centenarians that include living life with purpose and at lower stress levels, being socially connected, drinking wine with friends, eating until only partially full, eating mostly plant-based foods, and consuming meat only about once a week.

Expected cognitive changes are different from more serious cognitive changes such as dementia, a term describing confusion and cognitive impairment that may have a number of aetiologies, including Alzheimer's disease, Parkinson's disease, vitamin deficiencies and alcohol abuse. Symptoms include memory loss, language problems, paranoia and impaired judgement and reasoning. Diagnosis should be made only following a detailed multifactorial examination by appropriate medical practitioners.

Nutritional consequences of dementia are poor appetite, malnutrition and weight loss, for reasons including failure to recognise food, impaired swallowing ability, failure to recognise hunger or thirst, physical disabilities and forgetting to eat. Potential strategies to address these consequences are likely to be developed in a multidisciplinary environment with speech therapy, occupational therapy and physiotherapy in conjunction with a dietitian if the older person is in a hospital or residential care setting. Older adults living in the community can be assisted to continue living at home with support. Management plans might include texture-modified diets, exercise therapy to maximise physical function and aids such as alarm calls to remind patients to eat and drink at meal times.

As dementia is a progressive condition, the patient may become unable to or may refuse to eat, and tube feeding may be suggested, although there is limited evidence of benefit from this strategy and patients may repeatedly remove the tube due to discomfort. Inability to eat is common in late-stage dementia and signals that end of life is near; therefore, such invasive strategies are often futile and counterproductive if it causes distress, and comfort or palliative care is a more frequently used strategy.

Disability

Disability is a broad term used to describe any mental or physical condition that impairs an individual's ability to perform tasks commensurate with optimum function for a person of a similar age. Arthritis, frailty and cardiovascular disease (CVD) are major causes of disability among older adults in Australia. The decline in fat-free mass that occurs as a result of normal ageing may manifest as reduced mobility, reduced capacity to perform activities of daily living and functional decline that can necessitate a move to a higher level of care, with consequent loss of independence and increased nutritional risk.

The oldest-old are often frail and require assistance in activities of daily living either at home or in residential aged care. This can include the need for assistance in meal set-up and feeding as well as in personal care. Increased reliance on others is implicated in the development of depression, an independent risk factor for poor nutritional status. However, poor nutritional status is also a risk factor for depression; hence, it is often unclear which comes first. Minimising risk should be a major goal in the nutritional management of older adults, in order to preserve independence and reduce frailty as much as possible.

NUTRIENT REQUIREMENTS AND INTAKES

The New Zealand Food and Nutrition Guidelines for Healthy Older People are specific to healthy older adults aged 65 years and over, and the Australian Dietary Guidelines are suitable for healthy individuals up to the age of 70 years (Ministry of Health 2013; NHMRC 2013). It is likely that the Australian guidelines are appropriate recommendations for healthy adults over the age of 70 years, as are the NRVs for Australia and New Zealand (NHMRC & Ministry of Health 2006); however, there are some other specific considerations in the oldest-old that are detailed in this section.

Macronutrients

Protein and energy

Achieving adequate protein and energy intakes in older age is essential for maintenance of body weight and lean mass. Unintentional weight loss and loss of lean mass is common in the oldest-old and contributes to increased risk of disability, morbidity and mortality. However, appetite regulation is impaired in normal ageing; therefore, achieving adequate intakes (AIs) can be challenging. This is compounded by sensory changes that make food less appealing and poor dentition that makes food difficult to eat.

Strategies to improve intake should focus on eating small frequent meals of nutrient-dense foods and not simply energy-dense foods, making every mouthful count. The aim should be for weight maintenance in most cases, even weight gain if possible for the very undernourished, and rarely weight loss, as there is evidence of poor outcomes when energy intake is intentionally restricted in older adults.

Water

Dehydration resulting from inadequate fluid intake is a common nutritional risk factor in the oldest-old. Risk factors for dehydration in the elderly (**urinary tract infection**, vomiting, diarrhoea, diabetes, infection from **pressure sores**) may increase requirements for fluids. Sense of perception of thirst lessens with age. Studies have demonstrated that following water deprivation, older adults do not get as thirsty as younger people, therefore do not drink enough to rehydrate themselves and are subsequently at higher risk of dehydration. Poor cognition, dementia, poor mobility and texture-modification requirements in older adults may further impede access to or the ability to access adequate fluid intake.

For older adults in whom poor appetite and unintentional weight loss are common problems,

A **urinary tract infection** is an infection that can affect the urethra, bladder or kidneys.

Pressure sores are chronic ulcers of the skin caused by prolonged pressure.

CASE STUDY 13.1 MIGRANT HEALTH

Tania Thodis

Profile and problem

Mr N. is an 89-year-old Greek-born Australian male who has been widowed for three years. He lives alone in his own home. He generally speaks Greek with friends and family but is also fluent in English. His late wife prepared meals, and he has been experiencing difficulty with cooking meals daily and preparing food for himself. His son and daughter-in-law live nearby, and he has dinner with them and his grandchildren once or twice per week. His family are concerned that he has lost weight, and he is also concerned about his energy levels and ability to complete his daily living tasks. He used to enjoy gardening daily but now gets quite tired and also finds daily housework tiring. He walks for about fifteen minutes two to three times a week to the local shopping centre to meet with friends for lunch. He has had some episodes of forgetfulness, such as leaving the stove on, and his family have noticed that he occasionally forgets words when speaking, which he gets quite distressed about. His current weight is 69 kilograms, his height is 170 centimetres. Over the last twelve months he has lost about 8 kilograms. He takes warfarin and medication for hypertension daily.

Assessment

Mr N.'s current intake consists mainly of takeaway-type meals, such as fish and chips or fried rice, or noodles and grilled or battered fish for lunch when he eats out. At home, his usual breakfast is a boiled egg and a slice of wholemeal bread with a drizzle of olive oil, a cup of full-fat milk and a Greek coffee (no sugar) or porridge with full-fat milk and a teaspoon of honey. For dinner he cooks pasta, legumes in a soup, fried chicken or fish, roast vegetables or two-minute noodles or has yoghurt leftovers from dinner with his family. Snacks include one to two pieces of fruit per day, and he may have a couple of sweet biscuits with a cup of black tea in the evening.

Treatment

Mr N. has agreed to do the following:

- focus on nutrient-dense foods and nourishing fluids, and small frequent meals
- ensure adequate fluid intake, with water, milk, juice or tea
- eat dairy foods such as yoghurt more frequently and increase intake of cheese and milk, daily when possible
- spend time in the sunshine daily to minimise risk of vitamin D deficiency, possibly requesting a blood test from local general practitioner to assess status
- drink nourishing fluids on days when appetite is low to assist meeting estimated protein requirements and minimise weight loss, and eat ice cream for dessert, continue use of full-fat milk and perhaps use a protein powder added to milk yoghurt
- continue social activities to minimise social isolation from friends and family
- link into community health for provision of home help to assist with activities of daily living.

Plan

Mr N.'s general practitioner will refer him to a dietitian, who will be able to do a home visit to monitor his weight and dietary intake and assess his ability to prepare appropriate meals for himself. Mr N. will also be referred for an Aged Care Assessment Team visit to determine his need and eligibility for home help and/or delivered meals. As Mr N.'s general practitioner assessed his vitamin D status and found it to be low, he will commence vitamin D supplementation. Mr N. will be encouraged to attend local Greek clubs once a week with the support of a carer.

increased intake of fluids may be implicated in exacerbating poor nutritional intake through displacing food intake. In some groups, particularly malnourished older adults, provision of adequate energy and protein can be achieved through nourishing fluids that also contribute to fluid intake.

Micronutrients

Vitamin B12

Vitamin B12 supplementation is often prescribed for older adults, suggesting that deficiency is common, with prevalence increasing with age and as a consequence of **atrophic gastritis**. Deficiency may be a result of inadequate dietary intake; however, there are physiological factors associated with ageing that may impact on the ability to absorb sufficient quantities of this vital micronutrient. Vitamin B12 is not well absorbed in older adults, often due to reduced amounts of stomach acid that may occur as a result of medications such as antacids and proton pump inhibitors. Even in the presence of adequate stomach acid, vitamin B12 can be absorbed only with adequate amounts of intrinsic factor, a protein secreted by the stomach that carries vitamin B12 through the small intestine, where it is absorbed.

High intakes of folic acid can mask vitamin B12 deficiency, which is important to remember in older adults in the context of mandatory folic acid fortification in the Australian food system. Although the New Zealand authorities have not elected to enforce folate fortification, the voluntary fortification of some New Zealand foods may also impact on recognition of vitamin B12 deficiency. Vitamin B12 is also associated with homocysteine metabolism and therefore may have a role to play in the reduction of CVD.

Calcium

Adequate calcium intake is a major factor related to the protection from osteoporosis, which is the greatest cause of fractures in older adults. Due to reduced intestinal absorption, increased urinary excretion of calcium (especially in the presence of high sodium intakes) and increased bone losses estimated to be about 0.5–1 per cent per annum for both sexes (Jones et al. 1994), requirements for calcium are increased in old age (at over 50 years and over 70 years). Due to the high requirements for calcium, adequate consumption through dietary sources alone may not be possible; therefore, supplementation may need to be considered.

Vitamin D

Lack of exposure to sunlight is one the major reasons for vitamin D deficiency in older adults, as only 5–10 per cent of intake of this nutrient is obtained through dietary sources. It is not surprising that older community-dwelling adults, and particularly those in **residential care** facilities, are at high risk for vitamin D deficiency, which is recognised as a significant public health issue in Australia and New Zealand. Older adults also have less vitamin D substrate (7-dehydrocholesterol in the skin) and lower production rates of vitamin D from even high levels of ultraviolet radiation exposure. Routine vitamin D supplementation is appropriate for most older adults, importantly in combination with adequate calcium, lowering the risk of fall-related fractures (Nowson et al. 2012).

Atrophic gastritis is a chronic inflammation of the stomach mucosa that leads to the replacement of functional with non-functional cells.

Residential care is long-term care given to people who stay in a residential setting rather than in their own home. Care provided depends on the level of independence of the resident but generally includes support for activities of daily living and meal provision.

COMMON CONCERNS AND HEALTH CONDITIONS

Nutritional risk factors

Risk factors for poor nutritional status are numerous and have been highlighted throughout this text; however, there are some that are worthy of further discussion, as they are particularly important or prevalent in the oldest-old.

Hospital admissions

In Australia, 48 per cent of hospital inpatient days were attributed to those aged 65 years and over in 2012–13, with the oldest-old accounting for more of those days than any other group (AIHW 2014). This is reflected in New Zealand, where in 2009–10 the 85 years and over age group had the highest rate of publicly funded hospitalisations, with around 87,000 per 100,000 (Ministry of Health 2012). It is well documented that hospital admission is associated with a decline in nutritional status, particularly among older adults (Ross et al. 2011). This occurs for a number of reasons. Patients may be subject to repeated fasting associated with required tests and procedures, unfamiliar or culturally inappropriate foods and set meal times, as well as being in pain, nauseous, having poor or inadequate dentition or being unable to feed themselves. In addition, staff often report that they are not able to provide feeding assistance or encouragement, as they are stretched at meal times with other tasks, such as medication rounds, and are unable to attend to all the nutritional needs of their patients. McWhirter and Pennington (1994) published a seminal article on decline in nutritional status during hospital admission and reported a mean weight loss of 5.4 per cent at discharge. A more recent study reported weight loss of up to 10 per cent in two-thirds of older hip-fracture patients, with those exhibiting greater weight loss having longer mean lengths of stay, of nine days, demonstrating that hospital admissions remain a risk factor for decline in nutritional status two decades on (Reider et al. 2013).

Consequently, malnutrition risk screening is particularly important in the acute setting, especially in the oldest-old. A range of instruments have been developed and demonstrated to be useful in detecting malnutrition among inpatients, the most commonly administered in Australia and New Zealand being the Malnutrition Universal Screening Tool, the Malnutrition Screening Tool and the Mini Nutritional Assessment Short-form. (See Barker et al. 2011 for further discussion of these instruments.)

Residential care

The Evidence Based Practice Guidelines for the Nutritional Management of Malnutrition in Adult Patients across the Continuum of Care, endorsed by the Dietitians Association of Australia, highlight the prevalence of malnutrition in a range of settings and confirm that the figures are higher (estimated at 40–70 per cent) for those in a higher level of care, such as a nursing home (Watterson et al. 2009). One of the risk factors for admission to residential care is prior hospital admission, which may be due to deterioration in nutritional status during admission resulting in functional decline. Once in residential care, however, older adults appear to be at risk of further deterioration in nutritional status, largely due to systems issues, including lack of staffing to assist residents with feeding, dependence on food services for all meal provision, structured and shared meal times and lack of flexibility to cater for individual food preferences or dietary needs.

Despite the high level of poor nutritional status in aged-care residents, there is little research on the nutrient requirements of this group. While it is generally accepted among nutritionists and dietitians that general healthy eating guidelines are not appropriate for adoption at this stage of the lifespan, the message has not necessarily filtered through to this care setting; hence, restrictions of sugar and sodium in particular are still commonplace.

Bereavement and social isolation

Although it can happen at any age, bereavement is most common in older adults, as an older person's social network will be largely made up of those of a similar age. Bereavement has repeatedly been demonstrated to have a negative effect on mental and physical health. A systematic review found strong evidence of a link between bereavement and nutritional risk and unintentional weight loss as a result of impaired nutrient intake and poor eating behaviours (Stahl & Schulz 2013). These effects were reported to be strongest in the first year after bereavement. It is widely reported that there is a link between depression and malnutrition and it is likely that these negative outcomes following bereavement are a reflection of grief, depression and anxiety impacting on appetite and consequently reducing dietary intake.

When a longstanding partner passes away, **social isolation** can become a major challenge. Energy intakes are reportedly lower among those living alone than among those living with others (Hickson 2006). For older adults, maintaining independence in transport becomes more difficult due to physical and mental decline, and this can impact on maintenance of social networks, so enjoyment of food with others can become less feasible with advanced age. Bereaved people can also find themselves in a challenging situation if they have never been responsible for meal preparation. It is not uncommon for older men to have little knowledge about foods, nutrition and cooking, the result of which is an increase in risk of poor dietary intake and subsequent poor nutritional status. Financial pressures are also common in advanced age, and these can translate into issues of food insecurity as the food budget is often perceived as dispensable.

Polypharmacy

The combination of advances in chronic disease and reduced physiological function with increasing age commonly translates into increased need for medications, in terms of both dose and quantity. Many older adults are prescribed multiple medications, each with their associated side-effects and contraindications. There are a range of medications that are known to affect nutrient bioavailability, and therefore some recommended dietary intakes (RDIs) may be inadequate to maintain optimal health. Some medications can

Social isolation describes a state of little or no interaction with others outside an individual's residence.

Polypharmacy is the use of multiple medications, usually defined as five or more.

induce depression or interfere with hunger signals, which themselves are risk factors for poor nutritional health. Additionally, the quantity of medications may impact on nutritional status, with multiple medications being consumed at meal times given preference over food and appetite for food diminishing as a result of feeling full.

Unintentional weight loss

The prevalence of unintentional weight loss is estimated to be over 25 per cent in community-dwelling older adults and may be as high as 60 per cent in older adults in residential aged-care facilities (McMinn et al. 2011). Studies in very healthy adults in the eighth decade of life have reported that the rate of unintentional weight loss is low, at only 0.1–0.2 kilograms per year; therefore, any discernible weight loss should not be dismissed as normal in older people.

While rapid unintentional weight loss of 5 per cent in one month is considered clinically meaningful, an annual loss of more than 4 per cent of body weight is also an important predictor of mortality, and weight loss of only 3 per cent over one year may have a similar impact in frail older adults. Left untreated, unintentional weight loss is associated with poor outcomes, including depression, reduced quality of life, impaired activities of daily living, reduced physical function, depressed immune function, increased lengths of hospital stays, impaired wound healing and increased mortality.

Unintentional weight loss in advanced age can be defined by three primary syndromes: starvation, sarcopenia and cachexia, each of which has defining characteristics and effective treatment strategies or emerging treatment options.

Starvation

It is generally accepted that starvation occurs purely as a result of protein-energy deficiency. Globally, this is most often due to a lack of sufficient food availability, but in developed countries such as Australia and New Zealand starvation is generally due to imposed restrictions even in the presence of adequate food provision. Reduced appetite, swallowing difficulties, dementia and social isolation, all common for the oldest-old, are established predictors of inadequate protein and energy intake. Consequences of starvation are loss of both lean mass and fat mass. The major factor that distinguishes starvation from sarcopenia and cachexia is that it is reversible when adequate energy and protein intake is achieved (Thomas 2002). Some strategies for improving intake include liberalisation of the diet, appetite stimulants, modifying the texture of the diet and provision of nourishing fluids.

A liberalised diet is an approach to relax dietary modifications or prescriptions that are often implemented to manage common diet-related comorbidities and risk factors for diseases such as diabetes, CVD and hypertension. These prescriptions often focus on total energy, sugar and sodium. The implications of a liberalised diet for the oldest-old is an improved enjoyment of food, increased dietary intake and reduced risk of frailty, with some evidence suggesting these changes do not impact negatively on risk of chronic disease (Baulderstone et al. 2012; Darmon et al. 2010).

Strategies to increase dietary intake might include the use of appetite stimulants. The most studied pharmacological appetite stimulants are progestogens and corticosteroids, and these have been approved for the treatment of disease-related cachexia in some countries. Other agents actively researched include ghrelin, thalidomide and non-steroidal anti-inflammatory drugs; however, the evidence for these is not strong. The appetite can also be stimulated by some widely available items such as herbs and spices, which can make food taste more appealing; ginger, for example, has long been successfully used as a stimulant. Hot food is also more stimulating to the appetite, as it tends to have more of an aroma. Alcohol is also an appetite stimulant, so having a drink with a meal can increase intake.

Texture modification of food and fluid may be necessary in the oldest-old, to manage functional and disease-related dysphagia. While the modification of the foods may lead to easy and safe consumption, it is important to consider that the process of texture modification can result in **nutrient dilution**. Another important consideration is that a texture-modified diet can appear uniform in appearance and so may be visually unappealing. Fortification of meals with additional energy sources, specifically protein, is often necessary to avoid compounding unintentional weight loss.

Nourishing fluids are often provided for the nutritional management of unintentional weight loss, as part of a high-energy, high-protein diet. The evidence

Nutrient dilution is a reduction of nutrient density, the nutrient intake related to energy intake.

suggests that they can successfully increase energy and protein intake among older adults. Commercial oral nutritional supplements can be purchased from supermarkets and pharmacies and generally come either as powders that can be made up into drinks or puddings or as ready-made. While not generally nutritionally complete, these products contain a range of vitamins and minerals as well as being high in energy and protein. Powdered commercial supplements can be added to foods and drinks at home to supplement the nutrient content, or everyday products commonly found in larders can be used to make a homemade supplement. Milk can be fortified by the addition of full-cream milk powder or evaporated milk to create a low-cost, nourishing fluid, and this can be further fortified by the addition of ice cream and flavoured syrups. Unintentional weight loss that is not resolved by increased dietary intake is unlikely to be due to simple starvation and may be due to sarcopenia or cachexia.

Sarcopenia

Sarcopenia is the progressive loss of muscle mass that occurs with normal ageing. It is associated with increased frailty, loss of strength, reduced physical function and diminished capacity for exercise as a result of decreased muscle mass, increased deposition of fat within muscles and alterations to the structure of muscle tissue at the microscopic level (Cruz-Jentoft et al. 2010; Thomas 2007).

The ageing process has detrimental effects on the contractile and material properties of the muscles and on the function of the tendons. Changes are predominantly due to a reduced rate of protein synthesis rather than increased protein degradation, with poor-quality protein produced as a result of oxidative damage due to free radicals. This accumulation of oxidative damage over time leads to the loss of type 2 muscle fibres (fast) due to conversion to type 1 fibres (slow), meaning there is a build-up of clusters of slow-moving

CASE STUDY 13.2 AGED CARE

Profile and problem

Mrs M., an 86-year-old woman living in a residential aged care facility, recently had a fall and sustained a hip fracture. She was diagnosed with dementia twelve months ago and her cognition has been deteriorating. Care staff have noticed increasing uninterest in food, and as a result she has gradually lost weight. Her supportive family are concerned. She was admitted to hospital for seven days and has recently returned to the care facility without surgery on her fractured hip. Since returning from hospital she appears to be eating very little of her meals. She is 160 centimetres tall and weighs 48 kilograms; she has lost a total of 8 kilograms in the past twelve months.

Assessment

Staff at the care facility and Mrs M.'s family want to prevent Mrs M. from losing any more weight. At present she is eating less than half of her main meals and drinking diet cordial regularly. She is often distracted at meal times and is seen pushing the food around her plate. Her mobility has significantly reduced following her hip fracture, and she is bed/chair bound. Lately, staff have noticed Mrs M. coughing while eating.

Treatment

A discussion with Mrs M., care facility staff and family members resulted in the following strategies:

- Mrs M. will receive six small meals per day.
- Main meals will be presented on a side plate.
- Mrs M. will drink regular rather than diet cordial.
- Meals will be prepared for Mrs M. with reference to a list of her favourite foods, provided by the family.
- If Mrs M. refuses a meal, an alternative such as a sandwich or even a dessert will be offered.
- Staff will dedicate time to assist Mrs M. with meals.
- Mrs M. will be referred to the visiting speech pathologist.

Plan

Mrs M. will be weighed on a weekly basis until her weight stabilises. Care facility staff will monitor her food and fluid intake, and the visiting dietitian will review this on their regular visits. Mrs M. may require a modified-consistency diet or thickened fluids as prescribed by the speech pathologist. Mrs M. is at risk of malnutrition and therefore adverse outcomes if her weight loss continues.

fibres that results in a decrease in muscle function. The outcome of this is a loss of power that manifests as reduced ability to perform actions such as rising from a chair or climbing stairs. This is exacerbated by the loss of muscle mass and the infiltration of muscle with fat tissue to replace lost bulk. As a result of the loss of muscle mass, unintentional weight loss is common in those with sarcopenia; however, in the context of the obesity epidemic, there may be no weight loss. Sarcopenic obesity is becoming increasingly common and is also associated with considerable adverse outcomes.

While the treatment strategies identified above for starvation are also likely to be important for treatment of sarcopenia, alone they would be unlikely to address unintentional weight loss, as, although inadequate intake is known to exacerbate the condition, sarcopenia is thought to occur regardless of energy balance. Strategies that have been recommended for sarcopenia include progressive resistance training that can increase the rate of protein synthesis to a level that is greater than the rate of muscle atrophy. A review has indicated that no strategy for the treatment of sarcopenia has been as successful as combined nutrition and exercise therapies (Rolland et al. 2011).

Cachexia

Although cachexia is commonly associated with a number of chronic conditions, such as cancers, human immunodeficiency virus /acquired immune deficiency syndrome (HIV/AIDS) and chronic obstructive pulmonary disease, it also occurs in older adults. The literature suggests a strong association between geriatric cachexia and decreased quality of life, reduced functional capacity and increased morbidity and mortality, with a worsening prognosis for any underlying disease. Cachexia is driven by increased inflammatory processes resulting in levels of catabolism that see losses of both fat and muscle, with a disproportionate loss of muscle tissue, responsible for the unintentional weight loss.

The underlying causative factor for cachexia appears to be an excess of **cytokines**. The acute-phase response is activated following any insult, such as trauma or disease, resulting in an increase in positive acute-phase proteins such as C-reactive protein, which is a sensitive marker of the presence of systemic inflammation. Inflammatory cells produce excess cytokines, particularly tumour necrosis factor, interleukin 1 and interleukin 6, that are thought to be key in the development of cachexia. The resultant increase in resting metabolic rate increase energy requirements and disturbs the balance between muscle anabolism and catabolism. Furthermore, energy needs are unlikely to be met, as cytokine production also causes anorexia, thus reducing appetite and nutritional intake. Finally, excess cytokines can cause nausea, vomiting and altered gastrointestinal function, further limiting the ability to achieve estimated nutrient requirements.

Treatment strategies for geriatric cachexia are not well established and require further investigation. Treatment with **hypercaloric feeding** alone has essentially been found to be ineffective (Thomas 2007). Consequently, research has largely focused on pharmacological interventions to reduce inflammation or to stimulate appetite, in addition to the provision of adequate dietary energy and protein.

Progestogens are steroid hormones that can stimulate appetite and weight gain in some groups of patients. However, any benefit of appetite stimulants, including progestogens, is likely to be short lived, due to failure to address the underlying causes of cachexia. There is also evidence that any gain in body weight as a result of steroid hormone treatment is primarily due to increased fat mass and water retention as opposed to muscle mass; therefore, this would be unlikely to provide long-term benefits.

Omega-3 fatty acids are known to have anti-inflammatory effects. It is therefore plausible that they would be an effective treatment strategy for geriatric cachexia. There is some indication of improvements in weight status, body composition and inflammatory and oxidative markers in cachectic patients with cancer treated with fish oil, marine phospholipids or omega-3 fatty acids.

Other therapies that have been investigated for the treatment of cachexia include thalidomide and ghrelin. Although thalidomide is an anti-inflammatory agent, research into its use in the treatment of cachexia has yielded inconclusive results, and there are indications

Cytokines are small proteins that are released from cells and signal for surrounding cells to behave in a particular way.

Hypercaloric feeding is the provision of a high-energy (energy-dense) diet and may be accomplished through fortification of food and/or addition of homemade or commercial oral nutritional supplements. It is not necessarily achieved through consumption of increased volumes of food.

that it may be poorly tolerated in some patient groups (Reid et al. 2012; Wilkes et al. 2011). Conversely, studies investigating the use of ghrelin have shown some promising beneficial effects on weight status, body composition, appetite and pro-inflammatory cytokines in cachectic individuals. However, there are some concerns for the long-term efficacy of this recently discovered hormone (DeBoer 2008) and for the safety of treatment using it (Chopin et al. 2011).

Palliative care

Palliative care is provided to people who have active, progressive, advanced disease or who are at advanced age and aims to maximise quality of life for the patient and their carers and family. The nutritional care provided to palliative patients depends on the reason for their care, but appetite may be impaired and/or eating may be difficult. Large weight loss may result in loose-fitting dentition, and swallowing may be impaired, making it challenging for patients to consume adequate food and drink.

Some general strategies that may be considered include texture modification and high-energy, high-protein diets. General principles such as making every mouthful count, small frequent meals and the use of flavourings to stimulate appetite and increase the appeal of food should be adopted.

However, it is important to remember that palliative care patients will not always be advanced to the point of disability and may be under the care of palliative services for a number of years; hence, nutritional management may simply be a matter of relaxing dietary prescription to maximise intake and improve pleasure in food. Nonetheless, many palliative care patients will be nearing the end of life and may refuse food. While small amounts of food and fluids may still be offered, patients should not be forced to eat or drink, and in those cases the time may arrive where it is appropriate for the patient and/or their family to request the withdrawal of nutritional support.

SUMMARY AND KEY MESSAGES

Achieving an age in excess of 85 years is a milestone and should be celebrated. A more liberal dietary intake should mean food and eating are more enjoyable, with concern mainly for avoidance of exacerbation of existing chronic diseases and unintentional weight loss. A liberalised diet should not be feared, especially for those adults who enter this stage of the life cycle free from chronic disease. One of the most challenging issues is the appropriate recognition of unintentional weight loss and its underlying causes. While numerous malnutrition screening instruments exist, they do not differentiate between starvation, sarcopenia and cachexia as the likely aetiology, so effective treatment strategies may be overlooked. Add to this the challenge of the potential masking of sarcopenia in the presence of obesity and the situation becomes complex.

Dying with dignity is of utmost importance, and while food and nutrition may play a role for some older adults, it should be recognised that they may, for any number of reasons, be of less importance as the end of life approaches and comfort becomes priority.

- Nutrition risk factors in older age include living alone and in social isolation, admission to hospital, admission to higher level of care, bereavement, social isolation and polypharmacy.
- Physiological changes can increase the risk of poor nutritional status of older adults and may include changes in body composition and sensory function, and gastrointestinal and neurological alterations.
- Some key nutrients of concern for older adults include protein, vitamin B12, calcium and vitamin D. Inadequate energy and water intake can also be cause for concern.
- Any unintentional weight loss in older adults should not be assumed to be a normal part of the ageing process. Investigations to identify the presence of starvation, cachexia and sarcopenia are important for guiding the most appropriate treatment strategies.
- End-of-life care may require nutrition support, but equally the older adult and/or their family may deem it most appropriate to withdraw this support.

REFERENCES

ABS (Australian Bureau of Statistics) 2008. *Population Projections, Australia, 2006 to 2101*. Cat. no. 3222.0. Commonwealth of Australia. <www.abs.gov.au/AUSSTATS/abs@.nsf/Lookup/3222.0Main+Features12006%20to%202101?OpenDocument>, accessed 14 November 2014.

——2013. *Population Projections, Australia, 2012 (base) to 2101*. Cat. no. 3222.0. Commonwealth of Australia. <www.abs.gov.au/AUSSTATS/abs@.nsf/DetailsPage/3222.02012%20(base)%20to%202101?OpenDocument>, accessed 30 November 2014.

AIHW (Australian Institute of Health and Welfare) 2014. Admitted patient care: Overview. *Australian Hospital Statistics 2012–13*. Cat no. HSE 145. Commonwealth of Australia. <www.aihw.gov.au/WorkArea/DownloadAsset.aspx?id=60129547092>, accessed 14 November 2014.

Barker, L.A., Gout, B.S. & Crowe, T.C. 2011. Hospital malnutrition: Prevalence, identification and impact on patients and the healthcare system. *International Journal of Environmental Research and Public Health* 8: 514–27.

Baulderstone, L., Yaxley, A., Luszcz, M. & Miller, M. 2012. Diet liberalisation in older Australians decreases frailty without increasing the risk of developing chronic disease. *Journal of Frailty & Aging* 1: 174–82.

Boyce, J.M. & Shone, G.R. 2006. Effects of ageing on smell and taste. *Postgraduate Medical Journal* 82: 239–41.

Buettner, D. 2012. *The Blue Zones: 9 lessons for living longer from the people who've lived the longest* 2nd edn. National Geographic Society, Washington, DC.

Chopin, L., Walpole, C., Seim, I., Cunningham, P. et al. 2011. Ghrelin and cancer. *Molecular and Cellular Endocrinology* 340: 65–9.

Cruz-Jentoft, A.J., Baeyens, J.P., Bauer, J.M., Boirie, Y. et al. 2010. Sarcopenia: European consensus on definition and diagnosis; Report of the European Working Group on Sarcopenia in Older People. *Age and Ageing* 39: 412–23.

Darmon, P., Kaiser, M.J., Bauer, J.M., Sieber, C.C. et al. 2010. Restrictive diets in the elderly: Never say never again? *Clinical Nutrition* 29: 170–4.

DeBoer, M. D. 2008. Emergence of ghrelin as a treatment for cachexia syndromes. *Nutrition* 24: 806–14.

D'Souza, A.L. 2007. Ageing and the gut. *Postgraduate Medical Journal* 83: 44–53.

Hickson, M. 2006. Malnutrition and ageing. *Postgraduate Medical Journal* 82: 2–8.

Janssen, I. & Mark, A. E. 2007. Elevated body mass index and mortality risk in the elderly. *Obesity Reviews* 8: 41–59.

Jones, G., Nguyen, T., Sambrook, P., Kelly, P.J. et al. 1994. Progressive loss of bone in the femoral neck in elderly people: Longitudinal findings from the Dubbo osteoporosis epidemiology study. *British Medical Journal* 309: 691–5.

McMinn, J., Steel, C. & Bowman, A. 2011. Investigation and management of unintentional weight loss in older adults. *British Medical Journal* 342: d1732.

McWhirter, J.P. & Pennington, C.R. 1994. Incidence and recognition of malnutrition in hospital. *British Medical Journal* 308: 945–8.

Ministry of Health 2012. *Hospital Events 2008–09 and 2009–10*. New Zealand Government. <www.health.govt.nz/system/files/documents/publications/hospital-events-2008-09-2009-10.pdf>, accessed 14 November 2014.

Ministry of Health 2013. *Food and Nutrition Guidelines for Healthy Older People: A background paper*. New Zealand Government. <www.health.govt.nz/system/files/documents/publications/food-and-nutrition-guidelines-healthy-older-people-background-paper-jan2013.pdf>, accessed 16 March 2014.

NHMRC (National Health and Medical Research Council) 2013. *Eat for Health: Australian Dietary Guidelines; Providing the scientific evidence for healthier Australian diets*. Commonwealth of Australia.

<www.eatforhealth.gov.au/sites/default/files/files/the_guidelines/n55_australian_dietary_guidelines.pdf>, accessed 16 March 2014..

NHMRC (National Health and Medical Research Council) & Ministry of Health 2006. *Nutrient Reference Values for Australia and New Zealand: Including recommended dietary intakes*. Commonwealth of Australia & New Zealand Government. <www.nhmrc.gov.au/_files_nhmrc/publications/attachments/n35.pdf>, accessed 26 September 2014.

Nowson, C.A., McGrath, J.J., Ebeling, P.R., Haikerwal, A. et al. 2012. Vitamin D and health in adults in Australia and New Zealand: A position statement. *Medical Journal of Australia* 196: 686–7.

Reid, J., Mills, M., Cantwell, M., Cardwell, C.R. et al. 2012. Thalidomide for managing cancer cachexia. *Cochrane Database of Systematic Reviews* issue 4, art. no. CD008664. <http://onlinelibrary.wiley.com/doi/10.1002/14651858.CD008664.pub2/full>, accessed 5 October 2014.

Reider, L., Hawkes, W., Hebel, J.R., D'Adamo, C. et al. 2013. The association between body mass index, weight loss and physical function in the year following a hip fracture. *Journal of Nutrition, Health & Aging* 17: 91–5.

Rolland, Y., Onder, G., Morley, J.E., Gillette, S. et al. 2011. Current and future pharmacologic treatment of sarcopenia. *Clinics in Geriatric Medicine* 27: 423–47.

Ross, L.J., Mudge, A.M., Young, A.M. & Banks, M. 2011. Everyone's problem but nobody's job: Staff perceptions and explanations for poor nutritional intake in older medical patients. *Nutrition & Dietetics* 68: 41–6.

Schiffman, S.S. 2009. Effects of aging on the human taste system. *Annals of the New York Academy of Sciences* 1170: 725–9.

Stahl, S. & Schulz, R. 2013. Changes in routine health behaviors following late-life bereavement: A systematic review. *Journal of Behavioral Medicine* 37: 736–55.

Statistics New Zealand 2012. *National Population Projections: 2011 (base)–2061*. New Zealand Government. <www.stats.govt.nz/browse_for_stats/population/estimates_and_projections/NationalPopulationProjections_HOTP2011/Commentary.aspx>, accessed 14 November 2014.

Thomas, D.R. 2002. Distinguishing starvation from cachexia. *Clinics in Geriatric Medicine* 18: 883–91.

——2007. Loss of skeletal muscle mass in aging: Examining the relationship of starvation, sarcopenia and cachexia. *Clinical Nutrition* 26: 389–99.

Watterson, C., Fraser, A., Banks, M., Isenring, E. et al. 2009. Evidence based practice guidelines for the nutritional management of malnutrition in adult patients across the continuum of care. *Nutrition & Dietetics* 66: S1–S34.

Wilkes, E.A., Selby, A.L., Cole, A.T., Freeman, J.G. et al. 2011. Poor tolerability of thalidomide in end-stage oesophageal cancer. *European Journal of Cancer Care* 20: 593–600.

TEST YOUR UNDERSTANDING

Note: There is only one correct answer for each question.

1. It is projected that within 50 years the oldest-old segment of the population in New Zealand will do which of the following?
   a. Triple
   b. Halve
   c. Double
   d. Remain the same
   e. None of the above

2. Which of the following might be an effect of polypharmacy?
   a. Decrease in bioavailability of some nutrients
   b. Replacement of food intake
   c. Interference with hunger signals
   d. All of the above
   e. None of the above

3. Which of the following is not a usual trend in body composition among older adults?
   a. Increase in muscle mass
   b. Increase in fat mass
   c. Redistribution of fat mass
   d. Increase in total body weight
   e. All of the above
4. Which of the following BMIs is considered most appropriate for older adults?
   a. 20–5
   b. 22–5
   c. 22–7
   d. All of the above
   e. None of the above
5. Which of the following is considered normal cognitive change for older adults?
   a. Short-term memory loss
   b. Long-term memory loss
   c. Ability to multitask
   d. Learning new knowledge
   e. All of the above
6. Which of the following is not a nutritional consequence of dementia?
   a. Poor appetite
   b. Weight loss
   c. Malnutrition
   d. Alterations in calcium metabolism
   e. All of the above
7. Which of the following nutrient supplementations may need to be considered for optimising nutritional status of older adults?
   a. Vitamin B12
   b. Calcium
   c. Vitamin D
   d. All of the above
   e. None of the above
8. Which of the following strategies for treatment of sarcopenia in older adults is believed to be the most effective?
   a. High-energy, high-protein diet
   b. Exercise
   c. High-energy, high-protein diet and exercise
   d. Vitamin supplementation
   e. None of the above
9. Which of the following can cause unintentional weight loss in older adults?
   a. Cachexia
   b. Starvation
   c. Sarcopenia
   d. All of the above
   e. None of the above
10. Which of the following nutritional considerations may be important for end-of-life care?
   a. Small, frequent meals
   b. Appetite stimulants
   c. Modified-texture foods and fluids
   d. Liberalised diet
   e. All of the above

STUDY QUESTIONS

1. Discuss the nutritional risk factors which are of particular relevance to the 85 years and over age group.
2. Compare and contrast the malnutrition screening tools that are commonly used with the oldest-old in the acute setting.
3. What are considered to be normal cognitive changes in older adulthood?
4. What are the key changes that take place in the body composition of older adults?
5. Discuss the causes and impacts of vitamin B12 deficiency in adults aged 85 years and over.
6. Discuss the aetiology of unintentional weight loss in older adults.

PART III

REGIONAL PERSPECTIVES

14

Aboriginal and Torres Strait Islander nutrition and health

Amanda Lee and Kerin O'Dea

LEARNING OUTCOMES

Upon completion of this chapter you will be able to:

- explain the reasons for the very poor health outcomes of most Aboriginals and Torres Strait Islanders in Australia, particularly the impacts of poverty and social disadvantage
- discuss how poor-quality diet is a major contributor to poor health outcomes
- describe the characteristics of Aboriginal and Torres Strait Islander diets in different settings
- discuss ways in which the diet of Indigenous Australians can be improved.

Aboriginals and Torres Strait Islanders suffer a much greater burden of ill-health than other Australians, with the health gap accounting for 59 per cent of the total burden of disease for Indigenous Australians (AIHW 2011b, 2013). Poor nutrition, including both undernutrition and overnutrition, is an important risk factor underlying this excess burden (HREOC 2008). However, data quality issues hamper the ability of the health surveillance system to accurately monitor nutritional status of Indigenous peoples and the influence of health determinants. Aboriginals and Torres Strait Islanders constitute 2.5 per cent of the Australian population; most live in capital cities (32 per cent) and regional areas (43 per cent), with 25 per cent living in remote areas (AIHW 2011b). While the 25 per cent of Indigenous Australians residing in remote areas experience a disproportionate amount of the health gap (40 per cent), the majority of the health gap affects residents of non-remote areas (AIHW 2013). The Indigenous Australian population has a younger age profile, with a median age of 21 years compared with 37 years for the non-Indigenous population (AIHW 2011b).

AUSTRALIAN PEOPLES

Aboriginal people first settled in Australia at least 50,000 years ago, with recent genetic evidence suggesting it may have been as early as 75,000 years ago—making them one of the oldest continuous populations outside Africa. Over time, Aboriginal people settled all over the continent. It is likely that the fertile coastal areas in northern, eastern and southern Australia supported larger populations than the vast arid regions in the centre of the continent—much as they do today.

Relatively little is known of the traditional diets and lifestyles of the Aboriginal peoples of south-eastern Australia, as their land was the first to be appropriated by the new settlers for farming, and the population was decimated by violent conflict over land and resources and by diseases such as smallpox.

Torres Strait Islanders are originally Melanesian and are related to the people of Papua New Guinea. They have inhabited the islands of the Torres Strait for at least 4000 years and are a culturally distinct group who traded with Cape York Aboriginal people and Papuans for many centuries (NHMRC 2000).

TRADITIONAL DIETS AND LIFESTYLES

Much of our knowledge of the traditional hunter-gatherer diet and lifestyle of Aboriginal people has come from the study of groups who continued with that diet and lifestyle into the twentieth century in the most remote parts of central and northern Australia (O'Dea 1991). These ethnographic studies generated data specific to particular (usually small) groups. However, there were important common features that were functions of the nomadic hunter-gatherer lifestyle. All Aboriginal populations participating in these studies were omnivorous, consuming a wide range of uncultivated plant foods and wild animals. The composition of the diet was influenced by the geographical location and the season of the year. Aboriginal people had a strikingly detailed knowledge of their country, including the availability of water (especially important in arid regions) and the impact of the annual cycle of seasonal changes on the availability and lifecycles of particular plant and animal foods.

People tended to live in extended family groups numbering between 20 and 30, coming together in larger groups for ceremonies only when there was sufficient food in an area to feed them. Both men and women contributed importantly to food procurement. The women generally provided the subsistence diet of the particular season and geographical area: plant foods, honey, eggs, small mammals, reptiles, fish, shellfish, crustaceans and insects. They usually hunted in groups with their children, passing on their knowledge and skills to the next generation. Men sometimes took part in small family expeditions but prided themselves on hunting for larger game such as kangaroos, wallabies, emus, turtles, crocodiles and dugongs. While it has been claimed that hunter-gatherers spent less time (three to five hours per day) ensuring their livelihood than agriculturalists and the full-time employed today, food procurement and preparation was energy intensive, requiring much walking, digging and chopping with primitive implements (for yams, reptiles, eggs, honey grubs and other insects), winnowing and grinding seeds, digging pits for cooking larger game and collecting firewood for cooking and warmth.

Children were traditionally breastfed until they were three to four years old, the age of weaning usually depending on the arrival of another sibling. Solids were generally not introduced until teeth erupted. Responsibility for feeding tended to rest with the child, who was expected to express a desire for food and was fed on demand; older children had priority over infants for feeding (NHMRC 2000).

The concept of *terra nullius* (land owned by no one) was used by early European settlers in Australia to justify their acquisition of land occupied by Aboriginal people. While Aboriginal people did not practise agriculture in the conventional sense, numerous practices ensured continuing food abundance. Examples of this include firestick farming, the systematic burning of grasslands and undergrowth in treed areas. This was sometimes associated with ambushing game, but it also encouraged new growth and attracted game in that way. It facilitated the germination of some seeds and helped maintain the savannah grasslands in northern and central Australia, as well as minimising very destructive fires by preventing or reducing excessive fuel burden of undergrowth and grasslands. Fish trapping and poisoning were also practised, which ensured continuance of food, as did gathering the native millet before it was ripe and storing it to allow efficient harvesting of seeds when they dried and ripened, as they would otherwise have been widely dispersed. Keeping waterholes viable in the arid zones of Australia ensured the survival of the local ecosystem of plants, animals and birds that depended on it and which were sources of food.

All animals were potential food sources: mammals, birds, reptiles, insects and marine species. Everything edible on an animal carcass was eaten, including muscle, fat, organ meats, bone marrow and small intestine and stomach, the last of which were eaten after their contents were discarded. Large intestines and gall bladders were discarded well away from the campsite. One of the most striking features of wild animals is their leanness in comparison with meats from animals domesticated for human consumption, which have been bred for fatness since the advent of agriculture. Waygu beef is an extreme example. Domesticated animals have fat deposits under the skin, within the abdomen and within and between muscles.

Almost without exception, wild animals do not have subcutaneous or intramuscular fat. (Marine mammals are an exception.) They do have some intra-abdominal fat, primarily surrounding the gonads, kidneys and intestines. However, these deposits are usually small and needed to be shared, as they were highly prized. Most of the fat in muscle from wild animals is structural (in cell membranes) and therefore tends to be highly polyunsaturated, and rich in long-chain omega-3 fatty acids. Likewise, the fat in organ meats tends to be polyunsaturated (as well as relatively rich in cholesterol). Brain is a rich source of long-chain omega-3 polyunsaturated fatty acids, especially docosahexaenoic acid (DHA), very important in neural and retinal development and believed to be important in maintaining mental health and preventing depression.

Liver is an excellent source of many important nutrients, including omega-3 and omega-6 polyunsaturated fatty acids, cholesterol, fat-soluble vitamins (A and D), iron and zinc and, perhaps less widely appreciated, nutrients usually associated with plants, such as vitamin C and folic acid. Liver, therefore, may have been particularly important in those times of the year when little plant food was available. Furthermore, vitamin C is in a stable matrix in liver and is not oxidised by cooking. Thus, liver from free-living animals is highly nutritious. Insects (witchetty grubs and bogong moths) or insect products (honey and honey ants) were also seasonal important dietary components.

The detailed knowledge that Aboriginal people had of their traditional lands and ecosystems allowed them to draw on a wide range of wild plant foods, including tuberous roots, legumes, seeds, nuts, fruits, berries and nectars. Relative to their cultivated forms, wild plants tend to be high in fibre, with carbohydrate that is slowly digested, lower in starch and sugars and rich in a wide range of bioactive phytochemicals (for example, flavonoids and polyphenols) with anti-oxidant and anti-inflammatory properties. Plant foods contributed significant amounts of protein, fat, carbohydrate, fibre and vitamin C to the diet, and it has been estimated that they provided 20–40 per cent of total energy intake.

Food preparation generally resulted in minimal loss of nutrients. Many plant foods were consumed fresh and raw as they were collected (including fruit, bulbs, gums, flowers and nectar). Any processing was to render a food more digestible or palatable, by baking starchy tubers, grinding and roasting seeds and cooking meat. There were also examples of processing to detoxify potentially poisonous or bitter foods—for example, the nuts of the cycad palm, and cheeky yam (*Dioscorea bulbifera*).

The most common forms of cooking were roasting on coals and baking in an earth oven, or pit. Animals were cooked whole: small animals, such as reptiles, birds, fish, shellfish, crustaceans and grubs, on coals and large animals in an earth oven heated by hot coals or stones. Vegetables that needed cooking were baked whole on coals. Blood and other juices were collected in the peritoneal cavity and eaten as a soup along with the internal organs before the meat was distributed. Large animals were prepared more formally by the hunters, following a prescribed ritual for preparation, cooking and distribution of the carcass.

Food intake could vary greatly both daily and seasonally. Food was usually eaten when it was available, with little wastage. After a successful hunt, large quantities of meat or fish could be consumed (up to 2–3 kilograms per person over one long sitting). This capacity to gorge when food was available was clearly an important survival strategy, taking advantage of an abundant food supply when possible and storing excess energy as fat to help with survival in times of food shortage.

The most highly prized components of the traditional diet were the relatively few energy-dense foods: animal fat, bone marrow, organ meats, eggs, fatty insects and honey. Muscle provided most weight and energy from a carcass. Animals (terrestrial and marine) were actively hunted when their fat depots were biggest, but those depots were generally not large and had to be shared by many. Thus, higher fat foods and honey were available only on a seasonal basis, and their procurement was often associated with high energy expenditure. However, it is significant that the most highly valued foods were those with the highest energy density. This was an important survival strategy, along with minimising unnecessary energy expenditure. It is interesting to speculate whether these food preferences and behaviours underlie the current obesity epidemic in developed Western societies.

In remote Australia as recently as the 1980s, despite living a sedentary, Westernised lifestyle, many older Aboriginal people retained the knowledge and ability to live as hunter-gatherers. Collaboration with such people with diabetes provided the opportunity to document the impact on health of temporary reversion to the hunter-gatherer lifestyle. After only seven weeks of such reversion in their traditional country in the remote West Kimberley, the health of a small group of people with diabetes had improved dramatically:

they lost weight, and all of the metabolic abnormalities of diabetes and risk factors for cardiovascular disease (CVD) were ameliorated or greatly improved (O'Dea 1984).

Their diet was carefully documented over a two-week period. It was low in energy (1200 kilojoules per day), with 54 per cent of energy coming from protein, 33 per cent from carbohydrate and 13 per cent from fat. Despite this, the people participating in the study rarely complained of hunger (O'Dea 1984). This may have been due to the high content of protein (see the discussion of the protein leverage hypothesis below under 'Economic'). While reverting to the hunter-gatherer lifestyle is not an option for Aboriginal people today, in some remote communities people do regularly go hunting to supplement the foods they buy from the store. However, for the vast majority in towns and cities, it is not a realistic option. Nevertheless, the model of the hunter-gatherer diet and lifestyle can be used as a benchmark and as a guide to healthier patterns of eating and lifestyle.

It must also be acknowledged that, although living as hunter-gatherers was clearly a very healthy lifestyle, it was tough. The pattern of food intake varied widely both daily and seasonally. For many parts of the country the diet was probably best described as subsistence interspersed with occasional feasts. It is therefore not surprising that people frequently opted for life on cattle stations and missions, where at least some of their food was provided on a regular basis in the form, initially, of simple rations (flour, sugar, tea, tobacco, and meat if available).

The traditional Torres Strait Islander diet varied from island to island. Seafood such as fish and shellfish was an important component of the diet, and turtle and dugong had a central place in cultural life, as did the keeping of pigs. Gardening was essential, not only for subsistence but also to provide food for ceremonies and trade. Traditional crops include taro, yams, bananas and coconuts. Several plant foods were stored and preserved (Leonard et al. 1995; NHMRC 2000).

CONTEMPORARY DIETS AND LIFESTYLES

Although there is a general paucity of quality dietary data for Aboriginal and Torres Strait Islander communities in all areas of Australia, available data of contemporary dietary habits are striking in their contrast to those of traditional habits (Lee 1996). Since European occupation of Australia, from the late eighteenth century, the process of Indigenous Australian acculturation from a traditional to a contemporary diet and lifestyle has been underscored by many social, political and environmental factors (NHMRC 2000, 2013a). With the transition from a traditional hunter-gather lifestyle to a settled Westernised existence, Aboriginal and Torres Strait Islander peoples' diet has generally changed from a varied, nutrient-dense diet to an energy-dense diet that is high in saturated fat, white flour and bread, sugar and salt (see Table 14.1).

The best community-level dietary data are available for remote communities where there are a limited number of food outlets (sometimes only a single store). Lee and colleagues developed and validated the store turnover method to estimate total community diet (Lee, Bailey et al. 1994; Lee, O'Dea et al. 1994). Even without precise population numbers, this method can be used to describe the relative contribution of and expenditure on specific foods, food groups and beverages.

An early study using the store turnover method in remote Aboriginal communities in the Northern Territory showed that sugar, flour, bread and meat provided more than half the apparent total energy intake. Fatty meats contributed nearly 40 per cent of the total fat intake in northern coastal communities and over 60 per cent in central desert communities. In both regions, white sugar per se contributed approximately 60 per cent of all sugars consumed (Lee, O'Dea et al. 1994). Compared with relevant national data on apparent consumption in Australia, intakes of sugar, white flour and sugar-sweetened carbonated beverages were much higher and intakes of wholegrain cereals, fruit and vegetables were much lower in the Aboriginal communities.

These worrying results have been confirmed by more recent studies in northern communities (Brimblecombe et al. 2013; Brimblecombe & O'Dea 2009). The most striking observation over more than twenty years is how little the nutritional quality of diets in remote communities has changed (Brimblecombe & O'Dea 2009). In three communities in remote northern Australia, of total food expenditure, around 25 per cent is on non-alcoholic beverages, around 16 per cent on sugar-sweetened beverages, and only about 2 per cent on fruit and 5 per cent on vegetables (Brimblecombe et al. 2013). In Central Australia, despite improvements in food supply, recent studies have shown that the energy-dense, nutrient-poor discretionary foods and drinks that provided 35 per cent of energy intake in

Table 14.1 Characteristics of hunter-gatherer and contemporary diets and PALs of Aboriginals and Torres Strait Islanders

Diet and activity	Hunter-gatherer	Contemporary
PAL	High	Low
Density		
Energy	Low	High
Nutrients	High	Low
Intake		
Energy	Adequate	Excessive
Protein	High	Low–moderate
Animal food	High	Moderate
Plant food	Moderate	Low
Carbohydrate	Moderate (slowly digested)	High (rapidly digested)
Complex carbohydrate	Moderate	Moderate
Sugars	Low	High
Dietary fibre	High	Low
Fat	Low	High
Saturated fat	Low	High
Alcohol	Not available	Available
Sodium to potassium ratio	Low	High

Note: PAL: physical activity level.

Source: O'Dea (1991).

remote desert communities in the 1980s now provide 41 per cent of energy, 28 per cent of fat and 84 per cent of sugar intake in those communities (Lee 2013). While diets continue to be high in sugar, sugar-sweetened beverages, white bread and flour and very low in fresh fruit and vegetables, there now appears to be greater reliance on convenience, takeaway and fast foods. Where data are available, the dietary profile tends to be better in those remote communities where more traditional foods are consumed.

In remote and rural Australia social disadvantage of Indigenous groups is amplified by higher costs of food, particularly perishable foods such as meat, fish, eggs, dairy, fruit and vegetables. Over a three-month period in 2005, Brimblecombe and O'Dea (2009) explored the relationship between dietary quality and energy density of foods and energy costs for an Aboriginal population living in a remote region of northern Australia. They found that foods with high energy density (yet lower nutrient density) were associated with lower costs, contributing disproportionately to energy availability in the community. The authors concluded that the energy–cost differential influences the capacity of Aboriginal people living in remote communities to attain a healthy diet. Few other recent objective Indigenous Australian dietary data are available.

In Torres Strait communities, marine foods continue to make substantial contributions to the diet. Most men, women and children living in the Torres Strait are involved in different aspects of collecting shellfish, fishing and hunting. In 1990, those Torres Strait Islanders living on three outer islands were estimated to consume between 191 and 450 grams per person per day of seafood (including turtle and dugong), which is considerably more than the Japanese seafood intake (102 grams per person per day). Some concerns have been raised more recently about the heavy metal content in the organ meat of dugong and turtle (Haswell-Elkins et al. 2007). Production of traditional garden staples continues to be important for some ceremonial purposes in Torres Strait Islander communities (Leonard et al. 1995).

A study using the store turnover method was undertaken in a small island community in Torres Strait in the early 1990s. More than half the energy in the community diet was derived from white flour, white rice, tinned meat and vegetable oil. Compared to recommended intakes, apparent dietary intake of fruit and vegetables from the community stores was about 15 per cent and 30 per cent, respectively, of recommended levels. People who depended on store foods would not be able to meet their needs for vitamins A, C and E and folic acid (Leonard et al.

1995). More recent dietary intake data for the region are not available.

Only limited quantitative dietary data are available pertaining to Aboriginal people's diet in the urban setting (Lawrence 2013). Because of methodological difficulties, individual dietary studies have tended to focus on qualitative and semi-quantitative assessment of the diet and to reflect dietary patterns and preferences rather than actual habitual intake (Guest & O'Dea 1992). A comparison of the food habits of Aboriginal people and non-Indigenous Australians in a city and a country town showed that in both localities Aboriginal groups consumed takeaway meals and added salt more often than their non-Indigenous counterparts (Guest & O'Dea 1992), and recent work has confirmed high intakes of salt in a regional town (Lawrence 2013).

Dietary Guidelines for Aboriginals and Torres Strait Islanders

The general Australian Dietary Guidelines are relevant to Aboriginal and Torres Strait Islander groups. Two other specific recommendations are also important (NHMRC 2013a):

- Choose store foods that are most like traditional bush foods—for example, fresh plant foods, wholegrain cereal foods, lean meat, poultry and seafood.
- Enjoy traditional bush foods whenever possible.

Where non-Indigenous Australians have sufficient understanding of the traditional Indigenous Australian food supply and culture, the additional guidelines for Aboriginals and Torres Strait Islanders may also be useful in a wider context to improve the nutritional status of the population.

HEALTH AND NUTRITION

In the most remote areas of Australia, small groups of Aboriginal people continued to live as hunter-gatherers well into the twentieth century. The little quantitative data on the health of these Aboriginal people, who had little or no Western contact, indicate that they were extremely lean and physically fit, exhibiting no evidence of the chronic diseases that plague groups today. For example, adults living a traditionally oriented lifestyle in north-east Arnhem Land in 1985 had low blood pressure, very low body mass index (BMI) of under 20 and low fasting glucose and cholesterol. They also had serum folate levels in the healthy range, which is unusual in other similar groups and indicative of high dietary quality (O'Dea et al. 1988).

There is evidence that, in common with some other populations at high risk of diabetes and related conditions (such as South Asians), the healthy BMI range is lower for Aboriginal people than for Australians of European origin. Aboriginal people tend to have a linear body build, being narrow across the shoulders and hips and having relatively longer limbs and a shorter torso than Australians of European origin—and relatively more body fat for a given BMI. Furthermore, when gaining weight, fat tends to be deposited centrally in Aboriginal people, and even modest weight gain can be associated with increased cardiometabolic risk in insulin resistance, impaired glucose tolerance and dyslipidaemia. Conversely, the healthy BMI range of Torres Strait Islanders tends to be higher than for Australians of European origin (NHMRC 2013a).

The regular health survey of Aboriginals and Torres Strait Islanders reported by the Australian Institute of Health and Welfare collects only data on fruit and vegetable consumption by self-reported methods in non-remote areas, and little recent information is available concerning the range of nutrition issues as risk factors for poor pregnancy, infant growth or chronic disease outcomes (AIHW 2013). This is an ongoing challenge in efforts to address poor Indigenous Australian health. More current data is available from the Aboriginal and Torres Strait Islander component of the Australian Health Survey (ABS 2014).

In 2004–05, consumption of the recommended daily intake of fruit and vegetables was reported as 42 per cent and 10 per cent, respectively, by Indigenous Australian people aged fifteen years or over. There were no significant differences between the proportion of Indigenous and non-Indigenous people who reported meeting these dietary recommendations. However, the questions asked have not been tested for use with Indigenous Australian groups, and it is unclear to what degree social desirability may have led to response bias, particularly given the implementation of fruit and vegetable promotion programs such as the Go for 2 & 5 project in Indigenous Australian communities. Self-reported data do not match the low levels of intake described consistently through application of more reliable quantitative methods. The 2012–13 Australian Aboriginal and Torres Strait Islander Health Survey found again that 42 per cent of Aboriginal and Torres Strait Islander people over the age of 15 years consumed fruit at the recommended daily levels. The proportion of Aboriginal and Torres Strait Islander people over the age of 15 years consuming the recommended daily serves of vegetables had dropped to 4.8 per cent in 2012–13, and was significantly less than non-Indigenous people (ABS 2014).

Good maternal nutrition and healthy infant and childhood growth are fundamental to the achievement and maintenance of health throughout life. The potential intergenerational effects of poor health and nutrient status in early life, and also gestational diabetes mellitus (GDM), have been well described (see Chapter 4). During 2005–07, the mean number of births for Aboriginal and Torres Strait Islander women was 2.6, compared with 1.9 for all women in Australia, with teenage motherhood being much more common in Indigenous Australian births (21 per cent) than in all Australian births (4 per cent).

Babies born to Indigenous Australian mothers in 2005–07 were twice as likely to be of low birthweight (LBW)—that is, less than 2.5 kilograms—than babies born to other Australian mothers (13 per cent compared to 7 per cent, respectively; AIHW 2013). LBW may be a result of foetal growth restriction or preterm birth, or a combination of the two. Factors that contribute to LBW include socio-economic disadvantage, size of the parents, age of the mother, number of babies previously born, mother's nutrient status, smoking and alcohol intake and illness during pregnancy (NHMRC 2000).

In 2007, with regard to behavioural risk factors, 51 per cent of Indigenous Australian mothers reported to have smoked tobacco during pregnancy, compared with 15 per cent of non-Indigenous Australian mothers, and 20 per cent of Indigenous Australian mothers reported consuming alcohol during pregnancy, although the majority (83 per cent) reporting drinking less while pregnant. Prevalence of LBW in Aboriginal people varies from less than 7 per cent in Victoria to over 20 per cent in parts of northern Australia (AIHW 2013).

While infant mortality in Aboriginals and Torres Strait Islanders has declined over the last 30 years, in 2012 it was almost twice that of non-Indigenous Australian infants, at approximately 8 per 1000 live births, compared with 4 per 1000 live births. Aboriginal and Torres Strait Islander infant mortality rates vary across jurisdictions, from 6 per 1000 in South Australia to 13 per 1000 in the Northern Territory (AIHW 2013).

The gap in life expectancy between Indigenous and non-Indigenous Australians due to communicable diseases and maternal and perinatal conditions has reduced. However, prevalence of poor pregnancy outcomes and infant malnutrition remains high in many areas, and LBW, failure to thrive and poor child growth are serious concerns in many Indigenous Australian communities (Gracey 2007). Conditions such as pneumonia, inner ear infections and trachoma occur among Indigenous Australian children at several times the rate of non-Indigenous children (AIHW 2013). Anaemia among young Indigenous Australian children is common (NHMRC 2000).

Growth retardation among Aboriginal and Torres Strait Islander infants from around six months of age has been documented frequently since the 1960s (NHMRC 2000). This pattern of growth faltering is still common in several areas, particularly in the Northern Territory and Western Australia; but although there has been some improvement over the years (Burns & Thomson 2008), there is a paucity of recent national data.

Increasing rates of overweight and obesity have been observed among Indigenous Australian children over the last two decades, particularly among Torres Strait Islanders (NATSINAP 2001; NHMRC 2013a); again, there is a paucity of national data available. Anthropometric measurements of Indigenous Australian children were collected in the 2012–13 Australian Health Survey. Twenty per cent of 2–14-year-old Aboriginal and Torres Strait Islander children were overweight and 10 per cent were obese. There were more overweight and obese girls (32 per cent) than boys (29 per cent). Eight per cent of children were considered underweight with the remaining 62 per cent in the normal weight range (ABS 2013).

Breastfeeding

The nutritional and immunological benefits of prolonged breastfeeding are particularly important in populations with a high prevalence of infectious diseases and high prevalence of high-risk infants, such as those of LBW or babies of diabetic mothers. Breastfeeding is also especially important in communities that may have limited access to a quality water supply and limited financial resources (NHMRC 2012).

In remote areas in 2004–05 high proportions of Aboriginal and Torres Strait Islander infants were being breastfed: 85 per cent of those aged less than six months and 82 per cent of those aged six to twelve months. Seventy per cent of Aboriginal and Torres Strait Islander infants less than twelve months of age in remote areas were breastfed in 2008, compared with 50 per cent of those in non-remote areas; the proportion of children aged up to three years of age who had never been breastfed was also lower in remote areas (14 per cent compared with 25 per cent) (AIHW 2013).

In 2010, the Australian National Infant Feeding Survey found that at less than three months, 33 per cent of Indigenous Australian infants were exclusively breastfed compared to 48 per cent of non-Indigenous Australian infants; at less than six months the figures were 7 per cent and 16 per cent respectively (AIHW 2011a). Rates of any breastfeeding at all were similarly high (98–9 per cent) in the most advantaged quintile for both Indigenous and non-Indigenous Australian mothers, and similarly lower in the least advantaged quintile (93–4 per cent) for both groups, respectively. In 2010, mothers reported that 31 per cent of Indigenous Australian infants aged three months were fed solids, compared with 9 per cent of non-Indigenous Australian infants (AIHW 2011a).

Micronutrients

Nutrition status has been measured infrequently in Aboriginal and Torres Strait Islander populations, but there have been some studies in some groups and environments (NHMRC 2000). Numbers of people surveyed have generally been small and have often been selected from vulnerable groups in the community—infants and pregnant and breastfeeding women. Quantitative comparison of the prevalence of vitamin deficiencies may be misleading, since varying methods and 'normal' ranges have been used in these studies. Multiple deficiencies frequently cluster in individuals, suggesting generally poor nutrient status rather than a specific micronutrient problem. In particular, poor status of ascorbic acid, folate and beta-carotene consistent with the very low contemporary dietary intakes of fruit and vegetables has often been described (NHMRC 2013). Iron deficiency is common among young children (NHMRC 2000).

HEALTH CONDITIONS

Morbidity and mortality

Unlike other major health risk factors such as smoking and alcohol intake, the burden of disease due to poor diet has not been assessed in Australia for either the whole population or Aboriginal and Torres Strait Islander groups. Recent work in the United States has identified that poor diet was the leading risk factor in that country in 2010, contributing around 14 per cent to the burden of disease (IHME 2013). Estimates could be expected to be at least this high for Indigenous Australian groups.

Chronic conditions such as circulatory diseases (including heart disease), type 2 diabetes mellitus (T2DM), respiratory diseases, musculoskeletal conditions, kidney disease and eye and ear problems are responsible for about 80 per cent of the observed difference in the burden of disease between the Indigenous and the non-Indigenous Australian population (AIHW 2013). These conditions not only occur more frequently among Aboriginal and Torres Strait Islander groups than in the non-Indigenous Australian population, but in the former group the age of onset is earlier and the consequences more severe. The conditions cause the most damage in what should be the most productive years of life. For example, in the 20–50 years age group, the prevalence of diabetes and the mortality from cardiovascular disease (CVD) are both more than ten times higher in Indigenous Australians than in non-Indigenous Australians.

Sixty-six per cent of Indigenous Australian deaths occur before the age of 65 years, compared with 20 per cent of non-Indigenous deaths. In 2005–07, life expectancy at birth for Indigenous Australians was 67 years for men and 73 years for women, compared with 79 years for all Australian men and 84 years for all Australian women, a difference of 10–12 years (AIHW 2013).

In 2005–06, Aboriginals and Torres Strait Islanders were hospitalised for potentially preventable conditions at five times the rate of non-Indigenous Australians, including at fourteen times the rate of non-Indigenous Australians for renal care involving dialysis, and at three times the rate for endocrine, nutrition and metabolic diseases including T2DM (AIHW 2013).

Overweight and obesity

Overweight and obesity are major risk factors in the development of many health problems affecting Aboriginals and Torres Strait Islanders (AIHW 2011b). Of the individual risk factors assessed in 2003 (which did not include diet), high body mass was the second leading cause of the burden of illness and injury after cigarette smoking among Aboriginals and Torres Strait Islanders, accounting for 11 per cent of the total burden of disease and 13 per cent of all deaths (AIHW 2013).

Recent measured height and weight data for Aboriginal and Torres Strait Islander groups are lacking but are being collected in the 2012–13 Australian Health Survey due to be reported by 2014. Self-reported heights and weights are unreliable, and this is especially likely to be the case among Aboriginal and Torres Strait Islander groups (NHMRC 2013). In 2004–05, in Indigenous Australian people aged eighteen years and over in non-remote areas, 34 per cent of men and 24 per cent women were classified as overweight, and 28 per cent of men and 34 per cent of women were classified as obese (AIHW 2011b). Between 1995 and 2004–05 rates of overweight and obesity among Indigenous Australian people in non-remote areas aged fifteen years over increased from 48 per cent to 56 per cent. In 2004–05 rates of overweight and obesity were higher in older age groups, ranging from 37 per cent of people aged 15–24 years to 74 per cent of people aged 55 years and over.

Aboriginal and Torres Strait Islander adults were nearly twice as likely to be obese as non-Indigenous Australians (AIHW 2013). In 2004–05 overweight and obesity tended to be more common among Torres Strait Islanders aged fifteen years or over (61 per cent) than among Aboriginal people in that age range (56 per cent), and the level of overweight and obesity was particularly high among Torres Strait Islanders living in the Torres Strait area, where 86 per cent reported having a BMI of 25 or greater (Burns & Thomson 2008).

Aboriginals and Torres Strait Islanders aged 35 years and over who were overweight or obese were more likely than those who were a healthy weight to report having T2DM or high blood sugar levels (22 per cent compared with 10 per cent) and/or CVD (36 per cent compared with 23 per cent; AIHW 2013).

INFLUENCES ON HEALTH AND NUTRITION

Social

A range of social determinants underpin the poor nutritional health status of Aboriginals and Torres Strait Islanders (AIHW 2011b; NHMRC 2013). Among these are the following:

- poverty, lower income and unemployment
- lower levels of education
- disrupted family and community cohesion
- social marginalisation
- stress
- lack of control over circumstances
- substance abuse
- inadequate and overcrowded housing
- inadequate sanitation, water supplies and hygiene
- limited access to transport
- incarceration, discrimination and racism.

On all these measures Indigenous Australian people suffer substantial disadvantage (NHMRC 2000). Cultural factors can have both positive and negative influences on health and nutritional status. The relationship between social environment and poor health operates in both directions; poor health can increase the risks of deprivation through stigma and reduced earning capacity (NHMRC 2013).

While there is a small but growing well-educated middle class of Indigenous Australian people, most Aboriginals and Torres Strait Islanders remain disadvantaged, with a high proportion relying on some form of welfare. Available data shows that Indigenous Australians are less likely to have any post-school education compared with non-Indigenous Australians (25 per cent compared with 47 per cent), have higher unemployment rates (8.5 per cent compared with 3.3 per cent), have a lower household income ($460 per week compared with $740) and are less likely to own their home (31 per cent compared with 71 per cent; AIHW 2011b). Only 68 per cent of Indigenous Australian students in year 3 and 63 per cent in year 5 of schooling achieved the national reading benchmark, compared with 94 per cent and 93 per cent, respectively, of all Australian students. Of Indigenous Australian students, only 79 per cent in year 3 and 69 per cent in year 5 achieved the national numeracy benchmark, compared with 96 per cent and 94 per cent respectively of all Australian students. About one-third (32 per cent) of Indigenous Australian people reported year 10 as their highest year of school completion. Less than one-quarter (22 per cent) had completed year 12, compared with almost one-half (47 per cent) of non-Indigenous Australians (AIHW 2013).

Although most Indigenous Australians live in metropolitan and regional areas, a much larger proportion of the Indigenous than of the non-Indigenous Australian population live in remote or very remote areas (25 per cent compared with 2 per cent) (AIHW 2011b). For all Australians, death rates for those living in regional areas are on average 1.05–1.15 times and for those living in remote areas 1.2–1.7 times higher than death rates in major cities, and in these areas the average death rates for Indigenous Australian people are more than three times higher than for the non-Indigenous population (AIHW 2013). This is likely to be related to limited occupational and educational opportunities and their effect on income, rather than any special attributes of the physical environment. Poor-quality food supply in remote areas is an ongoing issue, and poor access to medical services is an additional factor (AIHW 2013).

People in rural and remote areas can pay up to 50 per cent more for basic healthy foods than people living in urban and metropolitan areas (NHMRC 2013). For example, in Queensland the mean cost of a basket of healthy foods increased by over 63 per cent ($192.36) from 2000 to 2010 across the state—more than the increased cost of unhealthy foods and about double the increase in consumer price index for food in Brisbane for that period. Basic food items are less available in the more remote stores, as are fresh vegetables and fruits and healthier nutrition choices.

Among factors contributing to the higher costs of foods in rural and remote areas are increased food transport costs, high store overheads (including capital costs of building and maintaining long-term storage facilities and high accountancy costs) and greater wastage of food stock (Altman & Jordan 2009). Commitment and partnership across a range of government and industry bodies are necessary if the factors contributing to the high costs and limited supply of nutritious foods in rural and remote regions are to be tackled (COAG 2009).

Economic

It is now recognised that one of the key drivers of people's pattern of expenditure on food is the economics of food choice, whereby people with a low income in developed economies maximise energy (kilojoules) per dollar, sacrificing quality for quantity (Drewnowski & Spencer 2004). This is particularly the case in remote Indigenous Australian communities, where the industrial globalised food system provides a plethora of energy-dense, nutrient-poor processed foods rich in fat, salt, sugar and refined starches, which provide energy at low cost. These foods have a long shelf life and are very profitable. Fresh foods (that is, fruit, vegetables, lean meats, fish, dairy and eggs) are much more expensive per unit of energy, usually have a short shelf life and are much less profitable for retailers.

Simpson and Raubenheimer (2005) have argued that the protein content of the diet is a much more powerful driver of overall energy intake than carbohydrate and fat. They claim that low-cost highly processed foods dilute the protein content of modern Western diets and lead to overconsumption of energy in the form of fat and carbohydrate in order to meet protein requirements. This provides a potential explanation for the obesity epidemic and is consistent with the strong social gradient of obesity in Western societies. There is no question that price of food influences consumption patterns, so economic levers (taxes and/or subsidies) are the most logical approach to restoring the balance towards protein sources and away from carbohydrate and fat. This is particularly important for disadvantaged sectors of populations, in which the economics of food choice has the most devastating impact.

Lack of food security (see Chapter 18) causes hunger and anxiety related to food shortage in the short term and serious health consequences related to malnutrition and undernutrition in the medium to long term, including those conditions mediated by obesity. While it may seem paradoxical that food insecurity is linked to overweight, as discussed above, these issues can arise because energy-dense, nutrient-poor foods are the lowest cost options, whereas diets based on nutrient-dense whole foods, such as lean meats, whole grains and fresh vegetables and fruits, are more costly. It has been estimated that in Northern Territory communities up to 36 per cent of the family income is needed to purchase food, which is at least double the proportion required in urban regions of Australia.

Young children and pregnant and breastfeeding women tend to be particularly vulnerable to the short- and long-term effects of food insecurity, which can impact on children's growth, physical and socio-emotional development and learning potential. High food costs, poor access to healthy foods, convenience of takeaway foods, budgeting issues, overcrowding and poor knowledge and skills have been identified as barriers to healthy eating and potential food insecurity.

Twenty-four per cent of Aboriginals and Torres Strait Islanders aged fifteen years and over in 2004–05 reported they had run out of food in the last twelve months, compared to 5 per cent of non-Indigenous Australians. While Aboriginals and Torres Strait Islanders living in remote areas were more likely to report having run out of food in the last twelve months (36 per cent), this figure was also high for those in non-remote areas (20 per cent) and ranged from 18 per cent in New South Wales to 45 per cent in the Northern

Territory compared to 3–6 per cent across all states and territories for non-Indigenous Australians (AIHW 2011b). In past national surveys around 30 per cent of Indigenous Australian adults reported they sometimes worried about going without food. As national surveys are unlikely to include data for the most vulnerable groups, such as people experiencing homelessness, these results are likely to underestimate the true extent of the problem.

One of the proposed health equity targets of the Close the Gap campaign launched in 2007 was that 90 per cent of Indigenous Australian families would be able to access a healthy food basket for under 25 per cent of their income by 2018 (COAG 2014; HREOC 2008). However, it was both disappointing and surprising that nutrition issues were not included in the Council of Australian Governments' National Indigenous Reform Agreement (COAG 2011).

HEALTH AND NUTRITION INTERVENTIONS

Community-based programs

Importantly, marked and sustained improvements in diet and related health indicators have been demonstrated in multi-strategy, community-directed nutrition programs in some remote communities, including Minjilang in the Northern Territory (Lee, Bailey et al. 1994) and Looma in Western Australia (Rowley et al. 2000). The strength of these two intervention programs was that they looked not only at change in composition of the food supply but also at biomarkers of dietary change in the populations. The biomarkers reflected the changes seen at the population level: reductions in sugar and saturated fat and increases in fresh fruit and vegetables were associated with reductions in triglycerides, cholesterol and blood pressure and increases in red cell folate, vitamin C and carotenoids.

Intervention strategies implemented by the Minjilang community focused both on the supply side (improving food quality and access to healthy food in remote communities) and on the demand side (promoting nutrition through behaviour change). Specific strategies included individual health assessment; feedback including practical advice and support; store interventions including point-of-sale promotions, shelf signage, cooking demonstrations, cross-subsidisation to decrease price of fruit and vegetables and freight subsidisation of fresh foods; development of infrastructure to promote physical activity (such as provision of fencing for the football field and night lights for the basketball courts); a bush tucker vehicle and one-way grocery delivery service; and provision of affordable, healthy meals through the women's centre (Lee, Bailey et al. 1994). Similar strategies were implemented successfully at Looma (Rowley et al. 2000).

These successful programs demonstrated that nutrition interventions can work when community members are involved in all stages of the development, implementation and evaluation. Communities need to be supported to make improvements, and to achieve this, community stores need to be seen as essential services, like health and education, rather than simply viewed as small businesses.

Among more recent work, Black et al. (2013) found that a fruit and vegetable subsidy program in three Aboriginal communities in New South Wales was associated with improvements in some indicators of short-term health status in some children. Several studies have attempted to use micronutrient supplementation as a 'magic bullet'. This approach does not address the underlying issues of food insecurity or the poor nutritional profile of the broader diet; none of these studies has been successful.

Policy initiatives

The impact of the Income Management component of the Northern Territory Intervention (now the Stronger Futures policy; DHS 2013) was assessed by analysis of store sales in ten remote Northern Territory communities over a three-year period, including an eighteen-month period before Income Management, a four- to six-month period after the introduction of Income Management, a three-month period that coincided with a government stimulus payment and the remaining Income Management period (Brimblecombe et al. 2010). Key outcome measures included trends in total store sales and sales of total food and beverage, fruit and vegetables, soft drink and tobacco. Other than during the significant increase in the rate of sales for all outcome measures during the three-month government stimulus payment period concomitant with Income Management, there was no apparent beneficial effect on sales of tobacco, cigarettes soft drinks, fruit or vegetables.

Future strategies

Systematic, widespread, sustained implementation of evidence-based nutrition interventions is required

The National Aboriginal and Torres Strait Islander Nutrition Strategy and Action Plan

From 2000 to 2010 the National Aboriginal and Torres Strait Islander Nutrition Strategy and Action Plan (NATSINSAP) provided a framework for national action, under the Eat Well Australia agenda (2000–10), to drive dietary changes to improve the nutritional status of Aboriginal and Torres Strait Islanders (SIGNAL 2001). The primary action areas in the plan are listed below:

- food supply in remote and rural communities
- food security and socio-economic status
- family-focused nutrition promotion, involving resourcing programs
- communicating and disseminating 'good practice'
- nutrition in urban areas
- the environment and household infrastructure
- the Aboriginal and Torres Strait Islander nutrition workforce
- national food and nutrition information systems (NATSINSAP 2001).

NATSINSAP was evaluated in 2010. Although the results were not released publicly, it is understood that key findings were that, despite limited resources for implementation, significant outcomes were achieved in improving the food supply in remote areas, disseminating good practice and increasing Indigenous Australian workforce capacity.

Some projects developed under NATSINSAP are continuing. For example, nine practical resources, including a freight-improvement toolkit and a buyer's guide, to assist remote stores to stock, promote and monitor the sale of healthy food are available under the Remote Indigenous Stores and Takeaways Project (Department of Health 2008). While the resources are well used and improvements in food supply have been seen where nutritionists have been involved in their application, insufficient time was allowed for implementation before impact evaluation (Gregoriou & Leonard 2010).

urgently in communities of Aboriginals and Torres Strait Islanders. In addition, there is a need to test innovative approaches, particularly economic interventions. Based on the issues and evidence presented in this chapter, potential solutions are highlighted in Table 14.2. To be effective, all interventions should be developed in full consultation with the community and implemented under local community control.

Table 14.2 Potential strategies to improve Aboriginal and Torres Strait Islander nutrition

Intervention area	Strategy	Comment
Improve food supply	Improve store management practices, transport and stocking of healthy food.	Use Remote Indigenous Stores and Takeaway resources.
	Reinvigorate local food gardens and traditional food procurement projects.	Applicability varies throughout the nation.
	Provide affordable, healthy breakfast and lunches in childcare centres, kindergartens and schools.	Prepared by local community members, community kitchens, restaurants and/or cafes.
	Provide affordable, healthy community meals or takeaways.	
Improve affordability of healthy food supply: economic interventions	Introduce/expand freight subsidies to transport basic healthy foods to remote areas.	Trial for widespread roll-out, if successful. Assess different models—e.g., for subsidisation, trial WIC-style program for pregnant and lactating women and their babies up to age five, linked to primary care sector, with 'prescriptions' for healthy food.
	Food supplementation/subsidisation for women, infants and children.	
	Subsidise provision of fruit and vegetables in remote schools and other settings.	
	Expand in-store cross-subsidisation—increase mark-up of less healthy items and lower price margins on healthier foods.	
	Expand the current national differential taxation system to further favour competitive retail pricing of basic healthy foods.	
	Develop pre-programmed credit cards to reward healthy purchases.	

Intervention area	Strategy	Comment
Increase demand for healthy food	Expand brief nutrition interventions and early interventions in primary care, including 'well persons' health checks and follow-up action.	Engage elders to build on traditional knowledge and stories.
	Expand prenatal, antenatal and postnatal nutrition programs, including: implementation of appropriate infant growth assessment and action programs and promotion of breastfeeding and appropriate introduction of solid foods.	Ensure culturally appropriate approaches to development, implementation, evaluation and dissemination.
	Implement school-based nutrition promotion projects.	
	Implement community food literacy and budgeting programs.	
Increase capacity	Train and employ a nutrition workforce for Aboriginals and Torres Strait Islanders.	
	Improve housing, including food storage, preparation and cooking facilities.	
	Develop a national coordinated food and nutrition monitoring and surveillance system that includes a component for Aboriginals and Torres Strait Islanders.	

Note: WIC: American Nutrition Service for Women, Infants and Children (www.fns.usda.gov/wic/women-infants-and-children-wic).

SUMMARY AND KEY MESSAGES

It is important to understand the socio-economic drivers of the poor health profile of Aboriginals and Torres Strait Islanders, particularly in the more remote parts of Australia. Poor diet is the single most important risk factor contributing to this burden of disease. Discretionary foods high in saturated fats, added sugars and salt, and highly processed foods, particularly sugar-sweetened beverages, provide significant contributions to energy intake. Intakes of fresh foods, including fruit and vegetables, meat, fish, milk products and plant-based alternatives, are low.

- Aboriginals and Torres Strait Islanders experience higher rates of poor health outcomes than non-Indigenous Australians. Much of this is due to poor diet, which impacts on both undernutrition in early life and high prevalence and early incidence of chronic diseases such as T2DM, CVD, renal disease and some cancers.
- The poor health of many Aboriginals and Torres Strait Islanders contrasts sharply with the good health of those who lived traditional lifestyles. The traditional diet was characterised by low energy density and high nutrient density.
- Discretionary and highly processed foods are generally low cost and provide much more energy per dollar than minimally processed fresh foods—a situation termed the economics of food choice. Good-quality food sources of protein can be particularly expensive.
- There is a need to focus on improving the quality of the whole diet rather than focusing on individual nutrients.
- Effective interventions require full consultation with the community and local community control.

REFERENCES

ABS (Australian Bureau of Statistics) 2013. *Australian Aboriginal and Torres Strait Islander Health Survey: First results, 2012–13*. Cat no. 4727.0.55.001. Commonwealth of Australia. <www.abs.gov.au/ausstats/abs@.nsf/Lookup/A07BD8674C37D838CA257C2F001459FA?opendocument>, accessed 24 November 2014.

——2014. *Australian Aboriginal and Torres Strait Islander Health Survey: Updated results, 2012–13*. Cat no. 4727.0.55.066. Commonwealth of Australia. <www.abs.gov.au/ausstats/abs@.nsf/Lookup/4727.0.55.006main+features12012-13>, accessed 13 November 2014.

AIHW (Australian Institute of Health and Welfare) 2011a. *2010 Australian National Infant Feeding Survey: Indicator results*. Commonwealth of Australia. <www.aihw.gov.au/WorkArea/DownloadAsset.aspx?id=10737420925>, accessed 13 November 2014.

——2011b. *The Health and Welfare of Australia's Aboriginal and Torres Strait Islander people: An overview*. Cat. no. IHW 42. Commonwealth of Australia. <www.aihw.gov.au/WorkArea/DownloadAsset.aspx?id=10737418955>, accessed 11 September 2013.

——2013. *Aboriginal and Torres Strait Islander Health Performance Framework 2012: Detailed analyses*. Cat. no. IHW 94. Commonwealth of Australia. <www.aihw.gov.au/WorkArea/DownloadAsset.aspx?id=60129543818>, accessed 5 October 2014.

Altman, J.C. & Jordan, K. 2009. Submission to the House of Representatives Standing Committee on Aboriginal and Torres Strait Islander Affairs Inquiry into Community Stores in Remote Aboriginal and Torres Strait Islander Communities. Centre for Aboriginal Economic Policy Research, Canberra. <http://caepr.anu.edu.au/sites/default/files/Publications/topical/Altman_Jordan_Stores_0409.pdf>, accessed 13 November 2014.

Black, A.P., Vally, H., Morris, P.S., Daniel, M. et al. 2013. Health outcomes of a subsidised fruit and vegetable program for Aboriginal children in northern New South Wales. *Medical Journal of Australia* 199: 46–50.

Brimblecombe, J. & O'Dea, K. 2009. Role of energy cost in food choices for an Aboriginal community in Northern Australia. *Medical Journal of Australia* 190: 549–51.

Brimblecombe J., Ferguson M., Liberato, S. & O'Dea, K. 2013. Characteristics of the community-level diet in remote Aboriginal northern Australia. *Medical Journal of Australia* 198: 380–4.

Brimblecombe, J.K., McDonnell, J., Barnes, A., Garnggulkpuy Dhurrkay, J. et al. 2010. Impact of income management on store sales in the Northern Territory. *Medical Journal of Australia* 192: 549–54.

Burns, J. & Thomson, N. 2008. *Review of Nutrition and Growth among Indigenous Peoples*. Australian Indigenous HealthInfoNet. <www.healthinfonet.ecu.edu.au/health-risks/nutrition/reviews/our-review>, accessed 15 August 2013.

COAG (Council of Australian Governments) 2009. *National Strategy for Food Security in Remote Indigenous Communities*. COAG. <www.coag.gov.au/sites/default/files/nat_strat_food_security.pdf>, accessed 20 August 2013.

——2011. *National Indigenous Reform Agreement (Closing the Gap)*. COAG. <www.federalfinancialrelations.gov.au/content/npa/health_indigenous/indigenous-reform/national-agreement_sept_12.pdf>, accessed 13 November 2014.

——2014. *Closing the Gap in Indigenous Disadvantage*. COAG. <www.coag.gov.au/closing_the_gap_in_indigenous_disadvantage>, accessed 2 February 2014.

Department of Health 2008. *Remote Indigenous Stores and Takeaway (RIST) Project*. Commonwealth of Australia. <www.health.gov.au/internet/main/publishing.nsf/Content/phd-nutrition-rist>, accessed 20 August 2013.

DHS (Department of Human Services) 2013. *Income Management*. Commonwealth of Australia. <www.humanservices.gov.au/customer/services/centrelink/income-management>, accessed 22 August 2013.

Drewnowski, A. & Spencer, S.E. 2004. Poverty and obesity: The role of energy density and energy costs. *American Journal of Clinical Nutrition* 79: 6–16.

Gracey, M. 2007. Nutrition-related disorders in Indigenous Australians: How things have changed. *Medical Journal of Australia* 186: 15–17.

Gregoriou, A. & Leonard, D. 2010. *RIST: Evaluation of the remote Indigenous stores and takeaways resources in Queensland and the Anangu Pitjantjatjara Yankunytjatjara lands of South Australia*. Australian Indigenous HealthInfoNet. <www.healthinfonet.ecu.edu.au/key-resources/bibliography?lid=19667>, accessed 20 August 2013.

Guest, C.S. & O'Dea, K. 1992. Diabetes in Aborigines and other Australian populations. *Australian Journal of Public Health* (now *Australian and New Zealand Journal of Public Health*) 16: 340–9.

Haswell-Elkins, M., McGrath, V., Moore, M., Soisungwan, S. et al. 2007. Exploring potential dietary contributions including traditional seafoods and other determinants of urinary cadmium levels among Indigenous women of a Torres Strait Island. *Journal of Exposure Science and Environmental Epidemiology* 17: 298–306.

HREOC (Human Rights and Equal Opportunity Commission) 2008. *Close the Gap: National Indigenous health equality targets; Outcomes from the National Indigenous Health Equality Summit, Canberra, 18–20 March, 2008.* Australian Human Rights Commission. <www.humanrights.gov.au/sites/default/files/content/social_justice/health/targets/health_targets.pdf>, accessed 5 October 2014.

IHME (Institute for Health Metrics and Evaluation) 2013. *The State of US Health: Innovations, insights, and recommendations from the Global Burden of Disease Study.* IHME. <www.healthdata.org/sites/default/files/files/policy_report/2013/USHealth/IHME_GBD_USHealth_FullReport.pdf>, accessed 5 October 2014.

Lawrence, C.G. 2013. The urgency of monitoring salt consumption and its effects in Aboriginal and Torres Strait Islander Australians. *Medical Journal of Australia* 198: 365–6.

Lee, A.J. 1996. Transition of Australian Aboriginal diet and nutritional health. In Simopoulos, A.P. (ed.), Metabolic consequences of changing dietary patterns. *World Review of Nutrition and Dietetics* 79: 1–52.

——2013. Unpublished store-turnover data. Nganampa Health. (Submitted for publication to the *Australia and New Zealand Journal of Public Health*).

Lee, A.J., Bailey, A.P.V., Yarmirr, D., O'Dea, K. et al. 1994. Survival tucker: Improved diet and health indicators in an Aboriginal community. *Australian Journal of Public Health* (now *Australian and New Zealand Journal of Public Health*) 18: 277–85.

Lee, A., O'Dea, K. & Mathews, J. 1994. Apparent dietary intake in remote Aboriginal communities. *Australian Journal of Public Health* (now *Australian and New Zealand Journal of Public Health*) 18: 190–7.

Leonard, D., Beilin, R. & Moran, M. 1995. Which way kaikai blo umi? Food and nutrition in the Torres Strait. *Australian Journal of Public Health* (now *Australian and New Zealand Journal of Public Health*) 19: 589–95.

NATSINAP 2001. *National Aboriginal and Torres Strait Islander Nutrition Strategy and Action Plan and First Phase Activities.* National Public Health Partnership, Canberra. <www.health.vic.gov.au/archive/archive2014/nphp/publications/signal/natsinsa1.pdf>, accessed 13 November 2014.

NHMRC (National Health and Medical Research Council) 2000. *Nutrition in Aboriginal and Torres Strait Islander Peoples. An information paper.* Commonwealth of Australia. <www.nhmrc.gov.au/guidelines/publications/n26>, accessed 13 November 2014.

——2012. *Infant Feeding Guidelines for Health Workers.* Commonwealth of Australia, Canberra.

——2013a. *Eat for Health: Australian Dietary Guidelines; Providing the scientific evidence for healthier Australian diets.* Commonwealth of Australia. <www.eatforhealth.gov.au/sites/default/files/files/the_guidelines/n55_australian_dietary_guidelines.pdf>, accessed 16 March 2014.

——2013b. *Infant Feeding Guidelines for Health Workers.* Commonwealth of Australia, Canberra.

O'Dea, K. 1984. Marked improvement in carbohydrate and lipid metabolism in diabetic Australian Aborigines after temporary reversion to traditional lifestyle. *Diabetes* 33: 596–603.

——1991. Traditional diet and food preferences of Australian Aboriginal hunter-gatherers. *Philosophical Transactions of the Royal Society B: Biological Sciences* 334: 233–41.

O'Dea, K., White, N.G. & Sinclair, A.J. 1988. An investigation of nutrition-related risk factors in an isolated Aboriginal community in northern Australia: Advantages of a traditionally-orientated lifestyle. *Medical Journal of Australia* 148: 177–80.

Rowley, K.G., Daniel, M. & Skinner, K. et al. 2000. Effectiveness of a community directed 'healthy lifestyle' program in a remote Australian Aboriginal community. *Australian and New Zealand Journal of Public Health* 24: 136–44.

SIGNAL (Strategic Inter-governmental Nutrition Alliance) 2001. *National Aboriginal and Torres Strait Islander Nutrition Strategy and Action Plan (NATSINSAP) 2000–2010.* Commonwealth of Australia, Canberra.

Simpson, S.J. & Raubenheimer, D. 2005. Obesity: The protein leverage hypothesis. *Obesity Reviews* 6: 33–142.

15

New Zealand and Māori nutrition and health

Carol Wham

LEARNING OUTCOMES

Upon completion of this chapter you will be able to:

- outline the contributing factors that shape the eating patterns of New Zealanders
- describe the cultural importance of food for Māori
- explain how a holistic view of health for Māori is important for dietary change
- identify the key diet-related causes of health loss
- review the progress of community nutrition intervention programs.

New Zealand is a multicultural society and is confronted by an ever-increasing variety of foods. Food and beverage production is at the cornerstone of New Zealand's prosperity and a major source of export earnings. Māori are the *tangata whenua* (people of the land) and the country's original inhabitants. A unique New Zealand food identity has developed from the fusion of colonial beginnings, traditional Māori cuisine and an ethnically diverse food supply. This chapter examines the factors that shape the eating and lifestyle practices needed for New Zealanders to live healthy independent lives. It also reviews population-specific background papers which provide the evidence base for nutrition policy advice. Food and nutrition guidelines are available for healthy infants and toddlers, children and young people, adults, older people, and pregnant and breastfeeding women.

The most reliable and up-to-date information about the nutritional health of New Zealanders comes from the 2008–09 Adult Nutrition Survey (University of Otago & Ministry of Health 2011). Compared to the previous survey, in 1997, key issues to emerge were an increase in the prevalence of overweight and obesity and a rise in the predominance of food insecurity.

In New Zealand, inequalities in health are evident especially for Māori, who are three to four times more likely to experience disadvantage and hardship than non-Māori. For Māori, political, cultural and historical influences are recognised factors related to health disparities. The burden of overweight and obesity is greater in Māori than in non-Māori, which in turn is reflected in their higher risk of cardiovascular disease (CVD) and diabetes (Ministry of Health 2013a).

Amid the abundance of food choice options available to New Zealanders, key dietary factors leading to poor health have emerged from the 2013 report of the New Zealand Burden of Diseases, Injuries and Risk Factors Study (Ministry of Health 2013a). Some of the preventative community nutrition interventions that have been developed to address these diet-related problems are reviewed in this chapter.

NEW ZEALAND PEOPLES

Originally home solely to Māori, who arrived in the thirteenth century, New Zealand is now a multi-ethnic society. In 2013, of the 4.47 million New Zealanders, approximately 69 per cent were of European descent, 14.6 per cent were Māori, 9.2 per cent Asian and 6.9 per cent Pacific people (Statistics New Zealand 2013). Geographically, over three-quarters of the population live in the North Island, with one-third of the total living in Auckland. More than half the Pacific people who live in New Zealand were born there. The population is young and concentrated in the Auckland region. About half of the Pacific peoples living in New Zealand are Samoan. The next largest groups are Cook Island Māori and Tongan. Many more Cook Island Māori, Niueans and Tokelauans live in New Zealand than on their home islands. Among Asian people living in New Zealand, the Chinese form the largest group, followed by South Asians. Within these two communities there are families that have lived in New Zealand for several generations.

As with other Organisation for Economic Co-operation and Development (OECD) countries, New Zealand has an ageing population, and those over 65 years (14 per cent in 2013) will comprise a quarter of the population by 2051. The over 65 years group itself is ageing, with the oldest (85 years and over) expected to increase six-fold by 2051 (Statistics New Zealand 2000). This has direct implications for health expenditure, because there is a significant rise in the prevalence of disability with age. Good nutrition is a key determinant of successful ageing, as food not only is critical to physiological wellbeing but also contributes to the social, cultural and psychological quality of life. Older people are known to be at disproportionate risk of malnutrition and have an increased risk of developing health problems as a result of inadequate food and nutrient intake.

Māori culture is a key element of the New Zealand identity. The Treaty of Waitangi (1840) is New Zealand's founding constitutional document and is further supported by the United Nations Declaration on the Rights of Indigenous Peoples (2008). The treaty clearly defines indigenous rights to access and preservation of (traditional) food sources.

The Māori population is younger than the population as a whole. In 2006, the median age of Māori was 22.7 years, compared with the median age of 35.9 years of the total population. Thirty-five per cent of Māori compared with 22 per cent of the total population were aged less than fifteen years, and 3 per cent of Māori are aged 65 and over, compared with 12 per cent overall.

The number and proportion of older Māori are rising, and by 2021 the Māori population aged 65 years and older is projected to have grown by more than double the rate of non-Māori. By 2021, it is predicted, 7.6 per cent of Māori will be over 65 years, compared with 25 per cent of non-Māori. The increase is especially influenced by growth of the Māori population over 65 years, who are healthier and living longer. By 2026 it is probable that Māori will comprise 9.5 per cent of the older people's population (Ministry of Health 2011).

Māori are disadvantaged compared to most other groups. They have lower life expectancy and living and housing standards, poorer health and lower educational attainments. They share these characteristics, which have their roots in history, with Pacific people. Māori have significantly higher rates of preventable morbidity and mortality, due to circulatory diseases at earlier ages, and ischaemic heart disease is the leading cause of death for older Māori (Ministry of Health 2011). More than half of Māori live in areas of highest deprivation, which impacts on food security, food choices and nutrition status. High body mass index (BMI), high blood pressure, high blood cholesterol and inadequate intake of fruit and vegetables are key risk factors that contribute to a higher nutrition-related mortality in Māori (47 per cent) than in non-Māori adults (39 per cent).

Whanau is at the foundation of Māori society and has a vital role to support Māori as a source of strength, security and identity. The overall aim of He Korowai Oranga: Māori Health Strategy (Ministry of Health 2014) is healthy families (*whānau ora*), whereby Māori families can be supported to achieve their maximum health. The holistic nature of health for Māori means that whanau experience physical, spiritual, mental and emotional health. In the Māori world view, there is a fundamental belief that understanding and being connected to the past are significant for both the present and the future. This is demonstrated by the importance placed on *tūpuna* (ancestors) and *whakapapa* (genealogical connections). For many Māori the importance of a healthy and sustainable environment is incorporated into the world view.

TRADITIONAL DIETS AND LIFESTYLES

For Māori, *kai* (food) has cultural importance, and traditional Māori foods differ from food usually eaten by non-Māori. Traditional Māori foods are generally compatible with the national Food and Nutrition Guidelines, and their inclusion into the diet is promoted within the Māori and general communities. Māori apply a whanau, or group, approach to the growing, procurement, cooking and eating of food using traditional food systems. The concept of *manaakitanga* (sharing) ensures that food is available for all. Māori elders who are able to access important traditional foods on a regular basis have a better nutrition status and need to be supported to increase their consumption of traditional foods (Wham et al. 2012).

The first settlers brought no cultivatable plant foods or domesticated animals and depended on what the sea, rivers, lands and forests could supply (Buck 1952). The native foods were mainly *ika* (fish), *kai moana* (seafood), *manu* (birds) and plant foods, including roots, pith, leaves, fruit and berries. The *aruhe*, the common bracken fern rhizome, was widely spread and the predominant staple all the year round. The rhizomes were dug up and stacked in short lengths on end to dry, stored, then cooked over coals. The rhizomes could be pounded with hardwood and worked with the hands in a wooden bowl to extract the starch, which was made into cakes providing a rich source of carbohydrate. The curling *pikopiko* (fern shoots) were cooked as a green.

Later, voyaging canoes from Polynesia introduced cultivated plant foods and domesticated animals such as the *kiore* (Polynesian rat), *kuri* (Polynesian dog) and *poaka* (Polynesian pig), and many varied tropical fruits and staple root crops of the Pacific. The cultivation of the kumara (sweet potato) required more attention than in Polynesia. When planting, a ritual with chants was developed to ensure the aid of specific kumara gods to offset any unfavourable effects of climate and weather and to promote a prolific crop. Successfully cultivated kumara was stored for eating out of season. The *hue* (gourd) was mainly grown to provide containers for water and for preserving birds. *Kamokamo* (marrow), *puha* (sow-thistle) and *wātakirihi* (watercress) were other vegetables grown, as green leafy varieties. Pre-colonisation, Māori cultivated crops that provided them with health benefits. They produced foods that contained protective antioxidant phytochemicals and anti-inflammatory agents.

Māori ate a wide variety of shoots and leaves as greens, such as puha, which were rich in vitamin C and phenolic anti-inflammatory agents. Kumara and *urenika* (Māori potatoes) provided complex carbohydrates and a rich source of anthocyanins. Berries from the karaka tree, seeds from the flax and miro provided good sources of fatty acids.

Kai moana was a significant part of the Māori diet in coastal areas. It included *kina* (sea urchins), paua, pipi, *koura* (crayfish), *ngaeti* (periwinkles), *tuna* (eel), *patiki* (flounder), *inanga* (whitebait), *kuku* (mussels) and *parengo* (a type of seaweed). Many species of fish were caught on lines or in nets, and shellfish (mussels, paua, pupu and pipi) were gathered from the shore. Seaweed provided a rich source of minerals, vitamins and carbohydrate. Birds were snared in the native forests, and eel and whitebait taken from inland or estuarine waters.

The traditional Māori diet was therefore high in protein and fibre and low in fat and provided a healthy nutrient balance. The provision of a year-round food supply required considerable planning and energy. Māori knowledge is now being sought to provide answers nationally and globally as to how to achieve sustainability and adaptability in ever-changing environments.

Māori traditionally practised cooking for large numbers in a hangi, or earth oven, in which hot stones created steam to cook wrapped food. Both Māori and non-Māori enjoy its food's distinctive taste. Other cooking methods included roasting and, in geothermal areas, boiling or steaming using natural hot springs and pools. Occasionally, food would be boiled in non-geothermal areas by putting hot stones into a bowl with water and the food; and some food was also cooked over the open fire. Some foods were preserved using smoke, air-drying or layers of fat, particularly mutton birds.

Colonisation brought changes for Māori in culture and society and in procurement, preparation, cooking and consumption of traditional foods. With the arrival of whalers, sealers and European colonial settlers, kai was an important trading commodity. Gardens and food gathering were highly regarded. Trading saw the introduction of new foods and the beginning of significant dietary change. The introduction of cattle, sheep and pigs by Europeans meant less difficulty was experienced in procuring meat to complement vegetable foods. A similar effect resulted from the introduction of the Irish potato, which grew mostly everywhere and was easy to cultivate.

The adoption of the potato as a replacement for kumara had overwhelming effects on Māori from a physiological, cultural and nutritional perspective. Kumara was significant as both a highly nutritious food and a spiritual and physical link to their ancestral homeland, Hawaiki. The growing and storage of kumara by whanau and *hapu* (sub-tribe) was labour intensive and, like other aspects of natural Māori society, required a whanau approach and involved a *tikanga* (correct procedure) and cultural value-based system. By contrast, the cultivation and storage of potato was less laborious. Potato grew in abundance in almost any soil type without the need for soil enhancement; it was less seasonal and was not supported by specific cultural practices.

Colonisation also saw a marked change in Māori cooking practices, through the introduction of metal pots. Boiling of foods, as mentioned, had previously been restricted to the heating of water in wooden bowls by adding heated stones from an earth oven (Buck 1952). With the introduction of three-legged iron pots, boiling became popular for family cooking as it was far easier and more rapid than earth oven cooking. Eventually, the bucket and tin pannikin superseded the gourd water bottle, and with the change of beverage from water to tea the cultivation of the gourd was negated.

Later introduction of large cooking ranges in community kitchens relegated the need for the earth oven, and this traditional method became reserved only for cooking in quantity for feasts and funerals. The adoption of English ways of serving food, with crockery and knives and forks, also led to a lapse in the making of flax *kōnae* (serving platters) (Buck 1952).

CONTEMPORARY DIETS AND LIFESTYLES

All of these post-colonial changes serve to illustrate the stripping of cultural food practices. It has been said that today the cultural value of food has been almost given away with the reliance on a stranger to grow, produce and perhaps cook and prepare food (Durie 2001).

In the 1960s a survey with Tūhoe people, a tribal group, and 73 whanau living in the Ruatahuna Valley, in the north-east of New Zealand's North Island, indicated that little of the people's typical diet consisted of traditional foods; instead, staple foods were bread, potatoes, sugar, butter, meat and green vegetables, possibly puha, watercress or *tī kōuka* (cabbage tree) (Prior et al. 1964).

A comprehensive insight into the modern-day eating patterns of Māori was found in 2004 by the Te Wai o Rona: Diabetes Prevention Strategy. Although restricted to the responses of Māori people from the Waikato and Southern Lakes areas of the central North Island and not nationally representative, this survey revealed the most frequently consumed traditional foods among 2669 Māori whose mean age was 48 plus or minus thirteen years:

- *Kai moana*—eaten by 55 per cent of respondents
- *Puha*—26 per cent
- *Wātakirihi*—24 per cent
- *Hāngī* (meat slow cooked in an earth oven with kumara and other root vegetables)—18 per cent
- *Parāoa parai* (fried bread, Māori bread, *rēwena*)—18 per cent. (Rush et al. 2010)

Many traditional foods were still consumed at *hui* (community gatherings), and there was a range of traditional foods obtained from the sea and the land. Māori whose *marae* (meeting house) was within 5 kilometres of the *moana* (sea) ate more kai moana at hui and more often regarded kai moana as a traditional food than those who attended hui on marae inland.

As the majority of New Zealanders today are of British descent (and referred to by Māori as pakeha), it is not surprising that contemporary dietary habits have evolved from British cuisine. In the early days of its colonisation New Zealand produced staples of the English diet, such as lamb, beef, butter and cheese. These were made in large quantities on a counter-seasonal cycle for shipment to Britain. Today, New Zealand cuisine is largely derived from local ingredients and seasonal variations, although ingredients for many ethnic dishes are widely available.

New Zealand's food and beverage manufacturing sector is large and accounts for more than half of its exports. With milk produced from grass-fed animals, dairy products are the biggest export earner. New Zealand supermarkets typically provide 80 per cent of the population's fresh perishables (for example, meat, seafood, fruit, vegetables and milk and dairy products) and 20 per cent of its consumer-ready shelf-stable products. In recent decades New Zealand's food and beverage industry has begun to transform into a producer of consumer ready-packaged goods.

Present-day Māori cuisine is a mixture of Māori traditional foods, old-fashioned English cookery and contemporary dishes. The boil-up is a distinctively Māori method of cooking that boils root vegetables such as kumara and potatoes, puha and spinach in a pork stock. Dumplings, also known as doughboys, usually accompany the meal.

Contemporary Māori value kai for its spiritual origins, healing properties and importance to enduring physical, mental, spiritual and social wellbeing. The holistic view of kai remains an important part of the dietary practices of many Māori. Kai has important social and cultural significance. The gathering of kai for the marae is particularly important, especially when there are *manuhiri* (visitors). As hosts, the whanau, hapu and *iwi* (tribe) are obliged to help with the gathering, preparation and cooking of kai. The abundance and type of kai given to manuhiri provide a measure of the host's success, while the demonstration of manaakitanga symbolises *mana* (prestige).

HEALTH AND NUTRITION

The health and wellbeing of New Zealanders was investigated in the New Zealand Health Survey 2012–13. A random selection of more than 13,000 adults aged fifteen years and over and 4000 children aged from birth to fourteen years were interviewed face to face in their own homes (for children, interviews were at home with their parent or caregiver). A survey questionnaire collected demographic, socio-economic and health information including lifestyle behaviours. Measurements of height, weight and waist circumference were recorded (Ministry of Health 2013b). The 2008–09 New Zealand Adult Nutrition Survey provides an assessment of self-reported food and nutrient intake for all New Zealanders. A total of 4721 adults aged fifteen years and over participated in the survey, including 1040 Māori and 757 Pacific people (Ministry of Health 2012a, 2012b; University of Otago & Ministry of Health 2011).

Two-thirds (66 per cent) of adults reported eating the recommended three or more servings of vegetables per day, while 60 per cent reported daily consumption of two or more servings of fruit. Bread was the main contributor of energy, protein, carbohydrate and fibre to the diet, and wholegrain bread was chosen by 63 per cent of adults most of the time. Almost 50 per cent of younger adults (fifteen to eighteen years) chose to eat white bread, but its popularity decreased with age in favour of wholegrain bread (University of Otago & Ministry of Health 2011).

Reduced-fat (1.5 per cent fat) or trim milk (0.5 per cent fat) was chosen most of the time by nearly half (48 per cent) of adults, and usage of this type increased with age. Red meat was eaten one to two times a week by 30 per cent and three to four times a week by 45 per cent of adults. Nearly two-thirds of all adults (62 per cent) regularly or always removed excess meat fat. Chicken was also a popular choice, consumed by 56 per cent of adults one to two times a week and by 24 per cent three to four times a week. The skin of chicken (with accompanying fat) was regularly or always removed by nearly half (48 per cent) (University of Otago & Ministry of Health 2011).

Butter was used as a spread most of the time by 20 per cent of the participants, while 69 per cent chose margarine. Oil, rather than margarine, dripping, lard or other alternatives, was most often used for cooking (90 per cent) (University of Otago & Ministry of Health 2011).

Nutrient intakes derived from repeat 24-hour recalls on a subsample (25 per cent) of participants were compared to the estimated average requirement (EAR) from the nutrient reference values (NRVs) for Australia and New Zealand (NHMRC & Ministry of Health 2006). The mean percentage contribution to daily energy intake of all adults was carbohydrate (46.6 per cent), fat (33.7 per cent) and protein (16.5 per cent). Nearly two-thirds (59 per cent) of New Zealanders did not meet the national recommendations for dietary calcium intake, and 25 per cent of those surveyed had an inadequate intake of zinc. In general, the prevalence of inadequate nutrient intakes was highest in younger adults (fifteen to eighteen years). Compared to all other age groups their diets were characterised by fewer servings of vegetables and wholegrain breads than are recommended and by the highest frequency of consumption of hot chips and sugar-sweetened drinks.

Māori men and women consumed 43.9 and 46.6 per cent of their energy from carbohydrate, 16.8 and 16.3 per cent from protein and 36.6 and 35.6 per cent from fat, respectively. Compared to non-Māori, consumption of mean percentage of energy from fat was higher, but after adjusting for age there was no difference in the contribution to energy from protein or carbohydrate. Similarly, there were no significant differences between Māori and non-Māori participants in the usual daily median intakes of vitamin A, riboflavin, vitamin B12, calcium, zinc or selenium (Ministry of Health 2012a).

The estimated proportions of Māori adults with inadequate nutrient intakes, according to the EARs, are listed below:

- *Vitamin A*—men: 17.8 per cent, women: 16.6 per cent
- *Riboflavin*—men: 3.8 per cent, women: 5.4 per cent
- *Vitamin B12*—men: 5.1 per cent, women: 12.45 per cent
- *Calcium*—men: 53.3 per cent, women: 71.4 per cent
- *Zinc*—men: 34.3 per cent, women: 14.7 per cent
- *Selenium*—men: 31.5 per cent, women: 53.3 per cent. (Ministry of Health 2012a; NHMRC & Ministry of Health 2006)

Key dietary habits associated with nutrition status revealed that over half of Māori men (52 per cent) and women (59 per cent) ate the recommended three or more servings of vegetables a day. Fifty per cent of men and 57 per cent of women consumed two or more servings of fruit daily. Of those who reporting consuming bread, approximately half selected wholegrain bread most often (52 per cent men and 49 per cent women) and just under half selected white bread most often (43 per cent men and 47 per cent women). Almost all Māori men (97 per cent) and more than half of Māori women (58 per cent) used whole milk most of the time, and a quarter of men (24 per cent) and around a third of women (36 per cent) used reduced-fat, trim or skim milk most of the time. About a third of Māori men (37 per cent) and women (36 per cent) ate fresh or frozen fish or shellfish at least once a week (Ministry of Health 2012a).

Breastfeeding

Exclusive breastfeeding until the age of six months, when solid food should be introduced, has many benefits for the infant and mother and is recommended by the New Zealand Ministry of Health. In 2006 breastfeeding rates from Plunket (a maternal and child health service) data indicated that 25 per cent of all New Zealand babies were exclusively or fully breastfed at six months (29 per cent for New Zealand European and other, 17 per cent for Māori babies) (Royal New Zealand Plunket Society 2010). Growing Up in New Zealand is a longitudinal study that provides an up-to-date population-relevant picture of what it is like to be a child growing up in New Zealand in the 21st century. For 6162 babies in the study's cohort, 2012 data indicate exclusive breastfeeding (without water, formula or other fluids) was stopped at a median of four months (mean 3.85 months) with over 90 per cent no longer exclusively breastfed at six months (Morton et al. 2012). Although New Zealand data have limitations due to inconsistencies in definition, age of collection and percentage of the population from whom the data were captured, published studies of selected population groups concur that babies are not breastfed for long enough.

The 2012–13 New Zealand Health Survey found that 10.2 per cent of children under the age of five years had been started on solid food before four months of age; this rate had fallen from 16 per cent in 2006–07. Sixteen per cent of Māori children aged under five years had been given solid food before four months in the 2012–13 survey—a fall from 22 per cent in 2006–07 (Ministry of Health 2013b). Provincial studies have indicated that some babies are inappropriately introduced to cow's milk as a drink before twelve months of age. This may be associated with iron deficiency anaemia, which is present in 4–14 per cent of New Zealand babies and toddlers (Grant et al. 2007). Babies fed cow's milk before one year of age may be at increased risk of developing depleted iron stores, because the concentration and bioavailability of iron is low. In preschool children the prevalence of iron deficiency is generally unknown, but there are concerns and these children may be at risk.

During childhood, optimal nutrition is important for the maintenance of growth and good health. National surveys indicate that the meal patterns of some New Zealand children may compromise nutrient intake. Although most children (94 per cent) aged two to four years eat breakfast at home every day, breakfast eating tends to decrease with increasing age. Only 81 per cent of children aged ten to fourteen years eat breakfast at home every day. In the two- to fourteen-year age group, breakfast is eaten at home every day by 82 per cent of Māori children and 83 per cent of Pacific children. Key determinants of breakfast-eating habits are the availability of breakfast foods and the breakfast-eating patterns of parents and caregivers (University of Otago & Ministry of Health 2011).

Most children (over 90 per cent) up to twelve years in age have something to eat for lunch at school; about 66 per cent of those aged thirteen to eighteen years eat lunch on four or more school days, and nearly half purchase their lunch from the school canteen. Pacific and Māori children are more likely than other children to make purchases from the school canteen (University of Otago & Ministry of Health 2011).

The majority of parents and caregivers (98 per cent) report that their children have their main meal sitting down with the rest of the family/whanau; however, this practice tends to decrease with age. Morning snack foods eaten by children tend to contribute nearly as much energy as breakfast but are low in nutrients. The most popular choices are potato chips and other extruded snacks, fruit including dried fruit and roll-ups, biscuits, muesli bars and crackers. Children and young people tend to eat foods high in fat, sugar and salt on a regular basis; potato chips, corn snacks or chips are eaten by 85 per cent of children, and chocolate, confectionery, fancy biscuits and soft drinks by 50 per cent of children, at least once a week (University of Otago & Ministry of Health 2011).

Key findings of the New Zealand Health Survey 2012–13 showed that of children aged two to fourteen years 63.6 per cent were of normal weight, 21.6 per cent were overweight and 11.1 per cent were obese. The rate of obesity had increased from 8 per cent in 2006–07. Obesity rates in 2012–13 were 11 per cent for children aged two to four years and 11.5 per cent for children aged five to nine and 10.7 per cent for ten to fourteen years. For adolescents and young people the obesity rate was 20.6 per cent (Ministry of Health 2013b).

Māori children (*tamariki* Māori) aged two to fourteen years and young people (*rangatahi* Māori) aged fifteen to eighteen years are at increased risk of obesity. The 2006–07 New Zealand Health Survey data indicated that among *tamariki* Māori 2 per cent were underweight, 26 per cent were overweight and 12 per cent were obese, while among *rangatahi* Māori less than 1 per cent were underweight, 27 per cent of boys and 29 per cent of girls were overweight, and 23 per cent of boys and 27 per cent of girls were obese (Ministry of Health 2012a).

Food insecurity issues are evident in 20–30 per cent of New Zealand households with five- to eighteen-year-olds. Māori experience lower levels of household food security compared to the total population. For households with *tamariki* Māori, 64 per cent report always being able to afford to eat properly, compared with 78 per cent of New Zealand households overall.

Micronutrients

In New Zealand, most soils are low in iodine, resulting in foods that contribute to low dietary iodine intake. In the early part of the twentieth century, iodine deficiency disorder was endemic in New Zealand, characterised by high rates of goitre. In 1924 iodised salt was introduced, and goitre was eradicated by the 1950s. In the 1990s iodine deficiency reappeared, with evidence of mild iodine deficiency in New Zealand children, infants and toddlers (Skeaff et al. 2002, 2005). The re-emergence of iodine deficiency seems to have resulted from a decline in the use of iodised salt in the home, greater use of manufactured foods made with non-iodised salt and a decline in the use of iodophors as sanitisers in the dairy industry, which markedly reduced the iodine content of its foods.

To address this problem Food Standards Australia New Zealand (FSANZ) introduced mandatory iodine fortification of bread in 2009. Similar to the iodisation concentration of table salt, a range of 25 to 65 milligrams of iodine per kilogram of salt is now added to all yeast-leavened bread (excluding organic and unleavened bread). This fortification level is expected to increase iodine intakes in the population by 30–70 micrograms per day to ensure the recommended dietary intake (RDI) of 150 micrograms per day is met. Even with mandatory fortification, pregnant and breastfeeding women and their infants may continue to be at risk of iodine deficiency because of their higher requirements. Hence, a subsidised iodine tablet of 150 micrograms per day is recommended for these women.

New Zealand soils are also low in selenium, so low levels of selenium are found in locally grown plant foods. Fish, seafood and poultry are the main contributors of selenium to the New Zealand diet. Fortunately, although the dietary intake of selenium is lower than in many other countries, severe deficiency, which results in **Keshan disease**, has never been reported in New Zealand, and currently there is no evidence that mild selenium deficiency has clinically significant effects.

HEALTH CONDITIONS

Overweight and obesity

Of the 13,000 New Zealand adults assessed in the 2012–13 New Zealand Health Survey (Ministry of Health 2013b), two-thirds (64.4 per cent) were

Keshan disease is a condition characterised by cardiomyopathy (which can be fatal) caused by deficiency of selenium. First observed in Keshan province in China, it is also found in other areas where the selenium level in the soil is low. Treatment involves selenium supplementation.

either obese or overweight (about 31.3 per cent were obese and a further 34.1 per cent overweight). Over the previous ten years the obesity rate increased considerably: in 1997 it was 19 per cent. In 2009 New Zealand adults had the third highest obesity rate of the member countries of the OECD. The obesity rate differed by ethnicity: 62 per cent of Pacific adults were obese, and 44 per cent of Māori, 26 per cent of Europeans or other ethnicity and 16 per cent of Asians.

Māori adults were almost twice as likely to be obese as non-Māori adults, and Pacific adults were 2.5 times as likely to be obese as non-Pacific adults after adjusting for age and sex; the obesity rates for Māori and Pacific adults had not changed since 2006–07. The obesity rate of young adults is of particular concern. Among those aged 15–24 years the proportion who were obese increased from 14 per cent in 2006–07 to 20.6 per cent in 2012–13. Obesity rates were much higher among people living in the most deprived areas (43.5 per cent) than among people living in the least deprived areas (26.5 per cent), even when accounting for differences in age, sex and ethnic group (Ministry of Health 2013b).

Diabetes

The New Zealand Health Survey 2012–13 (Ministry of Health 2013b) found the rate of diagnosed diabetes was 5.8 per cent for adults (6.4 per cent for men and 5.1 per cent for women). Diabetes rates were relatively high among Māori adults (7.3 per cent), Pacific adults (12.5 per cent) and Asian men (8.1 per cent), compared with the national average (5.8 per cent). Diabetes was much more common in people living in more deprived areas (8.6 per cent) than in people living in the least deprived areas (4.4 per cent).

Cardiovascular disease

CVD is the leading cause of death in New Zealand. Compared with 2006–07, in 2012–13 a higher proportion of adults were taking medication for high blood pressure (14 per cent and 16 per cent, respectively) and for high cholesterol (8 per cent and 10 per cent). Since 2006–07 there was a reduction in the rate of adults diagnosed with ischaemic heart disease (5.5 to 4.9 per cent); however, there was an increase for stroke (1.8 to 2 per cent). Generally, CVD affects adults 55 years and older. About half of adults over the age of 65 years took medication for high blood pressure, and a third took medication for high cholesterol. Nearly a quarter of people over the age of 75 years had been diagnosed with ischaemic heart disease, and 10 per cent in this age group had reported having a stroke. People who lived in more deprived areas had higher rates of ischaemic heart disease than those living in the least deprived areas (6 per cent compared with 4.8 per cent) and stroke (2 per cent compared with 1.8 per cent). Compared to other adults, Māori were more likely to have higher rates of medication for high blood pressure (1.4 times), to have higher rates of ischaemic heart disease (1.8 times) and to have had a stroke (1.4 times) (Ministry of Health 2013b).

INFLUENCES ON HEALTH AND NUTRITION

The New Zealand Health Survey 2012–13 (Ministry of Health 2013b) findings indicate that people living in socio-economically deprived areas have higher levels of smoking, obesity, psychological distress, ischaemic heart disease, diabetes, arthritis, chronic pain and tooth loss than people living in less deprived areas. These higher rates are not explained by differences in the ethnic, age or sex structure of the population. People in the most deprived areas were less likely to eat the recommended three or more servings of vegetables each day than people in the least deprived areas (58 per cent compared with 72 per cent).

Similarly, unmet need for health care is much more common in the most deprived areas, affecting one in three people (35 per cent) in these areas in the previous year. This may exacerbate existing health problems. Unmet need is more common for Māori (39 per cent) than for the whole population (27 per cent). For Māori and Pacific people, cost is a major barrier to accessing health services and to filling prescriptions, while a lack of transport is a further barrier to accessing services. Additionally, one in seven adults (14 per cent) living in the most deprived areas had not filled a prescription due to the cost in the previous year.

Inequalities between the health status of Māori and other New Zealanders are well documented and are greater for those in more deprived socio-economic groups. Māori experience disadvantage across indicators of health status and access to health services. They report similar dietary and physical activity behaviours as other adults, but are more likely to smoke. There is also a higher rate of diagnosis of health conditions such as ischaemic heart disease and stroke for Māori adults. In recent decades there has been much concern that Māori have adopted the worst of European New Zealand eating habits and as a result are disproportionately more likely to experience obesity, heart disease and diabetes.

Health loss, as a measure of how much healthy life is lost due to early death, illness or disability, has been reported for all New Zealanders in the New Zealand Burden of Diseases, Injuries and Risk Factors Study 2006–2016 (Ministry of Health 2013a). In 2006 dietary risk factor causes of health loss, estimated using **disability-adjusted life years**, included high salt and saturated fat intake, low fruit and vegetable intake, and excess energy intake (high BMI). Diet-related other causes included high blood pressure, blood cholesterol and blood glucose, and low bone mineral density and physical activity. The health losses attributable to these risk factors are listed below:

- *High salt (sodium) intake (1.7 per cent of health losses)*—Mostly (91 per cent) mediated through the effect of blood pressure on CVD: coronary heart disease (CHD), stroke and hypertension. Most (70 per cent) of the high salt intake burden falls on men, reflecting their higher intake of sodium from foods and higher rates of CHD.
- *High saturated (low in polyunsaturated) fat intake (1.2 per cent of health losses)*—All from CHD. Men sustain about two-thirds of the health loss, again reflecting their higher burden of CHD.
- *Low fruit and vegetable intake (1 per cent of health losses)*—Largely contributed by CHD, ischaemic stroke and cancer (lung, oesophageal, mouth and pharynx). Men sustain about 60 per cent of this health loss as a result of their higher burden of CHD and lung cancer.
- *High BMI (7.9 per cent of health losses)*—Main contributors are CHD, diabetes, osteoarthritis, certain cancers and stroke. The BMI burden is greater in Māori than in non-Māori, reflecting the former group's higher average BMI and higher risk of CVD and diabetes. High BMI is projected to overtake tobacco use as the leading risk factor cause of health loss by 2016.
- *High blood pressure (6.4 per cent of health losses)*—Mostly from CHD and stroke. About 60 per cent of this health loss is sustained by men as a result of their higher burden of CVD and slightly higher blood pressure levels.

Disability-adjusted life years is a calculation of the number of years lost due to disability, ill-health or death. It attempts to measure the disease burden in years.

- *High blood cholesterol (3.2 per cent of health losses)*—Largely as a result of CHD and stroke. Men sustain about 60 per cent of the health loss, which reflects their higher burden of CHD. This health loss estimate is projected to decrease by 2016 on the proviso that total cholesterol levels continue to decline. This will depend on continued improvements in dietary fat intake (replacement of saturated fat with polyunsaturated fat) and increased coverage of those at risk with lipid-lowering medications.
- *High blood glucose (4.6 per cent of health losses)*—As modelled indirectly from survey data on glycated haemoglobin levels. The overall health loss estimate included the impact of high blood glucose or glycated haemoglobin levels on CHD and stroke as well as the burden from diabetes itself.
- *Low bone mineral density (0.2 per cent of health losses)*—Nearly all (97 per cent) from hip fractures, with vertebral, rib and forearm fragility fractures accounting for most of the remainder. The burden of low bone mineral density increases steeply with age from 65–9 years and is higher in women than in men.
- *Low physical activity (4.2 per cent of health losses)*—Attributed for about one-quarter of CHD and diabetes, one-fifth of breast cancer and ischaemic stroke and 17 per cent of colon cancer. This health loss is determined partly through its effect on BMI and partly through its direct effect on blood lipids and blood pressure.

Figure 15.1 shows attributable disability-adjusted life years burden as a percentage of total disability-adjusted life years (for the whole population) for those risk factors common to the Global Burden of Disease Study country estimates for New Zealand (2010) (IHME 2013) and the New Zealand Burden of Diseases, Injuries and Risk Factors Study (2006) (Ministry of Health 2013a) estimates.

Overall, more than one-third (37 per cent) of the total health loss in 2006 from all causes was sustained by people over the age of 65 years, although they make up only 12 per cent of the population. The ageing of the population partly explains why health loss is projected to increase from 2006 to 2016 by 13.4 per cent. Total health loss in 2006 was almost 1.8 times higher for Māori than for non-Māori. For diabetes and vascular disorders in particular, the health loss for Māori was 2.5 times higher than for non-Māori.

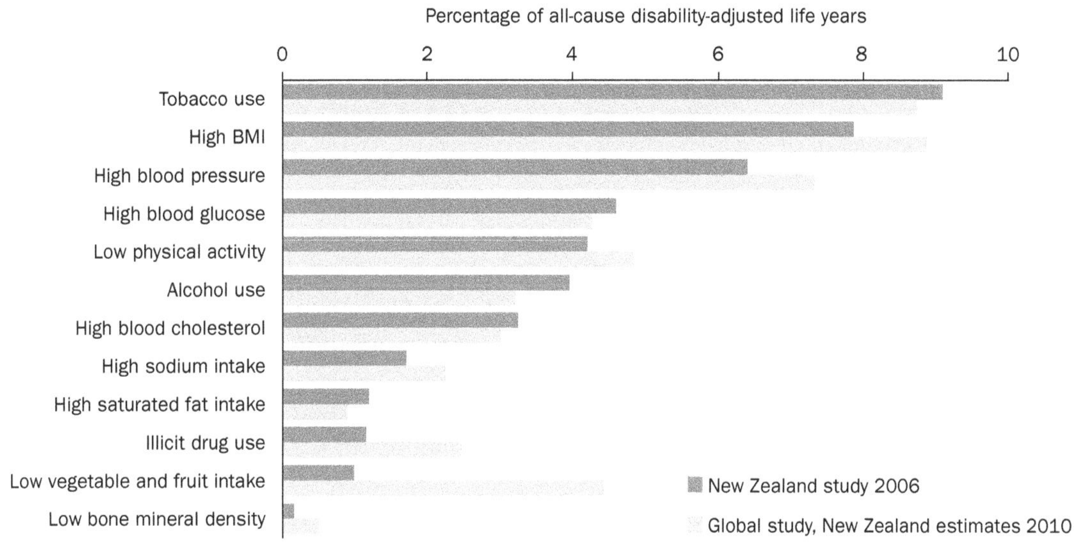

Figure 15.1 Attributable burden comparison, as a percentage of total disability-adjusted life years

Note: Ranking is according to the New Zealand study estimates (Ministry of Health 2013a).

Source: Adapted from GBD Profile: New Zealand (IHME 2013).

In 2006, using the health loss estimates in the New Zealand Burden of Diseases, Injuries and Risk Factors Study, health-adjusted life expectancy for men was 78.1 years, with 8.9 years (11 per cent) in poor health; for women it was 82.1 years, with 11.5 years (14 per cent) in poor health. Therefore, women can expect to live 1.4 years longer than men in full health.

HEALTH AND NUTRITION INTERVENTIONS

APPLE: A Pilot Program for Lifestyle and Exercise

The APPLE: A Pilot Program for Lifestyle and Exercise project was a controlled two-year obesity-prevention initiative in 730 primary school children aged five to twelve years between 2003 and 2005 (Taylor et al. 2013). The project involved four intervention and three control schools in the Otago region of the South Island. Most (83 per cent) of the children were New Zealand European, and 17 per cent were Māori. Measures of height, weight, waist circumference, blood pressure and diet were taken at baseline and follow-up, at year 1 and 2. Intervention schools partook in short bursts of activity in class as well as non-curricular activity at breaks, lunch time and after school. Initiatives included the provision of cooled water filters in each school, free fruit for six months and science classes which highlighted the adverse effects of sugar-sweetened beverages. Community activity coordinators assigned to each school facilitated activity programs for parents and community volunteers. Nutrition education focused on the reduction of sugar-sweetened beverage intake and the benefits of increased fruit and vegetable intake.

At year 2, findings showed that the mean BMI Z-score was significantly lower in intervention children than in control children after adjustment for baseline BMI and confounding variables. Waist circumference was significantly lower at year 2, and systolic blood pressure was reduced at year 1. The intervention children consumed fewer carbonated beverages and fruit juice or drinks and more fruit.

A two-year follow-up of the project showed that although the community activity coordinators were no longer present, continued benefits to BMI persisted in the intervention children. Children at follow-up in the whole group and in the group who underwent at least one or two years of intervention maintained significantly lower mean BMI Z-scores. Involvement of the school, parents and community as well as concurrent national initiatives are thought to have contributed to the sustained improvement (Taylor et al. 2013).

Fruit in Schools

Fruit in Schools is a Ministry of Health school-based initiative which started in 2006 in response to the 2002 National Children's Nutrition Survey finding that only 40 per cent of children ate the recommended two servings of fruit a day (Ministry of Health 2003). The program provides support to schools in low socio-economic communities to promote healthy eating through offering students a piece of fruit for each school day and assists such schools to promote healthy eating and physical activity as well as being sun smart. By 2008 the initiative had grown to involve approximately 470 schools from low socio-economic communities and over 95,000 students. A comprehensive evaluation of the program was conducted during 2005–09 (Boyd et al. 2009).

Overall findings indicated a positive impact on schools' and students' approaches to healthy lifestyles. Fruit in Schools was starting to positively impact on home behaviours, and through it the public health infrastructure had strengthened. The two-year tracking of a group of students before and after their school joined the program showed they were more likely to have increased their consumption of healthy foods such as fruit, vegetables and grains, increased physical activity and maintained sun-smart practices. The students also had more positive views about school than a group of comparison students.

Project Energize

Project Energize is a nutrition and physical activity intervention which aims to improve childhood obesity and cardiovascular risk factors in primary school children. It started as a randomised controlled trial in the period 2004–06 in 124 schools within the Waikato region of New Zealand. This region has a higher level of deprivation and a higher proportion of Māori children than the national average. By 2011, Project Energize had successfully engaged all but two of the 235 schools in the area.

The intervention is delivered by team of 'Energizers' who have trained expertise in physical activity or nutrition. Energizers work with eight to twelve schools each to implement a range of healthy eating initiatives and enhance physical activity. These include canteen makeovers, curriculum-based support for class teachers, lunchtime games and leadership training for students to lead physical activities before and after school. A home–school link program provides opportunities for parents to receive newsletters and attend information sessions.

An evaluation of Project Energize in 2011 compared indices of obesity and physical fitness (2474 younger and 2330 older primary school children with measurements undertaken in the 2004–06 trial (Rush et al. 2012). Findings showed the combined prevalence of obesity and overweight among younger and older children was lower by 31 per cent and 15 per cent, respectively, than that among control children. Physical fitness (time taken to complete a 550-metre run) was significantly higher in the 'Energized' children than in similarly aged children from another region (Rush et al. 2013). Enhanced physical activity has proved a popular and effective outcome measure.

Ngati and Healthy

Ngati and Healthy was a two-year community-led diabetes-prevention intervention aimed at reducing the incidence of insulin resistance in a predominantly Māori rural community on the East Coast of New Zealand. Residents in this region are predominantly from the Ngati Porou tribe, and community life is based around the hapu and whanau. Randomly selected residents aged 25 years and over enrolled on the Ngati Porou Haurora (Māori healthcare service) patient register were invited to participate. Surveys were undertaken before and after the intervention and included a 75-gram oral glucose tolerance test and anthropometric and blood pressure measurements.

The early involvement of community health workers, community members and local organisations helped to ensure the program became embedded into everyday community life. Community education classes included cooking skills, nutrition knowledge, food label reading, recipe swaps, exercise, and smoking cessation. There was encouragement for participants to monitor their own weight and for schools to develop policies for water only as well as healthy lunchboxes. Efforts were made for recommended foods to be available in local shops and supermarkets.

An evaluation between 2003 and 2006 showed a fall in overall prevalence of insulin resistance from 36 to 25 per cent. In 2006, minimum recommended exercise levels were achieved by 60 per cent of participants and wholegrain bread consumption increased to 65 per cent of participants compared with 45 per cent and 42 per cent, respectively, in 2003. The most substantial changes were in reductions in insulin resistance in women aged 25–49 years from 38 to 26 per cent. This group participated most actively in the program activities and reported the greatest lifestyle changes (Coppell et al. 2009).

SUMMARY AND KEY MESSAGES

New Zealanders enjoy a diverse range of foods; however, rates of overweight and obesity have increased since the 1980s, especially among Māori, Pacific peoples and those living in more socio-economically deprived areas. As a major risk factor for heart disease, type 2 diabetes mellitus (T2DM) and obesity-related conditions, high BMI is projected to be the leading cause of health loss in New Zealand by 2016 and is a major challenge for nutrition promoters.

Māori are disadvantaged compared to most other groups. Māori have a holistic view of health, and traditionally a whanau approach was applied to the growing, procurement and eating of food while maintaining sustainable food supplies. Traditional foods meet healthy goals, and Māori need support to access traditional foods to achieve their maximum health.

Nutrition promotion is required at all life stages across relevant settings to improve food and nutrition intake. A community-led diabetes-prevention intervention and several school-based nutrition and physical activity interventions have been successfully instigated to help reduce excess weight gain and risk of chronic disease. To curb the high rates of obesity, national policies are needed to increase the healthiness of food environments.

- New Zealand has an ethnically diverse populace and in line with other OECD member countries the population is ageing. The population of older Māori over 65 years is projected to grow in the next decade by more than double the rate of non-Māori.
- New Zealand soils are low in iodine. To prevent deficiency, bread flour (except organic) is iodine fortified. Supplementation with iodine is recommended for pregnant and breastfeeding women.
- Māori views on health take a holistic approach, and connections to the past are a significant aspect of health.
- Māori experience higher health disparity than non-Māori and have disproportionally higher rates of obesity, heart disease and T2DM.
- Traditional food and cultural eating practices are important for Māori. Emphasis needs to be placed on enabling Māori to access traditional food on a regular basis to improve nutrition status.
- High BMI is projected to be the leading risk factor cause of health loss by 2016. Nutrition promotion and lifestyle strategies are needed alongside social and environmental supports to reduce the burden of overweight and obesity.

REFERENCES

Boyd, S., Dingle, R., Hodgen, E., King, J. et al. 2009. *The Changing Face of Fruit in Schools: 2009 overview report*. New Zealand Council for Educational Research and Health Outcomes International. <www.nzcer.org.nz/system/files/changing-face-fruit-schools-2009-overview-report.pdf>, accessed 5 October 2014.

Buck, P. 1952. *The Coming of the Maori*. Whitcombe & Tombs Limited, Wellington.

Coppell, K.J., Tipene-Leach, D.C., Pahau, H.L.R., Williams, S.M. et al. 2009. Two-year results from a community-wide diabetes prevention intervention in a high risk Indigenous community: The Ngati and Healthy project. *Diabetes Research and Clinical Practice* 85: 220–7.

Durie, M. 2001. *Mauri Ora: The dynamics of Maori health*. Oxford University Press, Melbourne.

Grant, C.C., Wall, C.R., Brunt, D., Crengle, S. et al. 2007. Population prevalence and risk factors for iron deficiency in Auckland, New Zealand. *Journal of Paediatrics and Child Health* 43: 532–8.

IHME (Institute for Health Metrics and Evaluation) 2013. *GBD Profile: New Zealand*. IHME. <www.healthdata.org/sites/default/files/files/country_profiles/GBD/ihme_gbd_country_report_new_zealand.pdf>, accessed 13 November 2014.

Ministry of Health 2003. *NZ Food NZ Children: Key results of the 2002 National Children's Nutrition Survey*. New Zealand Government. <www.moh.govt.nz/notebook/nbbooks.nsf/0/658d849a 2bac7421cc256dd9006cc7ec/$FILE/nzfoodnzchildren.pdf>, accessed 14 November 2014.

——2011. *Tatau Kura Tangata: Health of Older Māori Chart Book 2011*. New Zealand Government. <www.health.govt.nz/system/files/documents/publications/health-of-older-maori-chart-book-2011-mar2011.pdf>, accessed 5 October 2014.

——2012a. *A Focus on Māori Nutrition: Findings from the 2008/09 New Zealand Adult Nutrition Survey*. New Zealand Government. <www.health.govt.nz/system/files/documents/publications/a-focus-on-maori-nutrition_final_1_may_2012.pdf>, accessed 8 October 2014.

——2012b. *A Focus on Pacific Nutrition: Findings from the 2008/09 New Zealand Adult Nutrition Survey*. New Zealand Government. <www.health.govt.nz/system/files/documents/publications/adult-nutrition-survey-pacific-nutrition_final_1_may_2012.pdf>, accessed 8 October 2014.

——2013a. *Health Loss in New Zealand: A report from the New Zealand Burden of Diseases, Injuries and Risk Factors Study, 2006–2016*. New Zealand Government. <www.health.govt.nz/publication/health-loss-new-zealand-report-new-zealand-burden-diseases-injuries-and-risk-factors-study-2006-2016>, accessed 21 November 2014.

——2013b. *The Health of New Zealand Adults 2012/13: Key findings of the New Zealand Health Survey*. New Zealand Government. <www.health.govt.nz/system/files/documents/publications/health-of-new-zealand-adults-2011-12-v2.pdf>, accessed 3 October 2014.

——2014. *He Korowai Oranga: Maori Health Strategy*. New Zealand Government. <www.health.govt.nz/our-work/populations/maori-health/he-korowai-oranga>, accessed 8 October 2013.

Morton, S., Atatoa Carr, P., Grant, C., Lee, A. et al. 2012. *Growing Up in New Zealand: A longitudinal study of New Zealand children and their families; Report 2, Now we are born*. University of Auckland. <https://cdn.auckland.ac.nz/assets/growingup/research-findings-impact/report02.pdf>, accessed 5 October 2014.

NHMRC (National Health and Medical Research Council) & Ministry of Health 2006. *Nutrient Reference Values for Australia and New Zealand: Including recommended dietary intakes*. Commonwealth of Australia & New Zealand Government. <www.nhmrc.gov.au/_files_nhmrc/publications/attachments/n35.pdf>, accessed 26 September 2014.

Prior, I., Rose, B. & Davidson, F. 1964. Metabolic maladies in New Zealand Maoris. *British Medical Journal* 1: 1065–9. <www.ncbi.nlm.nih.gov/pmc/articles/PMC1814418/pdf/brmedj02625-0023.pdf>, accessed 5 October 2014.

Royal New Zealand Plunket Society 2010. *Breastfeeding Data. Analysis of 2004–2009 data*. Royal New Zealand Plunket Society. <www.plunket.org.nz/assets/News--research/Plunket-Breastfeeding-Data-Analysis-of-2004-2009.pdf>, accessed 14 November 2014.

Rush, E., Hsi, E., Ferguson, L., Williams, M. et al. 2010. Traditional foods reported by a Maori community in 2004. *MAI Review* 2. <www.review.mai.ac.nz/index.php/MR/article/view/279/478>, accessed 5 October 2014.

Rush, E., Reed, P., Mclennan, S., Coppinger, T. et al. 2012. A school-based obesity control programme: Project Energize; Two-year outcomes. *British Journal of Nutrition* 107: 581–7.

Rush, E., Mclennan, S., Obolonkin, V., Vandal, A. et al. 2013. Project Energize: Whole-region primary school nutrition and physical activity programme; Evaluation of body size and fitness 5 years after the randomised controlled trial. *British Journal of Nutrition*. 111: 363–71.

Skeaff, S.A., Thomson, C.D. & Gibson, R.S. 2002. Mild iodine deficiency in a sample of New Zealand schoolchildren. *European Journal of Clinical Nutrition* 56: 1169–75. <www.nature.com/ejcn/journal/v56/n12/pdf/1601468a.pdf>, accessed 5 October 2014.

Skeaff, S.A., Ferguson, E.L., Mckenzie, J.E., Valeix, P. et al. 2005. Are breast-fed infants and toddlers in New Zealand at risk of iodine deficiency? *Nutrition* 21: 325–31.

Statistics New Zealand 2000. *Population Ageing in New Zealand*. Statistics New Zealand. <www.stats.govt.nz/browse_for_stats/people_and_communities/older_people/pop-ageing-in-nz.aspx>, accessed 13 November 2014.

—2013. *2013 Census QuickStats about National Highlights*. Statistics New Zealand. <www.stats.govt.nz/Census/2013-census/profile-and-summary-reports/quickstats-about-national-highlights/cultural-diversity.aspx>, accessed 13 November 2014.

Taylor, R.W., Mcauley, K.A., Barbezat, W., Farmer, V.L. et al. 2013. Two-year follow-up of an obesity prevention initiative in children: the APPLE project. *American Journal of Clinical Nutrition* 88: 1371–7. <http://ajcn.nutrition.org/content/88/5/1371.full.pdf+html>, accessed 5 October 2014.

University of Otago & Ministry of Health 2011. *A Focus on Nutrition: Key findings of the 2008/09 New Zealand Adult Nutrition Survey*. New Zealand Government. <www.health.govt.nz/system/files/documents/publications/a-focus-on-nutrition-v2.pdf>, accessed 2 October 2014.

Wham, C., Maxted, E., Dyall, L., Teh, R. et al. 2012. Korero te kai o te Rangatira: Nutritional wellbeing of Māori at the pinnacle of life. *Nutrition & Dietetics* 69, 213–16.

ADDITIONAL READING

Ministry of Health 2006. *Food and Nutrition Guidelines for Healthy Pregnant and Breastfeeding Women: A background paper*. New Zealand Government. <www.health.govt.nz/system/files/documents/publications/food-and-nutrition-guidelines-preg-and-bfeed.pdf>, accessed 18 September 2014.

——2008. *Food and Nutrition Guidelines for Healthy Infants and Toddlers (Aged 0–2): A background paper* 4th edn, partially revised December 2012. New Zealand Government. <www.health.govt.nz/system/files/documents/publications/food-and-nutrition-guidelines-healthy-infants-and-toddlers-revised-dec12.pdf>, accessed 16 March 2014.

——2012. *Food and Nutrition Guidelines for Healthy Children and Young People (Aged 2–18 years): A background paper* 1st edn. Wellington: New Zealand Government.

——2012. *The Health of New Zealand Children 2011/12: Key findings of the New Zealand Health Survey*. New Zealand Government. <www.health.govt.nz/system/files/documents/publications/health-of-new-zealand-child-2011-12-v2.pdf>, accessed 5 October 2014.

——2013. *Food and Nutrition Guidelines for Healthy Older People: A background paper*. New Zealand Government. <www.health.govt.nz/system/files/documents/publications/food-and-nutrition-guidelines-healthy-older-people-background-paper-jan2013.pdf>, accessed 16 March 2014.

16

Asia-Pacific nutrition and health

Mark L. Wahlqvist

LEARNING OUTCOMES

Upon completion of this chapter you will be able to:

- appreciate the extensive differences in food habits and patterns throughout the Asia-Pacific region and how they relate to health
- recognise the threats to food security and their causes in the Asia-Pacific region and consider how they might be addressed
- encourage food and health system approaches to the understanding of food-health relationships and their optimisation irrespective of regional differences
- demonstrate the interdependence of social, physical and eating activities for health
- identify the intergenerational and intragenerational features of nutritional status and how they may be incorporated into cross-culturally robust sustainable household and community health policy
- promote an understanding of the relevance of education and equity in nutrition-related health outcomes across a socio-economically disparate region.

The Asia-Pacific region is becoming the world's most economically dynamic. With this goes an increased demand for food, not only to deal with prior and intercurrent problems of nutrient inadequacy, but also as part of the quest for more animal-derived foods which seems to go with economic development, for more palatable foods and for nutritionally related wellbeing.

The ability of the region to produce more food in a sustainable and affordable way is limited. Imports into the region from other food-producing regions are at unprecedented levels. Food exports are also increasing where, as in Thailand, there is the ability to produce more rice and seafood, but this may not be sustainable as, for example, waterways on which rice depends are being compromised and fish stocks are being markedly depleted.

China, which first developed soy as a food, now imports 80 per cent of its soy, with much of it going to cooking oil production and the 'waste' to pig feed. As a consequence, the nutritional value of the soybean is degraded and even lost unless the soy is captured for traditional foodstuffs like tofu. The imported soy comes from the United States (largely genetically modified to resist weedkillers to which the weeds are becoming resistant and where there is loss of biodiversity), from Brazil (at the expense of the rainforests) and from Argentina (which suffers drought and recurrent inability to fill quotas). It might be thought that another green revolution, which increased staple crop production through superior plant breeds, could be repeated, but its lead plant breeder and Nobel prize winner, Norman Borlaug (2000), pointed out that this would not be possible because of, inter alia, the soil nutrient needs, increasingly dependent on finite fertiliser supplies, especially phosphate and potassium (with nitrogen less critical). This limitation is compounded by diminishing arable land, precarious supplies of safe water for food production and climate change with global warming, to which plants and animals are sensitive. It is clear that these phenomena set the stage for future food insecurity not only in the Asia-Pacific region, but also in the regions with which it trades.

The state of food regulation and safety in the Asia-Pacific region is varied. There are reliable and enforceable systems in Japan, with its advanced framework for foods for special health use, as well as in Singapore, Hong Kong and Taiwan where breaches are detected and prosecuted. In much of Southeast Asia food-borne illness remains common. In China fraud and contamination throughout the food system are all too common, depending on the effectiveness of local governance, the impact of rapid industrialisation and the use of fossil fuels. However, hazard analysis and critical control point protocols are being adopted, with evident improvements in local food systems.

Figure 16.1 shows the divergence of food energy available for consumption within the Asia-Pacific region and between regions as judged from food balance sheets (Peng et al. 2014). The trends have been progressively upward, which will have alleviated much hunger. Insofar as energy balance is struck between intake and expenditure, this may occur at different basal energy expenditures, which are dependent on body composition, especially lean mass (muscle and organs); this in turn depends mainly on physical activity, notably strength or resistance training of muscle, but also on the associated diet. It is of interest that at least Taiwan has begun to decrease the food energy available for consumption, which might reflect a matching decrease in physical activity as the hours of screen time and protracted hours of education supplant physical activity. Equally, such trends must be evaluated in terms of nutrient needs and how they are met as judged from national nutrition surveys.

The spectrum of foods available for consumption in the Asia-Pacific region is shown in Table 16.1, according to food balance sheets. The divergence is indicative of the range of food cultures in the region.

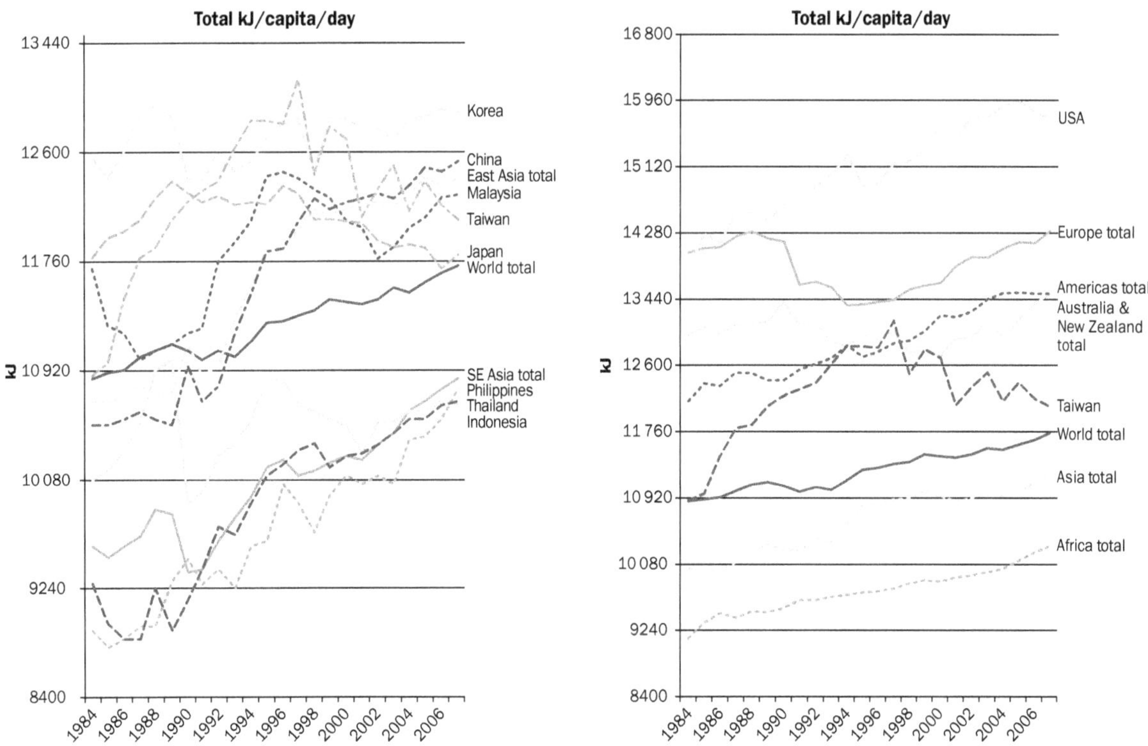

Figure 16.1a Disparate food energy for consumption within the Asia-Pacific region and by comparison with other regions: Energy supply

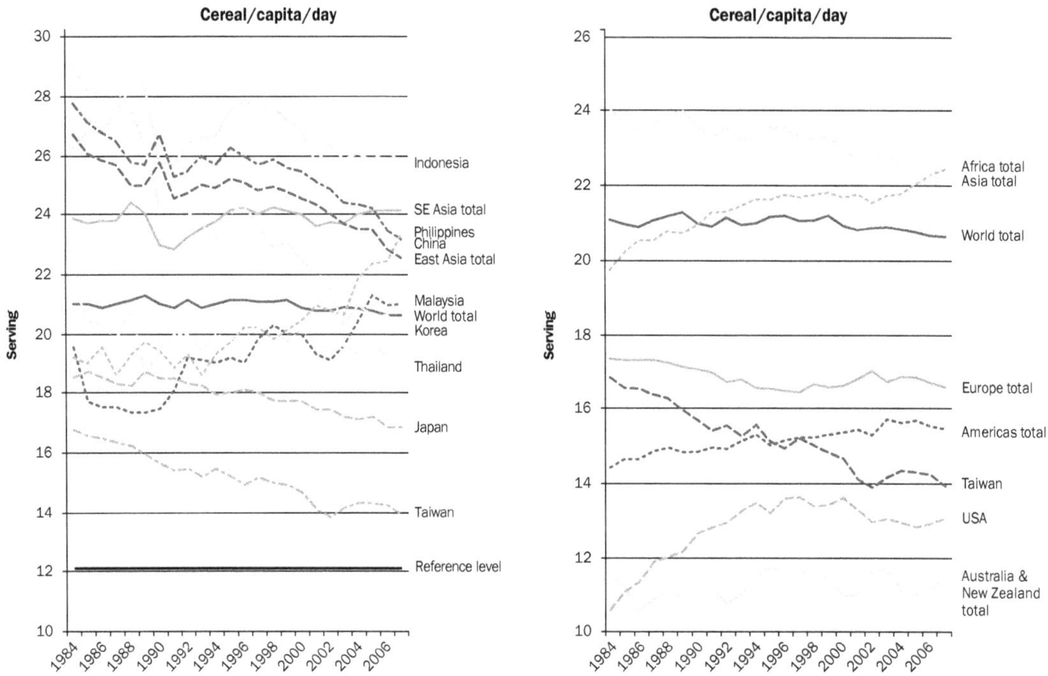

Figure 16.1b Disparate food energy for consumption within the Asia-Pacific region and by comparison with other regions: Supply of cereal group

Source: Adapted from Peng et al. (2014).

Table 16.1 Number of daily serves from each of the six food groups plus some important subgroups, by country or continent (2007)

Country/region	Cereal				Milk	Meat	Soybean	Livestock	
	Wheat	Rice	Rice:wheat	Total				Bovine meat	Pig meat
Indonesia	2.25	18.5	8.2:1	25.8	0.13	2.20	0.17	0.10	0.11
Malaysia	6.72	11.4	1.7:1	21.0	0.35	5.16	0.00	0.34	0.33
Philippines	2.07	19.3	9.3:1	23.3	0.21	3.34	0.01	0.21	0.81
Thailand	1.51	15.5	10.3:1	18.9	0.24	3.43	0.27	0.24	0.56
Southeast Asia total	*2.02*	*19.0*	*9.4:1*	*24.1*	*0.19*	*2.90*	*0.17*	*0.17*	*0.50*
Japan	5.12	9.20	1.8:1	16.8	0.95	7.63	1.23	0.49	0.77
Korea	5.55	12.5	2.3:1	20.2	0.34	6.14	0.79	0.56	1.23
China	8.15	11.9	1.5:1	23.2	0.34	4.99	0.50	0.24	1.43
Taiwan	4.88	6.89	1.4:1	13.9	0.63	7.99	2.11	0.17	1.96
East Asia total	*7.73*	*11.6*	*1.5:1*	*22.5*	*0.39*	*5.19*	*0.57*	*0.27*	*1.34*
Asia total	7.41	11.7	1.6:1	22.2	0.63	2.89	0.29	0.21	0.60
Europe total	11.46	0.74	0.1:1	16.6	2.43	5.43	0.01	0.93	1.39
Australia & NZ total	7.83	1.38	0.2:1	11.5	2.18	7.66	0.01	2.26	0.79
Americas total	6.45	2.92	0.5:1	15.4	1.78	5.43	0.09	1.64	0.69
Africa total	5.01	2.96	0.6:1	22.4	0.46	1.50	0.10	0.34	0.04
World total	7.38	7.98	1.1:1	20.6	0.96	3.33	0.20	0.51	0.61

Source: Peng et al. (2014).

HEALTH AND NUTRITION

All of the various life stages, from preconception to later stage elderly, are challenged in the Asia-Pacific region. There has been a decline in regional fertility rates, with populations of Chinese ancestry now having the lowest rates in the world (that is, Taiwan, Hong Kong, Singapore and China); this removes some of the potential population pressure on food and health systems. But as families get smaller (at least where religious constraints to family planning are not onerous) and life expectancy increases, there more people are in absolute numbers and relatively more elders.

Economic development is such that both parents often work and may spend significant time away from home base. Longer living grandparents become increasingly more important in the food beliefs, knowledge and skills of children. Indeed, there is intergenerational food habit sharing across at least three generations (Wahlqvist 2009). The sharing may operate from younger to older as well as the other way, especially as new information is generated among children by formal education or other means which include digital sources and advertising.

Food habits, as well as environmental exposures, of former generations may affect the way our genes work today—a process known as epigenetics. Thus, our progeny will likewise be affected by today's habits and exposures. We, therefore, have responsibility not only for our own health, but for that of our descendants. That children today may have all four grandparents alive provides a unique opportunity to take advantage of the collective nutritional experience of several generations. We are able to alter the potential outcomes of our inherited and acquired genetic (genomic) system by how we eat and live. This has become particularly evident in what is called the first 1000 days of life, from conception, in utero, until and including the second year of life. This is an especially vulnerable time for our genome and its expression; maternal diet before and during pregnancy and during lactation can affect lifelong health. In the Asia-Pacific region and globally, a program known as Scaling Up Nutrition aims to optimise this period of human development. In 2014, Scaling Up Nutrition was providing nutrition interventions in 54 countries around the world.

The problems of hunger and poverty in the Asia-Pacific region are declining, both numerically and in

Livestock (cont.)				Veg.	Fruits	Oil				Country/region
Poultry	Eggs	Fish, seafood	Total			Plant oils	Animal fats	Plant oil: animal fat	Total	
0.29	0.23	1.14	0.50	1.05	1.20	4.94	0.22	22.5:1	5.16	Indonesia
1.61	0.56	2.06	2.29	1.33	1.12	7.66	0.46	16.7:1	8.12	Malaysia
0.40	0.26	1.47	1.43	1.68	2.29	2.08	1.10	1.9:1	3.18	Philippines
0.51	0.43	1.30	1.31	1.16	1.60	3.76	0.42	9.0:1	4.18	Thailand
0.43	*0.24*	*1.21*	*1.10*	*1.52*	*1.35*	*3.84*	*0.50*	*7.7:1*	*4.34*	*Southeast Asia total*
0.87	0.89	3.19	2.13	3.01	0.82	8.32	0.84	9.9:1	9.16	Japan
0.63	0.47	2.21	2.41	6.24	1.35	8.22	1.26	6.5:1	9.48	Korea
0.54	0.79	1.03	2.21	7.91	1.13	5.18	0.92	5.6:1	6.10	China
1.37	0.72	1.55	3.50	3.00	2.12	11.50	2.18	5.3:1	13.69	Taiwan
0.57	*0.77*	*1.23*	*2.19*	*7.38*	*1.09*	*5.48*	*0.92*	*6.0:1*	*6.40*	*East Asia total*
0.37	0.39	0.77	1.19	4.17	1.09	5.02	1.04	4.8:1	6.06	Asia total
1.10	0.56	0.90	3.41	3.27	1.60	9.40	3.84	2.4:1	13.24	Europe total
1.94	0.29	0.84	4.99	2.88	1.82	9.76	4.08	2.4:1	13.84	Australia & NZ total
1.71	0.46	0.54	4.04	2.50	1.79	10.00	1.80	5.6:1	11.80	Americas total
0.23	0.09	0.34	0.61	1.83	1.50	4.66	0.34	13.7:1	5.00	Africa total
0.63	0.37	0.67	2.09	3.49	1.31	6.16	1.38	4.5:1	7.54	World total

prevalence as a percentage of the population affected (Figure 16.2); however, the numbers affected in South Asia remain greater than in other regions, being second only to sub-Saharan Africa. The improvements in the Asia-Pacific region account for much of a worldwide improvement.

Yet, at the same time, problems of overweight and obesity are increasing in the Asia-Pacific region (Gill 2006). This is evidence of energy dysregulation, which means that there are periods when there is a mismatch between energy intake and energy expenditure. This mismatch may be one simply of the energy (kilojoule) value of the amount of food eaten and the energy expended at rest by the tissues (especially muscle, brain, liver and blood vessels) and by physical activity. Our body composition, including the kind and distribution of fat, and our microbiota (the microorganisms in the gut, respiratory system, reproductive tract and skin), which constitute more than 90 per cent of our cells, also play a role in energy regulation. Increasingly, many health problems are being recognised as being to do with energy regulation, including obesity, diabetes cardiovascular disease (CVD), certain cancers and neurodegeneration like dementia and affective disorders (Hsu et al. 2011; Wahlqvist et al. 2010, 2012).

If linear growth, reflected in height or stature, is limited by nutrient inadequacy or episodes of infection, or both, since they often occur together, this is referred to as stunting. Stunting is used as an index of population nutrition status and to monitor progress with interventions or changes in nutrition policy (Acuin et al. 2011). It has become apparent that stunting and excess body fat can co-exist as a manifestation of the double burden of disease. A difficulty with the stunting concept is that its differentiation from healthy shortness depends on a knowledge of what has been eaten and of the episodes of infection. At the individual

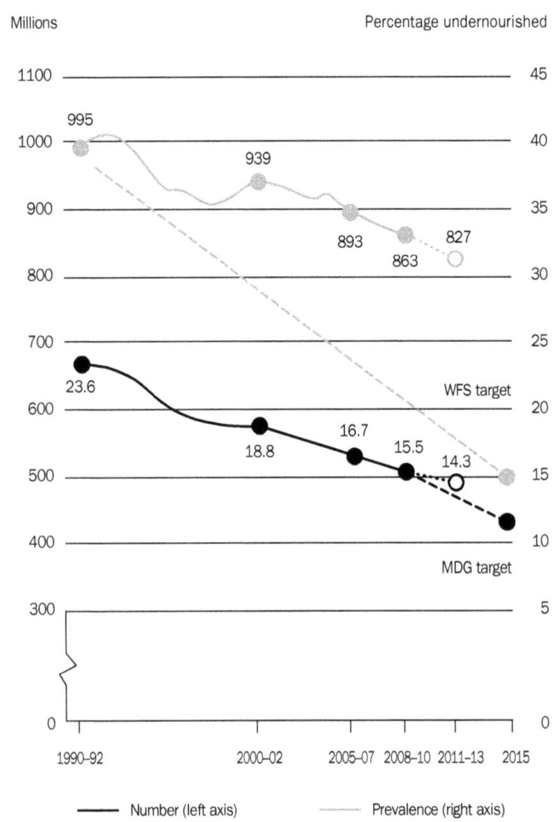

Figure 16.2a Undernourishment in the developing regions

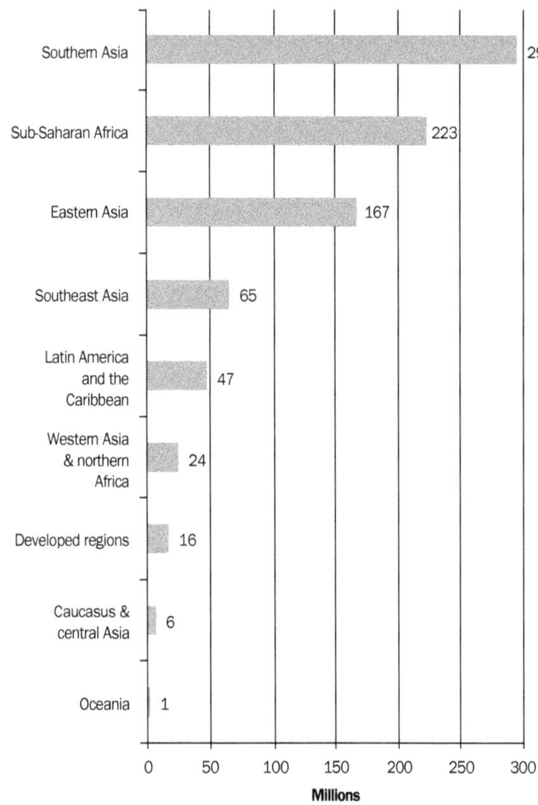

Figure 16.2b Undernourishment, by region, 2011–13

Note: MDG: Millennium Development Goal; WFS: World Food Summit. Total in Figure 16.2b is 844 million.

Source: Adapted from FAO (2013b).

level, this is a diagnostic process which can be followed by appropriate management of the underlying causes. At the population level, the complex pathogenesis of stunting may not be amenable to single or even multiple nutrient supplementation, even though some children may be stunted, say, on account of zinc deficiency. Indeed, the shortness may have already constituted an adaptation, and dealing with it rather than the associated problems like neurological or immunological development may create unwanted problems. For example, increased growth velocity is a risk factor for some cancers, like breast cancer, later in life. We are still not sure what optimal growth velocity and optimal height are. In Taiwan, schoolgirls overexposed to fast food outlets grow faster, while boys get fatter (Chiang et al. 2011). This same generation of Chinese girls is now experiencing increased breast cancer incidence.

HEALTH CONDITIONS

All body systems and organs are in some way affected by what and how we eat (see Table 16.2; Wahlqvist & Lee 2006). The extent to which one or other system or organ is compromised will depend on many factors, especially how well the genome works and how well organ functions are maintained. There must be favourable energy regulation (dependent on food intake and physical activity), an optimal supply of health-promoting bioactive chemicals from food and minimal exposure to damaging agents, be they chemical, physical or atmospheric (for example, actinic or climatic). These risk factors or scenarios vary remarkably throughout the Asia-Pacific region, and it is a testimony to human resilience that we can tolerate as much difference as we do. This resilience is now being put to the test much more, with migration, environmental change and profound changes in the multiple food systems of the region—for example, food production, transport, processing, packaging, marketing, sale, cooking and household eating practices.

It is difficult to say just how food and health secure peoples of the Asia-Pacific region were in pre-colonial times, but we can say that colonisation brought with it food and health adversity. For example, Pacific Island peoples were largely food sufficient, with at least seafood, green vegetables, root vegetables and coconut in abundance. It was similar for riverine and estuarine communities accounting for a large proportion of populations, since we have always needed access to water for survival. Traditional Chinese, Northeast Asian and Southeast Asian rice-based agriculture and aquaculture were also characteristically sustainable. Problems arose then and arise now through poor governance, inequitable social structures and conflict.

Health patterns that have nutrition determinants may reflect the settings in which food security is achieved. The school is one such setting where nutritious lunches may be provided in some countries in the Asia-Pacific region and not in others. This can make a difference to the child who comes from a disadvantaged background, but similarly, the child can carry nutrition messages back to the home from school. In Taiwan, socio-economically disadvantaged children may be vulnerable in the inter-semester breaks when the lunch is not available. This may compound the nutrient vulnerability seen in girls of lower birthweight (Chiang et al. 2013). Chen et al. (2009) noted that, in a study of some 750,000 Taiwanese primary school children, nutrient vulnerability increased with poverty and created a potentially vicious cycle of nutrition-related health problems, most impressively for mental health (see Figure 16.3).

INFLUENCES ON FOOD AND NUTRITION

The role of food to satisfy more than hunger can differ widely between cultures. In traditional societies, significant periods of time are usually spent eating together, whereas in high-income economies, where traditional connections are more often tenuous, eating together gives way to the costs of living and the pursuit

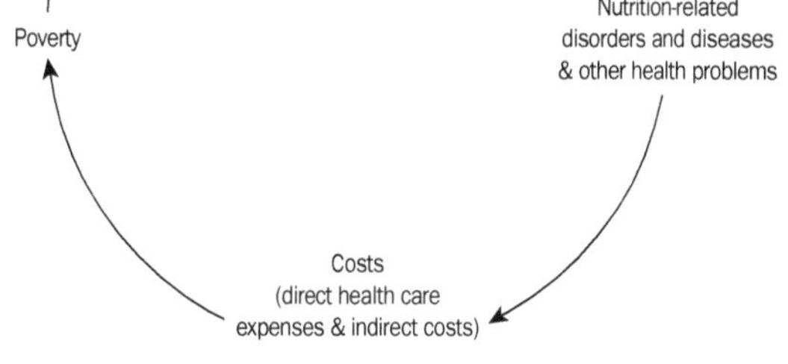

Figure 16.3 Disease cycle

Source: Adapted from Chen et al. (2009).

Table 16.2 Examples of bodily systems affected by diet

System	Examples	Nutritional factors
Neurological	Cognitive impairment at all life's stages	Deficiencies of iodine, iron, folate, vitamin B12, essential fatty acids; dyslipoproteinaemias dependent on apolipoprotein E status
	Cerebrovascular disease	Macrovascular disease risk factors
	Movement disorders (e.g. Parkinson's disease)	Oxidants: antioxidants
Reproductive	Spermatogenesis	Food antioxidant capacity
	Menstrual cycle/menopause	Phytoestrogens
Respiratory	Bronchoconstriction (asthma) Alveolar function: macrophages	Omega-3 (n-3) fatty acids Food antioxidants Vitamin D
Musculoskeletal	Inflammatory arthritides (e.g. rheumatoid)	Omega-3 (n-3) fatty acids, food antioxidants
	Bone health (osteoporosis)	Vitamin D, calcium, phosphorus, sodium effects on calcium excretion, homocysteine (through folate, vitamins B6 and B12), vitamin A (deficiency and toxicity), fruits and vegetables (phytochemicals), vitamin C
Gastrointestinal	Microflora and gut function: gastric (*Helicobacter pylori*); colonic (chronic inflammatory); bowel disease (neoplasia)	Prebiotics, probiotics, antibiotics
	Motility disorders	Caffeine, polyphenolics (culinary herbs), ginger, alcohol
	Hepatobiliary, pancreatic	Growth factors
Cardiovascular	Blood pressure Lipids Platelet function Endothelial function Glycaemic status Cardiac rhythm Abdominal fatness	Omega-3 (n-3) fatty acid sources (fish and plants) Sodium, potassium, magnesium, calcium Arginine, nuts Low-GI food Polyphenolics Alcohol Whole grains, fruits and vegetables (phytochemicals) including dietary fibre
Integument (skin)	Wrinkling (ageing) Skin cancer	Fat and fatty acids Tocoperol (vitamin E) Phytonutrients (fruits, tea)
Immunohaematological	Haemopoiesis	Micronutrients, essential fatty acid, energy and protein deficiency
	Lymphoma and leukaemia	Paternal nutrition, maternal nutrition
Endocrine	Thyroid	Iodine, anti-thyroid factors
	Insulin/pancreas	Energy balance, food patterns, intactness of foods
Special senses	Olfactory	Myriad of receptors: link to memory
	Taste	Preferences, polymorphisms, threshold with age and food components (sodium, caffeine)
	Auditory	Sound of eating (e.g. crunch, grind)
	Vision: retinal function (night blindness); macular function (maculopathy); lens health (cataract)	Vitamin A, carotenoids, zinc, alcohol, lutein, zeaxanthin, antioxidant foods, minimising ultraviolet damage
Mental health	Mood	Nutrient adequacy, social role of food

Note: GI: glycaemic index.

Source: Wahlqvist & Lee (2006).

of wealth. What was 'slow food' becomes 'fast food'. The health-giving properties of food are partly derived from its social role. They also derive from its quality, which requires that it remain relatively intact with favourable physicochemical characteristics as well as a broad spectrum of health-protective bioactive chemicals acting in concert. Countries with longevity generally have a food system which provides enough food, but from diverse sources. Australia is an example of food cultural pluralism on a background of ethnic diversity, where the supply is relatively secure. Other countries in the Asia-Pacific region with food security and diversity, in the top ten for longevity and disability-free life expectancies, include Japan (especially Okinawa), Hong Kong and Singapore. In each of these, food has an important social role.

Population-based studies in Taiwan illustrate how the food system itself is relevant to health outcomes, from production, transport, storage, retailing and cooking to consumption. Both more frequent shopping (Chang et al. 2012) and cooking (Chen et al. 2012) are predictive of longer lives, with men benefiting disproportionately. But, above all, a biodiverse diet, with relatively intact, unrefined or minimally processed foods or ingredients, provides the best guarantee that a full range of health-protective food components and structures will be obtained and that adverse components will be minimised by the dilution effect of diversity (Lee et al. 2011; Wahlqvist et al. 1989).

There are intrasocietal differences in food–health relationships, which are usually and unfortunately greatest between Indigenous and non-Indigenous peoples, then between urban and rural (rural health being less good) and between localities irrespective of urbanisation, in accordance with education, socio-economic status and gender (women living longer than men as long as maternal mortality in childbirth is not a problem). Habits which alter the associations between food and health are notably cigarette smoking, physical inactivity and alcohol excess. Of these, tobacco is a major problem in the Asia-Pacific region among men and a growing problem among women, thanks to young women being targeted by the tobacco industry as it exploits the anxieties about overweight. Social determinants of health are paramount, wherever people live, and they operate in part through the food system as reported by the World Health Organization (WHO).

Breakfast

Breakfast is an episode of eating with considerable differential across the Asia-Pacific region (Howden et al. 1993). Its expression depends on a number of factors. These include school and work schedules and distance from home to these settings (because of time to travel and food choices available on the way; Chiang et al. 2011), food culture and beliefs (as with rice, sweet potato or wheat-based diets, or preparedness to consume relatively expensive ready-to-eat processed breakfast cereals, which generally need added milk, to which non-dairy food cultures are unaccustomed) and trends towards convenient liquid breakfasts.

In turn, evidence is growing of impaired learning or cognition during the morning (Dunlop et al. 1981), with a large US study by Deloitte in 2013 showing that 'students who eat school breakfast on average attend 1.5 more days of school per year [and] score 17.5 per cent higher on standardized math tests' (Augustine-Thottungal et al. 2013), links to obesity in Fiji (Thompson-McCormick et al. 2010) and to compromised bone health in Japan (Kuroda et al. 2013), where breakfast is skipped. Yet, clearly, all breakfasts are not the same; nor is access to them comparable. So the place of breakfast in nutrition and health will vary.

Cultural beliefs and practices

There are many ideas about food habits and health that are current in the Asia-Pacific region which may run counter to present knowledge, may be useful adjuncts to it or may even challenge it.

Yin and Yang

The most pervasive and parallel belief system in the Asia-Pacific is about wellbeing and health in relation to 'tonic' or 'hot and cold foods' linked to the Yin–Yang concept, where Yin is cold and cool while Yang is warm and hot (Ho 1993). This is not about food temperature but about properties of foods which are defined by how they make people feel or how they should be used in various stages of life, climates or states of health. For example, some foods may or may not be used during menstruation, particular seasons or when certain pain syndromes are present. This belief system began in China and is still most in evidence there, but in diminishing degree with distance from China it is found throughout Northeast and Southeast Asia. The belief system is followed by many in conjunction with food-based dietary guidelines and may have some intrinsic merit, judged by studies conducted in accordance with scientific methodology (Ni et al. 2007).

Number of meals

Firm beliefs are held about the preferred number of episodes of eating. People may eat one, two, three or more times a day. The spread of nutrient intakes usually differs across the day, with more B group vitamins in the morning and more protein in the evening. Does this matter? We do not know, but it is conceivable that there are associations between daily profiles of eating episodes and diurnal rhythms.

Jenkins and colleagues (1989) in Toronto showed that nibbling or snacking through the day reduced the risk of coronary heart disease (CHD), and the same may be the case for the risk and management of diabetes, by not overloading the insulin-secretory system excessively at any one time, provided the snacks are of low glycaemic index (GI), which is to say in their the capacity to increase the blood glucose concentration.

In the Western Pacific region, hot meals are often preferred across the day. In some places, like Papua New Guinea, children may have a large breakfast and then snack throughout the day. European meal concepts are challenging these patterns, although not as much as has been assumed (Lipoeto et al. 2013).

Staple foods

Although most people in the Asia-Pacific region eat a staple food as their major energy source, this is second best to having a varied source of energy and, with it, a wide spectrum of bioactive and healthful food components. A staple-based diet is more prone to be nutrient deficient.

In the Asian region, rice is the major staple, with wheat being important in northern China and in much of India. Other grains, like millet and sorghum, and root vegetables, like sweet potato and cassava, are important in some areas.

Technically, wheat is more nutritious (nutrient dense) than rice, and it is thought that this may confer nutritional advantage in some populations, as with growth. Rice is also being targeted as a potential risk food for diabetes, because of some cultivars with high GIs. But it is worth remembering that staples are not usually eaten alone and that their net effect on human biology is the product of the entire diet and concurrent physical activity.

Historically, rice-based diets have not been associated with diabetes. While the cultivars used may have changed, this is unlikely to explain the increasing incidence of type 2 diabetes mellitus (T2DM) in the Asia-Pacific. There is another consideration: that wheat is more temperature sensitive than rice and therefore more likely to succumb to climate change than rice. It is also susceptible to types of rust, which are becoming difficult to control. If we discourage rice consumption in favour of wheat, we may contribute to future food insecurity among rice eaters.

In the Pacific, sweet potatoes are widely used. Not only may they be good sources of carotenoids, including those like beta-carotene, which is pro-vitamin A, but the leaves and stems serve as a most nutritious green vegetable. They manage to grow under a wide range of climatic conditions. At all times, it is important to consider not only the present nutritional value of a crop but also its sustainability.

Boiled water and tea

Boiling water has served people well from the point of view of microbiological safety. It is a deeply entrenched cultural practice of those of Chinese ancestry. It is, therefore, interesting that the world's second most commonly consumed beverage after water is tea, which requires boiled water, developed by the Chinese more than 1000 years ago.

Sugary drinks: barley tea

There is a marked difference between taste for sweetness on the one hand and for the savoury umami (a taste dependent on food nucleotides and typical of fermented soy sauce) on the other, from Northeast to Southeast Asia. Malaysian and Indonesian people fancy sweet foods and drinks. In Japan and Taiwan, even young people select savoury over sweet foods, with the exception of fruit; unsweetened barley tea is also enjoyed. As the evidence grows that sugary drinks play a role in the development of overweight and obesity, diabetes, CVD and certain cancers, these taste preferences are likely to become more health relevant.

Other considerations

Cooked or uncooked?

The human species is the only one that cooks, and the advent of cooking has played a key role in human evolution. The practice of cooking has been found to contribute to human health and longevity, probably for quite complex reasons, including the knowledge and control of one's food supply (Chen et al. 2012).

Cooking has the virtue of making food more edible and palatable—some foods, like beans, are difficult to eat without cooking. Cooking can also destroy certain anti-nutrients like antitrypsin (which reduces protein digestion), while breaking down essential heat-sensitive and water-soluble nutrients like thiamin, folic acid and vitamin C. It also kills potentially harmful microorganisms. In the Asia-Pacific region it is generally safer to eat cooked food and use boiled water or hot teas as a beverage for food safety reasons unless there is evidence to the contrary. The virtues of uncooked food can be flavour and smell, along with structural and nutrient retention.

If food is microbiologically safe, food variety can be enhanced by the consumption of both cooked and uncooked foods. For example, with tomatoes, cooking increases the bioavailability of the red colour lycopene, the most powerful antioxidant we eat, while the uncooked tomato retains more vitamin C.

Fruit and fruit juice

Years of research to discover how crucial food structure is in the delivery of food components from the gut to the circulation and beyond are culminating in evidence that while fruit can be protective against diabetes, fruit juice increases the risk of the disease (Muraki et al. 2013). Nevertheless, there is a good deal of difference between fruits, berries as fruit being one of the most protective against diabetes.

This is not to say that fruit juice does not have nutritional value in the provision of essential nutrients and other healthful phytonutrients, but for the adverse effects on glycaemic status to be offset, it is likely that other protective foods or physical activity might be required. We are likely to see different risk profiles for morbidity and mortality from diverse food patterns emerge across the Asia-Pacific region.

Non-nutritive drinks and diabetes

It has been somewhat surprising to learn that even beverages which are non-nutritive (having no energy value from sugars or other macronutrients) can increase the risk of diabetes. Could it be that such beverages inadvertently encourage the consumption of other energy-containing items? Or could it be that endocrine disruptors leached from the beverage container, like a plastic bottle or lid, may account for these observations? Such hormonal-like substances can increase resistance to insulin, the hormone which enables glucose to be assimilated by tissues as an energy source.

Affordability

Among food categories associated with longevity are fruits and vegetables. However, they are often relatively costly, especially fruits. So when people have marginal incomes, they tend to sacrifice fruits and vegetables in favour of staples. These nutritional economic considerations are increasingly important in national health policy and in the need to maintain this form of household management education (Lo et al. 2012).

Access to information services

Most nutrition and health policy centres on the food-based dietary guidelines formulated by the WHO and the Food and Agriculture Organization of the United Nations and promulgated at their Cyprus meeting in 1995, although they were consistent with national guidelines which had been evolving through previous decades. They embraced the importance of breastfeeding and, thereafter, the need for humans to eat a variety of foods and be physically active. Recognising the growing tide of chronic disease and the ongoing problems of food insecurity, they addressed trends which lowered the nutritional value and increased the risk to health of foods that come with refinement, added or excess salt, fat and sugar. There was encouragement to use relatively intact grains and other seeds like nuts, fruits and vegetables whose fats, sugars and other carbohydrates, were accompanied by many bioactive components and whose structure and physicochemistry were appreciably retained.

The guidelines were based on evidence from both nutrient and food science, but implementation was to be food based and localised in terms of sustainability and affordability. To this end, food-based dietary guidelines were produced by the Western Pacific office of the WHO in 1997 (Wahlqvist 1997). These provided a framework for national food and nutrition policies which of necessity engages the food, health and education systems. The difficulty has been their operationalisation at the community and household levels, to which the focus needs to shift (Wahlqvist 2009). The imperative to do so is crystallised by the increasing loss of habitable and sustainable food-producing ecosystems (Wahlqvist & Specht 1998).

The Millennium Development Goals of the United Nations system, developed in 2000 for realisation by 2015, have provided an added impetus to food and nutrition policy, since each of the eight goals has nutritional relevance (see box below). Of particular importance is how they articulate the importance of women and children in development; maternal literacy and primary education are prerequisites for socio-economic progress and health.

Millennium Development Goals

1. Eradicate extreme poverty and hunger.
2. Achieve universal primary education.
3. Promote gender equality and empower women.
4. Reduce child mortality.
5. Improve maternal health.
6. Combat HIV/AIDS, malaria and other diseases.
7. Ensure environmental sustainability.
8. Global partnership for development.

(Adapted from United Nations 2013)

In general, the first goal, to alleviate poverty and hunger, is making progress. This in part may be attributable to economic improvements, as evidenced in Bangladesh, where stunting has decreased as gross domestic product per capita has increased. However, the converse may also apply, which is that better nutrition contributes to economic development. Other nutrition-related problems have ameliorated in children, including stunting, vitamin A deficiency, anaemia and iodine deficiency (assessed by urinary excretion) by region.

From 2015, Sustainable Development Goals will complement the revision of the Millennium Development Goals. According to the International Social Science Council, these should cover six domains, which are all affected by the Earth's life support system, society and the economy (see box below).

Physical activity

The critical importance of regular activity in the Asia-Pacific region, as elsewhere, has been defined by a major **cohort study** of over 400,000 adults followed for an average of eight years in Taiwan. It showed that as little as fifteen minutes of regular moderate-intensity daily physical activity increased life expectancy of about three years, with a 14 per cent reduction in risk of all-cause mortality. Further benefit was obtained by up to about 40 minutes of exercise a day, when most chronic disease (cardiovascular, cancer and diabetes) was reduced. The benefits were seen for both women and men and at all ages (Wen et al. 2011).

Sustainable Development Goals as recommended by the International Social Science Council

1. Thriving lives and livelihoods
2. Sustainable food security
3. Secure sustainable water
4. Universal clean energy
5. Healthy and productive ecosystems
6. Governance for sustainable societies

(IGBP 2013)

A **cohort study** examines a group of individuals with a common characteristic. A longitudinal cohort study follows the group over a set period of time, sometimes for many years.

Being Indigenous or of an ethnic minority

Northeast Asian economies are having an increasing influence on food trade, but with resilience and extension of the major food cultures and habits of Northeast Asia (China, Japan and to some extent Korea), Indochina (Thailand and Vietnam), Southeast Asia (Indonesia, Singapore and Malaysia, with little impact from the Philippines) and South Asia (especially India) at the expense of minority cultures. Pacific Islanders and Papua New Guineans suffer from regional trade and policies which have increasingly compromised island food economies and locked them into imported foodstuffs of limited biological variety (for example, Australian rice in Papua New Guinea) and high energy density (for example, fatty meats from New Zealand). Australian mining of phosphate as fertiliser for phosphate-poor soils in Australia has irreparably damaged the ecosystems of nations like Nauru.

Support for education and training of minority in contrast to majority peoples from the Asia-Pacific region in food science, nutrition and health has been limited, with some exceptions, like support for tertiary institutions in Fiji and Papua New Guinea. Research into food habits and the health of minority groups is often reported, as for maternal and child health among the Orang Asli people in Malaysia and for diabetes among Pacific Islanders, but this is rarely accompanied by plans for follow-up of its application in the communities involved. This has left smaller nations and minority peoples vulnerable to rapid economic, food and health system change (Kuhnlein et al. 2013). At the same time, Australia and New Zealand have benefited greatly in food cultural diversification and the related health advantage as a consequence of their Asia-Pacific engagement and immigration from the region (Wahlqvist 2002).

CASE STUDY 16.1 LACTAGOGUE FOODS AND BREASTFEEDING AMONG BATAKNESE PEOPLE IN NORTH SUMATRA, INDONESIA

Mark L. Wahlqvist, Rizal Damanik and N. Wattanapenpaiboon

Breastfeeding policy and its success

On the basis of considerable evidence that in both early and later life children who are exclusively breastfed for the first six months of life have improved survival and less morbidity, encouragement to do so has become established WHO (n.d.) and United Nations policy. Benefits accrue to both mother and child, not only by way of ease, economics, safety of care and optimal nutrition, but also through bonding and psychosocial development.

There are immediate, medium-term and long-term biomedical and health advantages to the child. Less risk of infection is a particular early benefit, consistent with the general observation that effective nutrition is associated with better immunity. The breast microbiome, breast milk and breastfeeding are each contributory to the baby's resistance to infection. In childhood, there is more favourable energy balance and less respiratory illness. In adulthood, there is less likelihood of chronic disease.

Limiting factors for lactation

Although knowledge and support for breastfeeding and sanctions against inappropriate commercially driven formula feeding help mothers to succeed with breastfeeding, difficulties do arise. Where the healthcare system is in good stead, it can be possible to compromise, with exclusive breastfeeding for less than six months. Traditional food measures to promote lactation are found in many cultures, but they tend to be disregarded by health advisors as not scientifically based or, possibly, of risk to mother and child, when the composition and effects of herbal preparations are undocumented.

Lactagogue (or galactagogue) foods

There are endogenous (physiological) factors which activate and mediate lactation, notably oxytocin, prolactin and dopamine (which acts through prolactin). These may be altered pharmacologically with agents like metoclopramide and domperidone.

Traditional foods which are believed to assist lactation include goat's rue (*Galega officinalis*), fenugreek and fennel in European cultures. There is a much wider array of such foods or herbs in Asian cultures and elsewhere.

Bataknese women

Bataknese people live principally to the north of Lake Toba and around Simalungun in Sumatra. They trace their origins to Indochina and a migration to Sumatra in the 1200s. Indeed, ancestral graves of their leaders are extant from that time. They brought with them plants which were accorded healthful properties, among which is one named torbangun (*Coleus amboinicus* Lour.), which is cultivated but also grows wild.

Bataknese women take pride in their traditional foods, medicines and the gardens that provide them. They prevail upon their men to ensure successful cultivation. It is said by Bataknese people that women who consume the green leafy vegetable torbangun around delivery and during lactation not only have healthier babies, but remain more youthful in appearance.

One of us is Bataknese by birth (R.D.) and one (M.L.W.) has been conferred status as a Bataknese (and given the Bataknese name Purba) in a traditional ceremony instituted by the Simalungun Women's Association. Together we have been privileged to work to document and test the validity of beliefs about torbangun and its traditional uses.

Bataknese women impress by their organisational strength and by the depth and cohesion of their cultural narratives. Using focus groups which included women of all ages, grandmaternal and maternal, we found the beliefs about torbangun as lactogogue to be strongly and consistently held. The torbangun leaves are made into a soup, which is consumed from parturition for at least one month (Damanik 2009). It is also used by both men and women for respiratory and other ailments.

We then tested the beliefs about lactation in field trials in which villages were assigned to one of three different putatively lactogogue approaches and the women in that village adhered to the assigned approach (Damanik et al. 2006). The approaches were a commercial vitamin B12 supplement promoted in Indonesia as a lactogogue, fenugreek seeds (as used by European women for lactogogue reasons) and a traditional torbangun soup (prepared centrally). Distributions were by motorbike to the villages.

A woman assistant lived in with the mother after parturition and weighed the baby before and after feeding to measure the volume of breast milk ingested by the baby. Samples of breast milk were also saved for analysis and blood samples were taken from the mother to ascertain whether the hormones controlling lactation were altered by the foodstuff.

Torbangun

The increase in breast milk volume with torbangun, but not with vitamin B12 or fenugreek is shown in Figure 16.4. There was no dilution of nutrients in the breast milk with torbangun, which might have happened as volume increased. While we do not know what the active factors in torbangun are, we did find that prolactin remained higher in the women who had the torbangun soup than in the other groups.

Conclusions

It has been possible to confirm the traditional beliefs and practices of Bataknese women in regard to traditional food-assisted lactation success, with increased breast milk volume for at least two months (a month beyond the use of the soup) and net nutrient availability to the baby. No adverse effects of this practice have been observed, and there is some evidence that the babies are healthier. Longer term observations would be of value, as would a better understanding of mechanisms. In the meantime, the food-based approach to breastfeeding success contributes not only to maternal and child health, but also to sustainable communities with associated agriculture.

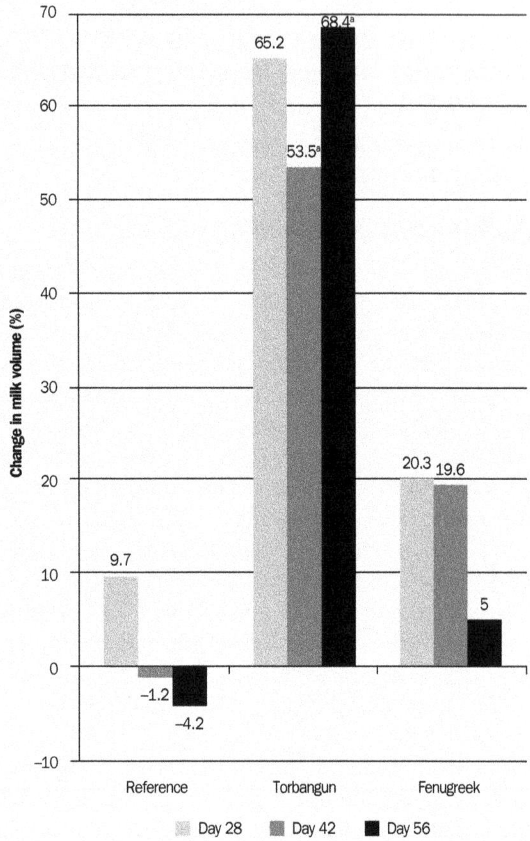

Figure 16.4 Torbangun effects on lactation in Bataknese women in Simalungun

a Significant difference from the reference group

Note: Change in breast milk intake during the two-month study period.

SUMMARY AND KEY MESSAGES

The Asia-Pacific region is rapidly growing both economically and in population. This has increased the demand for food volume, as well as a greater diversity of animal-derived and palatable foods. The region is strained to sustainably and affordably provide sufficient food for the population, and relies increasingly on imports from other regions.

- Diets in the Asia-Pacific region are traditionally rice-based. Diversification over time has improved nutrition-related health outcomes in this region.
- Food insecurity is a growing problem in the Asia-Pacific region and management of the food supply requires restrained use of resources and minimisation of food waste.
- Increasing population size will lead to a greater number of socio-economically disadvantaged and impoverished Asian citizens who emigrate from areas of drought and compromised food production. It is likely that this will contribute to large numbers of refugees experiencing food and water shortage and insanitary conditions.
- Improvements in food security must be addressed by all key stakeholders including the households and communities, all levels of government, non-government organisations, and the corporate sector.
- A focus on understanding ourselves as ecological creatures, being less anthropocentric (focused on our species) and caring about the locality or ecosystems in which we live and work may help change our otherwise destructive food–health trajectory in the Asia-Pacific region.

REFERENCES

Acuin, C.S., Khor, G.L., Liabsuetrakul, T., Achadi, E.L. et al. 2011. Maternal, neonatal, and child health in Southeast Asia: Towards greater regional collaboration. *Lancet* 377(9764): 516–25.

Augustine-Thottungal, R., Kern, J., Key, J. & Sherman, B. 2013. *Ending Childhood Hunger: A social impact analysis*. Deloitte and the No Kid Hungry Center for Best Practices. <www.nokidhungry.org/pdfs/school-breakfast-white-paper.pdf>, accessed 14 November 2014.

Borlaug, N. 2000. Ending world hunger: The promise of biotechnology and the threat of antiscience zealotry. *Plant Physiology* 124: 487–90.

Chang, Y.H., Chen, R.C., Wahlqvist, M.L. & Lee, M.S. 2012. Frequent shopping by men and women increases survival in the older Taiwanese population. *Journal of Epidemiology and Community Health* 66(7): e20.

Chen, L., Wahlqvist, M.L., Teng N.C. & Lu, H.M. 2009. Association of imputed food insecurity with disease and mental health in Taiwanese elementary school children. *Asia Pacific Journal of Clinical Nutrition* 18: 605–19.

Chen, R.C., Lee, M.S., Chang, Y.H. & Wahlqvist, M.L. 2012. Cooking frequency may enhance survival in Taiwanese elderly. *Public Health Nutrition* 15: 1142–9.

Chiang, P.H., Wahlqvist, M.L., Lee, M.S., Huang, L.Y. et al. 2011. Fast-food outlets and walkability in school neighbourhoods predict fatness in boys and height in girls: A Taiwanese population study. *Public Health Nutrition* 14: 1601–9.

Chiang, P.H., Huang, L.Y., Lo, Y.T., Lee, M.S. et al. 2013. Bidirectionality and gender differences in emotional disturbance associations with obesity among Taiwanese schoolchildren. *Research in Developmental Disabilities* 34: 3504–16.

Damanik, R. 2009. Torbangun (*Coleus amboinicus* Lour): A Bataknese traditional cuisine perceived as lactagogue by Bataknese lactating women in Simalungun, North Sumatera, Indonesia. *Journal of Human Lactation* 25: 64–72.

Damanik, R., Wahlqvist, M.L. & Wattanapenpaiboon, N. 2006. Lactagogue effects of Torbangun, a Bataknese traditional cuisine. *Asia Pacific Journal of Clinical Nutrition* 15: 267–74.

Dunlop, W.L., Wahlqvist, M.L., Rutishauser, I.H.E. & Nestel, P.J. 1981. The effect of a breakfast-oriented nutrition education program in food intake patterns and alertness of schoolchildren. *Proceedings of the Nutrition Society of Australia* 6: 104.

FAO (Food and Agriculture Organization of the United Nations) 2013. *The State of Food Insecurity in the World 2013*. FAO. <www.fao.org/publications/sofi/2013/en/>, accessed 24 November 2013.

Gill, T. 2006. Epidemiology and health impact of obesity: An Asia Pacific perspective. *Asia Pacific Journal of Clinical Nutrition* 15 Suppl.: 3–14.

Ho, Z.C. 1993. Principles of diet therapy in ancient Chinese medicine: 'Huang Di Nei Jing'. *Asia Pacific Journal of Clinical Nutrition* 2: 91–5.

Howden, J.A., Chong, Y.H., Leung, S.F., Rabuco, L.B. et al. 1993. Breakfast practices in the Asian region. *Asia Pacific Journal of Clinical Nutrition* 2: 77–84.

Hsu, C.C., Wahlqvist, M.L., Lee, M.S. & Tsai, H.N. 2011. Incidence of dementia is increased in type 2 diabetes and reduced by the use of sulfonylureas and metformin. *Journal of Alzheimer's Disease* 24: 485–93.

IGBP (International Geosphere—Biosphere Programme) 2013. *Sustainable Development Goals for People and Planet*. IGBP. <www.igbp.net/news/news/news/sustainabledevelopmentgoalsforpeopleandplanet.5.561163a13d60576e12c4.html>, accessed 6 December 2013.

Jenkins, D.J., Wolever, T.M., Vuksan, V., Brighenti, F. et al. 1989. Nibbling versus gorging: metabolic advantages of increased meal frequency. *New England Journal of Medicine* 321: 929–34.

Kuhnlein, H.V., Erasmus, B., Spigelski, D. & Burlingame, B. 2013. *Indigenous Peoples' Food Systems and Well-being: Interventions and policies for healthy communities*. FAO. <www.fao.org/docrep/018/i3144e/i3144e.pdf>, accessed 24 November 2013.

Kuroda, T., Onoe, Y., Yoshikata, R. & Ohta, H. 2013. Relationship between skipping breakfast and bone mineral density in young Japanese women. *Asia Pacific Journal of Clinical Nutrition* 22: 583–9.

Lee, M.S., Huang, Y.C., Su, H.H., Lee, M.Z. et al. 2011. A simple food quality index predicts mortality in elderly Taiwanese. *Journal of Nutrition Health and Aging* 15: 815–21.

Lipoeto, N.I., Geok Lin, K. & Angeles-Agdeppa, I. 2013. Food consumption patterns and nutrition transition in South-East Asia. *Public Health Nutrition* 16 1637–43.

Lo, Y.T., Chang, Y.H., Wahlqvist, M.L., Huang, H.B. et al. 2012. Spending on vegetable and fruit consumption could reduce all-cause mortality among older adults. *Nutrition Journal* 11: 113.

Muraki, I., Imamura, F., Manson, J.E., Hu, F.B. et al. 2013. Fruit consumption and risk of type 2 diabetes: Results from three prospective longitudinal cohort studies. *British Medical Journal* 347: f5001.

Ni, L., Lin, X. & Rao, P. 2007. Validation of a mathematical model for determining the Yin-Yang nature of fruits. *Asia Pacific Journal of Clinical Nutrition* 16(Suppl. 1): 208–14. <http://apjcn.nhri.org.tw/server/APJCN/16%20Suppl%201//208.pdf>, accessed 5 October 2014.

Peng, C.J., Lee, M.S. & Wahlqvist, M.L. 2014. Trends in the Asia Pacific food supply from 1984 to 2007 and their implications for food security and equity. *Asia Pacific Journal of Clinical Nutrition*. In press.

Thompson-McCormick, J.J., Thomas, J.J., Bainivualiku, A., Khan, A.N. et al. 2010. Breakfast skipping as a risk correlate of overweight and obesity in school-going ethnic Fijian adolescent girls. *Asia Pacific Journal of Clinical Nutrition* 19: 372–82.

United Nations 2013. We Can End Poverty: Millennium Development Goals and Beyond 2015 (website). United Nations. <www.un.org/millenniumgoals/>, accessed 6 December 2013.

Wahlqvist, M.L. 1997. *Food-based Dietary Guidelines for the Western Pacific: Nutrition in transition*. World Health Organization, Manila.

——2002. Asian migration to Australia: Food and health consequences. *Asia Pacific Journal of Clinical Nutrition* 11(Suppl. 3): S562–S568.

——2009. Connected community and household food-based strategy (CCH-FBS): Its importance for health, food safety, sustainability and security in diverse localities. *Ecology of Food and Nutrition* 48: 457–81.

Wahlqvist, M.L. & Lee, M.S. 2006. Nutrition in health care practice. *Journal of Medical Sciences* 26: 157–64.

Wahlqvist, M.L. & Specht, R.L. 1998. Food variety and biodiversity: Econutrition. *Asia Pacific Journal of Clinical Nutrition* 7: 314–19.

Wahlqvist, M.L., Chang, H.Y., Chen, C.C., Hsu, C.C. et al. 2010. Is impaired energy regulation the core of the metabolic syndrome in various ethnic groups of the USA and Taiwan? *BMC Endocrine Disorders* 10: 11.

Wahlqvist, M.L., Lee, M.S., Chuang, S.Y., Hsu, C.C. et al. 2012. Increased risk of affective disorders in type 2 diabetes is minimized by sulfonylurea and metformin combination: A population-based cohort study. *BMC Medicine* 10: 150.

Wahlqvist, M.L., Lo, C.S. & Myers, K.A. 1989. Food variety is associated with less macrovascular disease in those with type II diabetes and their healthy contols. *Journal of the American College of Nutrition* 8: 515–23.

Wen, C.P., Wai, J.P., Tsai, M.K., Yang, Y.C. et al. 2011. Minimum amount of physical activity for reduced mortality and extended life expectancy: A prospective cohort study. *Lancet* 378: 1244–53.

WHO (World Health Organization) n.d. *Breastfeeding*. WHO. <www.who.int/topics/breastfeeding/en/>, accessed 9 October 2014.

ADDITIONAL READING

Food and Agriculture Organization of the United Nations 2013. *The State of Food and Agriculture*. FAO. <www.fao.org/docrep/018/i3300e/i3300e.pdf>, accessed 6 December 2013.

Organisation for Economic Co-operation and Development & WHO 2012. *Health at a Glance: Asia/Pacific 2012*. OECD iLibrary. <www.oecd-ilibrary.org/social-issues-migration-health/health-at-a-glance-asia-pacific-2012_9789264183902-en>, accessed 5 October 2014.

Wahlqvist, M.L. 2011. *Food and Nutrition in Australia: Food and health systems in Australia and New Zealand* 3rd edn. Allen & Unwin, Sydney.

PART IV

EXTENSION KNOWLEDGE

17

Sports nutrition

Regina Belski

> **LEARNING OUTCOMES**
>
> Upon completion of this chapter you will be able to:
>
> - describe the sports nutrition and hydration recommendations for before, during and after training
> - list the nutritional needs and common challenges faced by endurance athletes, weight category and aesthetic sport athletes, young athletes and travelling athletes
> - explain ways to overcome some of the nutrition challenges faced by athletes.

Sports nutrition has become a significant topic of interest for many active people as well as recreational and elite athletes. Much research has been conducted over the last two decades, leading to a significantly improved understanding of nutritional needs for training and performance and more clear-cut recommendations. This chapter introduces the basics of sports nutrition recommendations for training and discusses the common nutritional concerns for athletes participating in selected sports.

NUTRIENT REQUIREMENTS

Athletes' eating habits are influenced by their age, gender, sport and body composition goals, as well as their level of training. Studies investigating the diets of elite athletes have demonstrated that male athletes and athletes in endurance sports tend to consume higher amounts of total energy and that endurance athletes consume the highest amount of carbohydrate—a key fuel for endurance sports—relative to athletes in other sports. Athletes involved in aesthetic and weight category sports were found to have the lowest carbohydrate and energy intakes during and following training; this is not surprising, given their body composition goals usually involve fat loss. Overall, it appears that while the athletes surveyed had reasonable overall diets, their training-specific nutrition was not optimal (Burke et al. 2003).

Athletes need appropriate amounts of energy, carbohydrate, protein and fat as well as vitamins and minerals to meet the demands of training, daily activities and physiological maintenance. In Australia and New Zealand the micronutrient recommendations for athletes are the same as those for the rest of the population; however, the macronutrient needs of athletes and very active individuals may be greater than average. Current recommendations are discussed below. It is common for macronutrient recommendations to be presented in grams per kilogram of body weight per day, reflecting that individuals of different sizes will have different daily total requirements; throughout this chapter, a measurement in grams per kilogram should be read as shorthand for grams per kilogram of body weight per day.

Protein

Some athletes have higher protein requirements than average; this includes endurance athletes undergoing heavy training (up to 1.6 grams per kilogram), resistance athletes in early training (1.5–1.7 grams per kilogram) and those in power sports (1.4–1.7 grams per kilogram). These increased requirements account for the additional repair and rebuild of muscle that is broken down during this type of training. Most other athletes can meet their requirements by consuming 0.8–1 gram per kilogram. It is not recommended for any athlete to consume more than 2 grams per kilogram of protein, as it does not appear to provide any performance benefit (Burke & Deakin 2010: 80).

Fat

Fat is an integral part of a healthy diet, functioning as an energy source and building block of hormones and cell membranes and supporting the absorption of fat-soluble vitamins. As per the nutrient reference values (NRVs) for Australia and New Zealand, the acceptable macronutrient distribution range (AMDR) for fat intake is from 20 to 35 per cent of total energy intake daily, and this is also appropriate for athletes (NHMRC & Ministry of Health 2006). Athletes should generally aim to include healthy fats in their diet and avoid excessive saturated and total fat intakes above the AMDR.

Occasionally, endurance athletes will engage in fat loading to obtain fat adaptation for endurance events. This practice involves athletes consuming a high-fat (65–70 per cent of energy), low-carbohydrate (less than 2.5 grams per kilogram) diet over a relatively short period of time, varying between five days and four weeks, with the aim to increase fat oxidation and reduce the use of muscle glycogen during the following submaximal exercise sessions. However, current evidence suggests that while this can have an impact on fat oxidation it does not appear to be beneficial to overall performance (Burke & Kiens 2006).

Carbohydrate

Carbohydrate is the key fuel for exercise and training with the body having only a very limited ability to store it as glycogen in the muscles and liver. Low glycogen stores as a result of insufficient or irregular carbohydrate intake can have detrimental effects on training and performance, including fatigue and higher risk of injury.

Carbohydrate requirements for fuel and recovery differ according to the nature of exercise or activity undertaken. Light activity requiring only low-intensity or simple skill-based exercise has a target carbohydrate intake of 3–5 grams per kilogram. A moderate exercise program lasting over one hour per day (for example, a 10-kilometre run at moderate intensity) has a target carbohydrate intake of 5–7 grams per kilogram. A high-endurance exercise program of moderate to high intensity lasting between one and three hours per day (for example, a 20-kilometre run at moderate or high intensity) has a target carbohydrate intake of 6–10 grams per kilogram. A very high exercise program of moderate to high intensity lasting more than four to five hours (for example, a half-ironman triathlon) has a target carbohydrate intake as high as 8–12 grams per kilogram (Australian Sports Commission 2011).

NUTRITION FOR TRAINING

Athletes want to go faster, harder and stronger, and the quality, quantity and timing of the fuel they put into their body in the form of food and drink play an important role in how successful they can become. Training nutrition is therefore an important area of interest and has received significant attention from researchers and athletes alike. Training nutrition can be split into three key time periods: pre-training, during training and post-training.

Pre-training

Pre-training nutrition refers to the foods and beverages consumed during the time immediately preceding the training session—that is, from two hours before training until the training begins. It has been shown to have a significant impact on an athlete's performance during training (Hargreaves et al. 2004).

Current recommendations suggest a low-fat, high-carbohydrate meal or snack two to four hours prior to training. This might be a breakfast-type meal such as a bowl of cereal with low-fat milk, or if training early (without one to two hours of waking time prior to training) then a small carbohydrate-rich snack (Australian Sports Commission 2009).

Large meals or meals high in fat are not recommended, as high fat content can slow absorption and can leave an athlete feeling uncomfortably full or nauseated when exercise begins.

Examples of suitable pre-training snacks include the following:

- A couple of slices of toast or crumpets with honey or jam
- Banana and honey sandwich
- Low-fat creamed rice
- Liquid breakfast substitutes
- Fruit and low-fat fruit yoghurt
- Cereal-based breakfast bar
- Fruit scroll.

Starting exercise well hydrated is another important pre-training strategy, as dehydration is one of the most common nutrition-related issues that can occur during exercise. In some elite sporting clubs athletes are asked to monitor their hydration via mid-stream urine tests measuring urine specific gravity on a daily basis, and in some clubs members can be fined if they turn up to training dehydrated.

If athletes do not have access to a urine-specific gravity test they can monitor how frequently they are urinating and the colour of the urine. If it is very bright yellow or dark, they are not well hydrated. The general rule of thumb is that the urine of properly hydrated individuals is clear to very pale yellow; for those moderately dehydrated it is light to bright yellow; and for those severely dehydrated it is dark yellow to almost orange. Some supplements and vitamin pills can impact on the colour and smell of urine and this can make using the colour method inaccurate.

Sawka et al. (2007) recommend that 'prehydrating with beverages, if needed, should be initiated at least several hours before the exercise task to enable fluid absorption and allow urine output to return toward normal levels. Consuming beverages with sodium and/or salted snacks or small meals with beverages can help stimulate thirst and retain needed fluids'.

During training

Research has revealed that for long training and exercise sessions, food and drink during training can have a very significant impact on performance (Coyle 2004). For training sessions or events of less than one hour duration, it appears that appropriate pre-event fuelling may be sufficient for athletes to obtain and maintain high performance. However, for events greater than one hour, adequate refuelling with carbohydrate and water appears to give significant benefits in regard to performance outcomes. Furthermore, when an appropriate amount of additional carbohydrate is consumed during endurance events, the importance of pre-training nutrition becomes less significant (Burke & Deakin 2010: 318).

Examples of suitable during-training snacks/and drinks include the following:

- Sports drinks
- Carbohydrate gels
- Sports bars
- Cereal-based breakfast bars
- Jelly lollies.

Carbohydrate

Exact recommendations vary according to the weight of the individual, the type of activity and the length of event; however, general guidelines are that under 45 minutes of exercise requires no extra carbohydrate; 45–75 minutes of exercise (sustained high-intensity) requires a small amount of additional carbohydrate, 1–2.5 hours of exercise (endurance) requires an extra 30–60 grams per hour of exercise, and 2.5–3 hours of exercise (ultra-endurance) requires up to 90 grams extra per hour of exercise (Australian Sports Commission 2011).

Carbohydrate has often been incorporated into rehydration fluids such as sports drinks to enable easy intake. This is discussed in further detail below.

Fluid

Rehydration via drinking plays an important role in preventing the dehydration and electrolyte losses that can occur during training or competition. These losses can have a detrimental impact on performance and include a loss of concentration, increased risk of injury and poorer overall performance.

Numerous studies have demonstrated that fluid replacement and rehydration during exercise can improve performance in a variety of settings and situations. These studies have also indicated that the addition of carbohydrate and electrolytes to fluids offers additional benefits (Millard-Stafford 1992).

As different sports place differing demands on athletes, and as individual athletes have differing sweat rates and sweat electrolyte concentrations, it is difficult to recommend a one-size-fits-all hydration plan. Fluid replacement guidelines come from the American College of Sports Medicine (Sawka et al. 2007). They differ from previous guidelines, in that they place additional emphasis on the importance of individualising hydration recommendations based on the athlete, sport and environment. Individual athletes should know about their own sweat losses, which can be estimated by measuring body weight before and after exercise where the loss of weight observed relates to fluid loss. It is this individual amount of weight loss that should be used to estimate the volume of fluid that needs to be consumed during and following exercise to replace all losses.

Electrolytes

It is well known that sweating results in the loss of water and electrolytes from the body. As exercise, especially if it is prolonged, can result in significant sweat losses, it becomes critical to replace not only the lost water but also the lost electrolytes. The main electrolytes lost in sweat include sodium, potassium, calcium, magnesium and chloride. However, despite the variety of electrolytes lost in sweat, research suggests that the only electrolyte that needs to be added to sports drinks is sodium (Maughan 1994). Therefore, manufacturers have worked together with sports scientists to create **isotonic sports drinks** to try to mimic the water to electrolyte ratios found in sweat. The isotonic nature of the drinks enables optimal rehydration in athletes, by speeding up **gastric emptying**. Typical sodium content of sports drinks is between 10 and 25 millimoles per litre.

Sports drinks usually combine water, electrolytes and carbohydrate in ratios that have been shown in research studies to enable effective rehydration. The addition of carbohydrate to sports drinks not only provides the muscles with the energy they need to continue working, but also plays a role in enhancing water absorption in the small intestine. Enhanced water absorption occurs only if carbohydrate is added at the right concentration. If the concentration of carbohydrate is too high, this will have a negative impact on rehydration, as it could lead to water retention in the gut and slow gastric emptying. Therefore, it is generally recommended that athletes neither over-concentrate nor over-dilute commercial sports drinks and that sports drink powders should be prepared according to package directions.

Isotonic sports drinks are designed to contain similar concentrations of electrolytes and carbohydrate as those found in human cells, to maximise rehydration following sweat loss.

Gastric emptying is the time it takes for fluid to go from the mouth, through the stomach and out of the body.

Taste and temperature of fluids have also been shown to play a role in rehydrating an athlete. A pleasant-tasting cold drink appears to be the best option. Researchers have shown that palatable fluids lead to higher consumption, hence better rehydration (Szlyk et al. 1989), and cool fluids appear to improve performance, especially in hot conditions (Lee et al. 2008).

It is important to note that every athlete will have individual differences in sweat losses, fuel use and rate of gastric emptying; therefore, it is critical that they are encouraged to focus on developing a personal approach to rehydration and sports nutrition that meets their individual needs. Sawka et al. (2007) recommend that 'individuals should develop customized fluid replacement programs that prevent excessive dehydration (less than 2 per cent body weight reductions from baseline body weight). The routine measurement of pre- and post-exercise body weights is useful for determining sweat rates and customized fluid replacement programs. Consumption of beverages containing electrolytes and carbohydrates can help sustain fluid–electrolyte balance and exercise performance'.

Post-training recovery

The key purposes of post-training nutrition are refuelling and recovery. Refuelling involves replenishing glycogen stores and replacing fluid and electrolytes.

Examples of suitable post-training drinks and snacks include the following:

- Sports drink and low-fat yoghurt
- Low-fat creamed rice
- Liquid breakfast substitutes
- Fruit and low-fat fruit yoghurt
- Pancakes with maple syrup and low-fat yoghurt
- Low-fat chocolate milk.

Carbohydrate

Research has consistently shown that the hours immediately following exercise are when muscle glycogen resynthesis is most rapid; hence, an immediate intake of carbohydrate is recommended. The type and amount of carbohydrate appear to be important, with the current recommendations being consumption of 1 gram of carbohydrate per kilogram of body weight immediately after exercise and repeated each hour until the next usual meal (Burke & Deakin 2010: 360, 382). Higher glycaemic index (GI) carbohydrate (see Chapter 1) consumed every fifteen to twenty minutes in the form of snacks may be particularly useful. This emphasises the importance of planning ahead and taking suitable snacks to consume following the training session, as otherwise suitable choices may not be available at the training or game venue.

The amount of carbohydrate required to refuel glycogen stores will depend on the intensity and length of the exercise session; therefore, a 30-minute walk will require less aggressive refuelling than a two-hour run at high intensity. Following intense and long training sessions the 24-hour recovery plan should aim to provide 7–12 grams of carbohydrate per kilogram. For shorter and less intense sessions not resulting in significant drops in glycogen stores, daily intakes of 3–5 grams per kilogram may be sufficient (Burke & Deakin 2010: 380).

Fluid

In the hours following a long and intense training session in which significant fluid loss has occurred, the thirst mechanism alone will not sufficiently encourage adequate fluid consumption to enable full rehydration. Therefore, a rehydration plan needs to be implemented.

If fluid loss has been assessed by weighing the athlete before and after training and weight loss exceeds 2 per cent of body weight, then 1.5 litres of fluid should be consumed for every kilogram of weight lost. To enable this to occur efficiently, beverages should be selected based on an individual's taste preferences and ideally at a cool temperature. The recommended electrolyte content of rehydration beverages, to maximise retention of ingested fluids, appears to be as high as 50–90 millimoles per litre; however, commercial sports drinks contain significantly lower amounts, so the additional sodium can come from consuming foods with sodium content (Burke & Deakin 2010: 381). Sawka et al. (2007) recommend that, 'if time permits, consumption of normal meals and beverages will restore **euhydration** ... Consuming beverages and snacks with sodium will help expedite rapid and complete recovery by stimulating thirst and fluid retention. Intravenous fluid replacement is generally not advantageous, unless medically merited.'

Euhydration is the normal state of body water, with no dehydration.

Alcohol

Alcohol is not recommended as a post-training beverage. While it is common for recreational athletes in Australia and New Zealand to have a beer after a game of cricket, football or rugby, it is not recommended, as it not only impairs rehydration, but is also likely to impact on the quality of the foods then consumed and may also impair glycogen synthesis (Burke & Deakin 2010: 379). Researchers in Queensland, Australia, have been working on a low-alcohol beer with added electrolytes that may be a more suitable alternative after exercise (Desbrow et al. 2013).

Protein

While post-training protein bars and shakes have been widely marketed over recent years as the ultimate muscle-building and recovery foods, the evidence for consuming protein immediately following training is not strong. The benefits of protein consumption immediately after exercise relate to snacks that contain both carbohydrate and protein. The amount of protein reported to be required in early recovery (the first 15–30 minutes after training) is quite small, at only 10–20 grams. It appears that total daily protein intake, amino acid profile of the foods consumed and the spread of the protein sources over the day play more integral roles.

SPECIFIC CONCERNS

Endurance sports

Endurance sports athletes, such as ironman triathletes and marathon runners, need to have optimal training nutrition as well as invest time in putting together race or event-day plans. The two key nutrition challenges they face relate to carbohydrate and fluid intake.

Adequate carbohydrate intake can become a challenge for endurance athletes, as total daily needs may be as high as 8–12 grams of carbohydrate per kilogram for ultra-endurance athletes on event and high training days (Australian Sports Commission 2011). These athletes need to accommodate the requirements within their day-to-day eating while not exceeding their energy requirements.

Incorporating pre- and post-training meals in day-to-day eating may simply involve a planned meal or snack having an appropriate amount of carbohydrate and protein to act as a refuelling snack. For example, an athlete could eat a low-fat breakfast before training that includes carbohydrate and a small amount of protein, then eat a fruit salad with low-fat yoghurt after training.

Weight category and aesthetic sports

Many athletes in weight category and aesthetic sports desire changes in body composition. In most cases this involves fat loss, with or without gain in muscle mass. This presents the challenge of meeting high nutritional needs for training while implementing an energy deficit for fat loss to occur.

In the case of weight category sports, weight is non-negotiable; in other words, if an athlete does not weigh in at the right weight on competition day, they are disqualified and do not compete. This strict regulation of weight category sports, including lightweight rowing, boxing, many martial arts and horse racing, to name a few, can often lead to some very undesirable and in some cases dangerous weight-loss approaches, often referred to as weight-making.

Weight-making usually involves trying to dehydrate the body so that the fluid loss will result in rapid weight loss and the lighter weight required for competition will be quickly reached. Current popular practices include using saunas, diuretic pills and laxatives, training in hot climates and wearing excessive clothing to induce extreme sweating. Athletes have also been reported to put themselves on food and fluid restrictions, consuming minimal amounts of fluids in the days leading up to competition. This not only presents significant health risks, but is also likely to impair performance if an athlete cannot be adequately rehydrated after weighing in and prior to their event.

Weight losses of 5–6 per cent of body weight using this approach are not uncommon and have been shown to decrease performance, impair cognitive function, increase susceptibility to overheating and result in abnormal bone metabolism (Burke & Deakin 2010: 152). Unfortunately, in the past, weight-making practices have also resulted in the death of athletes (Wilson et al. 2013). The practice is highly undesirable and discouraged. Athletes in weight category sports are advised to keep their weight as close as possible to their competition weight between competitions, or, where the option exists, to consider building up lean body mass to compete in a higher weight category, for which extreme weight-making practices will not be required.

For both weight category and aesthetic sports, athletes' weight-loss recommendations need to align with nutrition recommendations for health and wellbeing. Therefore, energy as well as macronutrient and micronutrient needs must be considered prior to nutrition advice being given. It may not be appropriate to encourage a young female gymnast with a body mass index (BMI) of 18.5 to lose any more weight even if this is a personal goal of the athlete, as it would result in her becoming underweight and is likely to lead to poorer long-term health outcomes.

Young athletes

The young athlete is not simply a small adult athlete and should not be treated as such. They present the additional challenge of meeting growth and development requirements on top of high training needs. Furthermore, young athletes are often very busy and, between school, homework and training, have little time to eat. The sheer volume of food required by a young growing athlete may also prove a challenge, as fitting the energy and nutrients needed into a relatively small body can be difficult. It is critical that the nutritional requirements for both growth and training be met, as inadequate intake can lead to delays in development and growth that can have both short- and long-term consequences.

Young athletes may be less efficient at recognising dehydration and therefore more at risk of heat-related illness. It is important that coaches and parents are aware of this and support their young athletes by providing and encouraging appropriate hydration practices.

With adolescent females, adequate energy and nutrient intakes are required for appropriate fat gain and body composition changes required for menses to commence. It is not uncommon for young females with high training loads to experience delays with the onset of menses; this is often related to energy intake inadequate to meet both growth and training needs and high training loads. It is not ideal for health and development; hence, an increase in food intake and reduction in training load may be required.

As parents are usually the food providers for young athletes it is important that they are aware of the high nutrient requirements of their children and provide suitable and adequate foods, drinks and snacks to meet their growth and training requirements. This includes appropriate composition of main meals as well as snacks provided around training and on competition days.

Travelling athletes

For elite and sub-elite athletes, travel becomes part of life, and the routines and strategies that work effectively at home may be disrupted by travelling to and competing in a different environment. Challenges faced may include jet-lag, increased risk of food-borne illness, and environmental stress. The availability of foods and drinks usually consumed by an athlete may be limited, impacting on their nutritional intake and status.

Athletes are encouraged to plan their nutrition and hydration strategy well in advance of a trip, to investigate what foods and beverages will be available at their destination, and to find out whether they need to pack particular foods or drinks to take on their travels. It is not uncommon for athletes to take sports drink powders and gels on trips to minimise any problems when they arrive. Furthermore, even with domestic travel throughout Australia and New Zealand, athletes should be aware of what is available at their destination. For example, the variety of supplements available in central Melbourne or Auckland may be greater than the variety available in more regional areas, so any athlete with very specific preferences travelling to compete should plan ahead.

SUMMARY AND KEY MESSAGES

Athletes generally have higher macronutrient requirements than the sedentary population, for protein and carbohydrate. The nature of the sport in which an athlete partakes and their goals in regards to body composition have significant impacts on the nutritional challenges they face.

- Nutrition and hydration practices before, during and after training and competition are critical to athletes' success and aid in reducing fatigue and risk of injury.
- Endurance athletes have high carbohydrate requirements and need to pay very careful attention to their hydration practices.
- Athletes in weight category and aesthetic sports are at risk of engaging in dangerous weight-making behaviours and need to maintain a healthy weight throughout both the off season and the competition season to avoid having to lose a lot of weight quickly.
- Young athletes face the additional challenge of meeting nutritional requirements for growth and development as well as for training, often making it difficult to obtain sufficient quality food at the right times.
- Athletes who travel as part of their sporting endeavours face additional nutrition challenges in finding suitable foods in other countries and having adequate food storage and preparation facilities.
- The importance of good nutritional intake that is suited to the lifestyle, training and personal goals of the individual athlete cannot be understated, as it is critical not only for optimal performance but also good health outcomes and injury prevention.

REFERENCES

Australian Sports Commission 2009. *Nutritional Preparation for Tournaments and Multiple Heat Competitions.* Commonwealth of Australia. <www.ausport.gov.au/ais/nutrition/factsheets/competition_and_training/nutritional_preparation_for_tournaments_and_multiple_heat_competitions>, accessed 1 November 2013.

——2011, *Carbohydrate: The facts.* Commonwealth of Australia. <www.ausport.gov.au/ais/nutrition/factsheets/basics/carbohydrate__how_much>, accessed 1 November 2013.

Burke, L.M. & Deakin, V. 2010, *Clinical Sports Nutrition* 4th edn. McGraw-Hill Australia, Sydney.

Burke, L.M, & Kiens, B. 2006. 'Fat adaptation' for athletic performance: The nail in the coffin? *Journal of Applied Physiology* 100: 7–8.

Burke, L.M., Slater, G., Broad, E.M., Haukka, J. et al. 2003. Eating patterns and meal frequency of elite Australian athletes. *International Journal of Sport Nutrition and Exercise Metabolism* 13: 521–38.

Coyle, E.F. 2004. Fluid and fuel intake during exercise. *Journal of Sports Sciences* 22: 39–55.

Desbrow, B., Murray, D. & Leveritt, M. 2013. Beer as a sports drink? Manipulating beer's ingredients to replace lost fluid. *International Journal of Sport Nutrition and Exercise Metabolism* 23: 593–600.

Hargreaves, M., Hawley, J.A. & Jeukendrup, A. 2004. Pre-exercise carbohydrate and fat ingestion: Effects on metabolism and performance. *Journal of Sports Sciences* 22: 31–8.

Lee, J., Shirreffs, S.M. & Maughan, R.J. 2008. Cold drink ingestion improves exercise endurance capacity in the heat. *Medicine and Science in Sports and Exercise* 40: 1637–44.

Maughan, R.J. 1994. Fluid and electrolyte loss and replacement in exercise. In Harries, M., Williams, C., Stanish, W.D. & Micheli, L.L. (eds). *Oxford Textbook of Sports Medicine*, 82–93. Oxford University Press, Oxford.

Millard-Stafford, M. 1992. Fluid replacement during exercise in the heat. *Sports Medicine* 13: 223–33.

NHMRC (National Health and Medical Research Council) & Ministry of Health 2006. *Nutrient Reference Values for Australia and New Zealand: Including recommended dietary intakes.* Commonwealth of Australia & New Zealand Government. <www.nhmrc.gov.au/_files_nhmrc/publications/attachments/n35.pdf>, accessed 16 March 2014.

Sawka, M.N., Burke, L.M., Eichner, E.R., Maughan, R.J. et al. 2007. American College of Sports Medicine position stand: Exercise and fluid replacement. *Medicine and Science in Sports and Exercise* 39: 377–90.

Szlyk, P.C., Sils, I.V., Francesconi, R.P., Hubbard, R.W. et al. 1989. Effects of water temperature and flavoring on voluntary dehydration in men. *Physiology and Behavior* 45: 639–47.

Wilson, G., Hawken, M.B., Poole, I., Sparks, A. et al. 2013. Rapid weight-loss impairs simulated riding performance and strength in jockeys: Implications for making-weight. *Journal of Sports Sciences* 32: 383–91.

ADDITIONAL READING

American College of Sports Medicine n.d. *Position Stands.* American College of Sports Medicine. <www.acsm.org/access-public-information/position-stands>, accessed 9 October 2014.

Australian Institute of Sport n.d. *Nutrition.* Australian Sports Commission, Commonwealth of Australia. <www.ausport.gov.au/ais/nutrition>, accessed 9 October 2014.

18

Poverty, disadvantage and food insecurity

Sharon Croxford

LEARNING OUTCOMES

Upon completion of this chapter you will be able to:

- use appropriate terminology when discussing disadvantage
- define income poverty and relative poverty and their relationship to disadvantage
- describe how health is related to disadvantage
- explain what it means to be food insecure
- outline strategies that may be employed to tackle food insecurity, listing advantages and disadvantages of each.

Education, employment, income and health are linked to disadvantage. Poverty and food insecurity arise from disadvantage, and the more disadvantaged a person, the more likely they are to have health problems. This chapter looks at disadvantage, poverty, food insecurity and health and introduces the reader to some policy and practice approaches to tackling some of the problems.

The level of disadvantage an individual or family experiences may vary throughout their life. It is possible that someone who shows no sign of disadvantage will find that through a change of circumstances they lose their job, home and family connections and find themselves in extreme disadvantage, experiencing homelessness without finances and even food. Typical measures of disadvantage include education, employment, income and health. Given these measures, it would seem logical that people who completed less formal education, are without work or enough work, are receiving a low income from any source and report poor health are the disadvantaged members of our society. Add to this **social exclusion** and lack of connection with the broader community and it is conceivable that someone that meets all these measures is the most disadvantaged in our society.

Social exclusion is a situation in which an individual or group is unable to participate in many regular activities and systems within society. Housing, education, health care and engagement in community and social activities are difficult to access due to a systematic failure to prevent people from becoming excluded or to keep them included. Social inclusion is the term used to describe the opposite situation.

Measures of Australia's Progress is a compilation of data from four key indicators that attempts to answer the question of whether life is getting better in Australia. Two of the key indicators, society and economy, can help to provide a general picture of how some measures of disadvantage are faring. In the 2013 report, the themes of health, and learning and knowledge (education), within the society indicator, showed improvement compared to the ten years prior. That is, people had a higher life expectancy at birth, and the proportion of people aged 25–64 years with vocational or higher education had increased within the past ten years (ABS 2013). The themes of close relationships, home and community connections and diversity showed no change. Within the economy indicator, opportunities (a measure of the number of educated people in skilled occupations) and jobs (unemployment rate) had improved.

In New Zealand, key findings in 2010 on progress towards sustainable development included the discoveries that while health expectancy had increased from 1996 to 2006 and disposable income had increased from 1996 to 2006 yet remained stable between 2006 and 2010, the annual unemployment rate had also increased in the twelve years prior to 2009. Income inequality had grown between 1988 and 2009, and economic hardship indicators had shown no change since 1988, although this indicator of the proportion of the population in households with low incomes saw a steep rise between 1988 and 1992 but had been steadily declining since then. The proportion of adults aged 25–64 years who completed at least secondary school increased between 2001 and 2009 (Statistics New Zealand 2011).

At first glance this may suggest that Australians are healthier, more educated and more gainfully employed and are socially and personally connected with friends, family and community. In New Zealand people are living longer and some have more money to spend, but there are fewer job opportunities despite more people leaving secondary school with a qualification. Importantly, the growing gap is clear between those who have access to more advantages, or at least options to meet needs, and those who do not.

Poverty has historically been described in relative and absolute terms. Income poverty looks at poverty from an economic perspective, with absolute poverty a measure of a person's ability to provide themselves with food, clothing and some sort of housing. Relative poverty attempts to describe a person's situation in relation to other segments of society and more general standards of living. Both definitions attempt to describe poverty in terms of tangible items—that is, items that are purchased or exchanged—but do not take into account other needs for a person to function in society, such as education and health care, cultural identity and participation in community.

Despite these difficulties with terminology it is a basic human right for a person to live in a society where they have the right to work and have an adequate income. In Australia, 13.9 per cent of all people and 17.7 per cent of children were living below the **poverty line** of $400 per week for single adults and $841 for a family of two adults and two children in 2011–12, taking into account housing costs. This was an increase of approximately 3 per cent since 2001. If the poverty line used to determine these figures were increased to 60 per cent, in line with Britain, Ireland and the European Union's calculation, then 22 per cent of all people and 25.5 per cent of children in Australia would be classed as living in poverty (ACOSS 2014).

Those who are living in poverty include people who are working but receiving low incomes and those with insufficient work; however, 61.2 per cent of people below the poverty line in 2011–12 were unemployed and 40.1 per cent had social security payments as their primary income. The groups in Australia who are at highest risk of poverty are households whose members are not employed, single adults under 65 years (that is, of an age at which they are able to work), households whose main income is from social security payments, people with disabilities and single-parent families (ACOSS 2014).

When the 60 per cent calculation is applied to New Zealand, one in four children was living in poverty in 2012. Twenty per cent of children were living in severe poverty (less than 50 per cent of median income after housing costs are considered), within families in income poverty, and 60 per cent of those in poverty had been in this persistent state for the seven-year period surveyed (2002–03 to 2008–09) (Craig et al. 2013). Income poverty rates for single-person working-age

The **poverty line** is a figure used widely by governments and other organisations and calculated on 50 or 60 per cent (depending on the country) of median household income—that is, half of the 'middle' income for all households. The lower level corresponds with an extremely simple standard of living.

households in New Zealand were 31 per cent in 2011 and 2012, with older single people at increased risk. The population poverty rate in 2012 was 14 per cent (MSD 2013).

It is important to note that **wealth** is distributed more unequally than income. In New Zealand, 25 per cent of gross income yet 50 per cent of total wealth belongs to those in the top income decile. In Australia, 45 per cent of wealth belongs to the top 10 per cent. This is similar to other Organisation for Economic Co-operation and Development member countries but significantly lower than the 70 per cent figure for the United States (MSD 2013).

FOOD INEQUALITY

Food security

The Food and Agriculture Organization (FAO) of the United Nations describes food security as a state 'when all people, at all times, have physical, social and economic access to sufficient, safe and nutritious food which meets their dietary needs and food preferences for an active and healthy life' (FAO n.d.). In other words, it is a situation for individuals and groups in which their needs concerning food access, food availability and food use are all met. The World Health Organization (WHO) describes these in detail:

- *Food access*—This is the ability to source and consume a nutritious diet. Access includes being able to do all that is required to buy, transport, store, prepare and cook foods. This implies that people have the finances, time and mobility to achieve the production of a healthy meal.
- *Food availability*—A food supply within a community, be it as small as a household or as large as a country, to meet the needs of its population, is termed food availability. Food outlets need to be within reach, and food within outlets needs to meet both the nutritional and the cultural needs of the community, to be available at the right price, and to be of high quality and sufficient variety. On a broader scale, food needs to be produced sustainably, with productivity enhanced.
- *Food use*—This includes knowledge and skills required to undertake the purchase, preparation and cooking of foods and to make appropriate choices. Adequate safe water and sanitation are required to ensure safe food. Issues of food waste and losses following harvest need to be considered. (WHO n.d.)

A further dimension to food security involves stability in all of the above.

Food insecurity

Food insecurity, a term used to describe the results of food inequality, is linked with a range of poor health outcomes. It is on many government and non-government agencies' agendas yet more often than not refers to global food security issues. At a local level food insecurity is a growing problem. More and more Australians and New Zealanders are facing food insecurity each year.

Research conducted in New Zealand showed that 15 per cent of respondents in the Survey of Families, Income, and Employment from 2000 to 2010 were deemed food insecure in 2004–05. Typical questions asked to determine food security status were based on food experiences within the past twelve months. In this study three questions were used relating to the use of **food relief**, the need to buy cheaper food in order to pay for other things and often going without fresh fruit and vegetables. A positive answer to any of these questions indicated food insecurity. People most likely to be food insecure included those of Māori or Pacific Island ethnicity, single-parent families, those who were unemployed, seeking work or in receipt of a benefit within the past twelve months and those at lower socio-economic levels (Carter et al. 2010).

In Australia an estimated two million people access food relief each year, with this number rising. Relief services report that many of their clients are return users, becoming semi-reliant on the support provided; yet at the same time financial aid and food aid are restricted to ensure there is enough to go around. Those most likely to experience food insecurity at much higher rates than the 5 per cent determined in Australian national surveys include Aboriginals and Torres Strait

Wealth in this context refers to money, possessions of value, property and other assets. These objects can be exchanged or have a monetary value. A person's total wealth can be calculated based on the monetary value of their objects.

Food relief is a type of local food aid or assistance offered to people in need. It can take the form of a meal, grocery or food parcel or voucher to spend on a meal or buy food.

Islanders (24 per cent), unemployed people and single-parent households (23 per cent), second lowest quintile earners and rental households (20 per cent) and young people (15 per cent) (Browne et al. 2009; Burns 2004).

Larger surveys in Australia typically use a single-item measure on food insecurity that asks a respondent if they have run out of food in the last twelve months and could not afford to buy more. Some researchers would suggest that this underestimates the numbers that are food insecure. A team in Queensland determined food security status using the single-item measure and a range of versions of the United States Department of Agriculture's Household Food Security Survey Module that had six, ten and eighteen items (Ramsey et al. 2011). All of the more detailed versions showed significantly higher rates of food insecurity, at 21.3, 22 and 24.6 per cent from the measures with six, ten and eighteen items, respectively, compared to 19.5 per cent from the single-item measure. Another study of more than 3000 people over 49 years of age found that people who reported food insecurity were more likely to be women, in rented accommodation and living on a pension only (Russell et al. 2014).

The seemingly incongruous link between food insecurity and obesity has been gathering momentum around the globe for many years. In 2004 it seemed that there was a strong link between women, food insecurity and obesity, with Burns (2004) stating that the risk of obesity was 20–40 per cent higher in those women who were food insecure. A study in New South Wales suggested that obesity was more prevalent among people who reported being food insecure and that higher consumption of energy-dense, nutrient-poor foods could be associated with both food insecurity and obesity when they occurred concurrently (Innes-Hughes et al. 2010). In the above study women were at higher risk than men. However, with the recent rapid increase in the number of obese socially disadvantaged men, especially in Australia, the trend may change. Many other studies have shown that buying cheaper foods results in meals and snacks that are high in added fats, sugars and refined grains, and that these foods are perceived as filling and are relied on by many who experience food insecurity.

Health and nutrition interventions

The fact that such inequalities in food, nutrition and health exist suggests changes in policies and practices could have a dramatic impact. Improvements in food security will address, at least in part, the nutrition and health inequalities. The objectives of policies and practices are therefore to ensure that every person:

- does not experience hunger as a result of running out of food and not being able to buy more
- does not need to rely on family, friends or food aid from emergency relief agencies to meet their food and nutrition needs
- does not suffer any anxiety about getting food
- has sufficient culturally appropriate food, including fresh fruit and vegetables at affordable prices
- has access to adequate food storage and cooking facilities
- has the knowledge and skills to select, store, prepare, cook and eat healthy food
- has an overall diet that is varied and meets their nutrient requirements.

There are a range of policy actions that would improve food security, but for policies and their tools to be written properly and to become successful they need to raise awareness of the right of each individual to have a diet that meets all their needs; identify the real barriers to food security for individuals and groups, which may be different within different sectors of the community and across the country; advocate for food security for all; and seek cooperation across sectors.

Food access

Access to food requires financial resources. An adequate income is a major, but not the only, barrier for many disadvantaged people when buying food. Even if food costs are covered, having time and ability to get to the shops must be available as well as adequate storage, preparation and cooking facilities. While solutions to many of these issues appear simple on paper (for example, provision of food storage and cooking equipment), there are many people for whom much more creative solutions are needed.

One example of such a solution is the Café Meals Program, which has been operating in different forms for many years, and through which eligible members are able to purchase food and drink using a voucher-like system. The cohealth (formerly Doutta Galla Community Health) and City of Melbourne Café Meals programs have operated within central Melbourne for some time.

The programs cater for people experiencing homelessness; a dietitian assesses referrals for eligibility and the appropriateness of the program, puts other supports in place to improve food and nutrition for those referred, and eligible members are provided with a membership card and can purchase meals up to an agreed value at any of the member cafes. Members must pay an agreed minimum amount for each meal. The programs also offer Home and Community Care clients an alternative to delivered meals when they would benefit from the social interaction gained through visiting local cafes.

Many participants in the programs have a history of disadvantage and have been socially excluded. The Café Meals provide more than financial support in sourcing nutritious meals by considering an individual's connection with the community, lack of which forms part of the broader definition of disadvantage.

The Australian Bureau of Statistics defines a person as homeless if where they are living:

- is in a dwelling that is inadequate; or
- has no tenure, or if the initial tenure is short and not extendable; or
- does not allow them to have control of, and access to space for social relations.

This definition is centred on the concept of 'home' as a place that is not just a roof but a place of 'security, stability, privacy, safety' (ABS 2012b).

Other definitions include primary, secondary and sometimes tertiary homelessness to describe various levels of 'without roof or home'. They take into account those who are sleeping rough, staying in refuges, shelters or institutions, 'couch surfing' (moving from place to place and sleeping on couches or similar), or living in boarding houses, caravan parks or unfit housing.

Food availability

There are many facets in the supply of food that can be improved to address food insecurity. The concept of a food desert has been talked about for decades. While the distances may differ slightly between surveys, a food desert is an area where the closest outlet that provides good-quality food is outside a reasonable distance to travel. If food is not easily available to people their food security may be affected, especially where physical mobility and transport become an issue.

Food waste is intricately linked to food availability yet is often overlooked. Much research and effort go into developing crops with higher yields and efficiencies in production, but food waste seems to be relatively neglected. Nearly half of all food produced is wasted before it gets from the farm to the kitchens of the world, with much fruit and vegetable produce discarded because it does not look good enough! Diverting this and other food waste from entering landfills could alleviate hunger for many, both internationally and locally.

Food rescue and recovery programs have been developed to retrieve food that would normally be wasted and to place it back in the system. By October 2014, the Australian organisation SecondBite had redistributed nearly 12.6 million kilograms of food since 2005 that would otherwise have ended up in landfill (SecondBite 2014). OzHarvest, another Australian organisation, diverted 1600 tonnes of food in 2012 alone (OzHarvest 2012). In Wellington, by October 2014 Kaibosh had rescued 295 tonnes of food since 2008 (Kaibosh 2014). Much of the food recovered through such programs ends up on the tables of those in need—that is, the disadvantaged in our societies. Many of those tables are not within household kitchens but at a community centre where a communal meal may be shared with others in need.

This and other forms of food relief are becoming keys to providing food for millions of people each year. The strain being placed on community agencies is increasing, with many running out of food and parcels provided becoming smaller with less variety.

Food use

Improving food security through food use is really about education. When food access and availability are met, knowledge and skills in all aspects of food, nutrition and cooking empower people to maintain food security. It is not always that simple in real life, however. Many disadvantaged people have competing priorities for their time and resources, so engaging in programs to improve food skills may be quite low on their list of priorities. A person experiencing housing stress will be focused on resolving their housing situation, or someone without work may be actively seeking work, and all their energies will be focused on this, so it is important that a full assessment of an individual's situation and needs is conducted before referral to any programs. A full social history may be more important than a dietary history.

People who have never had experience with food shopping and cooking may be anxious about getting involved in something as seemingly foreign as a cooking class. Programs that aim to tackle issues of food use include Foodcents, a program designed to help save money on grocery shopping yet provide a healthy diet; Australian Community Kitchens, a group of people who meet to socialise and cook affordable and nutritious meals; and the Australian Community Gardens Network, which organises places where people come together to grow fresh food, learn about growing food and socialise.

HEALTH CONDITIONS

Health and broader wellbeing are linked to advantage. Relatively disadvantaged members of the community live shorter lives and have higher rates of illness, disability and death when compared with their relatively advantaged counterparts (CSDH 2008). It is important to note that a person's position on the **socio-economic gradient of health** may falter. It is not a simple relationship between income, occupation and education that secures a position on the gradient. While it seems logical within this model that a fall in socio-economic status would result in a lower position on the gradient, those with seemingly good relative security who experience an illness (either physical or mental) or an injury can suffer socio-economic disadvantages resulting in a fall down the gradient. Geographical location within Australia also impacts on health, with rural and remote communities generally having higher rates of overweight and obesity and poorer health when compared to city communities.

Overweight, obesity and diabetes

Figures from 2011–12 showed that 63.4 per cent of all Australian adults were overweight or obese, with 28.3 per cent obese (ABS 2012a). Overweight and obesity are not evenly distributed across the Australian population, however. Aboriginals and Torres Strait Islanders are 1.2 times more likely to be overweight, 1.9 times more likely to be obese and over three times more likely to be morbidly obese than non-Indigenous Australians (see Chapter 14; AIHW 2008). Prevalence of obesity also increases with disadvantage for both men and women in Australia, with 30.5 per cent of women and 33.4 per cent of men who were experiencing the most disadvantage (of the lowest socio-economic status) being obese in 2007–08. For women this is 50 per cent greater and for men 43 per cent greater when compared to the most advantaged in society (AIHW 2003). Importantly, the gap between upper and lower socio-economic status is widening, and the rate of obesity in men experiencing disadvantage has overtaken that of women.

Socio-economic gradient of health describes the phenomenon of health improving as one moves up the socio-economic ladder. That is, the burden of disease sits with those who lack resources and/or are from minority groups.

The New Zealand Health Survey of 2012–13 found that 31 per cent of adults (fifteen years and over) were obese and an additional 34 per cent overweight, suggesting that the problem is slightly greater in New Zealand when compared to Australia. Forty-eight per cent of Māori and 68 per cent of Pacific Islander adults were obese. Adults living in deprived areas within New Zealand were 1.5 times more likely to be obese than those living in the least deprived areas (Ministry of Health 2013).

In Australia, cultural background and country of birth indicate likelihood of obesity, with migrants born in southern and eastern Europe and Oceania (excluding Australia) more likely to be overweight or obese and those from Southeast Asia less likely than non-migrant Australians (AIHW 2010). Pacific people in New Zealand have high rates of diabetes (12.5 per cent), as do Māori (7.3 per cent), who also have higher rates of asthma, ischaemic heart disease and stroke when compared with non-Māori (Ministry of Health 2013); see Chapter 15). Aboriginals and Torres Strait Islanders are three times more likely to have diabetes, at 12.8 per cent, than non-Indigenous Australians (4.2 per cent) (AIHW 2013).

The New Zealand Health Survey in 2012–13 found higher rates of most health conditions, especially psychological distress and diabetes, among people living in the most socio-economically deprived areas of the country. This same group of people showed greater levels of health risk behaviour, with higher smoking, hazardous drinking, inadequate fruit and vegetable intake and low physical activity than people of higher socio-economic status (Ministry of Health 2013).

This corresponds with the Australian situation, in which people living in lower socio-economic areas were more likely to be overweight or obese, less physically active and twice as likely to smoke compared with people living in higher socio-economic areas. Data from 2007–08 found the difference was minimal between the most disadvantaged and the least disadvantaged with regards to meeting fruit and vegetable recommendations, with both groups doing extremely poorly. Only 5 per cent of women and 4 per cent of men in the most disadvantaged groups and 8 per cent of women and 5 per cent of men in the least disadvantaged groups were eating enough fruit and vegetables (AIHW 2012). There are marked differences between consumption of fruit and consumption of vegetables, with more people in both groups doing better at reaching the fruit recommendation; however, those experiencing disadvantage consume markedly less fruit. The reasons for the high rates of obesity in disadvantaged groups are clearly more complex than fresh fruit and vegetable consumption.

Cardiovascular disease

Cardiovascular disease (CVD) and its associated risk factors affect more people of disadvantage than of advantage in both Australia and New Zealand. The data in Tables 18.1 and 18.2 show this general trend.

Table 18.1 Proportion of population groups with CVD and selected risk factors for all New Zealanders (%) and adjusted rate ratios for Māori, Pacific Islanders and disadvantaged

Population group	High blood pressure[a]	High cholesterol[a]	Ischaemic heart disease	Stroke	Obesity[b]
All New Zealanders	15.9	11.0	4.9	2.0	31.3
Māori	1.4	1.2	1.8	1.4	1.8
Pacific Islanders	1.3	1.2	1.4	1.4	2.4
Disadvantaged	1.4	1.1	1.7	1.4	1.5

a Taking medication.
b BMI of 30 or greater.

Note: BMI: body mass index; CVD: cardiovascular disease. The figures for Māori, Pacific Islanders and disadvantaged are adjusted rate ratios.

Source: Ministry of Health (2013).

Table 18.2 Proportion of population groups with CVD and selected risk factors for all Australians (%) and adjusted rate ratios for Aboriginals, Torres Strait Islanders and disadvantaged

Population group	High blood pressure	High cholesterol	Cardiovascular disease	Obesity[a]
All Australians	29.9[b]	47.9[b]	17.0[b]	28.3
Aboriginals and Torres Strait Islanders[c]	1.2	–	1.2	–
Disadvantaged[d]	32.6	49.5	17.0	32.0

a BMI of 30 or greater.
b Twenty-five years and over.
c Eighteen years and over. Age-standardised rate ratios. Age standardisation sees Aboriginals and Torres Strait Islanders more likely to have high blood pressure at rate ratio of 1.2 and CVD at 1.3.
d Figures for blood pressure and cholesterol relate to those who did not complete secondary school as a marker of socio-economic status. The lowest socio-economic groups have 17–18 per cent CVD prevalence compared to 14 per cent for the highest socio-economic groups.

Note: BMI: body mass index; CVD: cardiovascular disease. Empty cells indicate no data available.

Source: ABS (2012a, 2013); AIHW (2013).

SUMMARY AND KEY MESSAGES

Food insecurity results from disadvantage. Those who are food insecure have greater health problems. Food insecurity cannot be solved through food relief but through a broader policy and practice shift that ensures every person has access to education, employment and income in a socially inclusive society.

- The proportion of people moving into poverty and food insecurity each year is increasing.
- A higher proportion of people experiencing disadvantage are obese when compared with people experiencing advantage.
- Food recovery, rescue and relief are playing larger roles in getting food to people in need.
- Policy needs to address the growing inequalities in society.
- Practice needs to take into account each individual's situation.

REFERENCES

ABS (Australian Bureau of Statistics) 2012a. *Australian Health Survey: First results, 2011–12*. Cat. no. 4364.0.55.001. Commonwealth of Australia. <www.ausstats.abs.gov.au/ausstats/subscriber.nsf/0/1680ECA402368CCFCA257AC90015AA4E/$File/4364.0.55.001.pdf>, accessed 28 September 2014.

——2012b. *Information Paper: A statistical definition of homelessness*, 2012. Cat. no. 4922.0. Commonwealth of Australia. <www.abs.gov.au/ausstats/abs@.nsf/Latestproducts/E84E0BDA4DD31050CA257A6E0017FD0E?opendocument>, accessed 10 October 2014.

——2013. *Measures of Australia's Progress*, 2013. Cat. no. 1370.0. Commonwealth of Australia. <www.abs.gov.au/AUSSTATS/abs@.nsf/mf/1370.0>, accessed 9 November 2014.

ACOSS (Australian Council of Social Services) 2014. *Poverty in Australia 2014*. ACOSS. <http://acoss.org.au/images/uploads/ACOSS_Poverty_in_Australia_2014.pdf>, accessed 14 November 2014.

AIHW (Australian Institute of Health and Welfare) (O'Brien, K. & Webbie, K.) 2003. *Are All Australians Gaining Weight? Differentials in overweight and obesity among adults, 1989–90 to 2001*. Bulletin 11, cat. no. AUS 39, Commonwealth of Australia. <www.aihw.gov.au/WorkArea/DownloadAsset.aspx?id=6442453186>, accessed 24 January 2014.

——(Penm, E.) 2008. *Cardiovascular Disease and Its Associated Risk Factors in Aboriginal and Torres Strait Islander Peoples 2004–05*. Cardiovascular Disease Series 29, cat. no. CVD 41. Commonwealth of Australia. <www.aihw.gov.au/WorkArea/DownloadAsset.aspx?id=6442455071>, accessed 24 January 2014.

——2010. *Australia's Health 2010: The twelfth biennial health report of the Australian Institute of Health and Welfare*. Australia's Health Series 12, cat. no. AUS 122. Commonwealth of Australia. <www.aihw.gov.au/WorkArea/DownloadAsset.aspx?id=6442452962/>, accessed 10 November 2014.

——2012. *Australia's Health 2012: The thirteenth biennial report of the Australian Institute of Health and Welfare*. Australia's Health Series 13, cat. no. AUS 156. Commonwealth of Australia. <www.aihw.gov.au/WorkArea/DownloadAsset.aspx?id=10737422169>, accessed 6 October 2014.

——2013. *Populations of Interest*. Commonwealth of Australia. <www.aihw.gov.au/diabetes/populations-of-interest/>, accessed 10 January 2014.

Browne, J., Laurence, S. & Thorpe, S. 2009. *Acting on Food Insecurity in Urban Aboriginal and Torres Strait Islander Communities: Policy and practice interventions to improve local access and supply of nutritious food*. Australian Indigenous HealthInfoNet. <www.healthinfonet.ecu.edu.au/health-risks/nutrition/other-reviews>, accessed 24 January 2014.

Burns, C. 2004. *A Review of the Literature Describing the link between Poverty, Food Insecurity and Obesity with Specific Reference to Australia*. Victorian Health Promotion Foundation, Melbourne.

Carter, K.N., Lanumata, T., Kruse, K. & Dorton, D. 2010. What are the determinants of food insecurity in New Zealand and does this differ for males and females? *Australian and New Zealand Journal of Public Health* 34(5). <www.otago.ac.nz/wellington/otago020409.pdf>, accessed 10 January 2014.

Craig, E., Reddington, A., Wicken, A., Oben, G. et al. 2013. *Child Poverty Monitor 2013 Technical Report.* Updated 2014. New Zealand Child and Youth Epidemiology Service. <http://img.scoop.co.nz/media/pdfs/1312/2013_Child_Poverty_Monitor_Technical_Report_MASTER.pdf>, accessed 10 January 2014.

CSDH (Commission on Social Determinants of Health) 2008. *Closing the Gap in a Generation: Health equity through action on the social determinants of health; Final report of the Commission on Social Determinants of Health.* WHO, Geneva.

FAO (Food and Agriculture Organization of the United Nations) n.d. *Food Security Statistics.* FAO. <www.fao.org/economic/ess/ess-fs/en/>, accessed 9 October 2014.

Innes-Hughes, C., Thrift, A. & Cosgrove, C. 2010. *A Further Analysis of the Weight Status and Dietary Characteristics of People Reporting Food Insecurity in NSW: NSW Population Health Survey data 2007 and 2000.* University of Sydney. <http://ses.library.usyd.edu.au//bitstream/2123/9083/1/PANORG_Innes-Hughes_Monit_update_foodinsecurity_161210.pdf>, accessed 24 January 2014.

Kaibosh (website) 2014. <www.kaibosh.org.nz>, accessed 10 October 2014.

Ministry of Health 2013. *New Zealand Health Survey: Annual update of key findings 2012/13.* New Zealand Government. <www.health.govt.nz/system/files/documents/publications/new-zealand-health-survey-annual-update-2012-13-dec13.pdf>, accessed 6 October 2014.

MSD (Ministry of Social Development) 2013. *Household Incomes in New Zealand: Trends in indicators of inequality and hardship 1982 to 2012.* New Zealand Government. <www.health.govt.nz/publication/new-zealand-health-survey-annual-update-key-findings-2012-13>, accessed 10 January 2014.

OzHarvest 2012. *Annual Report 2012.* OzHarvest. <www.ozharvest.org/wp-content/uploads/2014/06/OzHarvest-Annual-Report-2012.pdf?380524>, accessed 10 October 2014.

Ramsey, R., Giskes, K., Gallegos, D. & Turrell, G. 2011. *Current measure of food insecurity used in the National Health Survey may be underestimating its prevalence.* Paper presented at 2nd National Food Futures Conference, Hobart, 22–23 November 2011.

Russell, J., Flood, V., Yeatman, H. & Mitchell, P. 2014. Prevalence and risk factors of food insecurity among a cohort of older Australians. *Journal of Nutrition, Health and Ageing* 18: 3–8.

SecondBite (website) 2014. <http://secondbite.org>, accessed 10 October 2014.

Statistics New Zealand 2011. *Key Findings on New Zealand's Progress Using a Sustainable Development Approach: 2010.* New Zealand Government. <www.stats.govt.nz/browse_for_stats/snapshots-of-nz/Measuring-NZ-progress-sustainable-dev-%20approach/key-findings-2010.aspx>, accessed 6 October 2014.

WHO (World Health Organization) n.d. *Food Security.* WHO. <www.who.int/trade/glossary/story028/en/>, accessed 11 November 2013.

ADDITIONAL READING

Australian City Farms and Community Gardens Network (website) 2014. <http://communitygarden.org.au/acfcgn/>, accessed 10 October 2014.

Australian Community Kitchens (website) 2014. <www.communitykitchens.org.au/>, accessed 10 October 2014.

Foodcents (website) 2014. <www.foodcentsprogram.com.au/>, accessed 10 October 2014.

19

Mental health

Adrienne Forsyth

LEARNING OUTCOMES

Upon completion of this chapter you will be able to:

- describe the prevalence and impact of mental disorders in Australia and New Zealand
- list the aetiology, symptoms and impacts of the most commonly occurring mental disorders
- understand the nutrition-related impacts of mental disorders
- highlight the key considerations for working with individuals with a mental disorder.

Mental disorders represent the most prevalent and burdensome group of health conditions globally. They have the ability to impact on work, study and social activities, as well as activities of daily living including the procurement and preparation of sufficient nutritious food. Emerging research is demonstrating an association between diet quality and several common mental disorders. For other mental disorders, dietitians play a key role in assisting individuals to manage nutrition-related symptoms or consequences.

The 2007 National Survey of Mental Health and Wellbeing estimated that 3.2 million Australians (equivalent to 20 per cent of the population between the ages of 16 and 85 years) experienced a mental disorder within the past year (ABS 2009). This survey also estimated that nearly half of the population aged 16–85 years (7.3 million people, or 45.5 per cent of the population) have experienced a mental disorder in their lifetime. In New Zealand, 16 per cent of adults had been diagnosed with common mental disorders at some stage in their life by 2011–12 (Ministry of Health 2012). Mental disorders are the leading cause of disability burden in Australia and are estimated to account for 24 per cent of total years lost due to disability (AIHW 2007). They account for 13 per cent of the total burden of disease in Australia, behind only cancer and cardiovascular disease (CVD) for morbidity and mortality. Individuals experiencing mental illness are more likely to also be experiencing comorbid illness or injury such as dementia, CVD, disability and alcohol abuse (AIHW 2012).

Mental disorders are formally diagnosed by a psychologist or psychiatrist using the criteria specified in the *Diagnostic and Statistical Manual of Mental Disorders* (fifth edition, known as *DSM-5*; APA 2013), although many people are treated following a simple screening with their general practitioner or other health professional.

Individuals experiencing acute severe mental illness may be admitted to inpatient hospital programs. Dietitians working in these environments are uniquely placed to impact on the nutrition of these patients from both a food service and an individual case-management perspective. A nutritional assessment including usual diet history, anthropometry and biochemistry should be conducted to determine the patient's nutritional status. Nutrient deficiencies should be corrected where applicable. Ongoing education should be provided on planning, shopping for and preparing healthy meals.

DEPRESSIVE DISORDERS

It is normal for all individuals to experience changes in mood. A depressive episode becomes problematic when it persists for an extended duration (two weeks or more) and interferes with normal work, family or social activities. Depression is characterised by depressed mood, diminished interest in daily activities, reduced or increased appetite, feelings of worthlessness, difficulty concentrating and in some cases suicidal ideation (thoughts of suicide or a desire to die, in some cases with plans for killing oneself). Depressive disorders include major depressive disorder and other classifications of depression, including persistent depressive disorder, disruptive mood dysregulation disorder and premenstrual dysphoric disorder.

More than 6 per cent of Australians will have experienced depression within the past twelve months (AIHW 2012), with 14 per cent of the New Zealand population having a diagnosis of depression in their lifetime (Ministry of Health 2012). The most common age of onset of depression is late teens to early twenties. Depression occurs more frequently in individuals with a family history of depression and may be triggered by stressful life events. Episodes of depression can last for weeks, months or even years and can reoccur. Traditional treatment options for depression include antidepressant medication and psychological therapies. There is substantial evidence to support exercise for the prevention and management of depression, and there is growing evidence to support dietary strategies.

A lack of energy and motivation coupled with changes in appetite can lead to poor dietary behaviours. Studies have demonstrated that individuals experiencing a depressive illness consume lower than recommended amounts of core food groups and higher than recommended amounts of 'junk' foods (Forsyth et al. 2012; Jacka et al. 2010; Kuczmarski et al. 2010). There is also a demonstrated association between depression and obesity (Luppino et al. 2010).

If poor mental health has impacted on an individual's motivation to participate in daily activities, they may experience difficulty with the planning, shopping, preparation and cooking required to produce healthy meals. In some cases, prolonged or recurrent mental illness including depression can impact on employment and financial means to purchase nutritious foods.

Health professionals seeking to modify the dietary intake of individuals with depressive illness need to consider the factors that influence their dietary habits. Patient-centred counselling strategies such as **motivational interviewing** are more time consuming but may be more likely to result in improved outcomes than diet prescription.

There are several dietary patterns that have been associated with a reduced incidence of depression, particularly traditional diets such as those from the Mediterranean and Norway (for a review, see Quirk et al. 2013). Eating patterns that follow basic healthy eating principles expressed in national dietary guidelines are generally associated with a reduced incidence of depression. Some specific foods and nutrients have also been investigated as possible factors in the prevention or management of depression, including fish oils, folate, magnesium and zinc. There is limited evidence to support the use of these nutrients in the management of depression, but insufficient evidence to recommend daily doses higher than the recommended dietary intakes (RDIs; see Chapter 1).

Prudent dietary advice for individuals experiencing depressive illness would include following the Australian Dietary Guidelines (NHMRC 2013) or the New Zealand Food and Nutrition Guidelines (Ministry of Health 2003), including the recommended serves of fruit, vegetables and whole grains as sources of folate and magnesium. Health professionals should assist patients to identify and overcome barriers to accessing sufficient nutritious food. Finally, given the substantial evidence supporting physical activity in the prevention and management of depression,

Motivational interviewing is a client-centred approach to behaviour change that is goal directed with the aim of exploring and resolving ambivalence.

patients should be encouraged to adopt a program of regular physical activity (in patients with comorbid physical illness or injury this may require the support of a general practitioner, exercise physiologist and/or physiotherapist).

ANXIETY DISORDERS

While some stress-induced anxiety is normal, an anxiety disorder is characterised by persistent excessive anxiety lasting six months or more. Anxiety disorders are characterised by excessive fear (emotional response to an imminent threat, either real or perceived) and anxiety (anticipation of future threat) with related behavioural disturbances. Anxiety disorders include generalised anxiety disorder, panic disorder, social anxiety disorder, specific phobias and other anxiety conditions, including those experienced by some children such as separation anxiety and selective mutism (a condition in which an individual does not speak in specific situations or to specific people or groups of people; most individuals with selective mutism also experience social anxiety).

The most common form of anxiety is generalised anxiety disorder, in which individuals experience persistent and excessive anxiety about various aspects of life, including work and school performance. Corresponding physical symptoms include restlessness, fatigue, difficulty concentrating, irritability, muscle tension and sleep disturbance.

It is estimated that 14.4 per cent of the Australian population will have experienced an anxiety disorder in the past year (AIHW 2012), and 6 per cent of the New Zealand population will experience an anxiety disorder in their lifetime (Ministry of Health 2012). Females are twice as likely as males to develop an anxiety disorder and, while anxiety typically begins in adolescence, it is most commonly experienced by individuals in their thirties and forties. The onset of anxiety occurs frequently after the experience of a severe stressor. Like depression, anxiety commonly runs in families, and many individuals experience both depression and anxiety. Anxiety disorders are often chronic, with symptoms persisting for several years.

Traditional treatment for anxiety involves anxiolytic medication and psychological therapies. There is evidence to support exercise in the short-term (daily) management of anxiety. There is a limited evidence base to support the benefits of specific foods, nutrients or dietary patterns in the management of anxiety. It would appear that the healthy dietary patterns beneficial in the prevention or management of depression may also be of benefit in the prevention and management of anxiety.

Dietary patterns of individuals experiencing an anxiety disorder often vary depending on the severity of the anxiety. People experiencing mild anxiety frequently describe periods of emotional eating in response to their stress, in which they consume 'comfort foods' in an attempt to manage their anxiety. While this behaviour may result in a short-term reduction in anxiety, it is problematic because non-hungry eating can lead to excess energy intake, weight gain, obesity and related metabolic conditions. Relying on emotional eating as a stress-management strategy also prevents individuals experiencing anxiety disorders from developing other stress-management techniques.

For individuals experiencing more severe anxiety, the related digestive upset and diminished appetite can lead to reduced dietary intake, weight loss and compromised nutritional status. This may further impact on both physical and mental health and compound existing symptoms of difficulty in concentrating and irritability. People experiencing severe levels of anxiety (or depression) should be managed by a psychiatrist, because they are likely to require treatment options beyond lifestyle intervention to effect a rapid improvement in their mental health.

When experiencing an anxiety disorder, individuals are more likely to disengage from their daily activities, including work, study and social commitments. This pattern may be more significant in those experiencing social anxiety disorder or specific phobias, such as a fear of crowds or of open spaces. Avoidance behaviours used to minimise anxiety may lead to social isolation and impact on ability to work. These factors combine to limit the financial, social and physical means of obtaining adequate nutritious foods.

SUBSTANCE USE DISORDERS

Substance use disorders are experienced by 5 per cent of the Australian population each year (AIHW 2012) and commonly co-occur with depression and/or anxiety. The New Zealand Mental Health Survey (Wells 2006) reported a twelve-month prevalence of 3.5 per cent. Substance use disorders may involve the use of alcohol, tobacco or caffeine, or over-the-counter, prescription or illicit drugs. Use of these substances is considered to be a substance use disorder when the use persists despite significant substance-related problems. There are several substance use disorders, with common characteristics of impaired personal control, inappropriate social behaviours, risky use and developed tolerance or withdrawal symptoms. Individuals with substance use disorders consume these substances in larger amounts or for longer time periods than they initially intended, and they may wish to reduce their substance use.

Obtaining and using substances, and recovering from substance use, can consume a large part of the day, and those with a substance use disorder may experience cravings for the substance. Substance use may lead to and continue in spite of problems at work, home or school. Individuals may withdraw from friends, families and social activities and give up occupational or recreational activities due to their substance use. Risky use of the substance may involve use in physically hazardous situations and continued use despite knowing that substance use has caused or exacerbated a persistent or recurrent physical or psychological problem.

Individuals may develop tolerance to the substance and require increasingly higher doses to achieve the desired effect. People experience withdrawal symptoms when blood or tissue concentrations of some substances decline, prompting them to use the substance to reduce the withdrawal symptoms.

Considerations for people with a substance use disorder are similar to those for people with depression and anxiety. In addition, it is important to consider potential nutritional deficiencies that can result from long-term substance use. Substance use may replace the desire for or financial and practical means to access sufficient nutritious food. In the case of alcohol, in addition to replacing food and limiting nutrient intake, it also impairs the body's ability to absorb some nutrients and over the long term can contribute to the risk of CVD and some cancers. One such nutrient deficiency is characterised by Wernicke–Korsakoff syndrome, in which a thiamin (vitamin B1) deficiency leads to poor muscle coordination, impaired memory and damaged nerves. For some individuals, food may be used to satisfy cravings or as a substitute for substance use during recovery. Patients should be assisted to develop other recovery strategies and appropriate long-term eating habits.

OTHER MENTAL DISORDERS

Other mental disorders, including bipolar disorder and schizophrenia, are experienced by a smaller proportion of the population and can have significant impacts on usual social, study and work activities. Like other mental disorders, the onset of these conditions is commonly during adolescence and early adulthood, and there seems to be a strong genetic component that may be 'activated' by environmental factors such as positive or negative life events. Treatment is often guided by a psychiatrist and involves a combination of medication and psychological therapies.

Schizophrenia is characterised by abnormalities in any of five domains, which are delusions, fixed and sometimes irrational beliefs held despite evidence to the contrary; hallucinations, perceiving things that do not exist through seeing, hearing, feeling, smelling or tasting; disorganised thinking or speech; grossly disorganised or abnormal motor behaviour, including catatonia (lack of or limited movement); and negative symptoms such as diminished emotional expression and reduced motivation. Bipolar disorder is characterised by episodes of mania that may be interspersed with depressive episodes. Mania is a persistent abnormally elevated or irritable mood, with goal-directed behaviour. Symptoms of mania include inflated self-esteem or grandiosity, reduced need for sleep, talking more than usual, the feeling that thoughts are racing, ability to be easily distracted, increased goal-directed activity or purposeless activity and participation in risky behaviour.

Management of psychotic disorders including schizophrenia typically involves atypical, or second-generation, antipsychotic medications, which have become more effective in managing psychotic symptoms but often have the undesired side effects of weight gain and metabolic illness, including an increased risk of developing type 2 diabetes mellitus (T2DM). Dietary strategies to employ when working with patients with psychotic illness involve healthy eating nutrition education based on the Australian Dietary Guidelines (NHMRC 2013) or the New Zealand Food and Nutrition Guidelines (Ministry of Health 2003), teaching portion control, providing dietary strategies to increase satiety and minimise energy intake and incorporating regular physical activity to increase energy expenditure and assist with the management of metabolic disease.

EATING DISORDERS

Eating disorders collectively refer to a group of conditions characterised by disturbed eating behaviours. There are several feeding and eating disorders listed in the *DSM-5* (APA 2013). The most common and well-known disorders include anorexia nervosa, bulimia nervosa and binge-eating disorder.

Anorexia nervosa

Anorexia nervosa is characterised by a persistent restriction of energy intake, intense fear of gaining weight or becoming fat and disturbed self-perception of weight or shape. The incidence of anorexia nervosa in males is approximately one-tenth of that in females. Onset tends to occur during adolescence or young adulthood, and rarely after the age of 40. The onset often coincides with stressful life events such as leaving home and is more frequent in those with a family history of anorexia nervosa and in those who demonstrated obsessional traits or anxiety during childhood.

Recovery from anorexia nervosa may take several years, with most individuals experiencing remission within five years. Individuals with extremely low body weight may be admitted to hospital to restore weight and address medical complications. Significant weight loss, fluid and electrolyte imbalance, anaemia, amenorrhea, reduced bone mineral density, altered enzyme and hormone levels, and heart and brain dysfunctions can occur as a result of starvation. Additionally, individuals with anorexia nervosa may complain of cold intolerance, abdominal pain, constipation, lethargy and excess energy.

Bulimia nervosa

Bulimia nervosa is characterised by an undue influence of body weight or shape on self-evaluation and recurrent episodes of **binge eating** with accompanying inappropriate **compensatory behaviours** to prevent weight gain. Like anorexia nervosa, bulimia nervosa is more common among females than males, and onset is typically in adolescence or young adulthood. Binge episodes often begin during or following an attempted weight-loss diet and can also be triggered by stressful life events. Binge eating, with or without compensatory behaviours, often persists for several years with interspersed periods of remission.

In many individuals, symptoms eventually subside even without treatment. Physiological impacts of bulimia nervosa can include menstrual irregularity, fluid and electrolyte disturbances, oesophageal tears, gastric rupture and cardiac arrhythmias.

Binge eating involves eating substantially more food than most individuals would consume under similar circumstances within a discrete period of time, such as a two-hour period, accompanied by a sense of lack of control over the type and amount of food consumed.

Compensatory behaviours are actions taken to prevent weight gain. They may include fasting, excessive exercise, self-induced vomiting and misuse of laxatives, diuretics, enemas or other medications.

Binge-eating disorder

Individuals with binge-eating disorder exhibit the binge-eating behaviours of bulimia nervosa without the accompanying compensatory behaviours. They experience distress about their binge eating and may feel embarrassed, depressed or guilty. Binge-eating disorder also runs in families, and triggers include low mood, stressful events, dieting behaviours, boredom and negative self-evaluation of body weight or shape. Individuals with binge-eating disorder may be normal weight, overweight or obese. Binge eating is not reliably associated with obesity. Symptoms of binge-eating disorder may persist for many years. Eventual improvement is likely both with and without treatment.

TREATING EATING DISORDERS

Each of the eating disorders described above involves a degree of distorted or undue influence of body shape and weight. The conditions may also involve other associated mental disorders such as depression, anxiety or bipolar disorder. Appropriate treatment will be individual and based on the disordered eating practices and/or compensatory behaviours displayed, as well as the underlying triggers for these behaviours.

The ultimate goals of dietary treatment for people with eating disorders are to correct any nutrient deficiencies, restore body weight to minimum healthy levels if underweight (body mass index [BMI] 18.5 for adults or using the appropriate BMI growth charts for children and adolescents) and assist with the development of healthy eating patterns. Appropriate eating patterns will vary for each individual, and while the size, frequency and types of meals will vary, all eating patterns should aim to meet the minimum recommendations (at least) for the core food groups in the Australian Guide to Healthy Eating (NHMRC n.d.) or the New Zealand Food and Nutrition Guidelines (Ministry of Health 2003) and to provide sufficient energy to maintain body weight within a healthy weight range. Eating disorders are complex and psychological in nature, so it is important that treatment involves a team of health professionals including at a minimum a general practitioner or physician and a psychologist or psychiatrist as well as a dietitian.

SUMMARY AND KEY MESSAGES

Nearly half of all Australians and one in seven adults in New Zealand will experience a mental disorder in their lifetime. Mental disorders can have a significant impact on dietary habits, and assisting individuals to develop healthy eating patterns can help to prevent, manage or treat mental disorders.

- Depression, anxiety and substance use disorder are the most common mental disorders experienced by Australians and New Zealanders.
- Symptoms of these disorders can impact on individuals' ability to procure, prepare and consume sufficient nutritious food.
- Following a healthy eating pattern such as the Australian Guide to Healthy Eating (NHMRC n.d.) or the New Zealand Food and Nutrition Guidelines (Ministry of Health 2003) is associated with lower levels of depression and anxiety.
- Eating disorders are complex psychological conditions that require support from a team of health professionals including a dietitian.

REFERENCES

ABS (Australian Bureau of Statistics) 2009. *National Survey of Mental Health and Wellbeing: Summary of results, 2007*. Cat. no. 4326.0. Commonwealth of Australia. <www.ausstats.abs.gov.au/ausstats/subscriber.nsf/0/6AE6DA447F985FC2CA2574EA00122BD6/$File/43260_2007.pdf>, accessed 6 October 2014.

AIHW (Australian Institute of Health and Welfare) 2007. *The Burden of Disease and Injury in Australia 2003*. (Begg, S. et al.) Cat. no. PHE 82. Commonwealth of Australia. <www.aihw.gov.au/WorkArea/DownloadAsset.aspx?id=6442459747>, accessed 6 October 2014.

——2012. *Australia's Health 2012: The thirteenth biennial report of the Australian Institute of Health and Welfare*. Australia's Health Series 13, cat. no. AUS 156. Commonwealth of Australia. <www.aihw.gov.au/WorkArea/DownloadAsset.aspx?id=10737422169>, accessed 6 October 2014.

APA (American Psychiatric Association) 2013. *Diagnostic and Statistic Manual of Mental Disorders* 5th edn (*DSM-5*). APA, Washington, DC.

Forsyth, A.K., Williams, P.G. & Deane, F.P. 2012. Nutrition status of primary care patients with depression and anxiety. *Australian Journal of Primary Health* 18: 172–6.

Jacka, F.N., Kremer, P.J., Leslie, E.R., Berk, M. et al. 2010. Associations between diet quality and depressed mood in adolescents: Results from the Australian Healthy Neighbourhoods Study. *Australian and New Zealand Journal of Psychiatry* 44: 435–42.

Kuczmarski, M.F., Cremer Sees, A., Hotchkiss, L., Cotugna, N. et al. 2010. Higher Healthy Eating Index—2005 scores associated with reduced symptoms of depression in an urban population: Findings from the Healthy Aging in Neighborhoods of Diversity across the Life Span (HANDLS) study. *Journal of the American Dietetic Association* 110: 383–9.

Luppino, F.S., De Wit, L.M., Bouvy, P.F., Stijnen, T. et al. 2010. Overweight, obesity, and depression: A systematic review and meta-analysis of longitudinal studies. *Archives of General Psychiatry* (now called *JAMA Psychiatry*) 67: 220–9.

Ministry of Health 2003. *Food and Nutrition Guidelines for Healthy Adults: A background paper.* New Zealand Government. <www.health.govt.nz/system/files/documents/publications/foodandnutritionguidelines-adults.pdf>, accessed 16 March 2014.

——2012. *The Health of New Zealand Adults 2011/12: Key findings of the New Zealand Health Survey.* New Zealand Government. <www.health.govt.nz/system/files/documents/publications/health-of-new-zealand-adults-2011-12-v2.pdf>, accessed 3 October 2014

NHMRC (National Health and Medical Research Council) 2013. *Eat for Health: Australian Dietary Guidelines; Providing the scientific evidence for healthier Australian diets.* Commonwealth of Australia. <www.eatforhealth.gov.au/sites/default/files/files/the_guidelines/n55_australian_dietary_guidelines.pdf>, accessed 16 March 2014.

——n.d. *Australian Guide to Healthy Eating.* Commonwealth of Australia. <www.eatforhealth.gov.au/guidelines/australian-guide-healthy-eating>, accessed 28 September 2014.

Quirk, S.E., Williams, L.J., Adrienne, O., Pasco, J.A. et al. 2013. The association between diet quality, dietary patterns and depression in adults: a systematic review. *BMC Psychiatry* 13: 175.

Wells, J.E. 2006. Twelve-month prevalence. In Oakley Browne, M.A., Wells, J.E. & Scott, K.M. (eds). *Te Rau Hinengaro: The New Zealand Mental Health Survey.* Wellington: Ministry of Health. <www.health.govt.nz/system/files/documents/publications/mental-health-survey-2006-12-month-prevalence.pdf>, accessed 31 October 2014.

ADDITIONAL READING

Beyondblue (website) 2014. <www.beyondblue.org.au/>, accessed 10 October 2014.

Black Dog Institute (website) 2014. <www.blackdoginstitute.org.au/>, accessed 10 October 2014.

Headspace (website) 2014. <www.headspace.org.au/>, accessed 10 October 2014.

Lifeline (website) 2014. <www.lifeline.org.au/>, accessed 10 October 2014.

Parletta, N., Milte, C.M. & Meyer, B.J. 2013. Nutritional modulation of cognitive function and mental health. *Journal of Nutritional Biochemistry* 24: 725–43.

20

Migrant health

Catherine Itsiopoulos and Antonia Thodis

LEARNING OUTCOMES

Upon completion of this chapter you will be able to:

- identify the factors that impact on the health and wellbeing of migrants to Australia and New Zealand
- list the major migrant groups in Australia and New Zealand
- describe the main nutrition-related health priorities of the major migrant groups in Australia and New Zealand
- discuss the health service needs of migrant groups in Australia and New Zealand.

Australia is considered to be a **culturally and linguistically diverse** community inhabited by rich multicultural populations who have chosen to migrate due to the social and economic opportunities that the country offers. Considered the 'lucky country' by early post–Second World War migrants, it continues to attract newcomers from all over the world. Migrants bring with them their diverse cultures, cuisines and health beliefs, and as they settle, raise their families and age they have specific health needs. Similarly, New Zealand has attracted a diverse migrant population which has shaped its multicultural society. This chapter is an extension of Chapter 2, which describes in detail the cultures, beliefs and food habits of different migrant groups in Australia and New Zealand, and focuses specifically on the health needs of migrants.

Post-colonial migration to Australia dates as far back as the late 1800s. Australia has often been described as a melting pot of different cultures, providing a richness of traditions. Migrants to Australia have shaped its culturally and linguistically diverse community and have created its multicultural cuisine and its society's internationalisation. The general health needs of individuals are common across most cultures, however some cultural differences must be identified and acknowledged as unique and these can impact upon the health of specific cultural groups.

Culturally and linguistically diverse, often shortened to the acronym CALD, is used to describe people who were born overseas and mainly speak a language other than English at home, although there may also be other categories.

The 1950s and 1960s saw a significant wave of migration to Australia by mostly young, work-ready and healthy Europeans from Greece, Italy, Malta, the United Kingdom, Scotland and Ireland, among other countries, who served to populate the country's labour force. Although some may have planned to work and earn a living in the lucky country then return to their homeland, a large majority of European immigrants became Australian citizens and now have become the elderly (elders) of their communities across Australia. The health needs of the ageing migrant population have become a major focus in recent times, due to their special needs.

Since the year 2000 migration patterns have changed, with Australia receiving larger numbers of temporary migrants (for education and short-term employment), primarily from the Asian region (Southeast Asia and South Asia). An important factor in the new pattern is an increased proportion of young adults and young families. The newer migrants will have different health needs from the ageing population of 1950s migrants.

At the end of 2012, the Australian Bureau of Statistics estimated the Australian resident population at 22,906,400, with a growth rate of 1.8 for that year, driven primarily by overseas migration. In June 2013, of the estimated resident population 27.7 per cent (6.4 million people) were born overseas (ABS 2013). For more details on immigration in Australia see Chapter 2.

New Zealand is home to over 200 ethnicities. The influx of migrants poses the country with a lot of challenges, because the experience of migration differs from one ethnicity to another.

INFLUENCES ON HEALTH

Health and wellbeing are significantly affected by **socio-economic factors**. This helps explain the health inequalities that exist within the Australian and New Zealand communities. Generally, disadvantaged people in society will have poorer health and less access to healthcare services. People born overseas who have migrated to Australia or New Zealand are more often affected by socio-economic factors. Language barriers form one of the major factors that affect employability and therefore income.

Socio-economic factors include level of education, type of occupation and level of income, and the interaction between the factors can impact upon a person's health status either positively or negatively.

Australia's immigration survey considers the socio-economic factors that influence overseas-born residents. The surveys for 2011–12 and 2012–13 reported that the unemployment rate for recent migrants was 7 per cent compared with 5 per cent for Australian-born people. Migrants who came from English-speaking countries (New Zealand, United Kingdom and Ireland) were more likely to be employed than those who came from non-English-speaking countries (81 per cent compared with 64 per cent, respectively). According to the Australian Bureau of Statistics, about one-third (35 per cent) of recent migrants reported experiencing some difficulty with finding their first job in Australia. Of the migrants who experienced difficulty:

- 64 per cent reported a lack of Australian work experience or references
- 33 per cent experienced language difficulties
- 23 per cent reported a lack of local contacts or networks
- 15 per cent said there were no jobs in their locality, line of work or at all
- 15 per cent had difficulty with their skills or qualifications not being recognised
- 8 per cent said they did not know how or where to apply for jobs
- 7 per cent had restrictions due to their visa type
- 7 per cent had difficulties with transport or had no driver's licence. (ABS 2011)

There is also a category for 'skilled migrants', of whom in November 2010 an estimated 90 per cent were in the labour force and 87 per cent were currently employed (ABS 2011). In September 2014 New Zealand's unemployment rate was 5.4 per cent, which indicates an increase over the last decade and has been linked to migration (Statistics New Zealand 2014).

The length of residency in New Zealand is a significant factor influencing migrant employment outcomes. Those who have lived in New Zealand for less than five years are more likely to have a different experience to long-term migrants. However, recent estimates indicate that 25 per cent of New Zealand's employed people were born overseas (New Zealand Statistics 2013). Compared with New Zealand-born people, migrants were more likely to have tertiary qualifications (21 per cent compared with 36.7 percent respectively) and this may be attributable to changes in New Zealand's immigration policies since the 1990s, which have selected for highly qualified and skilled migrants. Language difficulties and barriers to employment from overseas qualifications not being recognised in New Zealand do persist.

Unlike the early migrants, who were primarily labourers and had low levels of education, the majority of recent migrants to Australia have arrived with a good level of English and some education. Many come to study in Australia or on a skilled migration visa. Australia is viewed by overseas countries as a destination to obtain a qualification which will assist with future employment. While many migrants (65 per cent) arrive in Australia with non-school qualifications, almost one-third of recent migrants (31 per cent) achieved this qualification after settling in Australia, with half (46 per cent) of these migrants successfully completing at least a bachelor degree (ABS 2011).

Socio-economic status impacts on health in terms of access to health care and the proportion of household expenditure allocated to food. The latter will impact on the diet quality of an individual and ultimately upon health outcomes. The source of household income for migrants is related to visa type, with wages and salary the main income for those on skilled migration visas (92 per cent), and government pensions and allowance the main source for 57 per cent of people arriving on humanitarian visas (ABS 2011).

Male and female migrants from Asia are less likely to drink alcohol at high risk levels, as are females from Southern Europe. However, males and females from the United Kingdom and Ireland are more likely to drink alcohol at high risk levels compared with Australian-born males and females. Migrants from Southern Europe and Asia were both more likely to be inactive compared with Australian-born people (ABS 2013).

HEALTH CONDITIONS

People born overseas are generally in good health on their arrival in Australia because of the rigorous health checks they are required to undertake to be eligible for migration. This is often referred to as the healthy migrant effect, as those who are ill are unlikely or unable to migrate. Studies have shown that for many conditions migrants have better health statuses and use fewer health services than the Australian-born population. They generally have longer life expectancy, lower death rates and hospitalisation rates and lower rates of some chronic diseases (AIHW 2002).

For some health conditions the healthy migrant effect disappears as the length of time or years since migration increases and this has been noted for migrants to Australia and New Zealand. While the healthy migrant effect has been well documented in Australia and New Zealand, the underlying reasons for the decline in health associated with increased years since migration for some migrant populations and not for others are less well known. In Australia, for example, the Sudanese community has witnessed a decline in vitamin D status. People with dark skin require more exposure to sunlight, which is limited and seasonal in Australia. More research is required to identify the effect in different migrant groups, particularly in New Zealand (Hajat et al. 2010).

Overseas-born Australian residents have death rates that are 10–15 per cent lower than Australian-born people. During the 1990s, migrants born in Southern Europe, especially Greece, had one of the lowest all-cause and cardiovascular (CVD) death rates compared with Australian-born people and migrants from other countries (Young 1992). Elderly Greek migrants continue to show low rates of CVD mortality despite more than 60 years since migration (ABS 2011). Studies of food intake patterns of Greek-born migrants over the past 50–60 years have shown that the now elderly Greek-born migrants have retained significant elements of their traditional diet and lifestyle (Kouris-Blazos et al. 1996).

Acculturation, and more specifically **dietary acculturation**, may be contributing in part to the improved all-cause and CVD mortality profile of Greek migants, especially for conditions that have a diet and lifestyle component, such as CVD, type 2 diabetes mellitus (T2DM) and metabolic syndrome (MetS). In cases where migrants arrive in a country with a better health profile than the host country, a resistance to acculturation to the local diet and lifestyle practices may be beneficial or preferred. The Greek migrants to Australian are an example.

Acculturation can have a positive or negative impact on health. For example, in New Zealand, migrants from India have double the risk of being treated for high cholesterol and a four-fold increased risk for T2DM compared with New Zealanders from a European background. With dietary and lifestyle

Acculturation is the process of adjustment and changes that can occur after migration from the country of birth to a new country. Migrants have a choice as to whether they retain their culture of origin or embrace the new culture's beliefs and practices, and this choice can affect both psychological and physical health.

Dietary acculturation occurs when migrants adopt the eating patterns of their new host country. Individuals and groups resist dietary acculturation by striving to retain the eating patterns of their original culture, despite their new surroundings.

acculturation their high risk for T2DM may be reduced and approach the risk levels of the host population. The incidence of diabetes is increasing among Chinese and other Asian communities. Asians born or resident in New Zealand for more than ten years are more likely than recent Asian immigrants to be overweight or obese, drink and smoke (Scragg 2010).

More recent data show that migrants born in Asia have the lowest **death rates** from all causes, 28 per cent less than Australian-born males and 20 per cent less than Australian-born females (AIHW 2012). Female migrants from the United Kingdom and Ireland have higher death rates from breast cancer compared with Australian-born women, and both men and women from the United Kingdom and Ireland have higher death rates from lung cancer, which is associated with a higher prevalence of smoking in these migrants (AIHW 2012).

In New Zealand, between 1991 and 2001, the Asian population increased by 140 per cent. This heterogeneous Asian group is the fastest growing in New Zealand largely due to immigration, and consists of the Chinese, followed by Indians, Filipinos and Koreans (Statistics New Zealand 2014). Asian immigrants are often in better health than the New Zealand-born population on arrival but appear to be losing the health advantage over time. The growth and diversity of the New Zealand Asian population is likely to have implications for the health needs to be met by health services. However, there have been few investigations of Asian health and mortality at the national level; more research has been done on the health inequalities of the three major New Zealand ethnic groups: Māori, Pacific, and non-Māori/non-Pacific populations (Jatrana et al. 2014).

The Middle Eastern, Latin American and African group (MELAA) is also growing fast in New Zealand. Overall, a longer period of residence in New Zealand by migrant groups was associated with an increased likelihood of being overweight and obese, and drinking and smoking (Perumal 2010). Compared with Australian-born residents, those Australians born in Southern Europe, the Pacific Islands, Southeast Asia and South Asia have a higher prevalence of diabetes and diabetes-related deaths; those from South Asia (India), Southeast Asia, the Middle East and North Africa have a higher prevalence of gestational diabetes mellitus (GDM); those from Asia are less likely to be overweight or obese; and those from Southern Europe are more likely to be overweight or obese.

A **death rate** is a measure of the number of deaths (in general, or due to a specific cause) in a particular population, scaled to the size of that population, per unit of time.

A report commissioned by Australia's National Mental Health Commission claimed in 2013 that mental health issues (including suicide) were highly prevalent among migrant communities, although accurate data on prevalence within different community groups and information of risk factors and protective factors was not available (Minas et al. 2013). As many migrant groups are socially disadvantaged due to unemployment and low income, it is not surprising that these groups experience more mental health issues. For refugees and asylum seekers the risk of mental health issues is even greater, due to the major psychological trauma associated with the precipitating circumstances leading to their status and the prolonged detention experienced upon migration.

Generally, countries like Australia and New Zealand do not experience significant rates of infectious diseases because of effective and widespread vaccination programs. However, with the new waves of refugee migrants, health problems that had previously been eradicated are seeing a resurgence. This may be due to inadequate (or absent) vaccinations, severe and prolonged undernutrition leading to vitamin and mineral deficiencies, such as rickets in children through vitamin D deficiency, and dental caries caused by poor diet. Vitamin D deficiency is also prevalent among communities where the women are veiled and therefore not exposed to the sun.

HEALTH SERVICES

Who is responsible for the care of migrants in Australia and New Zealand? In Australia, general practitioners are the first point of contact and are relied upon to perform early health screening and assessment as well as ongoing care. This was improved somewhat by the addition of refugee and humanitarian entrants to the Medical Benefits Schedule to recognise the importance of general practitioners in caring for refugees (Department of Health n.d.). In New

Australia

- Mental Health in Multicultural Australia (<www.mhima.org.au>) focuses on building networks and collaborative partnerships to develop mental health reform strategies for Australia's culturally and linguistically diverse communities.
- The Centre for Culture, Ethnicity and Health (<www.ceh.org.au>) was set up in 1993 and advises healthcare institutions how to improve access to health services for culturally and linguistically diverse communities, given the recognition that migrant health services must be socially and culturally inclusive to meet the needs for equal access and making health care accessible to everyone despite their background or country of birth. It is important to educate staff not to make assumptions about the cultural backgrounds of individuals.

New Zealand

- The Migrant Action Trust (<www.migrantactiontrust.org.nz/>) assists migrants and refugees to settle in New Zealand, aiming to offer support throughout the often difficult time of migration, from people's arrival in New Zealand until they become fully integrated into New Zealand society.

Zealand, publicly funded services (hospital, general practitioner, medications in hospital, breast screening) are available to migrants who become New Zealand citizens, Australian citizens, those holding a work visa, scholarship and aid program recipients, and refugees and protected persons (Ministry of Health 2014).

Most recent migrants and a large proportion of longer settled migrants speak a language other than English at home. For longer settled migrants, Mandarin, Cantonese, Italian and Vietnamese were the most common languages other than English spoken at home, while for recent immigrants, Mandarin, Punjabi, Hindi and Arabic were most frequently spoken at home (ABS 2012). In addition to language, each different migrant group has cultural, religious and societal factors that impact health and wellbeing (as described in Chapter 2), and the challenge for health service providers is to ensure their services are accessible to the different migrant groups so that they are not disadvantaged.

Critics argue that health institutions are not meeting the needs of diverse multicultural groups because there is a lack of data on their health requirements. They claim that most of the research on health needs is focused on the English-speaking population and non-English speaking populations are often excluded (Minas et al. 2013).

Access to healthcare services is generally lower for all migrant groups, according to national hospitalisation data (AIHW 2002). For migrants in good health there may be less need for health services; however, for many socio-economically disadvantaged groups access to health care is poor. This may be due to an inability to afford private health care and the need to wait for services in the public health system or a lack of culturally appropriate services to meet the needs of specific migrant groups. Some of the issues relating to appropriate healthcare services for migrant groups are discussed below.

Barriers

Migrants from non-English-speaking countries are likely to have more difficulty accessing Australian and New Zealand health services. The key issues that affect health care in a multi-ethnic setting include lack of knowledge about the available services, language barriers and varying cultural attitudes to health and interaction with health professionals.

The healthcare system will likely be foreign to new migrants, and past experiences with their birth country's perhaps substandard system may impact upon their perception of health professionals. Therefore, health service providers need to be mindful of cultural backgrounds and the sensitivities of their patients born overseas. The perceptions of health and behaviour when individuals experience a medical issue are highly influenced by the cultural background.

Migrants who have limited English skills may be fearful presenting to their general or allied health practitioner and may not be able to explain symptoms or understand medical or technical jargon. They may feel less empowered and may be less likely to ask questions, or they may reply only yes or no to questions to avoid having further dialogue. This can impact on the accuracy of information that is obtained from a consultation, may reduce understanding of treatment and compliance to treatment, may discourage regular visits and may worsen the health of migrants. Interpreters are very important in the healthcare system, as are bilingual and multilingual health professionals.

Medicare is the Australian healthcare scheme funded by the federal government for all Australian citizens (irrespective of private health insurance) and some overseas visitors, which enables access to health services.

Limited knowledge of the health system may result in migrants delaying a visit to the doctor. Access to **Medicare** and hence free health care is available to permanent residents, those awaiting processing of their permanent residency claims and foreign citizens with **mutual care agreements** with Australia. However, migrants with student visas and tourists are required to pay the costs of treatment.

Limited research into migrant health in Australia and the underrepresentation of migrants can contribute to undermining equity in healthcare provision. The additional language or cultural barriers faced by migrants that are not experienced by English-speaking migrants or Australian-born patients create an obstacle to positive health outcomes. For older migrants, lack of engagement with the healthcare system may impact on the ability to self-manage chronic conditions such as heart disease and diabetes, and in this case effective communication, through bilingual health professionals or access to interpreters, is crucial to effective care.

Mutual care agreements are also known as reciprocal health agreements and exist between some countries, such as New Zealand and Australia, to allow individuals travelling overseas to access specific publicly funded healthcare services to the same extent as a national of the country visited. However, agreements do not provide full coverage, and travel and health insurance are still recommended.

Best practice model

A best-practice model for delivery of health services to migrant groups remains elusive, however, the Australian government's approaches to health care for migrants has been used as a model. The approach is based on a division of responsibility between the states and the federal government. Each state has its own arrangements for migrant care which recognise that not all migrants are the same or have the same needs. Access to health services is varied and may include standard health services such as community health centres or general practitioners or more specialised services, for example, services for refugees who have experienced torture or hardship brought about by war or dislocation. The current debate surrounding the admission of refugees to Australia continues with no definitive solutions but it is important to note that, on arrival, refugees are entitled to free medical assessment, unlike asylum seekers.

SUMMARY AND KEY MESSAGES

Australia and New Zealand can be described as culturally and linguistically diverse, given that a large proportion of residents in both countries were born overseas. The general health needs of individuals are common across cultures. However, the diversity of beliefs and traditions that immigrants bring with them to their host country means that there are factors at play, including socio-economic factors, that are unique within their specific cultures and can impact upon health needs and health status. These factors must be identified and acknowledged.

- Migrants to Australia and New Zealand generally have 10–15 per cent lower death rates from all causes than people born in those countries.
- Common morbidities of migrant groups include type 2 diabetes mellitus (T2DM), GDM, overweight and obesity and mental health issues.
- Common lifestyle risk factors in migrants include high-risk alcohol intake and physical inactivity.
- Migrants are more socio-economically disadvantaged due to language barriers, lower education, unemployment and lower income.
- Migrants have difficulty accessing health services due to language barriers and culturally inappropriate services.

REFERENCES

ABS (Australian Bureau of Statistics) 2011. *Characteristics of Recent Migrants, Australia*. Cat. no. 6250.0. Commonwealth of Australia. <www.ausstats.abs.gov.au/ausstats/subscriber.nsf/0/5DC8BB78350AC6FCCA25789900192D24/$File/62500_Nov%202010replacement.pdf>, accessed 10 October 2014.

——2012. *Reflecting a Nation: Stories from the 2011 census, 2012–2013*. Cat. no. 2071.0. Commonwealth of Australia. <www.abs.gov.au/ausstats/abs@.nsf/Lookup/2071.0main+features902012-2013>, accessed 10 October 2014.

——2013. *Australian Demographic Statistics, Dec 2012*. Cat. no. 3101.0. Commonwealth of Australia. <www.abs.gov.au/ausstats/abs@.nsf/Previousproducts/3101.0Main%20Features3Dec%202012?opendocument&tabname=Summary&prodno=3101.0&issue=Dec%202012&num=&view=>, accessed 10 October 2014.

AIHW (Australian Institute of Health and Welfare) 2002. *Australian Health Inequalities: 1 Birthplace*. (Singh, M. & de Looper, M.) Bulletin 2, cat. no. AUS 27. Commonwealth of Australia. <www.aihw.gov.au/WorkArea/DownloadAsset.aspx?id=6442453153>, accessed 6 October 2014.

——2012. *Australia's Health 2012: The thirteenth biennial report of the Australian Institute of Health and Welfare*. Australia's Health Series 13, cat. no. AUS 156. Commonwealth of Australia. <www.aihw.gov.au/WorkArea/DownloadAsset.aspx?id=10737422169>, accessed 6 October 2014.

Department of Health n.d. *Medicare Benefits Schedule Item 707*. Commonwealth of Australia. <www9.health.gov.au/mbs/fullDisplay.cfm?type=item&qt=ItemID&q=707>, accessed April 2014.

Hajat, A., Blakely, T., Dayal, S. & Jatrana, S. 2010. Do New Zealand's immigrants have a mortality advantage? Evidence from the New Zealand census—Mortality study. *Ethnicity and Health* 15: 531–47.

Jatrana, S., Richardson, K., Blakely, T. & Dayal, S. 2014. Does mortality vary between Asian subgroups in New Zealand: An application of hierarchical Bayesian modelling. *PLoS ONE* 9(8). <www.plosone.org/article/info%3Adoi%2F10.1371%2Fjournal.pone.0105141>, accessed 17 November 2014.

Kouris-Blazos, A., Wahlqvist, M.L., Trichopoulou, A., Polychronopoulos, E. et al. 1996. Health and nutritional status of elderly Greek migrants to Melbourne, Australia. *Age Ageing* 25: 177–89.

Minas, H., Kakuma, R., Too, L.S., Vayani, H. et al. 2013. *Mental Health Research and Evaluation in Multicultural Australia: Developing a culture of inclusion*. National Mental Health Commission, Commonwealth of Australia. <www.mentalhealthcommission.gov.au/media/80646/2093%20MHiMA%20CALD%20REPORT_06.pdf>, accessed 6 October 2014.

Ministry of Health 2014. *A Guide to Eligibility for Publicly Funded Health Services*. New Zealand Government. <www.health.govt.nz/new-zealand-health-system/eligibility-publicly-funded-health-services/guide-eligibility-publicly-funded-health-services-0>, accessed April 2014.

Perumal, L. 2010. *Health Needs Assessment of Middle Eastern, Latin American and African People Living in the Auckland Region*. Auckland: Auckland District Health Board. <www.adhb.govt.nz/healthneeds/Document/MELAAHealthNeedsAssessment.pdf>, accessed 9 November 2014.

Scragg, R. 2010. *Asian Health in Aotearoa in 2006–2007: Trends since 2002–2003*. Auckland: Northern DHB Support Agency. <www.asianhealth.govt.nz/Publications/Asian%20Health%20Trends%20Scragg%202010.pdf>, accessed 9 November 2014.

Statistics New Zealand 2013. *Employed Migrants in New Zealand Report High-quality Working Life*. New Zealand Government. <www.stats.govt.nz/browse_for_stats/income-and-work/employment_and_unemployment/snapshot-working-migrants.aspx>, accessed 17 November 2014.

——2014. *Household Labour Force Survey: September 2014 quarter*. New Zealand Government. <www.stats.govt.nz/browse_for_stats/income-and-work/employment_and_unemployment/HouseholdLabourForceSurvey_HOTPSep14qtr.aspx>, accessed 17 November 2014.

Young, C. 1992. Mortality, the ultimate indicator of survival: The differential experienced between birthplace groups. In Donovan, J. (ed.). *Immigrants in Australia: A health profile*. Canberra, Australian Government Publishing Service.

APPENDICES

APPENDIX 1

Clinical signs and symptoms of nutritional deficiencies

Antigone Kouris-Blazos

Assessment of nutritional status is the first step in planning and evaluating the nutritional care of individuals or groups and is determined on the basis of several different kinds of information that characterise different stages in the development of a nutritional problem.

A HOLISTIC APPROACH

Assessment of nutritional status and ultimately a nutritional diagnosis require a holistic approach because they should take into consideration the following:

- *Medical history and genetic predisposition*
- *Socio-demographic circumstances*—e.g., living arrangements, finances, exercise
- *Psychosocial circumstances*—e.g., stress, mood, anxiety, depression, grieving
- *Dietary history*—e.g., food distribution, cooking, eating out, cuisine/culture, food variety, nutrient intake
- *Nutritional blood and urine tests*—e.g., plasma zinc (ideally above 12 micromoles per litre); blood levels of magnesium pick up only severe deficiency, so clinical symptoms can be more useful for identifying borderline deficiency; serum B12 (ideally above 300 picomoles per litre); urine iodine (ideally above 100 micrograms per litre); serum selenium; vitamin D (ideally above 75 nanomoles per litre); iron (ferritin or iron stores ideally above 30 nanograms per millilitre but may be artificially raised due to chronic infection/disease)
- *Anthropometry*—e.g., weight history, fat distribution and composition
- *Clinical assessment*—e.g., assess skin, hair, nails, appetite, dental issues, tongue, headaches, mood, muscle pain, joint pain, cramps, sugar/salt cravings, fatigue, menstruation, sleep; bowel motions, in particular, can provide important information about nutritional status and problems with digestion and malabsorption of nutrients
- *Medications and potential adverse effects on nutritional status*—see Appendix 2
- *Vitamin/herbal supplement use and potential adverse or beneficial effects and interactions with medications*—see Appendix 2.

These assessments together are necessary to detect clinical (overt) and subclinical (borderline) nutrient deficiencies. The most important part of a clinical assessment is the socio-demographic, medical and dietary information obtained about the individual that will help identify the patient at risk of nutritional deficiencies. This is history-taking.

RISK FACTORS FOR NUTRITIONAL DEFICIENCIES

An individual can have an excellent diet and yet experience nutritional deficiencies if what they are eating is not properly digested and absorbed; that is, 'you are what you absorb'. Therefore, patients at highest risk of deficiency will be patients experiencing problems with their stomach, such as **hypochlorhydria**. Also, patients with bowel problems or altered bowel motions

> **Hypochlorhydria** refers to states where the production of gastric acid in the stomach is low. It is associated with various other medical problems.

will also be at high risk of nutritional deficiencies. Below is a list of risk factors for nutritional deficiency:

- *Digestive problems*—hypochlorhydria, **coeliac disease**, **Crohn's disease**, thyroid disease
- *Being elderly*—hypochlorhydria more common, polypharmacy, especially in those who are disabled or chronically ill
- *Medications that affect absorption, metabolism or excretion of nutrients*—see Appendix 2
- *Chronic diseases*—e.g., diabetes can increase the excretion of zinc, magnesium, chromium; and conditions that affect food intake—e.g., chronic fatigue syndrome
- *Allergies or sensitivities to particular foods*
- *'Crash dieting' or chronic low-energy diets*
- *Poor or altered food intake*—e.g., due to restricted food budget, food supply or food storage, rising food prices, poor cooking skills
- *Vegetarianism*
- *Sedentary lifestyles*—especially with limited sun exposure
- *Pregnancy and breastfeeding, being female and having young children and excessively bleeding during menstruation*
- *Drinking more than one to two (women) or more than two to four (men) standard alcoholic drinks a day*
- *Smoking cigarettes and using illicit drugs.*

This information provides clues to the likely natures of nutritional problems and why they occur. Symptoms (manifestations reported by the individual) and signs (observations made by a qualified examiner, such as a medical practitioner or dietitian) of nutrition-related conditions can be valuable aids in the detection of nutritional problems. However, because they often occur late in the development of the nutritional problem, a diagnosis of a nutritional deficiency cannot usually be made solely on the basis of a clinical examination. Many nutrition-related signs and symptoms are non-specific and occur for non-nutritional reasons, so these need to be considered first. Usually, the presence of a group of related clinical signs and symptoms is a better indication than a single sign or symptom. The tissues with the fastest turnover rates are the most likely to show signs of nutrient deficiencies or excesses, such as the hair, skin and tongue (an indirect reflection of the status of the villi of the gut).

Coeliac disease is an autoimmune disorder of the small intestine that occurs in genetically predisposed people of all ages from middle infancy onward. It is caused by exposure to gluten (a protein in wheat) and similar proteins in other cereals such as barley, rye and oats. Exposure to gluten causes inflammation in the bowel, leading to damage of the epithelium of the bowel.

Crohn's disease is an inflammatory bowel disease caused by a combination of environmental, immune and bacterial factors in genetically susceptible individuals. The inflammationcan occure in any part of the bowel from the mouth to the anus and can lead to severe abdominal pain, diarrhoea, bowel obstruction and weight loss.

GOOD NUTRITIONAL STATUS

Good nutritional status begins in the stomach with hydrochloric acid. Adequate hydrochloric acid is needed to extract nutrients from food. The secretion of the acid from the parietal cells in the stomach is one of the first in a sequence of events that must occur for digestion and nutrient absorption to take place. Optimal digestion and absorption of nutrients takes place only when digestive hormones, hydrochloric acid, enzymes and bile are released at just the right times and in the right amounts to properly prepare food for each phase of digestion. Hydrochloric acid also destroys harmful bacteria and prevents the overgrowth of bacteria, yeast, fungi and parasites in the digestive tract.

The lining of the stomach consists of mucus comprising 95 per cent water as well as bicarbonate to protect it from the hydrochloric acid. This lining is replaced every three days and acts as a barrier against hydrochloric acid and other stomach secretions.

Zinc is essential for the production of hydrochloric acid because it is a cofactor for the enzyme carbonic anhydrase that provides the ionic form of hydrogen (H^+) for hydrochloric acid. Zinc deficiency can therefore adversely affect the first stage of digestion. Zinc is also needed for the maintenance of a healthy stomach and intestinal mucosal lining (which is lost every two weeks), so a zinc deficiency can adversely affect later stages of digestion and absorption as well. A zinc deficiency can flag overall poor nutritional status due to compromised overall digestion and absorption.

The pH of the stomach needs to reach around 2 for the proper breakdown of nutrients, especially minerals, vitamin B12, folate, fat-soluble vitamins and protein. Stomach pH is usually at its lowest when the stomach is empty; it is maintained between 1 and 3 prior to ingestion of food and rises thereafter. As food is ingested, the stretching of the stomach lining signals the secretion of gastrin, which regulates the release of hydrochloric acid (from parietal cells) and pepsinogen. Pepsinogen is then converted to the enzyme pepsin, which in turn breaks down protein into peptides and amino acids. Gastrin also stimulates the production of stomach mucosa and stimulates the stomach muscles to contract and churn, to mix the food.

Hydrochloric acid provides the 'acid trigger' to the digestive cascade. When the stomach contents have been churned and bathed in the acid, the acidified chyme (partially digested food) sends a signal to the pyloric sphincter that the chyme is ready to enter the small intestine for the final phases of digestion. The acid chyme is released into the duodenum and comes into contact with the lining of the gut, which stimulates the release of the hormones secretin and cholecystokinin. Secretin signals the pancreas to discharge bicarbonate ions to increase the alkalinity and pancreatic enzymes (which only work in an alkaline environment) to finish off the digestive process. Cholecystokinin activates the liver to release bile for fat digestion.

Low hydrochloric acid, or hypochlorhydria, is commonly seen with advancing age (from 40, but with significant declines after 60) but also with hypothyroidism, gastritis, *Helicobacter pylori* infection and zinc deficiency, and especially in patients who have been on long-term reflux medication and antacids. Many studies have pointed to impaired acid secretion in relation to increased age. This relationship is mainly seen in people with gastrointestinal symptoms. According to a report by Segal and Samloff (1973) of 1590 patients, the incidence of achlorhydria was 19 per cent in the fifth decade of life and 69 per cent in the eighth decade of life. The increased rate of achlorhydria was also associated with a rise in the frequency of gastric cancer. These findings may be explained by the higher prevalence of *H. pylori* infection in older individuals. Similarly, a study from Denmark by Christiansen (1968) on healthy people showed that the incidence of achlorhydria in patients increased rapidly from 1.8 per cent in the fifth decade to 18.5 per cent in the eighth decade. A consensus conference on hypochlorhydria in the elderly also reported high rates (20 per cent) in people over the age of 70 years (Holt et al. 1989). The rising levels of pH due to age continue to trigger specialised stomach cells to release more and more gastrin, to stimulate parietal cells to produce greater amounts of hydrochloric acid, resulting in hypergastrinaemia and an increased risk of stomach and bowel cancer.

In hypochlorhydria, the elevated pH delays gastric emptying and causes the lower oesophageal sphincter to open, resulting in food rising up into the oesophagus and causing the symptoms of gastro-oesophageal reflux. About 15 per cent of cases of gastro-oesophageal reflux disease are caused by low hydrochloric acid. Elevated pH in the stomach prevents the solubilisation of minerals (especially iron and calcium), fat-soluble vitamins, thiamin and folate and prevents the break-down of protein and release of vitamin B12 (which is attached to protein). It can cause bacterial overgrowth in the stomach and small bowel (also called small intestinal bowel overgrowth), contributing to bad breath, belching and stomach discomfort after eating. Small intestinal bowel overgrowth is the most common cause of malabsorption among older adults. Competition between bacteria and the human host for ingested nutrients leads to malabsorption and considerable morbidity due to micronutrient deficiency.

The elevated pH also causes the pyloric sphincter to open and release partially digested food into the small intestine prematurely. Since the chyme exiting the stomach is now at a higher pH, the pancreas and liver will not be properly stimulated to release digestive enzymes and bile (also known as pancreatic/bile insufficiency). All of these secretions are critical to the complete digestion process that allows food to be fully broken down into its smallest components so that nutrients can be properly absorbed in the small intestine and utilised by the body. As a result, food transits through the bowel at a much faster rate, resulting in looser pale or green-coloured stools which may float (due to fat malabsorption) and have an offensive smell (see Table A1.1).

Partially digested food entering the colon will stimulate bacterial fermentation, causing bloating, wind, offensive odours and potentially an imbalance of the bowel flora, favouring pathogenic over beneficial bacteria. Also, large particles of undigested proteins can make their way into the blood stream, causing health problems ranging from allergies to autoimmune diseases. Correcting a zinc deficiency may help improve hydrochloric acid levels and digestion, and consumption of acidic foods (lemon or cranberry juice) may also facilitate digestion. The bacterial overgrowth may need to be treated with antibiotics, probiotics and antimicrobial foods—for example, garlic, oregano and thyme.

Interpreting stools

The following stool chart (Table A1.1) is commonly used in practice to assist with the diagnosis of altered bowel motions.

Table A1.1 Bristol Stool Chart and assessments

Type	Appearance	Description	Assessment
1	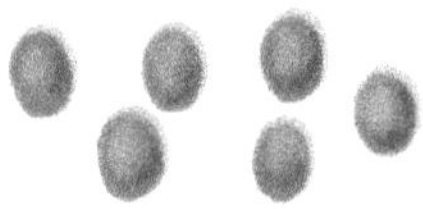	Separate hard lumps, like nuts (hard to pass)	Slower transit time—e.g., too little fibre, dysbiosis/poor bowel flora, hypothyroid, iron supplements, food intolerance
2	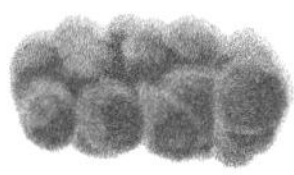	Sausage shaped but lumpy	
3		Like a sausage but with cracks on the surface	
4		Like a sausage or snake, smooth and soft	Normal
5		Soft blobs with clear-cut edges	Faster transit time—e.g., fibre, dysbiosis/poor bowel flora, food intolerance, hypochlorhydria, pancreatic/bile insufficiency, malabsorption of minerals/fat-soluble vitamins, sign of zinc/vitamin B12 deficiency
6	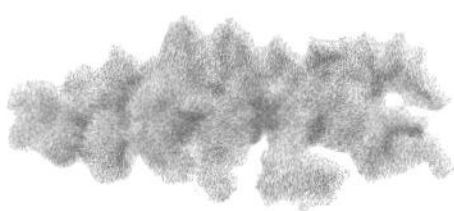	Fluffy pieces with ragged edges, a mushy stool	
7		Watery, no solid pieces, entirely liquid	

Source: Adapted from Heaton & Lewis (1997).

Interpretations of different stool colours are given below:

- *Light brown*—Normal healthy stool
- *Yellow, floating, greasy, foul smelling*—Faster transit time, commonly caused by hypochlorhydria. Undigested fat in the stools, pancreatic and/or bile insufficiency (potential malabsorption of fat-soluble vitamins/minerals)
- *Green*—Rapid transit time, dysbiosis/poor bowel flora
- *Grey or clay*—Faster transit time, commonly caused by hypochlorhydria. Undigested fat in the stools, pancreatic and/or bile insufficiency (potential malabsorption of fat-soluble vitamins/minerals)
- *Dark red–maroon*—Intestinal bleeding in the colon
- *Black, tarry, foul smelling*—Intestinal bleeding higher up in the gastrointestinal tract (stomach or upper small intestine).

NUTRITIONAL DIAGNOSIS AND THERAPY

Table A1.2 describes groups of signs associated with different kinds of nutrient imbalances. A holistic assessment will enable the practitioner to make a nutritional diagnosis that in turn will inform the most appropriate short-term nutritional therapy and diet therapy (which may need to be continued long term). Short-term nutritional therapy may involve the use of supplements, fortified foods and drinks or specific nutrient-dense foods to correct the deficiency. If the deficiency is marked, a supplement may be the best therapy to provide quick relief of symptoms.

Long-term dietary therapy needs to be implemented at the same time as the nutritional therapy and may need to continue until there is resolution of symptoms—for example, the low **FODMAP** diet. Once symptoms have been resolved, the patient should be advised of an appropriate long-term diet plan for their condition and genetic profile in order to maintain corrected nutrient levels. If the patient is unable to maintain the diet plan, the practitioner needs to consider the continued use of supplements.

FODMAPs are a collection of poorly absorbed, short-chain carbohydrates that occur naturally in many foods. The acronym FODMAP describes: the **O**ligosaccharides fructans and galacto-oligosaccharides (GOS) present in foods such as wheat, onions, garlic and legumes; the **D**isaccharide lactose, present in mammal milk and some milk products; the **M**onosaccharide fructose (when consumed in excess of glucose), present in honey, apples, pears and high fructose corn syrup; **A**nd **P**olyols, including sorbitol and mannitol, present in apples, pears, stone fruit and many artificially sweetened gums and confectionary (Shepherd et al. 2013).

Table A1.2 Clinical signs and symptoms and corresponding possible nutritional deficiency

Signs and symptoms	Nutrient that may be inadequate	Other possible causes
General		
Poor growth	Protein, energy, iron, zinc, iodine, vitamin D, calcium, essential fatty acids, vitamin A	
Poor immunity	Zinc, selenium, vitamin C, vitamin D, vitamin E	
Poor appetite	Zinc, magnesium, iron, thiamin, niacin, folate, vitamin B12	
Fatigued, irritable	Protein, zinc, iron, chromium, pyridoxine, vitamin B12, folate, vitamin C	
Moody, depressed	Protein, zinc, iodine, magnesium, thiamin, niacin, pyridoxine, vitamin B12, vitamin C, vitamin D	
Migraine, headache	Riboflavin, vitamin B complex, magnesium	
Sleep disturbance	Magnesium, calcium, zinc, iron, pyridoxine, vitamin C, vitamin D	
Poor dream recall	Pyridoxine	
Impaired memory/cognition	Iron, zinc, iodine, thiamin, niacin, vitamin B12, folate	
Bone/joint pain	Calcium , zinc, iodine, boron, vitamin D, vitamin C, essential fatty acids	Inflammation caused by bacteria
Irregular periods	Iodine, iron, zinc, vitamin E, essential fatty acids	Thyroid
Mastalgia	Vitamin A, vitamin E, essential fatty acids	
Premenstrual syndrome, dysmenorrhoea, menorrhagia	Pyridoxine, calcium, magnesium, essential fatty acids	

continues

Table A1.2 Clinical signs and symptoms and corresponding possible nutritional deficiency *continued*

Signs and symptoms	Nutrient that may be inadequate	Other possible causes
Hair		
Hair loss, dry/brittle hair, slow hair growth	Protein, zinc, iron, iodine, riboflavin, biotin, essential fatty acids, vitamin A	Thyroid, excess vitamin A
Prematurely greying hair	Copper, biotin, vitamin B12	
Dry, flaking scalp	Zinc, magnesium, selenium, biotin, vitamin A	
Eyes		
Dark under-eye circles	Iron	Allergy, poor/inadequate sleep
Pale conjunctiva	Iron	
Impaired night vision	Zinc, vitamin A	Ageing
Twitching eye lid, facial spasms	Magnesium, calcium	Poor/inadequate sleep
Nose, taste, tongue, mouth		
Poor smell (anosmia)	Zinc, vitamin A	Sinusitis
Impaired taste	Zinc, vitamin A, iron	
Pale tongue	Iron	
Magenta/blue tongue	Riboflavin, biotin	
Large/swollen pale tongue	Iodine, iron	Hypothyroid
Smooth, bright-red tongue	Riboflavin, pyridoxine, vitamin B12, iron, folate, biotin	
Raw, painful, dark-red tongue	Niacin, vitamin B12, folate	
Berry-like red tongue	Vitamin B complex	
Cherry-tip tongue	Niacin, pyridoxine	
Burning sensation in mouth/throat	Vitamin D	
Lip cheilosis (burning/soreness)	Riboflavin, niacin, pyridoxine	
Angular stomatitis	Iron, vitamin B complex	
Bleeding gums	Riboflavin, vitamin C, vitamin K	Gingivitis, poor dental hygiene
Nails		
Vertical corrugations (pronounced)	Protein, B complex zinc, iron	
Pronounced central ridge	Protein, iron, folate	
Horizontal grooves	Protein, zinc, selenium, calcium	Past severe illness
Leukonychia (white spots)	Zinc	
White half-moon at base	Pyridoxine	
Dry, thin, brittle	Protein, essential fatty acids, iron, calcium	
Peeling, splitting	Protein, calcium	
Spoon shaped, brittle	Iron	
Soft, bendy,	Protein, zinc	
Growth arrest, thickened	Protein, zinc	
Yellow	Vitamin E	
Egg shell	Vitamin A	
Stools[a]		
Diarrhoea	Zinc, iron, niacin, vitamin B12, folate, biotin	Excess magnesium, excess vitamin C, dysbiosis/poor bowel flora
Constipation	Water, dietary fibre, iodine iron, niacin, dysbiosis/poor bowel flora	Thyroid
Mucus in stool		Bowel inflammation (inflammatory bowel disease)

Signs and symptoms	Nutrient that may be inadequate	Other possible causes
Muscles		
Muscle pain/ache	Calcium, magnesium, potassium, selenium, thiamin, vitamin C, vitamin E, vitamin D, coenzyme Q10	
Cramps	Calcium, magnesium, iron, potassium, sodium, vitamin D, vitamin C	
Twitching, spasms, restless legs	Calcium, magnesium, iron, potassium, pyridoxine, vitamin D	
Calf muscle tenderness	Vitamin B complex, vitamin E	
Weakness, wasting	Protein, thiamin, vitamin D, potassium, calcium	
Decreased muscle reflexes	Vitamin B complex, vitamin E	
Skin		
Excessive ageing, wrinkles	Vitamin E, essential fatty acids, iodine, selenium	Sun damage
Perifollicular hyperkeratosis (toad skin)	Zinc, vitamin A, vitamin B complex, vitamin C, essential fatty acids	
Shark skin (dyssebacia)	Riboflavin	
Oily, scaly, seborrhoeic dermatitis (nasolabial folds, eyebrows, forehead)	Riboflavin, niacin, pyridoxine, biotin, copper, essential fatty acids	
Dry, scaly, coarse, itchy dermatitis	Zinc, iodine, niacin, essential fatty acids, biotin, vitamin A, vitamin C, vitamin E	
Dry, fish scale/flaky paint, especially on the legs	Vitamin A, zinc	Hypothyroid
Hyperpigmented, non-scaly macules (patches)	Vitamin A, zinc, vitamin C, niacin, pyridoxine, vitamin B12, folate, essential fatty acids	Insulin resistance
Hyperpigmented, scaly dermal patches on face/limbs (pellagra)	Thiamin, niacin, biotin, zinc, essential fatty acids	
Liver spots (ceroid accumulation)	Vitamin E	
Eczema	Biotin, zinc, essential fatty acids	
Psoriasis	Vitamin D, vitamin A, essential fatty acids, zinc	
Acne	Essential fatty acids, zinc, vitamin A, vitamin C	
Poor wound healing	Protein, essential fatty acids, vitamin A, pyridoxine	
Nerves		
Impaired coordination/balance, disorientation, ataxic gait	Thiamin, vitamin B12 (neurological changes can occur without haematologic changes), niacin, vitamin E	
Neuropathy (weakness, ataxia, pins and needles, paraesthesia), foot/wrist drop, reduced tendon reflexes, numbness	Thiamin, riboflavin, pyridoxine, vitamin B12, carnitine, folate, magnesium, calcium, potassium, essential fatty acids, chromium, iron, vitamin E	
Postural hypotension	Vitamin B6, vitamin B complex, iron	

a See also Table A1.1.

Note: Some of these signs and symptoms can occur for non-nutritional reasons, which need to be ruled out first.

Source: Wahlqvist & Kouris-Blazos (2011).

REFERENCES

Christiansen, P.M. 1968. The incidence of achlorhydria and hypochlorhydria in healthy subjects and patients with gastrointestinal diseases. *Scandinavian Journal of Gastroenterology* 3(5): 497–508.

Heaton, K.W. & Lewis, S.J. 1997. Stool form scale as a useful guide to intestinal transit time. *Scandinavian Journal of Gastroenterology* 32: 920–4.

Holt, P.R., Rosenberg, I.H. & Russell, R.M. 1989. Causes and consequences of hypochlorhydria in the elderly. *Digestive Diseases and Sciences* 34: 933–7.

Segal, H.L. & Samloff, I.M. 1973. Gastric cancer: Increased frequency in patients with achlorhydria. *American Journal of Digestive Diseases* 18: 295–9.

Shepherd, S.J., Lomer, M.C.E. & Gibson, P.R. 2013. Short-chain carbohydrate and functional gastrointestinal disorders. *American Journal of Gastroenterology* 108: 707–17.

Wahlqvist, M. & Kouris-Blazos, A. 2011. Nutrition assessment and monitoring. In Wahlqvist, M. (ed.). *Food and Nutrition: Food and health systems in Australia and New Zealand*, 683–709. Allen & Unwin, Sydney.

ADDITIONAL READING

Gibson, R. 2005. *Principles of Nutritional Assessment* 2nd edn. Oxford University Press, New York.

Geobel, L. & Yousef, G.M. 2014. Hidden clues to diagnosing nutritional deficiencies. *Medscape*. <http://reference.medscape.com/features/slideshow/nutrition-def>, accessed 12 November 2014.

Heimburger, D.C. & Ard, J.D. 2006. *Handbook of Clinical Nutrition* 4th edn. Elsevier, Philadelphia, PA.

Kohli, R.D. & Katz, J. 2013. Achlorhydria. *Medscape*. <http://emedicine.medscape.com/article/170066-overview>, accessed 25 August 2013.

Kotsirilos, V., Vitetta, L., Sali, A. & Kouris-Blazos, A. 2011. Nutritional assessment and therapies. In Kotsirilos, V. et al. *A Guide to Evidence-based Integrative and Complementary Medicine*, 14–47. Elsevier, Sydney.

Kouris-Blazos, A. 2011. *Food Sources of Nutrients*. Lulu, Raleigh, NC.

McLaren, D.S. 1992. *A Colour Atlas and Text of Diet-related Disorders* 2nd edn, Wolfe & Mosby—Year Book Europe Ltd, London.

——1999. Clinical manifestations of human vitamin and mineral disorders: A resume. In Shils, M., Olson, J.A., Shike, M. & Ross, C. *Modern Nutrition in Health and Disease* 9th edn, 485–503. Williams & Wilkins, Baltimore, MD.

Newton, M.J. & Halsted, C.H. 1999. Clinical and functional assessment of adults. In Shils, M., Olson, J.A., Shike, M. & Ross, C. *Modern Nutrition in Health and Disease* 9th edn, 895–902. Williams & Wilkins, Baltimore, MD.

APPENDIX 2

Drug interactions with foods, herbs and nutrients

Antigone Kouris-Blazos

Pharmaceuticals have both beneficial and adverse effects, although there tends to be a focus on the benefits. Furthermore, drug–drug interactions are generally integral to decision-making by medical practitioners, yet the impacts of drug–food/herb and drug–nutrient interactions are rarely acknowledged or mostly deemed clinically insignificant. Even though an individual may have a good diet providing adequate amounts of vitamins and minerals, they may be tipped into nutritional deficiency due to their medications. The elderly are at particular risk of nutritional deficiencies due to reduced appetite and consequently reduced food intake compounded by the use of multiple drugs. If these deficiencies are not corrected through diet (and supplements if indicated) they may further complicate the management of the current health condition or create new health problems.

A combination of dietary assessment, nutritional blood tests and physical examination for clinical signs and symptoms of nutritional deficiencies should be carried out to determine if a clinical or subclinical deficiency is present (see Appendix 1). Further complicating dietary management and the achievement of health goals, some medications can have the following effects:

- facilitate weight gain—e.g., angiotensin II antagonists, beta blockers, calcium channel blockers, anticonvulsants, antipsychotics, benzodiazepines, antidepressants, non-steroidal anti-inflammatory drugs, oral contraceptive pill, hormone replacement, corticosteroids, sulfonylureas, glitazones, some proton pump inhibitors
- increase blood sugar levels—e.g., oral contraceptive pill, antidepressants, diuretics, beta blockers, calcium channel blockers, fenofibrate, phenytoin, some antibiotics, terbutaline, theophylline
- increase cholesterol levels—e.g., allopurinol, antidepressants, diuretics, beta blockers, angiotensin II antagonists.

Herbal supplements with clinically proven effects can be safely combined with medications if the patient is under the care of a knowledgeable healthcare practitioner. Many clinical scenarios benefit from using herbal products and drugs together. For example, people with diabetes may benefit from taking *Panax ginseng*, bitter melon or goat's rue, because these herbs can lower blood sugars and might allow patients to rely on less medication. The herb valerian can be combined with benzodiazepines and might enable patients to use less medication.

However, interactions between herbs and drugs can sometimes result in adverse clinical outcomes. If combined with selective serotonin reuptake inhibitors (antidepressants), St John's wort can result in a life-threatening medical emergency known as serotonin syndrome. The herb also affects a liver enzyme responsible for metabolising most medications so should never be taken with most medications. St John's wort also interacts mildly with over-the-counter painkillers, like non-steroidal anti-inflammatory drugs, by increasing the risk of sunburn (photosensitivity). Hawthorn berries can improve heart contractile function so can be taken with heart medications but can increase the risk of bleeding if combined with blood-thinning medications.

THE ROLE OF THE DIETITIAN

The prevalence of the use of complementary medicines (that is, vitamin, mineral and herbal supplements) by the Australian adult population is high, ranging between 52 and 66 per cent.The use of complementary medicine in New Zealand has been reported in a 2007 study to be even higher at 70 per cent (Wilson et al. 2007).

Of interest, a 2008 National Prescribing Service study found that over 60 per cent of general practitioners surveyed were not aware of the risk of side effects and drug interactions with some commonly used complementary medicines (Williamson et al. 2008). A survey conducted in 2010 on 1121 pharmacy customers located in metropolitan and regional areas of Australia found that complementary medicine use was very high, with 72 per cent of respondents reporting use in the past 12 months. The most popular complementary medicines used were multivitamins, fish oils, vitamin C, glucosamine, vitamin B complex, probiotics, echinacea, coenzyme Q10, *Ginkgo biloba*, St John's wort and valerian. The majority of complementary medicine users (70 per cent) reported that they found the products effective. Complementary medicines were self-prescribed (by 42 per cent of the pharmacy customers surveyed) or had been recommended to them by a medical doctor (32 per cent), family member or friend (20 per cent), naturopath (20 per cent), pharmacy assistant (13 per cent), health food store staff member (7 per cent), pharmacist (10 per cent) or someone else (6 per cent) (Braun et al. 2010).

There are hundreds of complementary medicines being sold in supermarkets, pharmacies and health food shops, so the potential for consumer confusion is high. Understandably, the consumer needs assistance. While research indicates that patients prefer to discuss their use of complementary medicines with their general practitioner and pharmacist, they are increasingly asking dietitians and other health professionals about the appropriate use of these products. Interestingly, a recent study of pharmacy customers reported that some participants felt that pharmacists were not adequately skilled to counsel them about complementary medicines, as they did not refer to pharmacists as an information source (Braun et al. 2010). This correlates with reports that pharmacists often feel uncomfortable dealing with complementary medicine queries due to insufficient knowledge and training. Dietitians have in-depth training in nutritional diagnosis, medical nutrition therapy and increasingly in drug–nutrient/herb interactions and are therefore an invaluable resource to both the patient and the general practitioner.

DRUGS AND NUTRITIONAL PROBLEMS

Proton pump inhibitors and histamine receptor antagonists

These drugs (especially proton pump inhibitors), used for reflux, shut down production of acid, resulting in hypochlorhydria and elevated gastrin levels. This affects digestion and absorption of protein, fat, minerals (calcium, iron, magnesium, zinc, selenium and chromium) and vitamins (B1, B2, B12, folate, C, D and E). The reduced protein digestion can result in preservation of the protein structure and its cell epitopes and can thus stimulate immunoglobulin E induction and manifest as food allergy, oeosinophilic gastroenteritis and oesophagitis.

Hypochlorhydria can increase the risk of small intestinal bowel overgrowth, so a probiotic food or supplement or antimicrobial herb may be indicated. The drugs can also alter bowel motions and may not be tolerated by patients with fructose malabsorption or irritable bowel syndrome. They can alter taste, impair appetite and increase weight. They may impair the absorption of iron; the risk of developing iron deficiency anaemia is compounded in patients with a low intake of haem iron.

Long-term use is associated with hypomagnesaemia and hypocalcaemia, which in turn can cause muscle cramps, arrhythmias, tremors, seizures and increased risk of fractures. Supplementation of calcium- and magnesium-dense foods may be needed with long-term therapy. Magnesium hydroxide and calcium carbonate supplements require stomach acid for digestion and absorption so may be less well absorbed unless taken with a meal; citrate versions are preferable (except if taking oral hypoglycaemic medication—see below). Long-term use is also associated with lower zinc stores and plasma levels, which can reduce stomach acid levels even further, because zinc is a cofactor for the enzyme carbonic anhydrase needed to make the ionic form of hydrogen (H+) for hydrochloric acid production. Supplementation or zinc-dense foods may be needed with long-term therapy.

Doses greater than 20 milligrams per day may decrease vitamin B12 absorption to less than 1 per cent of baseline. Stomach acid is needed to release vitamin B12 from dietary protein. Supplementation or vitamin B12–dense foods may be needed with long-term therapy. (Vitamin B12 supplements do not need stomach acid for absorption.)

A multivitamin is sometimes recommended with these drugs, especially in the elderly with poor food intake. It is important to advise patients to separate all supplements from medications by two to three hours to minimise interactions. Cranberry juice and other acidic foods have a beneficial interaction with these drugs, by increasing absorption of vitamin B12 and minerals. Licorice as a supplement also has a beneficial interaction, by enhancing ulcer healing, and garlic has a beneficial interaction, by inhibiting the growth of *Helicobacter pylori*.

Antacids containing magnesium hydroxide can block absorption of these drugs, so they should be taken apart, separated by two to four hours. St John's wort reduces the drugs' effectiveness.

Antacids

Antacids neutralise stomach acid, and their high levels of calcium can interfere with the absorption of iron, zinc, chromium, copper, vitamins A, B1, B12, folate, D, E and K. Aluminium in some antacids can bind dietary phosphates, leading to calcium depletion and osteomalacia. They can alter taste and impair appetite. Long-term use of antacids can adversely increase serum magnesium levels, which can negatively affect heart, nerve and muscle function. Elderly patients should not take antacids at meal times or with other dietary supplements.

Calcium or magnesium citrate, vitamin C supplements, citrus juices and milk can increase aluminium absorption; doses should be taken apart, separated by at least two hours. Iron, zinc and fibre supplements and foods high in oxalates (for example, tea and wheat germ) and phytates (for example, bran and oats) can reduce the absorption of antacids.

Laxatives

Laxatives can cause steatorrhoea with chronic use, reducing absorption of fat-soluble vitamins (A, D, E and K) and increasing excretion of sodium, potassium, calcium, iron and magnesium. To maximise absorption of ingested nutrients that are affected by laxatives, the consumption of dairy foods, and calcium and magnesium supplements are not recommended within two hours of taking a laxative. Aloe vera outer leaf can have additive (adverse) effects due to anthraquinones. It may be prudent to recommend a multivitamin with chronic use of laxatives.

Aspirin and other blood-thinning drugs

Aspirin can reduce blood sugar levels, cholesterol and inflammation (high-sensitivity C-reactive protein). It can increase the risk of gastrointestinal bleeding with chronic use, is associated with significantly lower mean serum ferritin and is a contributor to iron deficiency in the very elderly. It can also reduce the absorption of iron, zinc, potassium, calcium, vitamin B12, folate and vitamin C. Psyllium husk can reduce absorption of aspirin, and doses should be separated by at least one hour.

Risk of bleeding may be increased by a high intake of fish oil (over 3000 milligrams per day of eicosapentaenoic acid [EPA] plus docosapentaenoic acid [DPA]), flaxseed oil (over 30 grams per day), evening primrose oil (over 1 gram per day) or vitamin E (over 1000 international units [660 milligrams]). Other blood-thinning herbs and foods that have the potential to increase the blood thinning effects of aspirin, warfarin or clopidogrel include aloe vera, andrographis, baical skullcap, bilberry, carnitine, celery seed, chamomile, chondroitin, cinnamon, coenzyme Q10 (unlikely at under 150 milligrams per day), cranberry, devil's claw, dong quai, feverfew (unlikely), garlic supplement (over 7 grams per day), ginger supplement (over 10 grams per day), ginkgo (unlikely), ginseng, glucosamine (with warfarin), goji (with warfarin), guarana, grape seed extract, horseradish (unlikely), licorice (with warfarin), meadowsweet, myrrh, green tea, krill oil, policosanol (over 10 milligrams per day), red clover (with warfarin, but unlikely), rosemary (with warfarin), saw palmetto, St John's wort (with warfarin and clopidogrel), turmeric and willow bark (over 240 milligrams per day).

Oral hypoglycaemics

These drugs (including metformin, thioglitazones and sulfonylureas) can decrease absorption of vitamin B12 and folate and increase homocysteine. They can also alter taste and impair appetite. Metformin reduces blood sugar, low-density lipoprotein cholesterol and triglycerides and reduces the loss of lean body mass. Sulfonylureas can affect thyroid function (and cause weight gain) by reducing the uptake of iodine by the thyroid.

Magnesium supplements can increase the absorption of these drugs. Quercetin can enhance the effects of thioglitazones (exercise caution), while St John's wort can reduce the effects of sulfonylureas. The drugs' therapeutic effect may also be reduced by potassium or magnesium citrate supplements. Vitamin and mineral supplements that can have an additive effect (requiring adjustment of drug dose) include vitamin E, magnesium, chromium, zinc, coenzyme Q10, lipoic acid, biotin and inositol. Foods and herbal supplements that can have an additive effect (requiring adjustment of drug dose) include aloe vera, andrographis, bilberry, bitter melon, cinnamon, damiana, elderberry, eucalyptus, fenugreek, garlic, ginger, ginkgo, ginseng, goat's rue, green tea, guar, guggul, gymnema, horse chestnut, myrrh, milk thistle, olive leaf extract, psyllium, resveretrol and turmeric.

Antihypertensives

Ace inhibitors and angiotensin II antagonists

These drugs attach to zinc and can cause zinc deficiency by increasing urinary zinc excretion. This may account for some side effects (impaired appetite, altered taste and skin numbness or tingling). It is important to monitor zinc status. An increased zinc intake may be required with long-term therapy.

Iron supplements may reduce a dry cough caused by ace inhibitors; doses should be separated by at least two hours. Potassium supplements should be avoided; high-potassium foods are contraindicated only if serum potassium is elevated. These drugs contain magnesium, so high-dose magnesium supplements (over 300 milligrams per day) should be used with caution. Licorice (in supplements or herbal teas) can reduce the drugs' effects; that is, high-dose glycyrrhizin taken long term can lead to increased blood pressure. St John's wort and dong quai can also reduce the drugs' effects and increase the risk of sunburn.

Foods and herbal supplements that can have an additive effect (requiring adjustment of dose) include arginine, coleus forskohlii, evening primrose oil, garlic, hawthorn, olive leaf, oats, olive oil, omega-3 fatty acids (fish oil) and omega-6 fatty acids and stinging nettle.

Loop and thiazide diuretics

These drugs may increase the risk of glucose intolerance and diabetes and may increase low-density lipoprotein cholesterol and triglycerides. They increase the excretion of sodium, potassium, magnesium, zinc, calcium (with loop), iodine (with thiazide), vitamins B1 (with loop), B6, B12 and folate, meaning that supplements may be indicated; it is important to monitor blood and clinical signs and symptoms. They can alter taste and impair appetite. The drugs can increase homocysteine levels; vitamin B12 and folate may help lower these levels. High-zinc foods and supplements may be indicated.

Long-term use (over six months, mainly with loop) may lead to magnesium deficiency, which in turn can increase loss of potassium and vitamin B1, further complicating existing heart conditions. Magnesium is needed for potassium (and sodium and calcium) homoeostasis. Magnesium is an essential cofactor in the activation of thiamin. Magnesium supplements (about 300 milligrams per day) or magnesium-dense foods may be considered if the patient has unexplained low blood potassium, calcium or sodium. Strict sodium restriction is not advisable, as it may cause hyponatraemia.

High-potassium foods or supplements are frequently prescribed (except in potassium-sparing diuretics). Potassium supplements are better absorbed and mobilised in the presence of magnesium. Magnesium supplements or magnesium-dense foods may be considered if the patient is not responding to potassium supplements.

Vitamin B1 deficiency (mainly with loop) can aggravate congestive heart failure, oedema, muscle pain, poor appetite, mental confusion and risk of falls. Vitamin B1 supplements or vitamin B1–dense foods may be considered if the patient's heart condition deteriorates.

Thiazide diuretics can reduce coenzyme Q10 serum levels; a supplement may be indicated with long-term use. They can also increase blood levels of calcium by decreasing excretion and, indirectly, by affecting vitamin D metabolism. Calcium and vitamin D supplements should therefore be used with caution. Signs of hypercalcaemia should be sought.

Herbal supplements that can have an additive effect (requiring adjustment of dose) include dandelion leaf, stinging nettle, elderberry and green tea (see also ace inhibitors). Herbal supplements that can reduce the drugs' effects (requiring adjustment of dose) include: black cohosh, dong quai, guarana, horsetail, juniper, licorice (over 100 milligrams glycyrrhizin per day), St John's wort (see also ace inhibitors).

Thyroxine (inactive T4)

This is used in the treatment of hypothyroidism. It contains inactive T4, which needs to be converted to active T3 in the body. Secretion of thyroid stimulating hormone, production of T4 and conversion of endogenous or exogenous inactive T4 to active T3 in the thyroid, liver and other tissues require an adequate intake (AI) of iodine, iron, selenium, zinc, magnesium, omega-3 fatty acids, vitamins A and E and tyrosine. Correcting deficiencies of these nutrients may improve subclinical hypothyroidism (thyroid stimulating hormone 2–4 milliunits per litre), reducing the need for medication. These nutrients can also have an additive effect on thyroid function in medicated patients, potentially resulting in a need for a reduced dose of thyroxine. This may be desirable, since thyroxine therapy can have side effects—for example, it potentiates glucose intolerance.

The drug does not cause nutrient deficiencies, but its absorption and efficacy is affected by food, mineral supplements and some medications. Absorption requires normal gastric acid secretion. Thyroxine should be taken on an empty stomach, ideally one hour before food, two hours after food or late in the evening, to maximise absorption. Mineral supplements or mineral-fortified foods, especially calcium, magnesium, iron, zinc, selenium and chromium, can reduce absorption of thyroxine and should be separated from the dose by four hours. Foods and supplements high in fibre and/or soy should be separated from thyroxine by several hours.

Sulfonylureas and diuretic thiazides can reduce iodine uptake by the thyroid gland, and beta blockers, lithium and oestrogen can reduce conversion of inactive T4 to active T3. Hypochlorhydria increases the drug's dose requirements.

Iodine is essential for the production of endogenous T4 and T3. Mild iodine deficiency re-emerged in Australia between 2000 and 2010, with 43 per cent of the population having inadequate iodine intakes (Savige et al. 2011). Good food sources of iodine include kelp, seaweed, fish and iodised salt. Iodine deficiency can be detected by way of several fasting urinary iodine tests. If iodine deficiency is identified, low-dose iodine supplement approaching the recommended dietary intake of 150 micrograms daily may be necessary with a concomitant reduction in thyroxine dose if thyroid stimulating hormone levels drops. High-dose iodine supplements should be avoided, as they can block thyroid hormone synthesis and create an underactive state. Kelp supplements should therefore be avoided, as it is usually not clear how much iodine is in the product.

A T4:T3 ratio over three may suggest selenium deficiency. Selenium is essential for conversion of T4 to T3 and breakdown of T3 and can help lower antibody levels to the thyroid. However, since both iodine and selenium deficiencies can co-exist, iodine deficiency must be corrected first to enable the thyroid to manufacture enough T4.

Foods and herbal or nutrient supplements that may increase thyroid function or the effects of thyroxine (potentially requiring adjustment of dose) include low-dose iodine, kelp or seaweed, iron, selenium, zinc, vitamin A, vitamin E, tyrosine, brahmi and withania. Foods and herbal or nutrient supplements that may reduce thyroid function or the effects of thyroxine (potentially requiring adjustment of dose) include aloe vera, bugleweed, carnitine, celery seed, fenugreek, horseradish, high-dose iodine, kelp or seaweed, isoflavones, lemon balm, red rice yeast extract and s-adenosylmethionine.

Goitrogenic foods that may reduce thyroid function through reduced utilisation of iodine include uncooked broccoli, cauliflower and cabbage, garlic, onion, linseed, rapeseed, lima beans, soy, peanuts, swede, sweet potato, millet and cassava. However, of these, only cassava has been proven to cause hypothyroidism in humans. Soy foods have been observed to reduce absorption of thyroid medication, so they should be separated from the dose by four hours. Soy foods and isoflavone supplements may also directly inhibit the function of the thyroid gland, although this may be significant only in people who are deficient in iodine.

PRACTICE TIPS

Health professionals need to be knowledgeable and vigilant regarding nutrition-related clinical symptoms that may be caused by pharmaceuticals. If a patient is having trouble losing weight or has elevated blood sugars or lipids despite good dietary management, their medications may be contributing.

Clinical symptoms of nutritional deficiencies combined with laboratory data are needed to validate poor nutritional status or deficiency before supplementation is prescribed. If patients are already taking vitamin or mineral supplements these should be stopped for several days before a nutritional blood test to avoid bias.

Short-term nutritional therapy may require a patient to take a high-dose supplement for a couple of months to correct the nutritional deficiency caused by the effects of medications, poor diet or a medical condition. The patient's general practitioner should be notified if the patient starts taking any supplements, as it may affect drug action, requiring adjustment of dose (which is usually carried out by the general practitioner). Long-term diet therapy aims to maintain levels of nutrients through diet (after short-term therapy has been completed), but in some cases a low-dose multivitamin may be indicated and can make an important contribution to the health of patients on long-term prescription medications.

Grapefruit juice, pomegranate juice, echinacea,

black cohosh and milk thistle interact with some drugs by inhibiting drug-metabolising intestinal and hepatic cytochrome P450 enzymes, especially cytochrome P450 3A4, for up to 72 hours. This significantly increases the bioavailability of many drugs (for example, statins, antidepressants, beta blockers, calcium channel blockers, hormone replacement therapy, warfarin, anticonvulsants and antipsychotics) and can raise blood levels into toxic ranges.

St John's wort adversely reacts with most drugs by inhibiting liver enzymes. Many herbs, some vitamins (high-dose vitamin E and C) and some foods (cranberry, garlic and fish oil) have blood-thinning effects; therefore, use should be recommended with caution with blood-thinning medications. Blood-thinning herbs, vitamins, foods and oils should be stopped two weeks before an operation or a procedure that may cause bleeding. Separate all herb, vitamin, mineral, fibre and slippery elm supplement doses from medication doses by two to four hours.

REFERENCES

Braun, L., Tiralongo, E., Wilkinson, J., Spitzer, O. et al. 2010. Perception, use and attitudes of pharmacy customers on complementary medicines and pharmacy practice. *BMC Complementary and Alternative Medicine* 10: 38.

Savige, G., Kouris-Blazos, A. & Wahlqvist, M.L. 2011. Iodine and thyroid function. In Wahlqvist, M.L. (ed.). *Food & Nutrition: Food and health systems in Australia and New Zealand* 3rd edn, 671–86. Allen & Unwin, Sydney.

Williamson, M., Tudball, J., Toms, M., Garden, F. et al. 2008. *Information Use and Needs of Complementary Medicines Users*. National Prescribing Service. <www.nps.org.au/__data/assets/pdf_file/0010/66619/Complementary_Medicines_Report_-_Consumers.pdf>, accessed 6 October 2014.

Wilson, K., Dowson, C. & Mangin, D. 2007. Prevalence of complementary and alternative medicine use in Christchurch, New Zealand: Children attending general practice versus paediatric outpatients. *New Zealand Medical Journal* 23: 120 (1251): U2464.

ADDITIONAL READING

Braun, L. & Cohen, M. 2010. *Herbs and Natural Supplements: An evidence-based guide* 3rd edn. Elsevier, Sydney.

Brown, C.M., Barner, J.C. & Shah, S. 2005. Community pharmacist's actions when patients use complementary and alternative therapies with medications. *Journal of American Pharmacists Association* 45: 41–7.

Brown, J., Morgan, T., Adams, J., Grunseit, A. et al. 2008. *Complementary Medicines Information Use and Needs of Health Professionals: General practitioners and pharmacists*. Updated April 2009. National Prescribing Service. <www.nps.org.au/__data/assets/pdf_file/0020/66620/CMs_Report_-_HP_-_Apr_09.pdf>, accessed 6 October 2014.

Chang, Z.G., Kennedy, D.T. & Holdford, D.A. 2000. Small RE: Pharmacists knowledge and attitudes toward herbal medicine. *Annals of Pharmacotherapy* 34: 710–15.

Coleman, Y. 1998. *Drug–Nutrient Interactions: The manual*. Nutrition Consultants Australia, Melbourne.

——2013. *Medications and Nutrition: Favourite fifty*. Nutrition Consultants Australia, Melbourne.

Easton, K. 2007. *Complementary Medicines: Attitudes and information needs of consumers and healthcare professionals*. National Prescribing Service Limited. <www.nps.org.au/__data/assets/pdf_file/0007/26872/ComplementaryMedicinesReport.pdf>, accessed 6 October 2014.

Kotsirilos, V., Vitteta, L. & Sali, A. 2010. *A Guide to Evidence-based Integrative and Complementary Medicine*. Elsevier, Sydney.

Kotsirilos, V., Kouris-Blazos, A. & Phelps, K. (eds) 2010. Integrative medicine: What is the role of the dietitian? Letter to the editor. *Nutrition and Dietetics* 67: 303–5.

Kotsirilos, V., Vitetta, L., Sali, A. & Kouris-Blazos, A. 2011. Herb–nutrient–drug interactions. In Kotsirilos, V., Vitetta, L. & Sali, A. *A Guide to Evidence-based Integrative and Complementary Medicine*, 835–48. Elsevier, Sydney.

Kouris-Blazos, A. 2009. Drug–nutrient interactions. *Geriatric Medicine in General Practice* 2 (July): 26–7.

MacLennan, A., Myers, S. & Taylor, A. 2006. The continuing use of complementary and alternative medicine in South Australia: Costs and beliefs in 2004. *Medical Journal of Australia* 184: 27–31.

Naidu, S., Wilkinson, J.M. & Simpson, M.D. 2005. Attitudes of Australian pharmacists toward complementary and alternative medicines. *Annals of Pharmacotherapy* 39: 1456–61.

Test your understanding answers

Chapter 3

1. d
2. c
3. c
4. b
5. a
6. d
7. e
8. e
9. b
10. e

Chapter 4

1. b
2. d
3. d
4. c
5. a
6. d
7. e
8. c
9. c
10. e

Chapter 5

1. c
2. b
3. e
4. b
5. a
6. c
7. b
8. c
9. e
10. a

Chapter 6

1. c
2. b
3. b
4. b
5. a
6. d
7. e
8. d
9. e
10. e

Chapter 7

1. c
2. b
3. a
4. d
5. c
6. a
7. c
8. c
9. d
10. e

Chapter 8

1. d
2. a
3. d
4. e
5. d
6. a
7. e
8. c
9. a
10. e

Chapter 9

1. a
2. d
3. e
4. b
5. a
6. c
7. b
8. c
9. e
10. d

Chapter 10

1. e
2. d
3. d
4. e
5. c
6. b
7. d
8. d
9. b
10. e

Chapter 11

1. a
2. d
3. c
4. d
5. b
6. d
7. d
8. c
9. d
10. b

Chapter 12

1. d
2. b
3. d
4. d
5. e
6. c
7. e
8. c
9. a
10. e

Chapter 13

1. c
2. d
3. a
4. c
5. b
6. d
7. d
8. c
9. d
10. e

Index

Numbers and letters that precede the names of chemicals are ignored in filing; for example, N-acetyl sugars can be found under acetyl sugars. Page numbers in **bold** indicate the major treatment of a topic. Page numbers in *italics* refer to figures.

For Product Safety Concerns and Information please contact our EU
representative GPSR@taylorandfrancis.com
Taylor & Francis Verlag GmbH, Kaufingerstraße 24, 80331 München, Germany

www.ingramcontent.com/pod-product-compliance
Ingram Content Group UK Ltd.
Pitfield, Milton Keynes, MK11 3LW, UK
UKHW052213190926
475702UK00006B/1005

* 9 7 8 0 3 6 7 7 1 8 1 0 7 *

FOOD & NUTRITION THROUGHOUT LIFE